A TEXTBOOK OF
MANUFACTURING TECHNOLOGY–I

[For the students of B.E. / B.Tech Anna University & other Technical Universities of India]

Dr. P.C. SHARMA
B.Sc. Engg. (Mech.) (Hons.)
M.Sc. Engg. (Mech.) (Distinction)
Ph.D. (I.I.T. Delhi)
MISTE, LMISME
Visiting Professor
Department of Mechanical Engineering
S.U.S. College of Engineering and Technology
Tangori (Mohali), Punjab
Formerly of Punjab Engineering College
CHANDIGARH

S.CHAND
PUBLISHING

S Chand And Company Limited
(ISO 9001 Certified Company)

S Chand And Company Limited
(ISO 9001 Certified Company)
Head Office: Block B-1, House No. D-1, Ground Floor, Mohan Co-operative Industrial Estate, New Delhi – 110 044 | Phone: 011-66672000
Registered Office: A-27, 2nd Floor, Mohan Co-operative Industrial Estate, New Delhi – 110 044 Phone: 011-49731800
www.**schandpublishing.com**; e-mail: **info@schandpublishing.com**

Branches

Chennai : Ph: 23632120; chennai@schandpublishing.com
Guwahati : Ph: 2738811, 2735640; guwahati@schandpublishing.com
Hyderabad : Ph: 40186018; hyderabad@schandpublishing.com
Jalandhar : Ph: 4645630; jalandhar@schandpublishing.com
Kolkata : Ph: 23357458, 23353914; kolkata@schandpublishing.com
Lucknow : Ph: 4003633; lucknow@schandpublishing.com
Mumbai : Ph: 25000297; mumbai@schandpublishing.com
Patna : Ph: 2260011; patna@schandpublishing.com

First Edition 2007
Reprints 2008, 2010, 2011, 2012, 2013, 2014, 2015, 2017, 2018, 2020

Reprint 2022

ISBN: 978-81-219-2821-2 **Product Code:** H3MFT62MAFT10ENAA07R

PRINTED IN INDIA

By Vikas Publishing House Private Limited, Plot 20/4, Site-IV, Industrial Area Sahibabad, Ghaziabad – 201 010 and Published by S Chand And Company Limited, A-27, 2nd Floor, Mohan Co-operative Industrial Estate, New Delhi – 110 044.

PREFACE TO THE FIRST EDITION

The book has been written to meet the need of First Year Mechanical Engineering students of Anna University, for the course "Manufacturing Technology–I". The author hopes that the matter will come up to the expectations of both the students and the teachers.

Suggestions for the improvement of the text shall be highly appreciated and duly obliged.

P.C. Sharma

CONTENTS

1

Introduction

1.1. INTRODUCTION

The main aim behind advances in engineering and technology has been to raise the standard of living of man and to make his life more comfortable. The major role in this direction has been played by manufacturing science. Manufacturing is an essential component of any industrialised economy. The word 'Manufacturing' is derived from the Latin, *manus* = hand and *factus* = made, that is, the literal meaning is "made by hand". However, in the broader sense, the word 'Manufacturing' means the making of goods and articles by hand and/or by machinery. Thus 'Manufacturing engineering' or "Production Engineering" can be defined as the study of the various processes required to produce parts and to assemble them into machines and mechanisms. Production or Manu- facturing is a critical link in the design cycle, which starts with a creative idea and ends with a successful product, Fig. 1.1.

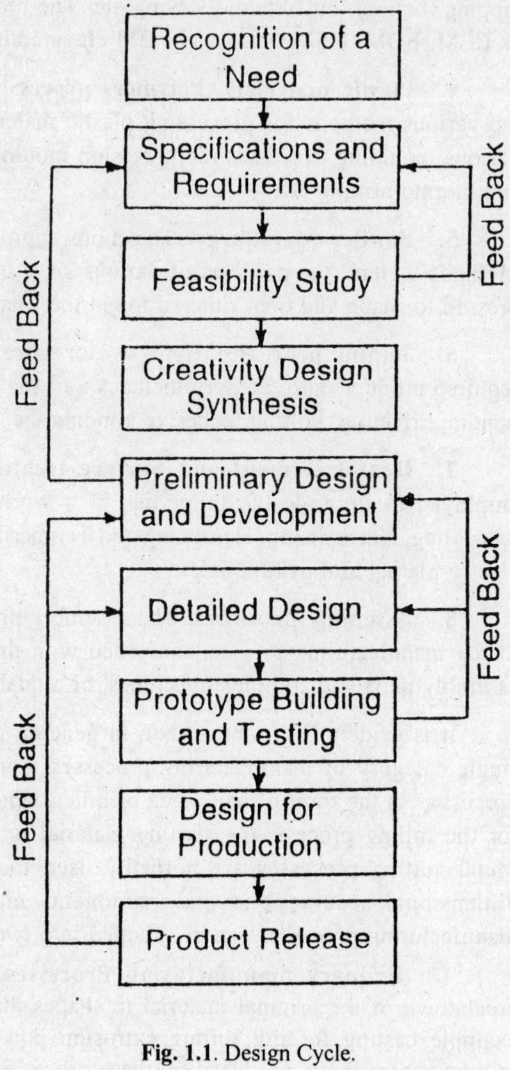

Fig. 1.1. Design Cycle.

1.2. CLASSIFICATION OF MANUFACTURING PROCESSES

When one thinks as to how the various components of machines are produced, many techniques come to the mind, for example, casting, forging, rolling, machining, welding etc. The manufacturing processes are so varied that there is no simple and universally accepted criteria of classifying these. However, all the manufacturing processes may be grouped into the following main categories:

1. Casting Processes. Here, the metal in the molten state is poured into a mould and allowed to solidify into a shape. The mould may be expendable or permanent.

The examples are : Sand casting, Permanent mould casting, Die casting, Precision investment casting and centrifugal casting etc.

2. Deformation Processes. In these processes, the material is plastically deformed (hot or cold) under the action of an external force, to produce the required shape. No material is removed, but is only displaced and deformed to get the final shape. This category includes metal working/ forming processes such as : forging, rolling, extrusion and drawing etc. and also sheet metal working processes such as deep drawing and bending etc. The unconventional forming processes such as High Energy Rate Forming (HERF) and High Velocity Forming (HVF) methods also fall under this category.

3. Machining Processes. In machining processes, also known as Metal cutting or chip forming processes, material is removed from a work piece to get the final shape of the product. The processes include : Turning, milling, drilling, broaching, shaving, grinding, polishing, lapping, honing, buffing and sawing etc. The modern unconventional machining processes such as ECM, EDM, USM, AJM and LBM etc. are also included in this category.

4. Plastic materials / Polymers processing methods. Under this category are included the various methods for processing plastic materials/ polymers, for example, shape casting, the various moulding processes (compression moulding, injection moulding, transfer moulding etc.) and thermoforming etc.

5. Powder Metallurgy. The more appropriate name should be "Particulate Processing Methods". Here, the particles of various sizes of metals, ceramics, polymers and glass etc. are pressed to shape and then sintered to get the final product.

6. Joining processes. Here, two or more components are joined together to produce the required product. The category includes : all the welding processes, brazing, soldering, diffusion bonding, riveting, bolting, adhesive bonding etc.

7. Heat treatment and Surface treatment processes. Heat treatment processes are employed to improve the properties of a work piece. The category includes the processes : Annealing, Normalising, Hardening and Tempering methods. Surface treatment processes include electro-plating and painting etc.

8. Assembly processes. The assembly process for machines and mechanisms is the part of the manufacturing process concerned with the consecutive joining of the finished parts into assembly units and complete machines, of a quality that meets the manufacturing specifications.

It is evident from above that, in general, no component can be produced entirely by one single category of manufacturing processes. For example, the starting material for the forging process is in the form of bar, stock or billet, which are the end-products of rolling process. And, for the rolling process, the starting material is 'ingot', which is produced by casting process. Metal cutting processes are normally used to give the final shape and size (surface finish, dimensional accuracy) to the components made by other processes. Accordingly, all the manufacturing processes can be grouped into two main categories :-

(*i*) **Primary manufacturing Processes.** These processes are involved in the initial breakdown of the original material to shapes that are then processed for the final product ; for example, casting, forging, rolling, extrusion, powder metallurgy and plastic technology. However, in many cases, parts produced by these processes do not need other processes for producing the final shape and size, for example, precision investment casting, die-castings, precision forgings, powder metallurgy parts and parts produced by plastic processing methods.

(*ii*) **Secondary Manufacturing Processes.** These processes take the products of some primary process and change their geometry and properties to the semifinished or finished stage, *e.g.*, all metal removal processes and the rest of the metal forming processes, that is, drawing, spinning, swaging, coining, stretching, bending, deep drawing, wire / rod / tube drawing, sheet metal forming and rubber forming etc.

1.3. VARIOUS KINDS OF PRODUCTION

Depending upon the scale of production and the type of components being produced, the production or manufacturing can be classified as : piece or job-lot production; medium lot, moderate lot, or batch production; and mass production.

1. Piece or Job-lot production. Here, the parts are produced in small quantities to order, which are not periodically repeated. Companies engaged in piece production employ mainly general purpose equipment. Standard cutting tools and universal measuring facilities are used. The principle of complete interchangeability is not complied with. Hence, fitting is resorted to in assembly operations. The labour employed is more skilled. Lot sizes usually vary from 10 to 500 parts per lot, for example, manufacture of aeroplanes and oil field equipment, machine tools, giant hydroturbines, rolling mills and other heavy equipment etc.

2. Medium Production. It is characterized by the manufacture of parts in repeating lots or batches and to order. Mostly, the equipment used is general purpose type, equipped with universal, adjustable and sectional built up jigs, fixtures and tools. This enables the labour input and cost of production to be substantially reduced. Principles of interchangeability are to be strictly complied with. Lot size may range from a few parts to over 300 parts per lot and the yearly production may range from 2500 to 100,000 parts. An example of medium production is printing of books. Other examples are : machine tools, pumps, compressors, stationary I.C. engines etc.

3. Mass Production. It is characterized primarily by an established and stable object of production. The parts are produced in large quantities either intermittently or continuously, but are not dependent upon individual orders. The quantity is usually over 100,000 parts per year. A mass production enterprise deals with standardized products of limited variety, such as consumer durable and products for industrial use. Other features typical of mass production are : an extensive use of specialized (usually permanently set up) and single purpose machine tools, and mechanization and automation of production processes, with strict compliance of the principle of interchangeability. The latter greatly reduces the time required for assembly operations, Examples of mass production are :- nuts, bolts, screws, washers, pencils, matches, engine blocks, automobiles, bicycles, electric motors, sewing machines and tractors etc.

The most advanced form of mass production is 'Continuous-flow production', whose main feature is that the time required for each operation of the production line is equal to or a multiple of the set standard time all along the line. This enables work to be done without providing stock piles and in strictly definite intervals of time. Examples include : oil refineries and continuous chemical plants.

In mass production type of production, semi-skilled or even unskilled workers are needed to operate the machines. This type of production is capital-intensive and the unit cost is low.

1.4. COMPUTERS IN MANUFACTURING

Computers are being increasingly used in the design / production cycle of a part.

Computer-aided design (CAD) and computer aided manufacturing (CAM) are performing greater role in manufacturing industry. Computers are being used in every aspect of manufacturing, for example,

(*i*) for the control of production processes and machines with the aid of NC (numerical control), CNC (Computer Numerical Control) and DNC (Direct Numerical Control)

(*ii*) Computer-Aided Process Planning, (CAPP)

(*iii*) Production planning and control

(*iv*) Inventory management and for material management planning (MRP).

(*v*) Computer-Aided Quality Control, (CAQC).

(*vi*) Flexible Manufacturing Systems, (FMS).

(*vii*) Industrial Robots

(*viii*) Group Technology, (GT).

(*ix*) Computer Integrated Manufacturing System, (CIMS), and so on.

The use of computers in manufacturing has resulted in : improved productivity, quality, equipment utilization, reduced inventory and faster delivery.

1.5. SELECTION OF A MANUFACTURING PROCESS

From the above, it is thus clear that a manufacturer has the choice of selecting one of the many manufacturing processes for producing a given component. There are many factors involved in this exercise, but the selection of a suitable process can be made on the basis of the following considerations :-

(*i*) Type and nature of the starting material

(*ii*) Volume of production

(*iii*) Expected Quality and properties of the components.

(*iv*) Technical viability of the process.

(*v*) Economy

Other factors influencing the choice of a manufacturing process are : Geometrical shape, Toolings, jigs, fixtures and gauges needed, available equipment and delivery date.

The selection of material for a component should be based on the processing factors as well as the functional requirements of the component (Refer to Tables 1.1 to 1.3). The cheapest material meeting these requirements should be selected, because higher the alloy content, the higher the cost. Also, standard alloys should be selected. Selection of many different alloys should be avoided to reduce inventory costs.

The type of material is also an important factor. Hard and brittle materials cannot be mechanically worked conveniently, whereas these can be cast or machined by many methods.

A manufacturing method usually alters the properties of the material. For example, cold working renders a material stronger, harder and brittle (less ductile). A very good surface finish and close tolerances can be obtained by cold working operations as compared to hot working operations. However, they will be needing higher capacity equipment.

Size, shape and shape complexity are also important factors. Flat components with thin cross-sections cannot be cast conveniently, whereas complex parts cannot be mechanically worked. However, these can be conveniently cast or fabricated from individual parts. Complex shapes can be produced by forging and subsequent machining and finishing operations, and they have a thoughness that is far supperior to that of casting and P/M parts. Landing gear of a jet liner is manufactured by forging and machining processes. Very thin sections can be produced by cold rolling.

The geometrical shape of the component dictates the shape of the starting material and the processing method. For simple shaped components, for example, straight shafts, bolts, rivets etc., the selection of the shape and the starting material and the manufacturing method is easily made. As the complexity of the shape increases, there is a wide choice of the shape of the starting material and the manufacturing process. For example, a small pinion can either be machined from a bar stock or from a precision forged gear blank. The final choice will be dictated by economics. For this, volume of production is an important factor. Larger the volume of production, more economical it will be to invest in special equipment so as to reduce the cost per unit. Similarly, higher the production rate, economical it will be to select special purpose/automatic equipment. They will also be capable of meeting the delivery dates.

Production volume depends upon the product. For example, paper pins, paper clips, nuts, bolts, washers, speark plugs, ball bearings and ball point pens are produced in very large quantities. Whereas, jet engines for large commercial air crafts, diesel engines for locomotives etc. are manufactured in limited quantities. Production volume also plays a very significant role in determining, Economic order Quantity (EOQ) and Economic Lot or Batch quantity.

As for the production rate, processes such as Die Casting, P/M, sheet metal working processes and roll forming etc. are high production– rate processes. Whereas, processes such as sand Casting, conventional machining process, Unconventional machining and forming processes and the various adhesive and diffusion bonding processes are relatively slow processes. The production rates can, however, be increased by automation or by the use of multiple machines.

Lead time also gereatly influences the choice of a process. It is defined as the time to start production. Process such as rolling, forging, extrusion, die casting, roll forming and various sheet metal working process, require extensive and expensive dies, tools and equipment, resulting in long lead-time. On the other hand, the various conventional machining processes are very flexible and can be adapted to most requirements in a relatively short time. Lead time is also very small in the case of machining centres, flexibe Manufacturing cells (FMC) and Flexible Manufacturing Systems (FMS). They can respond quickly and effectively to product changes (both in desing and quantity).

It is clear from Table 1.3 that each manufacturing process has the capability of manufacturing a component within a certain tolerance range and surface quality. As the tolerance decreases and surface quality increases, the cost/unit increases. Therefore the widest tolerances and minimum surface quality that will meet the functional requirements of the component, should be specified.

Each manufacturing process has its own limitation of producing complex shapes and maximum and minimum size of the component (Table 1.3).

It is not always an easy task to select the best manufacturing process. Sometimes a product can be made by more than one competitive processes. So, in the final selection, cost is a very important factor. The cost of manufacturing a component by these methods should be compared and the optimum process selected. While doing these evaluations, along with the cost of processing the material to the finished product, the other important considerations are : material utilization factor, the effect of processing method on the material properties and the subsequent performance of the component in service.

The various manufacturing processes have been compared in Tables 1.1 to 1.4 on the basis of joinability, metal forming, design and cost considerations.

Table 1.1. Manufacturing Process : Joinability considerations

Material	Welding			Brazing	Soldering	Adhesive bond [TS, TP, EN][1]	Adhesive bond[2]	Thread Fastening	Riveting
	Arc	Oxy-Acet.	Resistance						
Cost Iron	C	R	O	D	O	(TS) C, (TP) O	C	C	R
Carbon steel	R	R	R	R	D	(TS) C, (TP) O	C	R	R
Stainless steel	R	C	R	R	C	(TS) C, (TP) O	C	R	C
Aluminium-Magnesium	C	C	C	C	O	(TS) R, (TP) O	R	C	C
Copper	C	C	C	R	R	(TS) C, (TP) O	C	C	C
Nickel	R	C	R	R	C	(TS) C, (TP) O	C	R	C
Titanium	C	X	C	D	O	(TS) O, (TP) X	C	C	X
Lead	C	C	X	X	R	R	R	R	R
Zinc	C	C	O	X	R	R	R	R	R
Thermoplastics	R	R	C	X	X	C	O	O	R
Ceramics	X	O	X	X	X	R	R	C	O
Leather	X	X	X	X	X	(EM) R, (TS) C	R	O	R
Fabric	X	X	X	X	X	(EM) R	R	O	R

R : Recommended ; C: Common : D : Difficult ; O : Seldom used ; X : Not used [TS : Thermoset ; TP : Thermoplastic : EM : Elastomeric][1] ; [2](Modified Compression Epoxy)

Table 1.2. Manufacturing Processes : Metal forming considerations

Manufacturing Process	*Irons*	*Low C-steel alloys*	*Heat and corrosion resistant alloy*	*Aluminium alloy*	*Copper alloy*	*Lead alloy*	*Magnesium alloy*	*Nickel alloy*	*Precious Metals*	*Tin alloy*	*Titanium alloy*	*Zinc alloy*
Sand casting	A	A	A	A	A	B	A	A	-	B	-	B
Shell mould casting	A	B	B	A	A	-	-	B	-	-	-	-
Permanent mould casting	A	B	-	A	B	B	A	B	-	B	-	B
Die casting	-	-	-	A	B	A	A	-	-	B	-	A
Plaster mould casting	-	-	-	A	A	-	-	-	-	-	-	-
Investment casting	-	A	B	A	A	-	A	B	B	-	-	-
Centrifugal casting	A	A	A	B	B	-	-	B	-	-	-	-
Drop forging	B	A	A	B	B	-	B	B	-	-	B	-
Press forging	-	A	A	B	B	-	B	B	-	-	B	-
Upset forging	-	A	A	B	B	-	B	B	-	-	B	-
Cold headed parts	-	A	B	A	A	B	-	B	B	-	-	-
Stampings, Drawing	-	A	B	B	A	-	B	B	B	-	B	B
Spinning	-	A	B	A	A	B	B	A	B	-	B	B
Screw machining	B	A	B	A	A	-	B	A	B	-	B	B
Roll forming	-	A	-	B	B	-	B	-	-	-	B	B
Extrusion	-	B	-	A	A	B	A	B	-	B	B	-
Powder Metallurgy	A	A	B	B	A	-	-	B	B	-	B	-
Electro forming	A	-	-	B	A	B	-	A	A	B	-	B

A : Materials most frequently used ; B : Also materials currently used

Table 1.3. Manufacturing Process : Design Considerations

S. No.	*Manufacturing Process* ↓	*Material choice*	*Complexity of part*	*Maximum Size*	*Minimum Size*	*Mechanical Properties*	*Precision and Tolerance*	*Special Structural Properties*	*Surface Smoothing*	*Surface Detail*	*Remarks*
1.	Sand casting	Wide ; Ferrous, Non ferrous Light Metals	Considerable	Largest	3 mm	Fair to High	± 0.6 to 0.3 mm per mm	Good Bearing structure	Poor	Poor	Usually require some machining
2.	Shell Mould casting	Wide	Moderate	Best for smaller parts	1.5 mm	Good	± 0.003 to 0.005 mm per mm	High Quality for cast metals	Good	Good	Best of low costing methods.
3.	Permanent Mould casting	Restricted : brass bronze aluminium	Limited	Moderate	2.5 mm	Fair good	± 0.6 mm per mm	None	Good	Good	Economical for mass production only.
4.	Plaster Mould casting	Narrow : Brass, bronze, aluminium	Considerable	Moderate	0.75 mm	Fair	± 0.25 to 0.125	Non	Good	Good	Little finishing required.
5.	Investment casting	Wide	Considerable	Moderate	0.75 mm	Good	± 0.125 mm per mm	None	Excellent	Excellent	Best for too complicated parts.
6.	Die casting	Narrow : Zinc, aluminium, brass, magnesium	Considerable	Large	0.06 mm	Fair to Good	± 0.025 mm per mm	None	Good	Good	Most economical where applicable
7.	Drop Forging	Medium	Moderate	Large	Small	High	± 0.25 to 0.75 mm	Toughness	Fair	Fair	Used for high strength
8.	Press Forging	Medium : best for non ferrous alloys	Limited	Moderate	Small	High	Medium better than drop forging	Toughness	Fair	Fair	Greater complexity than drop forging.
9.	Upset Forging	Medium : many ferrous and non-ferrous alloy	Limited	Medium	Moderate	High	Medium	Toughness	Medium	Medium	Best suited to small parts.
10.	Welding	Wide	Considerable	Unlimited	Moderate	Variable	Medium, not highly precise	Toughness	Depends upon components	None	Versatile process.

Table 1.4. Manufacturing Processes : Cost considerations.

S. No.	*Manufacturing Process* ↓	*Raw Material Cost*	*Tool and Die Costs*	*Direct Labour Cost*	*Finishing Cost*	*Scrap loss*
1.	Sand Casting	Low to Medium depending on metal	Low	High	High	Moderate : Scrap can be remelted.
2.	Shell Mould Casting	Low to Medium	Low to Moderate	Low	Low	Low
3.	Permanent Mould Casting	Medium	Medium	Moderate	Low to Moderate	Low
4.	Plaster Mould Casting	Medium	Medium	High	Low	Low
5	Investment Casting	High : best for special and costly alloys	Low to Moderate	High	Low	Low : Scrap is remelted.
6.	Die Casting	Medium	High	Low to Medium	Low	Low : Scrap can be remelted.
7.	Drop forging	Low to Moderate	High	Medium	Medium	Moderate
8.	Press forging	Low to Moderate	High	Medium	Medium	Moderate
9.	Upset forging	Low to Moderate	High	Medium	Medium	Medium : Lowest of forging processes.
10.	Welding	Low to Moderate	Low to moderate	Medium to High	Medium	Low : Practically none.

Production technology has become much more sophisticated since World War – II, due to technological innovations and advances in computerisation and quality management. The economic implications of new technology for manufacturing companies are complex. Changes in Production Technology, such as the development of CAD/CAM, have allowed the manufacturing of high quality, tailor made goods at low cost. Changes in Production Technology create whole classes of highly skilled positions to operate the machines and co-ordinate more complex organizational structures that originate from the technologies.

PROBLEMS

1. Discuss the importance of manufacturing science.
2. Define the term "Manufacturing".
3. Define "Manufacturing Engineering".
4. Classify the Manufacturing Process.
5. Discuss the various types of production.
6. Discuss the importance of Computers in Manufacturing
7. Discuss the various factors for selection of a manufacturing process for a given product.
8. Consider a product and describe how its production volume would affect selection of economical manufacturing process for it. (PTU).
9. Give two broad classes of manufacturing processes. Explain differences in these. (PTU).

2

The Casting Process

2.1. GENERAL

Casting is probably one of the most ancient processes of manufacturing metallic components. Also, with few exceptions, it is the first step in the manufacture of metallic components. The process involves the following basic steps :

1. Melting the metal.
2. Pouring it into a previously made mould or cavity which conforms to the shape of the desired component.
3. Allowing the molten metal to cool and solidify in the mould.
4. Removing the solidified component from the mould, cleaning it and subjecting it to further treatment, if necessary.

The solidified piece of metal, which is taken out of the mould, is called as "Casting". A plant where the castings are made is called a "Foundary". It is a collection of necessary materials, tools and equipment to produce a casting. The casting process is also called as "Founding". The word "Foundry" is derived from Latin word "fundere" meaning "melting and pouring".

Types of Foundries. All the foundries are basically of two types :

(*i*) **Jobbing Foundries.** These foundries are mostly independently owned. They produce castings on contract, within their capacity.

(*ii*) **Captive Foundries.** Such foundries are usually a department of a big manufacturing company. They produce castings exclusively for the parent company. Some captive foundries which achieve high production, sell a part of their output.

2.1.1. Advantages of Casting Process.

1. Parts (both small and large) of intricate shapes can be produced.
2. Almost all the metals and alloys and some plastics can be cast.
3. A part can be made almost to the finished shape before any machining is done.
4. Good mechanical and service properties.
5. Mechanical and automated casting processes help decrease the cost of castings.
6. The number of castings can vary from very few to several thousands.

However, casting imposes servere problems from the point of view of material properties and accuracy. Also, a complicated sequence of operations is required for metal casting. Again, the geometric complexity of the final product may be such that this process is of no use.

2.1.2. Applications. There is hardly any machine or equipment which does not have one or more cast components. The list is very long, for example, automobile engine blocks, cylinder blocks of automobile and airplane engines, pistons and piston rings, Machine tool beds and frames, mill rolls, Wheels and housings of steam and hydraulic turbines, Turbine vanes and aaircraft jet engine blades, Water supply and sewer pipes, Sanitary fittings and agricultural parts etc. In machine tools, internal combustion engines, compressors and other machines, the mass of castings may be as great as 70 to 85% of the product's total mass.

2.1.3. Classification of Casting Process. The list of the various casting processes is very long. However, there is one convenient way of classifying these processes. It is according to whether the moulds, patterns (used to make mould cavities) and cores (used to produce internal details in a component) are permanent or expendable (disposable).

1. Expendable Mould Casting. In this process, the mould cavity is obtained by consolidating a refractory material (moulding material) around a pattern. The mould has to be broken to take out the casting from the mould cavity. So, such moulds are *one casting moulds*. The moulding material can be sand or some other refractory material. The main drawback of sand mould casting process is that the dimensional accuracy and surface finish of the castings do not satisfy in many cases the requirements of modern machine building and instrument making industry. However, if the moulding material is used in the form of slurry (Slurry moulding), better surface finish and dimensional accuracy can be obtained.

The pattern used in this process can be permanent pattern (which can be used again and again and is made of wood, metal or plastic) or expendable pattern (Full mould process, lost wax method).

2. Permanent Mould Casting. In this process, the mould is used repeatedly and is not destroyed after the solidification of the casting. The moulds are adaptable to the production of tens and thousands of castings. Generally, *the process is practical for making parts of small and medium mass from light non-ferrous alloys. The castings produced by this method have smooth surface and increased accuracy of dimensions*. Due to the high cost of permanent moulds, the use of this method is limited to mass or quantity production.

3. Semi-permanent Mould Casting. These moulds are prepared from high refractory materials, for example, based on graphite. These moulds are not as durable as permanent moulds. So, these cannot be used for mass or quantity production, but for only a few tens of castings.

In all the above three methods, the cores used may be permanent (metallic) or expendable (made of core sand or of some other suitable material).

An iron foundry may have the following six prominent sections :

(*a*) Moulding and core making.
(*b*) Metal melting.
(*c*) Metal handling and pouring.
(*d*) Knockout.
(*e*) Fettling.
(*f*) Miscellaneous.

2.2. SAND - MOULD CASTING

This process accounts for about 80% of the total output of cast products. As mentioned above, the sand moulds are single-casting moulds and are completely destroyed for taking out the casting, after the metal has solidified in the mould cavity. The moulding material is sand, which is mixed with small amounts of other materials (binders and additives) and water to improve the cohesive strength and mouldability of sand. For making the mould, the moulding material will have to be consolidated and contained around the pattern. The metallic container is called as *flask*. There can be one flask or more than one flasks. The most common design is two flasks system. In the assembled position, the upper flask is called "Cope" and the bottom one "Drag". In three flasks system, the central flask is called "Cheek". One flask design is used in 'Full mould process' or in 'Pit moulding', where it is used as cope, the pit acting as the drag.

Depending upon the type of pattern used, the sand mould casting process is of two types :

1. Permanent or Removable pattern process. Here, the pattern is removed from the mould cavity, before the molten metal is poured into the mould cavity. This is the most common sand mould casting process.

2. Expendable or Disposable pattern process. Here, the pattern is not removed from the mould cavity, before the molten metal is poured into the mould cavity. It gets melted and forms a part of the final casting. This process will be discussed later.

2.2.1. Steps in making a casting by the expendable mould, "removable pattern sand moulding process".

1. Make the pattern. The material of the pattern can be : wood, metal or plastic.
2. With the help of patterns, prepare the mould and necessary cores.
3. Clamp the mould properly with cores placed properly in the mould cavity.
4. Melt the metal or alloy to be cast.
5. Pour the molten metal / alloy into the mould cavity.
6. Allow the molten metal to cool and solidify. Remove the casting from the mould. This operation is called "Shake out".
7. Clean and finish the casting. The operation is called as "fettling".
8. Test and inspect the casting.
9. Remove the defects if any and if possible (Salvaging the casting).
10. Stress relieve the casting by heat treatment.
11. Again inspect the casting.
12. The casting is ready for use.

2.2.2. Types of Sand Mould Casting process. Sand moulds can either be made by hand or on moulding machines. Hand moulding is done for piece and small lot production, whereas machine moulding is employed in large lot and mass production.

Depending upon the nature of the work place, hand moulding process can be classified as:

1. Bench Moulding. This is done only for small work.

2. Floor Moulding. This process is done on the foundary floor and is employed for medium sized and large castings.

3. Pit Moulding. This method is used for very large castings and is done on the foundary floor. However, a pit dug in the floor acts as the lower flask (drag) and the top flask (cope) is placed over the pit to complete the assembly. The walls of the pit are brick-lined and plastered with loam sand and allowed to dry. Sometimes, for large and tall castings, the bottom of the pit is rammed with a 50 to 80 mm layer of coke to improve the permeability of the mould. Vent pipes are run from this layer to the surface (at the sides) and the coke is covered with blacking sand.

In "Machine Moulding", the operations done ordinarily by hand, are done by machine. The operations include : Compacting the sand, rolling the mould over and drawing the pattern from the mould etc.

2.2.3. Types of Sand Moulds. As already discussed, a mould is an assembly of two or more flasks (metallic) or bonded refractory particles, with a primary cavity which is a negative of the desired part. It contains secondary cavities (pouring basin, sprue, runners and gates) for pouring and channeling the liquid metal into the primary cavity. If necessary, it also has a large cylindrical cavity for storing the molten metal (riser or feeding head) for feeding into the primary cavity to compensate for the shrinkage of molten metal in the primary cavity, on cooling and solidification.

According to the material used in their construction, the moulds are of following types :

1. Green Sand Moulds. A green sand mould is composed of a mixture of sand (silica sand, SiO_2), clay (which acts as binder) and water. The word "green" is associated with the condition of wetness or freshness and because the mould is left in the damp condition, hence the name "green sand mould". This type of mould is the cheapest and has the advantage that used sand is readily reclaimed. But, the mould being in the damp condition, is weak and cannot be stored for a longer period. Hence, such moulds are used for small and medium sized castings.

2. Dry Sand Moulds. Dry sand moulds are basically green sand moulds with two essential differences : the sand used for dry sand moulds contains 1 to 2% cereal flour and 1 to 2% pitch, whereas the sand mixture for green sand moulds may not contain these additives. Also, the prepared moulds are baked in an oven at 110 to 260^{0}C for several hours. The additives increase the hot strength due to evaporation of water as well as by the oxidation and polymerization of the pitch. So, dry sand moulds can be used for large castings. They give better surface finish and also reduce the incidence of the casting defects such as gas holes, blows or porosity that may occur as a result of steam generation in the mould (when the molten metal is poured into the green sand mould cavity). However, due to the greater strength of these moulds, tearing may occur in hot-short materials.

3. Skin - dry Sand Moulds. Here, after the mould is prepared, instead of entirely drying it out, the mould is partially dried around the cavity (to a depth of about 25 mm). This can be done in two ways :

(*i*) About 12.5 mm around the pattern, the proper moulding sand (as described under dry sand moulds) is used, the remaining mould contains ordinary green sand.

(*ii*) The entire mould is made of green sand and then the surface of the cavity is coated with a spray or wash of linseed oil, gelatinized starch or molasses water etc. The advantages and limitations of such moulds are the same as of the dry sand moulds.

4. Loam Sand Moulds. Loam sand consists of fine sand plus finely ground refractories, clay, graphite and fibrous reinforcements. It differs from ordinary moulding sand in that the percentage of clay in it is very high (of the order of 50%). This sand is used in pit moulding process for making moulds for very heavy and large parts (engine bodies, machine tool beds and frames etc.).

5. Cemented - bonded Moulds. Here, the moulding sand contains 10 to 15% of cement as the binder. Such a mould is stronger and harder. Such moulds are made in the pit moulding process and develop their strength by air drying and are used for large steel castings. However, it is very difficult to break away the sand from the casting.

6. CO_2 Moulds. The CO_2 moulding process is a sand moulding process in which sodium silicate ($Na_2O.\,x\,SiO_2$), that is, water glass is used as a binder (2 to 6%), rather than clay. After the mould is made, CO_2 gas is made to flow through the mould, the sand mixture hardens due to the following reaction.

$$Na_2O\,.\,x\,SiO_2 + n\,H_2O + CO_2 \rightarrow Na_2CO_3 + \underbrace{x\,.\,SiO_2\,.\,n\,(H_2O)}_{\text{Stiff gel.}}$$

here, x = 1.6 to 4, most often 2.

This reaction is very rapid and takes about 1 minute, which is very much less than the several hours needed to produce a dry sand mould. Such moulds can be used for producing very smooth and intricate castings, because the sand mix has a very high flowability to fill up corners and intricate conturs.

7. Resin - bonded Sand Moulds. Here, the green sand mixture is mixed with thermosetting resins (polymers) or an oil, such as, linseed oil or soyabean oil. During baking of the mould, the resin or oil oxidises and polymerizes around the sand particles, thus bonding them together. The strength of the polymerized resin is greater than that of pitch used in dry sand moulds. So, the moulds produced are stronger. Such a sand mixture is commonly used for making cores.

Baking is usually needed to make strong moulds or cores. Many times, the required strength of moulds is obtained without baking. These moulds are known as "Furan-no bake" moulds, "oil-no bake" moulds etc. Furan is a generic term denoting the basic structure of a class of chemical compounds. The resins used in the "no-bake" systems are compounds of furfuryl alcohols, urea and formaldehyde. A very low water content (less than 1%) is used in the above moulds. The synthetic liquid resin is mixed with sand and the mixture hardens at room temperature.

8. Dry Sand Core Moulds. When the moulding flasks are too large to fit in an oven (for baking) or when it costs too much to dry a large mass of sand, moulds are made from assemblies of sand cores. A sand core is usually prepared from core sand mixtures (discussed later) and is baked at 175 to 230°C for 4 to 24 h, depending upon sand preparation and mass.

9. Cold-box Mould Process. Here, various organic and inorganic binders are blended into sand to bond the grains chemically, imparting greater strengths to the mould. These moulds are dimensionally more accurate than green sand moulds, but are more expensive.

10. Composite Moulds. These moulds are made of two or more different materials, such as shells, plaster, sand with binder and graphite. These moulds combine the advantages of each material. They are used in shell moulding and other casting processes, generally for casting complex shapes, such as turbine impellers. These moulds result in : increased mould strength, improved dimensional accuracy and surface finish of castings and reduced overall costs and processing times.

The above mentioned types of moulds are compared in Table 2.1.

Table. 2.1. Comparison of Moulds

Type of Mould	*Advantages*	*Disadvantages*
a. **Green Sand Mould**	1. Least expensive	1. Sand control is more critical than in dry-sand moulds.
	2. Less distortion than in dry sand moulds, because no baking is required.	2. Erosion of mould is more common in the production of large castings.
	3. Flasks are ready for reuse in minimum time	3. Surface finish and dimensional accuracy deteriorate as the weight of the casting increases.
	4. Dimensional accuracy is good across the parting line.	
	5. Less danger of hot tearing of Castings than in other types of castings.	
b. **Dry Sand Mould**	1. Stronger than green sand moulds, thus are less prone to damage in handling.	1. Castings are more prone to hot tearing.
	2. Overall dimensional accuracy is better than for green sand moulds.	2. Distortion is greater than for green sand moulds because of baking.
	3. Surface finish of castings is better, mainly because dry sand moulds are coated with a wash.	3. More flask equipment is needed to produce the same number of finished pieces because processing cycles are longer than for green sand moulds.

			4. Production is slower than for green sand moulds.
c.	**Dry Sand Core Moulds**	1. Exceptionally good dimensional accuracy can be maintained. 2. Dry sand core moulding is adaptable to green sand foundaries that do not have large drying facilities.	1. Extreme care must be taken in setting the cores.
d.	**Furan, oil, CO_2 Moulds**	1. Sands are free flowing, therefore, ramming is eliminated or reduced. 2. Tensile strengths of moulds are higher than those of conventional moulds. This permits reduction of mould weight and easier handling of large moulds. 3. Moulds can be made without flasks. 4. Production rates are high. 5. Most of these moulds can produce castings to closer tolerances than are obtainable in green sand moulds.	1. Sands must be used immediately

Note. 1. More than 85% of all metal castings are poured in sand moulds, the balance are made in ceramic shell or metal moulds.

2. Majority of castings are poured in green sand moulds.

3. In a foundry, upto 90% of the moulding sand can be reprocessed to make new moulds.

2.2.4. Preparing a Sand Mould. The sequence of operations performed in the making of a sand mould is outlined below. For this, a green sand mould and a split pattern have been chosen.

The appropriate split pattern is made which is split into two equal parts at the parting plane and joined together with dowel pins. We will use a two flask system, (Fig. 2.1).

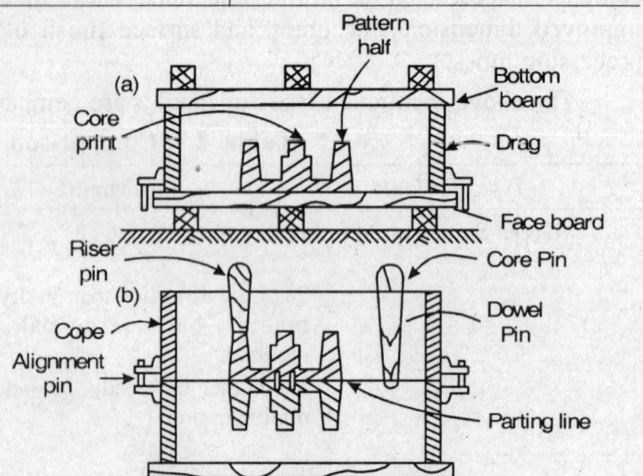

Fig. 2.1. A Green Sand Mould.

A. Hand Moulding Process.

1. The drag half of the pattern, that is, the half with dowel holes rather than dowel pins, is placed with the flat parting plane on a flat board called "Moulding board".

2. The drag is placed over the moulding board with the alignment or locating pins downwards.

3. A parting material is dusted over the pattern and the moulding board to facilitate both the removal of the pattern from the mould and the separation of the two mould halves.

4. The drag is filled with moulding sand and it is packed and rammed around the pattern. The ramming is done manually with hands and with hand rammers (wooden or iron). The sand should be properly rammed, that is, neither too hard nor too soft. If it is too soft, the mould will fall apart during handling or during pouring and if it is too hard, gases produced on pouring will not be able to leave it. Pneumatic ramming or mechanical ramming can be used for large moulds.

5. After the ramming, the excess sand is scrapped off with a straight bar called a "strike rod".

6. Vent holes are pierced in the sand (within 15 to 20 mm of the pattern surface) with "vent wires", for the gases to escape through.

7. A second moulding board is placed on the moulded drag half and clamped if the mould is too heavy to be turned over conveniently by hand. The mould is then turned over 180° and the original moulding board is removed.

8. The cope is mounted onto the drag and the two halves are properly aligned with the help of alignment pins.

9. The cope half of the pattern is properly positioned over the drag half of the pattern with help of dowel pins and dowel holes.

10. For making the sprue and riser, the sprue pin and riser pin are placed approximately 25 mm on either side of the pattern (usually along the parting line passing through the alignment pins).

11. Steps 3 to 6 are repeated.

12. A pouring basin is cut adjacent to the sprue and then the sprue and riser pins are with drawn.

13. The cope is carefully lifted off the temporarily separated from the drag and placed on one side.

14. To take out the split pattern from the drag, "draw spikes" are drawn into the pattern and the pattern is loosened from the sand by rapping them lightly in all directions with a wooden hammer called a "mallet". Then the pattern is lifted off with the help of draw spike. Before with drawing the pattern, the sand around it is moistened with a "Swab" so that the edges of the mould remain firm when the pattern is withdrawn.

15. The gate and runner are cut in the drag or both cope and drag, connecting the mould cavity and the sprue opening. Sometimes, the gate and runner are automatically made with the help of extensions on the pattern. If needed, all the cavity edges are repaired. Dirt remaining in the mould cavity is blown off with a stream of air. If cores are to be used, they are properly placed in position in the drag.

16. The mould is now assembled, the cope being carefully placed over the drag so that the locating pins fit into the holes.

17. If the lifting force on the cope due to the hydraulic pressure of the molten metal is greater than the weight of the cope, the cope must either be clamped to the drag or else weights must be placed on the top of the cope.

18. The mould is now ready for pouring.

Some terms used above are defined below :

(*a*) **Core.** It is made of core sand (or of some other suitable material and even of metal) and is used to make holes in the casting.

(*b*) **Core Prints.** Core prints are the projections on a pattern and are used to make recesses (core seats) in the mould to locate the core.

(*c*) **Pouring Basin.** It is a reservoir at the top of the sprue (in the cope) that receives the stream of molten metal poured from the ladle.

(*d*) **Sprue.** A sprue or downgate is a vertical channel that connects the pouring basin with runners and gates. It is made somewhat tapered downward for ease of moulding and more importantly to have a decreasing cross-sectional area corresponding to the increase in velocity of the molten metal as it flows down the sprue hole. This prevents turbulent flow and hence the drawing in of air along with the liquid into the mould cavity.

(*e*) **Sprue base or Well.** It is a reservoir at the bottom end of the sprue. It prevents excessive sand erosion when the molten metal strikes the runner at the sprue base. Also, there is considerable loss of velocity in the well.

(*f*) **Runner.** The runner is generally a horizontal channel whose functions are to trap slag and connect the sprue base with the gates (ingates), thus allowing the molten metal to enter the mould cavity.

(*g*) **Gates.** The gates (ingates) are the channels through which the incoming metal directly enters the mould cavity.

(*h*) **Risers.** The risers or feed heads are a part of the feeding system. These are reservoirs of molten metal that feed the metal in the casting proper as it solidifies, to prevent shrinkage cavities in the casting.

Bars and Gaggers. With large cope flasks, added support is normally required to keep the moulding sand from sagging or falling out when the cope is raised to remove the pattern. This support is provided by two elements, namely bars and gaggers. The bars subdivide the large cope flask areas into smaller areas which are able to support themselves, (Fig. 2.2). The bars should extend downward to within 25 to 50 mm of the pattern or parting surface. If added support is needed, gaggers should also be used. A gagger is an L-shaped steel rod (6.35 to 12.5 mm in diameter). Gaggers extend vertically downward from the top of the bar to within 25 to 50 mm of the pattern or parting surface. The gaggers are generally dipped in a clay slurry to improve the bonding of the mould sand to the gagger. Or their surfaces should be rough to help them obtain a better grip in the sand. They are placed in position after a depth of about 25 to 50 mm of rammed sand is placed over the pattern and parting surface. Their lower ends are then pressed

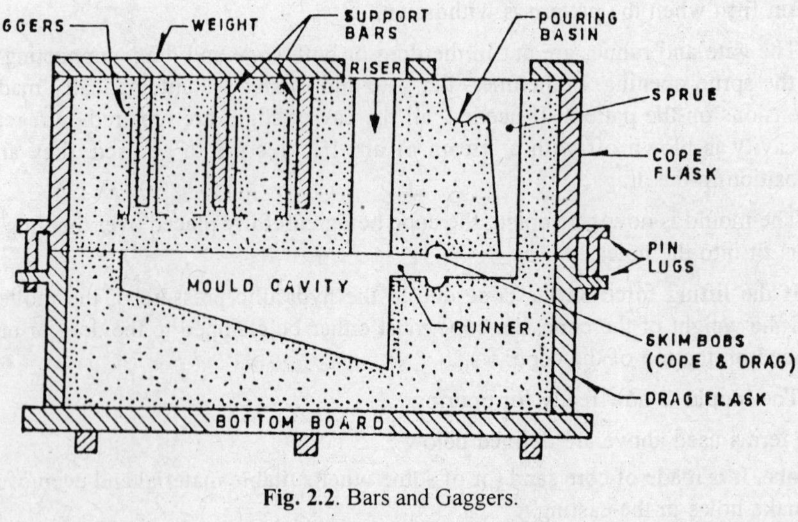

Fig. 2.2. Bars and Gaggers.

into the first layer of moulding sand with their upright portions against bars for support. The function of 'skim bob' in the figure is to trap foreign matter and slag in the molten metal, so that these do not enter the mould cavity. The portion of the skim bob in the cope traps lighter impurities and the heavier impurities are trapped is the portion in the drag.

Soldiers. In some cases, additional support may be secured by the use of nails or soldiers. Soldiers are wooden or steel pins with rough surfaces. These are placed in the moulding sand, as it is rammed, where reinforcement of the mould is needed.

Three Flask Mould. Sometimes a casting has re-entrant sections which make it more convenient to use a three or more flask mould rather than a two flask mould. The flasks between the cope and drag flasks are referred to as "cheeks". Fig. 2.3 illustrates the use of a cheek to cast a wheel having a groove in the rim (for example a V-belt pulley or a rope pulley).

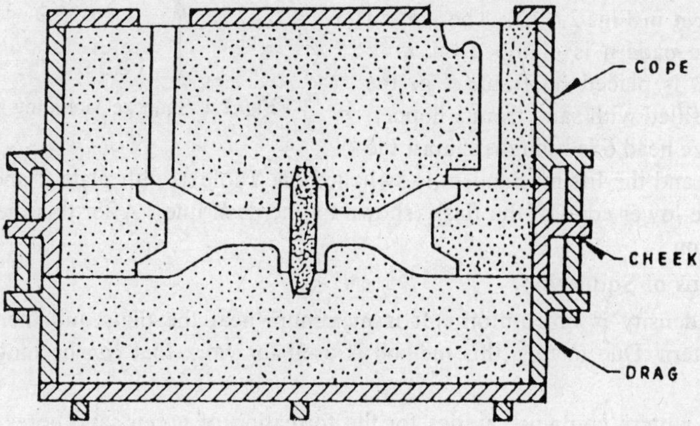

Fig. 2.3. Three Flask Mould.

(B) Machine Moulding. The hand moulding process discussed above is suitable only when the number of moulds to be made is small. But, when large quantities of castings are to be produced, the moulding is done on a machine. Moulding machines pack the sand and draw the pattern from the mould. The use of a moulding machine results in the following advantages :

1. The time required to make a mould is greatly reduced. This reduces the overall costs.
2. The labour productivity is greatly increased, owing to the mechanization of laborious operations of sand ramming and pattern removal.
3. More accurate castings are produced, because the method employs more accurate and less tapered patterns, dispenses with rapping for removing the pattern and secures more exact location of the cope and drag. This results in smaller machining allowances.
4. A higher quality of product is maintained.

A moulding machine consists of a large number of interconnected parts and mechanisms, which transmit and guide various motions in order to prepare a mould.

According to the method in which the sand is compacted around the pattern to make a mould, the moulding machines are classified as :

(*i*) Squeeze Moulding Machines.

(*ii*) Jolt Machines

(*iii*) Sand Slingers

(*a*) **Squeeze Moulding Machines.** These machines are operated by compressed air at a pressure from 5 to 7 atm. A schematic diagram of a top squeeze machine is shown in Fig. 2.4 (*a*). The pattern plate with pattern 2 is clamped on the work table 1 and flask 3 is placed on the plate. Then the sand frame 4 is placed on flask 3. The flask and frame are filled with moulding sand from a hopper located above the machine. Next the table lift mechanism is switched on and the flask together with the sand frame and pattern is lifted up against platen 5 of the stationary squeeze head 6. The platen enters the sand frame and compacts the moulding sand down to the upper edge of the flask (shown by a dash line). After the squeeze, the work table returns to its initial position.

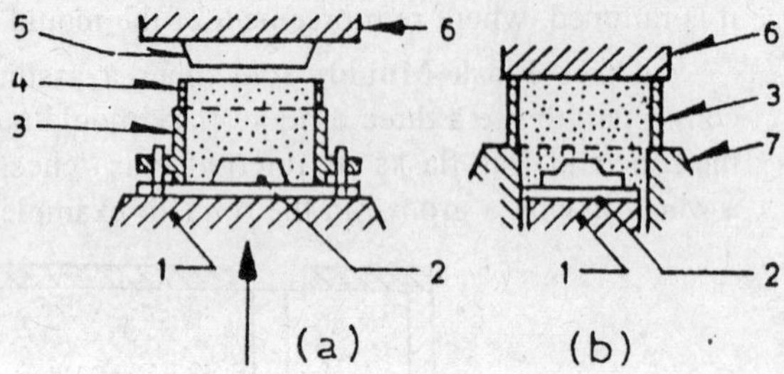

Fig. 2.4. Squeeze Moulding Machines.

The principle of a bottom squeeze machine is shown in Fig. 2.4 (*b*). The pattern plate 2 with the pattern is clamped on work table 1. Flask 3 is placed on frame 7 of the machine and is filled with sand from a hopper. Next, the squeeze head 6 is brought against the top of the flask and the lift mechanism is switched on. Table 1 with plate 2 and the pattern are pushed up to the lower edge of the flask (shown by the dash line). After this the table returns to the initial position.

Limitations of Squeezing.

1. Sand density is not uniform. It is maximum near the plate and then falls gradually towards the pattern. Due to this, this method is used for work that can be moulded in shallow flasks.

2. If the pattern contains cavities for the formation of green sand cores, squeezing does not make sand flow into the cavities effectively and get it packed properly.

(*b*) **Jolt Machine.** In the jolt moulding machine, the pattern and flask are mounted on a mould plate and the flask is filled with sand. The entire assembly is raised a small amount by means of an air cylinder and is then dropped against a fixed stop. The compacting of sand is achieved by the decelerating forces acting on it.

The working of a jolt moulding machine is shown in Fig. 2.5. The table 1 with moulding sand, is lifted by plunger 4 to a definite height (about 5 cm) when compressed air is admitted through pipe 5 and channels 6. Next the table drops since the air is released through hole 7. In falling, the table strikes the stationary guiding cylinder 8 and this impact packs the moulding sand in the flask. Springs 9, by cushioning the table blows, reduced noise and prevent destruction of the mechanism and the foundation. About 20 to 50 drops are needed to compact the sand, and the average machine operates at about 200 strokes per minute.

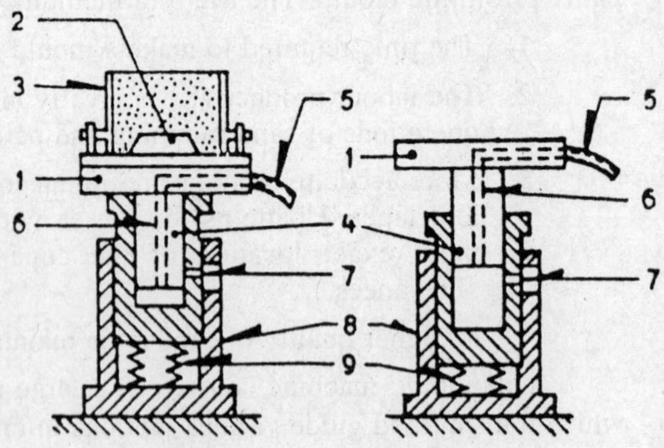

Fig. 2.5. Jolt Moulding Machine.

The drawback of the method is that the density of the sand in the mould is not uniform. It is greatest in the layers next to the pattern plate and lowest near the top of the mould, because in the course of impact, every upper layer acts on the lower layer. Also, there is high level of noise produced by the jolt machines in operation and there is considerable load on the foundation.

(*c*) **Jolt Squeeze Machine.** This machine is similar in appearance to a jolt machine. In addition, it has a vertical column rising above and behind the table to which is fastened a rigid support that overhangs the table. Also, the base contains two concentric air cylinders ; the smaller one to provide the jolting and the larger one to squeeze the mould against the overhead support. On this machine, the pattern can be attached to the match plate. The jolt squeeze method is free of the basic limitations of the squeeze method, that is, weak compaction near the bottom of the mould and that of the jolting method-weak compaction near the top of the mould. The method combines jolting with squeezing, which gives the mould a high and uniform density and increased strength and enables the production of accurate castings of high quality. Jolt squeeze machine now find most widespread use in the foundary practice.

(*d*) **Jolt Squeeze Strip Machine.** The jolt squeeze strip or jolt squeeze vibrate machine is similar to the jolt squeeze machine except that it has an air hose attached to the match plate so that the pattern can be vibrated while it is being with drawn from the mould. In this machine, both the cope and drag halves of the mould are made on one machine. First the drag half of the mould is completed and then the cope half is made with the match plate (with its attached air hose) is assembled between the cope and drag. The operator raises and turns aver the mould halves by hand.

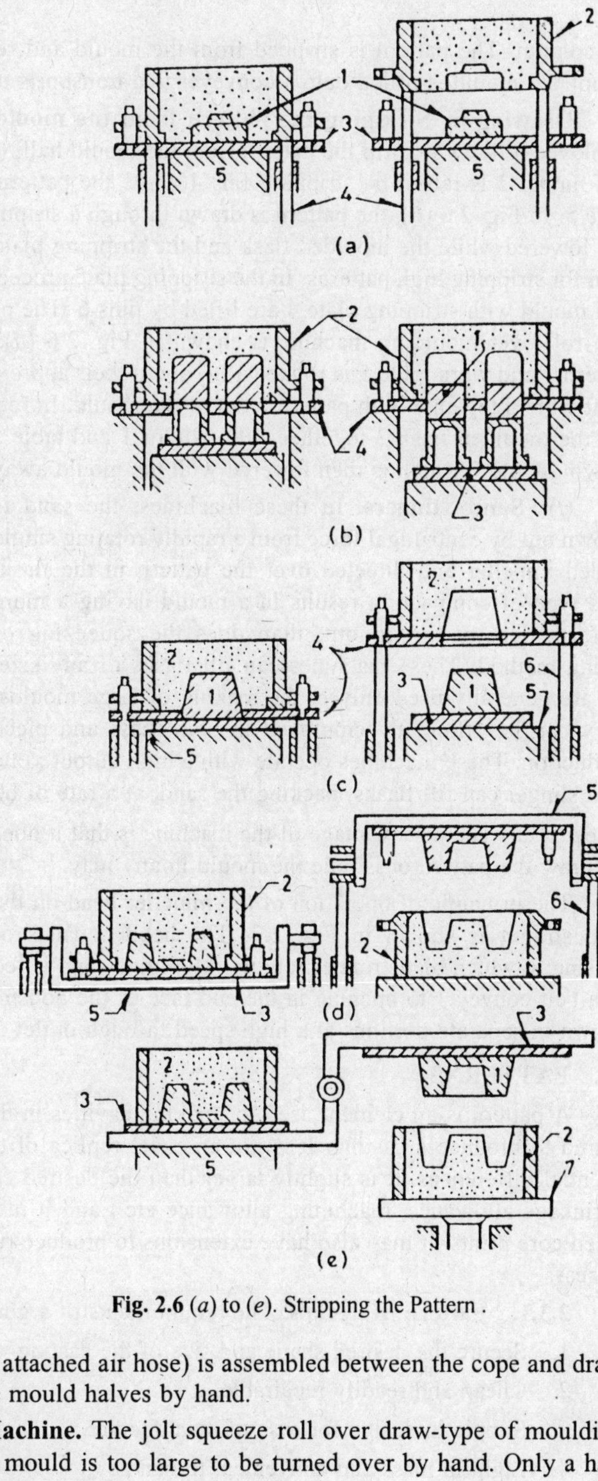

Fig. 2.6 (*a*) to (*e*). Stripping the Pattern .

(*e*) **Jolt Squeeze Roll-over Machine.** The jolt squeeze roll over draw-type of moulding machine is used when the size of the mould is too large to be turned over by hand. Only a half of the mould is made at one time on this machine.

The mould is compacted by jolting on the table in the foreground to which it is clamped. The table is raised and the mould is rolled over the centre column onto the roller table in back of the column. The pattern is stripped from the mould and returned to the jolting table while the completed mould is rolled onto a conveyer that transports it to the assembly and pouring area.

Drawing or Stripping the pattern from the mould halves. There are various methods employed to draw or strip the pattern from the mould half, (Fig. 2.6). In Fig. 2.6 (*a*), the rammed half mould 2 is raised by stripping fins 4 while the pattern 1 with pattern plate remains on the table 5. In Fig. 2.6 (*b*), the pattern is drawn through a stripping plate. Pattern 1 with pattern plate 3 is lowered while the moulded flask and the stripping plate 4 remain stationary. This method is used for stripping high patterns. In the stripping plate procedure shown in Fig. 2.6(*c*), the finished half mould with stripping plate 4 are lifted by pins 6. The pattern drawing principle incorporated in a roll over moulding machine is shown in Fig. 2.6 (*d*). The moulded flask 2 together with pattern 1 and work table 5 is rotated 180^0 about their approximate centre of gravity and then pins 6 lift table 5 together with pattern 1 out of the mould. In rock-over pattern draw method, Fig. 2.6 (*e*), the moulded flask 2 together with pattern 1 and table 5 is swung over by the arm onto the drawing table 7 which is then lowered with the mould away from the pattern.

(*f*) **Sand Slingers.** In these machines, the sand is thrown out by centrifugal force from a rapidly rotating single bladed impeller and directed over the pattern in the flask. This type of compaction results in a mould having a more uniform density throughout, than does the squeezing or jolting method. These machines can fill flasks of any size, but are generally used only in the making of large moulds. They can be efficiently employed in both mass and piece production. These machines operate with a high output ; one sand slinger can fill flasks, packing the sand, at a rate of 60 m^3 per hour. The disadvantage of the machine is that it does not draw the pattern or handle the mould in any may.

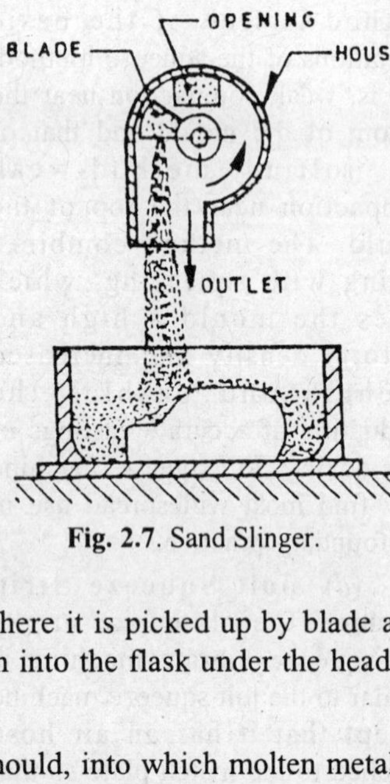

Fig. 2.7. Sand Slinger.

The principle of operation of the impeller head on the sand slinger is shown in Fig. 2.7. The head consists of housing in which blade rotates rapidly. Moulding sand is fed by a belt conveyer to opening in the end face of the housing where it is picked up by blade and thrown in separate portions at a high speed through outlet down into the flask under the head.

2.3. PATTERNS

A pattern is an element used for making cavities in the mould, into which molten metal is poured to produce a casting. It is not an exact replica of the casting desired. There are certain essential differences. It is slightly larger than the desired casting, due to the various allowances (shrinkage allowance, machining allowance etc.) and it may have several projections or bosses called core prints. It may also have extensions to produce runners and gates during the moulding process.

2.3.1. Pattern Materials. The requirements of a good pattern are :

1. Secure the desired shape and size of the casting.
2. Cheap and readily repairable.
3. Simple in design for ease of manufacture.
4. Light in mass and convenient to handle.
5. Have high strength and long life in order to make as many moulds as required.
6. Retain its dimensions and rigidity during the definite service life.
7. Its surface should be smooth and wear resistant.

8. Able to withstand rough handling.

The common materials used in pattern making include "wood, metal, plastic and quick setting compounds. Each material has its own advantages, limitations and field of application. Also, the required accuracy, strength and life of a pattern depend on the quantity of castings to be produced. Based on the above factors, we can choose the pattern material as follows :

(*i*) *Piece and short run production.* Wood

(*ii*) *Large scale and mass production.* Metal, being more durable than wood, though costlier.

(*iii*) *Batch production.* Plastics, for example, epoxy resins and also from gypsum and cement.

(*a*) **Wood.** The wood used for pattern making should be properly dried and seasoned. It should not contain more than 10% moisture to avoid warping and distortion during subsequent drying. It should be straight grained and free from knots.

Advantages.

1. Light in weight.
2. Comparatively inexpensive.
3. Good workability.
4. Lends itself to gluing and joining.
5. Holds well varnishes and paints.
6. Can be repaired easily.

Limitations.

1. Inherently non-uniform in structure.
2. Possess poor wear and abrasion resistance.
3. Cannot withstand rough handling.
4. Absorbs and gives off moisture, so that it varies in volume, warps and thus changes its mechanical properties. These drawbacks, however, can be remedied by drying and seasoning it and then giving coats of water proof varnishes and paints.

The following types of wood are commonly used for pattern making :

(*i*) **White Pine.** It is the most widely used wood, because of its straight grain and light weight and because it is soft, easy to work and unlikely to warp.

(*ii*) **Mahogany.** It is harder and more durable than white pine. Can be worked easily if straight grained. It is less likely to warp than some of other woods.

(*iii*) **Maple, Birch and Cherry.** These woods are harder and heavier than white pine. They tend to warp in large sections, so should be used for small patterns only. They should be carefully treated, because, they pick up moisture readily.

The other common wood materials are : Teak, Shisham, Kail and Deodar.

(*b*) **Metal.** A metal pattern can be either cast from a master wooden pattern or may be machined by the usual methods of machining. Metal patterns are usually used in machine moulding.

Advantages.

1. More durable and accurate in size than wooden patterns.
2. Have a smooth surface.

3. Do not deform in storage.
4. Are resistant to wear, abrasion, corrosion and swelling.
5. Can withstand rough handling.

Limitations.

1. Expensive as compared to wood.
2. Not easily repaired.
3. Heavier than wooden patterns.
4. Ferrous patterns can get rusted.

The common metals used for pattern making are :

(*i*) **C.I.** With fine grain can be used as a pattern material. It has low corrosion resistance unless protected. Heavier and difficult to work. However, it is cheaper and more durable than other metals.

(*ii*) **Brass.** May be easily worked and built up by soldering or brazing. It has a smooth, closed pore structure. It is expensive, therefore, generally used for small cast parts.

(*iii*) **Aluminium.** It is the best pattern material, because it is easily worked, light in weight and is corrosion resistant. It is, however, subject to shrinkage and wear by abrasive action.

(*iv*) **White Metal.** It has low shrinkage, can be cast easily, has low melting point, is light in weight and may be built up by soldering. However, it is subject to wear by abrasive action of sand.

(*c*) **Plastics.** The use of plastics for pattern material results in following advantages :

1. Facilitates the production process.
2. Makes it more economical in cost and labour.
3. Plastic patterns are highly resistant to corrosion, lighter and stronger than wood patterns.
4. Moulding sand sticks less to plastics than to wood.
5. No moisture absorption.
6. Smooth surface of patterns.
7. Strong and dimensionally stable.

Various plastics make good materials for the production of patterns. These are the compositions based on epoxy, phenol formaldehyde and polyester resins ; polyacrylates, polyethylene, polyvinylchloride, and others. In most wide use are cold-curing plastics based on epoxy resins and acrylates.

Plastic patterns are made by one of the following methods :

(*i*) By injecting a plastic material into a die.

(*ii*) Utilizing laminated construction by building up successive layers of resin and glass fibre.

(*iii*) By pouring a plastic material into a plaster mould.

Laminated plastic patterns with surface as smooth as glass are more durable than wood patterns. They draw more easily, are not prone to wear and scratches in service, and can be repaired quickly when damaged. A laminated fibre glass core box can resist tremendous pressures and rough treatments in foundry practice. Various combinations of apoxys, often with lay-on fibre glass for added strength, are employed. Metal grains are also added to the resin, to provide added strength to those areas likely to be abused by foundry methods. The polyurethanes exhibit outstanding resistance to damage, both from abrasion and indentation.

Gypsum patterns are capable of producing castings with intricate details and to very close tolerances. The two main types of gypsum are soft "plaster of paris" and hard metal casting plaster. However, soft plaster does not have the strength of hard plaster. Gypsum can be easily formed, has plasticity and can be easily repaired.

Patterns are also made of a combination of materials used for special purposes. To improve wear qualities and strength, metal inserts are often used, as well as resin-impregnated materials.

Practically, all aluminium match plates and cope and drag plates are cast in plaster moulds. This results in smooth pattern surfaces and accurate reproduction of fine details in the pattern. Due to this castings will have close dimensional tolerances.

Soft wood patterns : upto 50 pieces of medium size castings.

Hard wood patterns : for 50 to 200 castings.

Metal patterns with hardened steel wear plates : 200 to 5000 castings.

Note. Sometimes, if a pattern is first made in wood and then in some other metal by casting, double shrinkages allowances are provided.

2.3.2. Finishing of Patterns. After the patterns are made, they should be finished by sanding so that tool marks and other irregularities are erased. Then they should be applied with 2 to 3 coats of shellac. Shellac fills up the pores and imparts a smooth finish.

The finish of the casting depends on the finish of the pattern. If the pattern is to be preserved for a long period and if a colour scheme is to be used, a good quality enamel paint should be selected to spray or brush paint it.

2.3.3. Pattern Allowances. The difference in the dimensions of the casting and the pattern is due to the various allowances considered while designing a pattern for a casting. These allowances are discussed below :

1. Shrinkage Allowance. Since metal shrinks on solidification and contracts further on cooling to room temperature, linear dimensions of patterns are increased in respect of those of the finished casting to be obtained. This is called the "shrinkage allowance". It is given as mm/m. Typical values of shrinkage allowance for various metals are given below :

C.I., Malleable iron	= 10 mm/m
Brass, Cu, Al	= 15 mm/m
Steel	= 20 mm/m
Zinc, Lead	= 25 mm/m

While laying out a pattern, the dimensions are taken from a Pattern maker's rule, Called "Shrink scale", which is longer than a standard scale by the shrinkage value for the appropriate metal.

2. Machining Allowance. Machining allowance or finish allowance indicates how much larger the rough casting should be over the finished casting to allow sufficient material to insure that machining will "clean up" the surfaces. This machining allowance is added to all surfaces that are to be machined.

The amount of finish allowance depends on the material of the casting, its size, volume of production, method of moulding, configuration of the casting, the position the wall surface occupies in the mould and during pouring. Machining allowance is larger for hand moulding as compared to machine moulding. The largest allowances are taken for the surfaces located in the cope half of the mould, since they are liable to contamination due to slag. Typical machining allowances for sand casting are given in Table 2.2. The allowances are in mm per side. For internal surfaces such as bores, the allowance is about 0.8 mm greater and is negative.

Table 2.2. Typical Machining Allowances for Sand Casting

Material Cast	*Over all length of external surfaces, cm*			
	0 to 30	*30 to 60*	*60 to 105*	*105 to 150*
Al alloys	1.6	3.2	3.0	4.8
Brass, Bronze	1.6	3.2	3.0	4.8
C.I.	2.4	3.2	4.8	6.4
C.S.	3.2	4.8	6.0	9.6

3. Pattern draft or Taper. Pattern draft also termed is "drawn" is the taper placed on the pattern surfaces that are parallel to the direction in which the pattern is withdrawn from the mould (that is perpendicular to the parting plane), to allow removal of the pattern without damaging the mould cavity, Fig. 2.8 (*a*). The draft depends upon the method of moulding, the sand mixture used, the design and economic restrictions imposed on the casting. The common draft is 1^0 to 3^0 After applying the draft, the largest cross-section of the pattern will be at the parting line for external surfaces and reverse will be for internal surfaces, (Fig. 2.8 (*b*).)

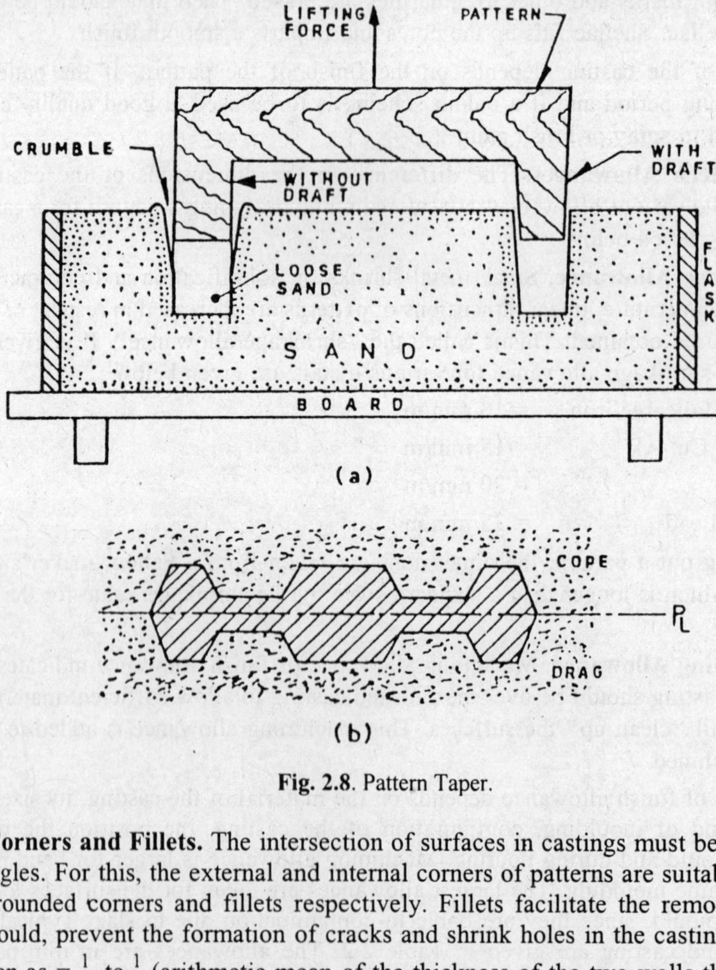

Fig. 2.8. Pattern Taper.

4. Corners and Fillets. The intersection of surfaces in castings must be smooth and form no sharp angles. For this, the external and internal corners of patterns are suitably rounded. They are called rounded corners and fillets respectively. Fillets facilitate the removal of the pattern from the mould, prevent the formation of cracks and shrink holes in the casting. The radius of a fillet is given as $= \frac{1}{5}$ to $\frac{1}{3}$ (arithmetic mean of the thickness of the two walls that form the angle in the pattern).

5. Rapping or Shake Allowance. To take the pattern out of the mould cavity it is slightly rapped to detach it from the mould cavity. Due to this, the cavity in the mould increases slightly. So, the pattern is made slightly smaller.

6. Distortion Allowance. This allowance is considered only for castings of irregular shape which are distorted in the process of cooling because of metal shrinkage.

2.3.4 Types of Patterns. Patterns may be classified as temporary and permanent patterns, depending upon the material used for the pattern. Temporary patterns of soft wood are easily made. However, they soon wear, warp or crack and so have a short life. Hardwood patterns are used more than any other type of patterns. The portions that wear may be protected by sheet metal. Permanent patterns are made of metals (usually Al or brass) or plastics that are easily cast and machined. Patterns may be classified from their utility point of view. The following factors affect the choice of a pattern.

(*i*) Number of Castings to be produced.

(*ii*) Size and complexity of the shape and size of casting.

(*iii*) Type of moulding method to be used.

The common types of patterns are discussed below :

1. Loose Pattern. A loose pattern is simply a replica of the desired casting. It is slightly larger than the casting (due to the allowances discussed above) and it may have several projections called core prints that the resulting casting does not have. Loose patterns get their name because they are not attached or mounted on a plate or frame. These patterns may be made of wood or metal depending upon the volume of production. The gates, runners and risers are added during moulding. This makes mould slow and labour intensive. Due to this, loose patterns are used when the number of castings to be made is small, say, upto 100. Loose patterns are of two types :

(*a*) One piece or solid pattern

(*b*) Split pattern

(*a*) **One piece or Solid Pattern.** This is the simplest type of pattern, exactly like the desired casting. For making a mould, the pattern is accommodated either in cope or drag. The moulding process is quite inconvenient and time consuming. So, such patterns are used for producing a few large castings, for example, stuffing box of steam engine.

(*b*) **Split or Parted Pattern.** These patterns are split along the parting plane (which may be flat or irregular surface) to facilitate the extraction of the pattern out of the mould before the pouring operation. Moulding with a split pattern has already been explained under Art. 2.2.4. Fig. 2.9 (*a*) shows a split pattern for casting a bush. The two parts of the pattern are joined together with the help of dowel pins. For a more complex casting, the pattern may be split in more than two parts.

2. Loose Piece Pattern. When a one piece solid pattern has projections or backdrafts which lie above or below the parting plane, it is impossible to with draw it from the mould. With such patterns, the projections are made with the help of loose pieces. A loose piece is attached to the main body of the pattern by a pin or with a dovetail slide. While moulding, sand is rammed securely around the loose piece. Then the pins are removed. The sand is then packed and rammed around the total pattern. When the main pattern is drawn, the loose pieces remain in the mould. These are then carefully rapped and drawn as shown in Fig. 2.9 (*b*). One drawback of loose pieces is that their shifting is possible during ramming.

Drawbacks. Another technique to make a mould with a one piece solid pattern (with projections) is the use of drawbacks. A drawback is a portion of the mould, which can be drawn

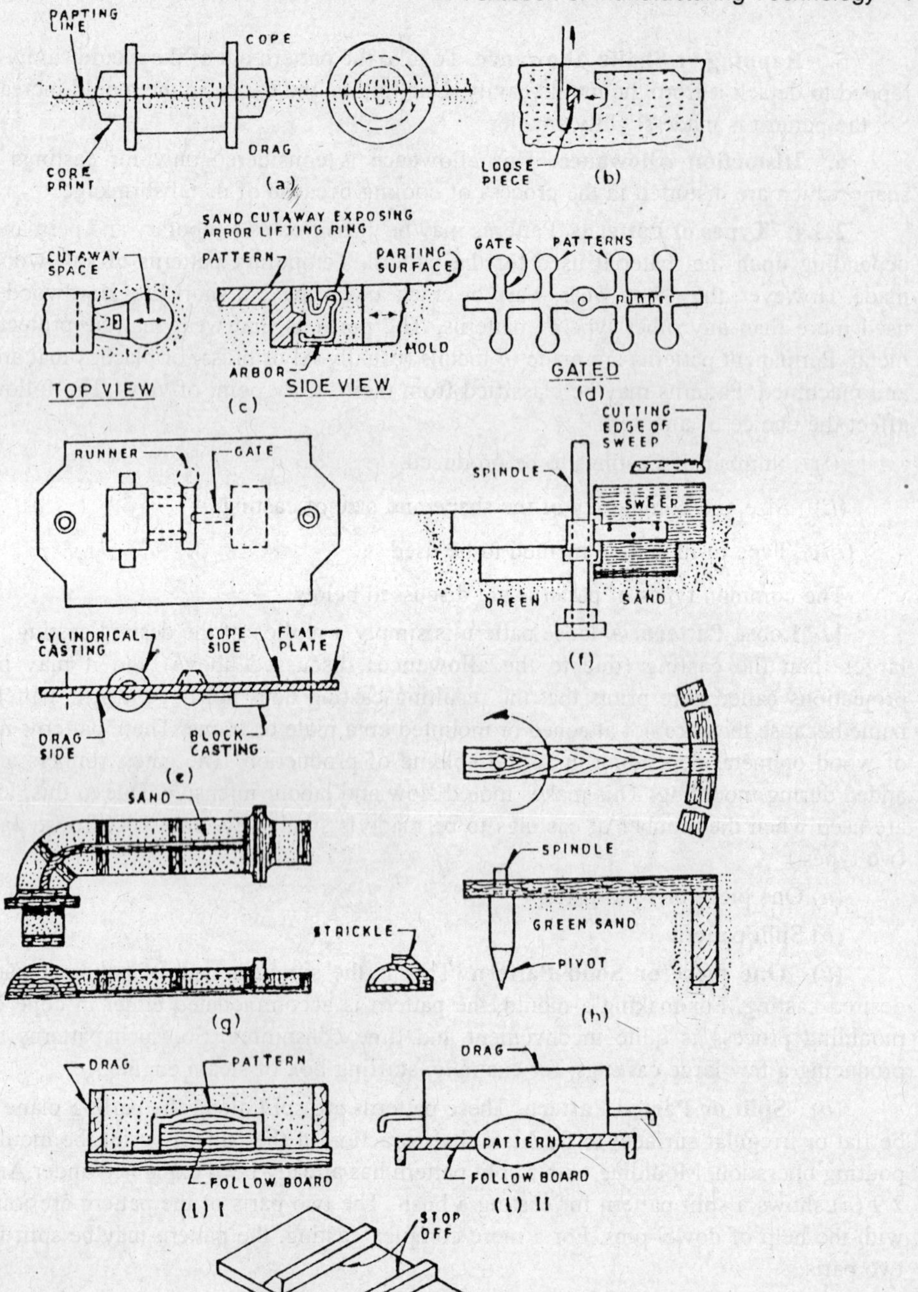

Fig. 2.9. Types of Patterns.

back horizontally in order to allow removal of the pattern. It may be rammed around a rigid support called an "arbor" to facilitate moving it, (Fig. 2.9 (*c*).)

3. Gated Patterns. A gated pattern is simply one or more loose patterns having attached gates and runners, (Fig. 2.9 (*d*)). Since the gates and runners are not to be cut by hand, gated patterns reduce the moulding time somewhat. Because of their higher cost, these patterns are used for producing small castings in mass production systems and on moulding machines.

4. Match Plate Pattern. A match plate pattern is a split pattern having the cope and drag portions mounted on opposite sides of a plate (usually metallic), called the "match plate" that conforms to the contour of the parting surface. The gates and runners are also mounted on the match plate, (Fig. 2.9 (*e*)), so that very little hand work is required. This results in higher productivity. This type of pattern is used for a large number of castings. Several patterns can be mounted on one match plate if the size of the casting is small. The patterns need not all be for the same casting. When match plate patterns are used, the moulding is generally done on a moulding machine. Piston rings of I.C. engines are produced by this process.

5. Cope and Drag Pattern. A cope and drag pattern is a split pattern having the cope and drag portions each mounted on separate match plates. These patterns are used when in the production of large castings, the complete moulds are too heavy and unwieldy to be handled by a single worker. The patterns are accurately located on the plates, so that when the two separately made mould halves are assembled together, the mould cavity is properly formed. For a higher rate of production, each half of the pattern is mounted on a separate moulding machine, one operator working on the cope part of the mould and the other on the drag part of the mould.

6. Sweep Patterns. A sweep is a section or board (wooden) of proper contour that is rotated about one edge to shape mould cavities having shapes of rotational symmetry, (Fig. 2.9 (*f*)). This type of pattern is used when a casting of large size is to be produced in a short time. A complete pattern is not necessary and would be very expensive for a very large casting where the tolerances are large. The moulds are made manually, either in a pit or on the foundary floor. Thus, this type of moulding is referred to as pit and floor moulding. The moulding sand used is the "loam sand". A framework of brick or wood supports the loam. Once the mould is ready, the sweep pattern and the post about which it rotates, are removed before pouring the mould cavity. Large kettles of C.I. are made by sweep patterns.

7. Skeleton Patterns. For large castings having simple geometrical shapes, skeleton patterns are used. Just like sweep patterns, these are simple wooden frames that outline the shape of the part to be cast and are also used as guides by the moulder in the hand shaping of the mould. This type of pattern is also used in pit or floor moulding process. Again, a complete pattern is not necessary and would be quite expensive for a very large casting where the tolerances are large, and when only a small number of castings is to be made.

Fig. 2.9 (*g*) shows a skeleton pattern for casting a pipe. It is a ribbed construction of wood which forms an outline of the pattern to be made. The framework is filled with loam sand and rammed. The sand must be true and smooth. Hence, a strike off board known as a "Strickle board" is used to shape the sand according to the desired shape of the skeleton pattern. For round shapes, the pattern is made in two halves which are joined together by means of screws or with glue etc. This type of pattern is used for casting water pipes, turbine casings etc.

8. Segmental Pattern. The segmental pattern is functionally similar to a sweep pattern in the sense, that both employ a part of the pattern instead of a complete pattern, for getting the required shape of the mould. The segmental pattern is in the form of a segment, (Fig. 2.9 (*h*)), and is used for moulding parts having circular shapes. To create the mould, it is rotated about the post in the same way as in sweep pattern. But it is not revolved continuously about the post to prepare the mould. Rather, it prepares the mould by parts. When one portion of the mould is completed, the pattern is lifted up and moved to the next portion to make the next segment of the mould. This process is continued until the entire mould is completed. Big gears and wheel rims etc. are produced with the help of segmental patterns.

9. Follow Board Pattern. A follow board is not a pattern but is a device (wooden board) used for various purposes as explained below :

(*i*) It is used for supporting a pattern which is very thin and fragile, which may break or collapse under the pressure when the sand is rammed above it. Such a pattern is set on a follow board which is shaped to the surface of the pattern, Fig. 2.9 *i* (*a*). After ramming the sand over the pattern and inverting the flask, the follow board is withdrawn. The other flask is mounted over the first one, filled with sand which is then rammed. During this operation, pattern remains in the drag and gets support from the rammed sand under it.

(*ii*) It helps to establish a parting plane with ease in a pattern that has an irregular shape, Fig. 2.9 *i* (*b*). A hole is cut into the follow board. The pattern is placed in the follow board and then it is used in the same manner as a match plate. A follow board is used with patterns that do not have a flat side and these are one piece solid types. Without the help of a follow board, it will be time consuming and costly to cut out an irregular parting surface and skilled workmen will be required for this.

(*iii*) Gated patterns are frequently set on a follow board which is shaped to the parting surface of the mould. This serves as an interface where the mould separates, which enables the pattern to be removed after the mould is prepared.

Stop-off. This is the portion of the pattern which is added for its own strength only if the pattern is fragile, (Fig. 2.9 (*j*)). It forms a cavity in the mould when the moulder withdraws the pattern, which is refilled with sand before pouring. Stop-offs are wooden pieces used to reinforce some portions of the pattern which are structurally weak, especially from the standpoint of repeated handling. They have no connection with the completed casting.

2.3.5. Colour Scheme for Patterns. Frequently, a print of the finished part is not furnished with a pattern. As a result, the foundary man is not able to take the necessary precautions to produce the best results. Many mistakes may be eliminated by indicating the functions of the various parts of the pattern with proper colours :

(*a*) A loose piece may get lost and unless the pattern is marked to indicate the seat of the loose piece, it is quite possible that the casting will be made from the incomplete pattern.

(*b*) With properly marked core prints, the moulder is constantly reminded that cores must be set in the mould before it is closed.

(*c*) Patterns with stop offs should be marked to remind the moulder to fill the mould cavity made by the stop off.

(*d*) If a moulder knows what surfaces are to be machined, he will, if possible mould the pattern in a position to produce a surface more nearly free of impurities.

The foundaries may use their own colour schemes. A common colour scheme is given below :

1. Surface as cast : Black
2. Machined surface : Red
3. Core prints and seats : Yellow
4. Loose pieces : Yellow/Red diagonal stripes
5. Stop-off : Yellow/Black diagonal stripes

As per Indian Standards, the colour scheme is given in: IS 1513 - 1971

2.3.6. Stack Moulding. Stack moulding is used to make small castings. An advantage of this process is that it requires much less floor space in the foundary. An upright stack moulding process is shown in Fig. 2.10. From 10 to 12 flask sections are arranged one above the other, having a common sprue through which all the moulds are poured.

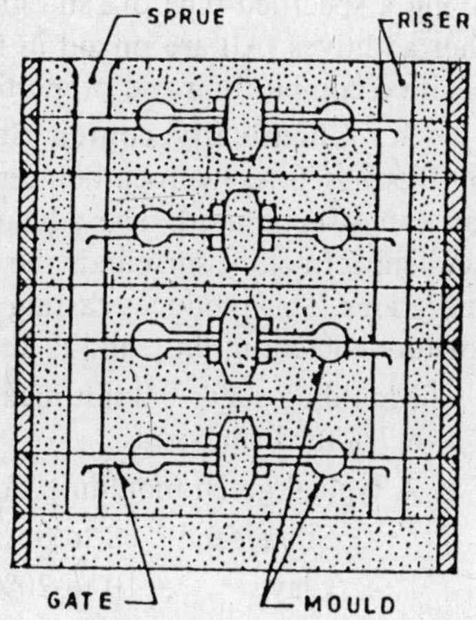

Fig. 2.10. Stack Moulding.

2.4. MOULDING MATERIALS

Various types of materials are used in foundaries for the manufacture of casting moulds and cores. These materials are divided into two groups : Basic and Auxiliary. Basic moulding materials include : Silica sands, which form the base and the various binders. The auxiliary group includes various additives which impart desired properties to the moulding and core sands.

The essential constituents of a moulding sand are : Silica sand, Binder, Additives and water. Silica sand as the widely used moulding material has the following advantages :

1. It is cheap, plentiful and easily available.

2. It is characterised by a high softening temperature and thermal stability, that is, highly refractory.

3. Easily moulded, reusable and capable of giving good details.

Silica sand forms the bulk of the moulding sand.

Binder (most common is clay) may be present in natural sand or added to silica sand separately. In combination with water, it is the bonding agent in a green sand. Although moist sand particles do adhere to one another slightly, but coating with moist clay, the strength (tensile and shear) of the sand mix is increased about three-fold. Clay imparts cohesiveness and plasticity to the moulding sand in the moist state and increases its strength after drying. When the moulding sand has the proper amount of water added to it to give a high strength with sufficient plasticity, the sand is said to be "at temper" or "tempered".

Additives impart to the moulding sand special properties (strength, thermal stability, permeability, refractoriness, thermal expansion etc.)

Sand. According to the amount of clayey matter they contain, the moulding sands are classified as :

1. Silica Sand : Upto 2% clay.
2. Lean or weak sand : 2 to 10% clay
3. Moderately strong sand : 10 to 20% clay
4. Strong sand : Upto 30% clay
5. Extra Strong sand (Loam sand) : Upto 50% clay.

There are three types of sand used for making moulds : natural, synthetic and chemically coated.

1. Natural Sand. A natural sand is the one which is available from natural deposits. Only additives and water need be added to it to make it satisfactory for moulding. The clay content of most natural sands is slightly higher than desired so that new sand can be continuously added to the used sand to replenish that which is lost.

2. Synthetic Sand. A synthetic sand is prepared by mixing a relatively clay free sand having a specified type of sand grain, with a specified type of clay binder as well as water and other additives. All are mixed in the foundary. This sand has the advantage that sand grains of specified composition and properties can be prepared on the basis of the metal being cast.

3. Chemically coated Sand. Clean silica grains are sometimes coated with a non-thermosetting hydrocarbon resin, which acts as a binder. An additional binder in the form of clay can also be used. The advantage of this sand is that the carbon in the resin which is an excellent refractory surrounds the sand grains and does not allow the molten metal to reach the sand grains. This produces casting with clean surfaces as the sand does not get fused in them. The moisture content in this sand is kept to above 3%.

Moulding sands exhibit maximum strength at a moisture content of 4% for lean sands and of 6 to 7% for loam sands.

A typical green moulding sand for gray iron moulding are given below :

Silica sand = 68 to 86%
Clay = 10 to 20%
Water = 3 to 6%
Additives = 1 to 6%

Binders. Binders used in a foundary are : Inorganic and Organic binders. Organic binders are mainly used for core making and will be discussed later.

Clay binders are the most common inorganic binders. Clays are formed by the weathering and decomposition of rocks. The common types of clay used in moulding sand are : Fireclay, Kaolinite, Illite and Bentonite. Kaolinite and Bentonite clays are most popular, because, they have high thermochemical stability.

1. Fire - clay. Fire clay is a refractory clay usually found in the coal measures.

2. Kaolinite. Its general composition is $Al_2O_3 . 2SiO_2 . 2H_2O$. It is one of the decomposition products of the slow weathering of the granite and basalt (a kind of black rock). It is the main constituent of China fire clay. Its melting point is 1750 to 1787°C.

3. Illite. This clay has the approximate composition :

$K_2O . Al_2O_3 . SiO_2 . H_2O$. It is formed from the weathering of mica rocks. Its particle size is about the same as the kaolinite clay and has similar moulding properties.

4. Bentonite. Its general composition is $MgO.Al_2O_3 . SiO_2 . H_2O$. It is formed from the weathering of volcanic or igneous rocks. It is creamy white powder. Its melting point is 1250 – 1300°C.

The basic constituent which gives refractoriness to a clay is alumina, Al_2O_3. Of all clays, bentonite is the most commonly used clay. It needs smaller amount of water to obtain a given degree of plasticity. This will result in less steam generation when the molten metal is poured into the mould, permitting a lower permeability in the moulding sand.

Other binders can be : Portland cement and Sodium silicate.

The percentage of binder in the moulding sand is of great importance. The bond must be strong enough to with stand the pressure of sand erosion by the melt, yet it must be sufficiently weakened by the heat of metal to allow shrinkage of the casting and finally removal of the sand without damage to the solid casting. However, bond must not destroy the permeability of the sand so that gases present in the melt or produced by the heat of the melt in the binder itself can escape.

Organic Binders. These binders are most frequently used in core making. Cereal binders are obtained from wheat, corn or rye; Resins; Drying oils, for example, linseed oil, fishoil, soyabean oil, and some mineral oils; pitch and molasses.

Additives. Additives are the materials added in small quantities to the moulding sand in order to enhance its existing properties and to impart to it special properties. These additives may be necessary to give a good surface finish to the casting or to eliminate casting defects that arise from either the expansion of the moulding sand as it is heated or the contraction of the casting as it cools in the mould. There may be some overlapping between an additive and a binder, because many people include organic binders also in the category of additives.

Metal penetration or burning-on is penetration of metal between sand grains and also strong adhesion of the fused or sintered sand to the metal of the casting surface. The burnt-on sand involves difficulty in cleaning operations and is responsible for rapid wear of cutting tools used to machine the castings. Some common additives used to prevent the above mentioned defects and to improve the quality of the casting, are discussed below :

1. **Sea Coal.** Sea coal or coal dust is finely ground soft coal (pulverised coal). It is added to moulding sand used to make ferrous castings. It tends to obtain smoother and cleaner surfaces of castings and also reduces the adherence of sand particles to the casting. It also increases the strength of the moulding sand. It is added upto 8%. Also, when molten metal fills the mould, coal dust burns and gives off volatile substances containing the gases CO and CO_2 which form a gas spacing between the mould walls and metal. This "gas jacket" not only prevents interaction between the metal of the casting and the sand and thus prevents the metal penetration into the sand, but also makes the mould more collapsible when the metal shrinks.

Other carbon rich materials which are sometimes substituted for sea Coal are : finely ground coke, Pitch and Asphalt (2%), graphite (0.2 to 2%).

2. **Cereals.** Foundary cereal is finely ground corn flour or corn starch. It (0.25 - 2%) increases the green and dry strengths of the moulding sand. Since the cereal is organic, it is "burned out" when hot metal comes in contact with it. This gives rise to space for accommodating the expansion of silica sand at the surfaces of mould cavity, without buckling these. Because of their low density, about 1% is generally sufficient.

3. **Saw Dust.** It increases the gas permeability and deformability of moulds and cores. It must be dry. Instead of saw dust, one can use peat that contains about 70 - 73% volatile substance, not over 5 or 6% ash and upto 25 to 30% moisture.

4. **Wood Flour.** It is ground wood particles or other cellulose materials such as grain hulls. They serve the same purpose as cereals except that they do not increase green strength as much. When required, about 1% is added.

5. **Silica Flour.** It is a very fine ground silica. It is generally mixed with about twice as much conventional moulding sand to make a "facing sand" and is used to surround the pattern. It thus improves the surface finish of the casting, and because of its purity, it increases the hot strength of the mould face. It also resists metal penetration and minimizes sand expansion defects.

Some other common additives are :

(*i*) **Fuel Oil.** It improves the mouldability of sand.

(*ii*) **Iron Oxide.** It develops hot strength.

(*iii*) **Dextrin.** It increases air setting strength, toughness and collapsibility and prevents sand from drying rapidly. During pouring of molten metal, it gasifies, thus producing extra space between the grains without any grain distortion.

(*iv*) **Molasses.** It is the by product of sugar industry. It enhances the bench life of sand and imparts high dry strength and collapsibility to moulds and cores. Due to high temperature in the mould; it develops CO_2 which sets up an hardening action of the mould.

Facing Materials. These materials when added to the moulding sand tend to obtain smoother and cleaner surfaces of castings, help easy peeling of sand from the casting surface during shake out and prevent metal penetration. Sea coal, pitch (distilled from soft coals), asphalt, graphite and silica flour discussed above, act as facing materials.

Cushion Materials. These materials when added to the moulding sand burn and form gases when the molten metal is poured into the mould cavity. This gives rise to space for accommodating the expansion of the sand at the mould cavity surfaces. Wood flour, cereals, cereal hulls and cellulose etc. discussed above are called as 'Cushion materials'.

2.4.1. Other types of Moulding Sands.

1. Facing Sand. This sand is used directly next to the surface of the pattern and comes into contact with the molten metal when the mould is poured. As a result, it is subjected to the severest conditions and must possess, therefore, high strength and refractoriness. This sand also provides a smoother casting surface and should be of fine texture. It is made of silica sand and clay, and some additives without the addition of used sand. The layer of facing sand in a mould usually ranges from 25 to 50 mm.

2. Backing Sand. This is the sand which is used to back up the facing sand and to fill the whole volume of the flask. Old, repeatedly used moulding sand is mainly employed for this purpose.

3. System Sand. In mechanised foundries, where machine moulding is employed a so called "system sand" is used to fill the whole flask. Since the whole mould is made up of this system sand, the strength, the permeability and refractoriness of the sand must be higher than that of backing sand.

Facing sand is always used to make dry sand moulds while system sand is frequently used for green sand moulding.

4. Parting Sand. This sand is used to prevent adhering of two halves of mould surfaces in each moulding box when they are separated. Thus, to ensure good parting, the mould surface (at contact of cope and Drag) should be treated with parting sand or some other parting material. It is also sprinkled or applied on the pattern surface (before the moulding sand is put over it) to avoid its sticking and permit its easy withdrawal from the mould. The parting sand is fine dry sand. There are other parting materials also used in foundary.

Dry parting materials. These are applied by dusting. These include : charcoal, ground bone and limestone, lycopodium (a yellow vegetable matter), tripolite (a silicate rock), ground nut shells, Talc (Magnesium silicate, $3MgO, 4SiO_2.H_2O$) and Calcium phosphate.

Wet parting materials. These are not used with wooden patterns, but are used mostly in machine moulding when metal patterns are used. They are wax based preparations, petroleum jelly mixed with oil, paraffin and stearic acid.

2.4.2. Mould Surface Coatings. Mould surfaces are coated (after the pattern is drawn out) with certain materials possessing high refractoriness. It eliminates the possibility of burn-on and enables castings with smooth surface to be obtained. However, the permeability of the mould gets reduced. Therefore, the coatings should not contain gas forming materials. Mould surface coatings which are also known as facings, dressings, washes, blackings or whitening, may be applied dry (by dusting) or wet in the form of thin cream. The various mould surface coating materials include : Coal dust, pitch, graphite, china clay, Zircon flour or French chalk (Calcium oxide).

2.4.3. Moulding Sand for Non-ferrous Castings. The melting point of non-ferrous metals is much lower than that of ferrous metals. Therefore, the moulding sands for non-ferrous castings may be less refractory and permeable. Also, a smooth surface is desirable on non-ferrous castings. Due to all this, the moulding sands for non-ferrous casting contain a considerable amount of clay and are fine grained.

2.4.4. Typical Moulding Sand Composition (By Weight) :

1. For Steel Castings

(a) *Green sand moulding facing sand :*

Silica Sand	= 95%
Bentonite	= 4%
Cereal type binder	= 1%
Water	= 3 to 4%

(*b*) *Green sand moulding backing sand*

Floor Sand	= 95%
Bentonite	= 5%
Water	= 2.5 to 4%

(c) *Dry sand moulding facing sand :*

(*i*) *For heavy section castings :*

Silica Sand	= 70%
Silica Flour	= 23%
Bentonite	= 5.5%
Mollases	= 1.5%
Water	= 6 to 7%

(ii) *For light section castings :*

Silica Sand	= 75%
Silica flour	= 20%
Bentonite	= 4%
Mollases	= 1%
Water	= 6 to 7%

2. For Iron Castings :

(a) *Green sand moulding facing sand :*

Old Silica Sand	= 60%
New Silica Sand	= 34 to 37%
Bentonite	= 1 to 4%
Coal dust	= 2%
Water	= 3 to 4%

Typical Core Sand Composition :

1. For steel Castings :

(a) *For Light section castings :*

Silica Sand	= 90%
Silca flour	= 6%
Core oil	= 2%
Bentonite	= 1%
Cereal binder	= 1%

(*b*) *For heavy section castings :*

Silica sand	= 43.5%
Silica Flour	= 48.5%

Core Oil = 4%

Bentonite = 4%

(c) *Small cores in average metal sections*

Silica sand = 80%

Silica Flour = 16%

Core Oil = 2.5%

Bentonite = 1.5%

2. For Iron-Castings :

(*a*) *For average castings :*

Silica sand = 93%

Core Oil = 5%

Bentonite = 2%

(*b*) *For heavy section castings :*

Silica Sand = 83%

Silica flour = 9.5%

Core Oil = 5%

Bentonite = 2.5%

A tipical composition of core sand used with CO_2 method, when water glass (sodium silicate) is used as a binder is :

Silica Sand = 80 to 90 kg

Sodium Silicate = 9 to 11 kg

Wood flour = 1.5 to 2.5 kg

Coal dust = 3.5 to 4 kg

2.4.5. Properties of Moulding Sand.

The success of the casting process depends to a large extent on the making of a satisfactory mould. For this, the moulding properties of the sand have to be controlled. These properties include : Porosity or Permeability, Strength, Refractoriness, Plasticity, Collapsibility and Adhesiveness etc. These properties are influenced, not only by the chemical composition, but also by the amount of binder in the sand, by its moisture content and by the shape and size of the silica sand grains. Often, the above variables affect the properties of moulding sand in contradictory manner. For example, the strength and plasticity increase with percentage of clay but this will reduce the porosity of the sand, and so on. The various variables are decided so as to get optimum moulding properties of the sand for a given situation. The properties of the moulding sand are briefly discussed below :

1. Permeability. Permeability or porosity of the moulding sand is the measure of its ability to permit air to flow through it.

Molten metal always contains a certain amount of dissolved gases which try to leave it when the metal solidifies. Also, when the mould is poured, water vapours and other gases are generated when the molten metal comes in contact with the moulding sand and the cores. If all these gases and vapours are not able to escape completely through the walls of the mould, they may penetrate the liquid metal where, after solidification, they form gas holes and pores. To avoid these defects, the moulding sand should have good gas permeability.

Permeability of the moulding sand is influenced by : quality and quantity of clays and quartz in the mixture, moisture content, degree of compaction etc. The higher the percentage of silica content of a moulding sand mixture (that is lower clay content), the better its permeability and vice versa. The larger the grain size, the better its permeability and vice versa. The distribution of grain size in a particular sand mix, greatly influences the permeability. If all the sand particles in a mix are of the same size, then the bulk density is low and the permeability is higher because of the voids between the sand particles. Angular grains that are randomly oriented will have a lower bulk density than spherical grains and thus higher permeability. A sand mixture containing grains of all sizes, from the largest that can be used to the smallest ones, will have a bulk density approaching the true density of the sand. In this case, the permeability of the mixture will be very low, because the smaller particles will fill up the interspaces of bigger grains.

Again, higher the silt content of sand, the lower its gas permeability. If the mould is rammed too hand, its permeability will decrease and vice versa.

2. Strength or Cohesiveness. It is defined as the property of holding together of sand grains. A moulding sand should have ample strength so that the mould does not collapse or get partially destroyed during conveying, turning over or closing. This property also enables the pattern to be removed without breaking the mould and to stand the flow of molten metal when it rushes inside the mould. Also, when the mould cavity is filled with molten metal, its bottom and walls are subjected to a high metallostatic pressure. If the sand is not strong enough to resist this pressure, the mould cavity gets enlarged and the castings obtained are oversize.

Strength of the moulding sand depends on the grain size and shape of sand particles, moisture content and density which varies with the content of clay binders in the sand mixture. The strength of the moulding sand grows with density, clay content of the mix and decreased size of sand grains. So, it is clear that as the strength of the moulding sand increases, its porosity decreases.

3. Refractoriness. It is the ability of the moulding sand mixture to withstand the heat of melt without showing any signs of softening or fusion. This property is greatly influenced by the purity of the sand particles and their size. It increases with the grain size of sand and its content and with the diminished amount of impurities and silt.

4. Plasticity or Flowability. It is the measure of the moulding sand to flow around and over a pattern during ramming and to uniformly fill the flask. This property may be enhanced by adding clay and water to the silica sand.

5. Collapsibility. This is the ability of the moulding sand to decrease in volume to some extent under the compressive forces developed by the shrinkage of metal during freezing and subsequent cooling. This property is especially important for cores. This property permits the moulding sand to collapse easily during shake out and permits of the core to collapse easily during its knockout from the cooled casting. Lack of collapsibility in the moulding sand and core may result in the formation of cracks in the casting. This property depends on the amount of quartz sand and binders and their type.

6. Adhesiveness. This is the property of sand mixture to adhere to another body (here, the moulding flasks). The moulding sand should cling to the sides of the moulding boxes so that it does not fall out when the flasks are lifted and turned over. This property depends on the type and amount of binder used in the sand mix.

7. Co-efficient of Expansion. The sand should have low co-efficient of expansion.

2.4.6. Other Refractory Materials. In addition to silica sand, some other refractory materials which are used for special purposes are :

1. Zircon : $Zr\,Si\,O_4$ 2. Chromite : $Fe\,Cr_2\,O_4$

3. Olivine : $((Mg\,Fe)_2\,Si\,O_4)$

2.4.7. Sand Testing :– Moulding sands should be of proper quality to meet the required level of accuracy and surface finish of castings. Proper quality of the moulding sand will result in sound castings that will decrease the cost/unit and increase the production.

The following tests have been recommended by B.I.S.

1. Moisture content test
2. Clay content test
3. Permeability test
4. Fineness test or Sand grain size test (Sieve analysis)
5. Strength test
6. Mould hardness test.

1. Moisture content test: Moisture is an important constituent of moulding sand. Its content in the moulding sand can be estimated as given below:–

(*a*) **By the loss of weight after evaporation:** Test sample is carefully weighed 50 gm of moulding sand. It is heated for 2 hours at 105°C to 110°C. It is then allowed to cool to room temperature on a desicator, after which it is reweighed. The difference in weight divided by the weight of the sample (50 gm) and multiplied by 100 gives the percentage moisture content.

(*b*) **By Moisture teller :–** In this method, the sample is placed on the teller pan which has 500 mesh screen in the bottom. Hot air is blown through the sand. Moisture gets removed in only about 5 minutes and so the method is quite fast. The sand sample is weighed directly in the teller pan. and the loss in weight is accordingly determined.

(*c*) **By moisture teller based on chemical reaction :** In this method, a measured weight of sand (10 gm) in a cap is placed in the moisture teller along with a container containing a measured amount of CaC_2 (Sand and CaC_2 should not come into contact with each other). The reactions of CaC_2 with moisture in the sand will produce C_2H_2,

$$CaC_2 + 2H_2O \longrightarrow C_2H_2 + Ca\,(OH)_2$$

The amount of C_2H_2 produced is proportional to the content of moisture in the sand. The moisture content is directly measured from a calibrated scale on the instrument.

2. Clay content test : Under this test, the clay content is measured by washing it off the sand sample. The standard dried sand sample (50 gm) is put in a mixing device along with distilled water and 1% NaOH solution. The mixture is stirred for 5 minutes and is then allowed to settle for 10 minutes. The sand settles down at the bottom. The dirty water is siphoned off from the top. The operation is repeated until the water above the sand in the mixer is clean. This water is drained off. The sand is dried and weighed. The loss of weight gives the clay content.

3. Permeability test : As already defined, permeability or porosity of the moulding sand is the measure of its ability to permit air to flow through it. It is measured in terms of "Permeability number." In this test, a test specimen of moulding sand (50.8 mm dia. × 50.8 mm long) is placed in a tube. Time taken for 2000 cm^3 of air at a pressure of 10 gm/cm^2 to pass through the specimen is noted. Then, the permeability number is given as:

$$P = \frac{V \times h}{p \times A \times t}$$

Where,
P = Permeability number
V = Volume of air passing through test specimen
h = Height of specimen
p = pressure of air
A = Cross-sectional area of specimen
t = time taken by air to flow through the specimen

Understandard conditions, the formula becomes

$$P = \frac{2000 \times 5.08}{10 \times \frac{\pi}{4} \times (5.08)^2 \times t}$$

$$= \frac{50.128}{t} \text{ (t is in minutes)}$$

or

$$P = \frac{501.28}{p \times t}$$

4. Fineness Test : Fineness test of the sand specimen determines the size of grains and the distribution of grains of different sizes in the moulding sand. Their effect on the properties of moulding sand has been discussed in Art.2.4.5. The test is performed on completely dry and clay free sand. The test apparatus consists of a set of graded sieves placed one above the other, with the coarsest sieve at the top and the finest sieve at the bottom, next to the pan. The sand is put in the top sieve and the set of sieve is shaken continuously for 15 minutes with the help of an electric motor. At the end of the shaking operation, the sieves are taken out and the sand retained on each is carefully weighed. Each weight is converted to a percentage basis. Each percentage is multiplied by a weighting factor and these are added to get the sum of products. Then the grain fineness number (GFN) is given as,

$$\text{GFN} = \frac{\text{Sum of products}}{\text{Sum of percentages of sand retained on each sieve and pan}}$$

As per AFS (American Foundrymen's Society), the various sievemesh numbers and the corresponding weighting factors (multiplying factors) are given below:

Sieve number	6	12	20	30	40	50	70	100	140	200	270	pan
Weightage factor	3	5	10	20	30	40	50	70	100	140	200	300

Example: Let the weight retained on each sieve are given as :

0.5, 1.0, 1.5, 2.0, 2.5, 4.5, 10, 15, 7.5, 3.5, 1.5 and 0.5 gm. GFN is calculated as explained below:

Sieve number	*Retained Sample (gm) Fi*	*Retained percentage Pi*	*Weightage factor Wi*	*WiFi*	*WiPi*
6	0.5	1	3	1.5	3
12	1.0	2	5	5.0	10
20	1.5	3	10	15.0	30
30	2.0	4	20	40	80
40	2.5	5	30	75	150
50	4.5	9	40	180	360
70	10	20	50	500	1000
100	15	30	70	1050	2100
140	7.5	15	100	750	1500
200	3.5	7	140	490	980
270	1.5	3	200	300	600
Pan	0.5	1	300	150	300
	Σ = 50	Σ = 100		Σ 3556.5	Σ 7113

$$\therefore \quad \text{GFN} = \frac{3556.5}{50} = 71.13$$

or $$\text{GFN} = \frac{7113}{100} = 71.13$$

5. Strength Tests : These tests are carried out to check the holding powers of various binders used in Green moulding sand, dry moulding sand and core sand. These tests are performed on a U.T.M. The various tests carried out include:–

(*i*) Compressive strength test

(*ii*) Shear strength test

(*iii*) Tensile strength test

(*iv*) Bending strength test

Out of these, compressive strength test is most important.

The test specimen is 5.08 cm $\phi \times$ 5.08 long. The test is performed by applying a uniformly increasing load on sand specimen until it breaks. The compressive strength is read directly on the scale of the machine. The other tests are also similarly carried out by using suitable jaws.

The green strength of moulding sands ranges from 30 to 160 k Pa

The dry strength of moulding sands ranges from 140 to 1800 k Pa

The green shear strength of moulding sands ranges from 10 to 50 k Pa.

6. Hardness Test : The hardness of the prepared mould surface is tested to check the ramming density of the sand. The instrument resembles a dial indicator, and carries a spring loaded steel ball of 5 mm . The ball is pressed into the rammed surface. The depth of penetration of the ball is indicated on the dial of the indicator (which is divided into 100 parts) in terms of hardness units. The values of hardness units for sands rammed to different degrees are approximately:–

40 for lightly rammed sand

50 for medium rammed sand

and 70 for hard rammed sand

2.5. CORES

2.5.1. Definition. A core is a body made of refractory material (sand or metal, metal cores being less frequently used), which is set into the prepared mould before closing and pouring it, for forming through holes, recesses, projections, undercuts and internal cavities.

2.5.2. Essential qualities of a Core. The cores are subject to much more severe thermal and mechanical effects than the mould, because, they are surrounded on all sides (except for the ends) by molten metal. Consequently, core sands should meet more stringent requirements. The dry strength and surface strength of a core must be higher than that of a mould. Core sands must have high refractoriness, good permeability and sufficient collapsibility.

The strength of a core will depend on the kind of sand and binding material used. Their bond strength must be greater to allow handling. Also, because they generally are subjected to bending forces produced by the hydrostatic pressure of the liquid trying to lift the core. The bond strength can be increased by using sand with sharper grains which bond together better.

As already discussed, permeability will depend on grain size and distribution of grains in the core sand mix. It will also depend upon the type and content of the binder.

Refractoriness or thermal stability of core can be increased by giving a thin coating of graphite or similar material to the surface of core.

A core should be capable of collapsing shortly after the molten metal has solidified around it. A lack of this ability to decrease in volume may lead to the formation of cracks in the casting as the metal cools. Collapsibility is enhanced by using oil binding materials.

2.5.3. Core Manufacture. The various methods employed for the manufacture of cores will depend on : shape and size of core, type of the binder used and the technique of imparting strength to the cores. The common core making methods are discussed below :

Core Sand: The core sand mainly consists of silica sand and an organic binder, with very little, if any, clay content, as noted under Art 2.4.4. The presence of clay in core sand reduces its permeability and collapsibility. The core sand may contain small percentages of other constituents also, to enhance its properties. The common organic binders used in core sands include: core oils, dextrin, cereals, sulphite liquor, mollases, resins and other commercial binding materials. The main ingredients of core oils are vegetable oils, for example linseed oil and corn oil. To these may be added: fish oil, petroleum oils and coal tar. Core oils impart sufficient strength and good collapsibility. In the CO_2 method of core making, water glass (sodium silicate) is used as the binder. The resin binders are discussed ahead.

1. Conventional Method :-

Cores are made manually or with machines. Cores are made by hand in core boxes or by using sweeps ; only core box methods are used for machine coremaking.

(*A*) **Hand Coremaking.** The manual method of making cores in core boxes is the most widespread method, because it is rather simple and effective. However, the method is useful for piece and small-lot production of castings only.

Coremaking consists of the following sequence of operations :

(*i*) Moulding a green core

(*ii*) Baking or curing

(*iii*) Finishing

(*iv*) Coating

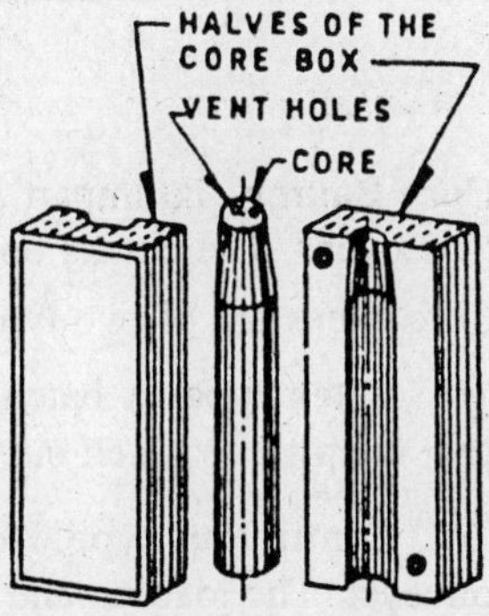

Fig. 2.11. Core box.

For moulding a green core, usually a split wooden core box is used, Fig. 2.11. If a core is made of two or more pieces, they are assembled together after baking by pasting or other methods. Refer to Fig. 2.11. The two halves of the core box are fastened together with clamps or dowel pins and the box is placed vertically on the bench. The operator then rams the sand, layer by layer, into the box, trims it off level along the top face.

Reinforcement of Cores. Some cores require internal reinforcing within the sand to increase their strength. This is done by embedding wires, rods or arbors within the core sections while they are being moulded.

Reinforcement (core irons) is most often made of annealed low carbon steel wire upto 8 mm in diameter. Heavy cores with a large cross-section may be reinforced with cast iron grids. It is important to properly locate the core iron in the core. It should never be too close to the surface, otherwise a thin layer of sand may break off.

Core Venting. While pouring the mould with molten metal, mould walls and cores heat up rapidly and give off a large volume of gases. These gases must be vented out through core prints in order to prevent casting defects such as "Blow holes". Vent holes may be made :

(*i*) By piercing the sand with a wire from the core print end after the sand has been rammed into the core box.

(*ii*) Additional vent holes are made with steel rods or pipes moulded into the cores and subsequently removing these after core moulding, leaving straight vent holes.

(*iii*) By inserting wicks, which form holes after wax melting and chord removal.

(*iv*) In two part cores, before they are pasted together, vent channels are cut along the parting surface.

(*v*) Large cores (round cores over 500 mm long and of a diameter upto 60 mm and more) are made hollow by using sweeps or templates or with cinder or coke filled centres to help provide proper venting.

After the core has been reinforced and vented, the core maker gently knocks on the box with a mallet to facilitate the core removal. The core box is laid in a horizontal position. The clamps or dowel pins are loosened, one box half is lifted off and the core is withdrawn from the other half.

Core Dryer. The cores must have adequate support while in the green state before they are baked or cured. For this, the cores are placed on a drying board or plate known as 'Core dryer', (Fig. 2.12), for transporting the cores to the oven. If a shaped core dryer is not available for a core of complex configuration, a sand bed can do well.

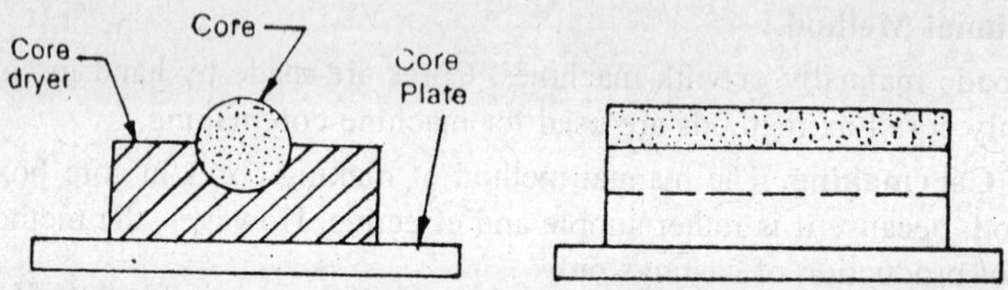

Fig. 2.12. Core dryer.

Core Baking. To impart proper strength to the cores, these are baked or cured in gas, oil or electric ovens. The curing operation drives off the moisture and hardens core binders.

Core ovens are of two types :

(*i*) **Batch type.** A batch of cores on the drying plate are placed in the oven. When they are cured, the plate is taken out and next batch of cores is placed in the oven.

(*ii*) **Continuous type.** Here, the cores on the drying plate move slowly through the oven, on a conveyer. The loading and unloading is continuous and the curing time is controlled by the rate of travel of the conveyor.

Curing temperature is in the range of 230°C, which should be closely controlled by suitable controlling instruments.

Dielectric Core Baking. This method of core baking is used for materials which are non-conductors of electricity. An A.C. field is establishes between two parallel plates (conductors) which function as an electric condenser. The material to be heated (here cores) is placed between these two conductors. When the high frequency A.C. (15 MHz) is switched on, heat is created within the molecules of the cores. A great advantage of the method is that the cores are heated uniformly throughout. Linseed oil (which requires oxidation for imparting strength to the cores) is not suitable for this method. The thermosetting synthetic-resin binders which do not need oxidation and cure at about 110°C are well suited for this method. Small sized cores can be baked only in a few seconds, while the larger ones will take only a few minutes for complete curing.

Plastic or Synthetic Resin Core Binders. These binders have got some important advantages : Approximately half the baking time as compared to linseed oil binders, and excellent collapsibility which will reduce casting defects like hot tears and also reduce the core knockout effort. Urea formaldehyde binders burn out faster and, therefore, collapse at lower temperature as compared to phenol formaldehyde binders. Thus, urea formaldehyde binders are used with low melting point metals such as Aluminium, Magnesium and thin section brass or bronze castings. On the other hand, phenol formaldehyde binders are used for cast iron, steel and thicker section brass and bronze castings.

(*B*) **Machine Core Making.** In the mass and large lot production of castings, cores are made on core making machines, which at present are progressively introduced into foundaries producing castings in small lots.

Coremaking machines increase the productivity of labour, make easier the work of operators, and produce cores of high accuracy, largely from sands of lowered green strength, which flow readily into deep pockets of the core box.

There are various types of core making machines used in the foundary : core blowing, jolting, squeezing, slinging and screw feed (extrusion) types. The principles of jolting, squeezing and slinging machines have already been discussed under Art. 2.2.4. Below, we shall discuss core blowing machines and screw feed machine.

(*a*) **Core Blowing Machines.** These machines can produce small and medium sized high quality cores of any shape, by forcing the core sand into a core box. The core box is made of metal, desirably of Aluminium, with numerous holes provided for the escape of air. These small holes (vented plugs) are covered by a fine screen to prevent the escape of sand. The compressed air at a pressure of about 700 kPa passes through a valve into a container (filled with core sand) and forces the core sand into the core box and compacts it (Fig. 2.13). The core box is filled in a few seconds. The drawback of the method is that the core sand has to be of low green strength, which involves difficulties in core transportation, removal of cores from boxes and so on.

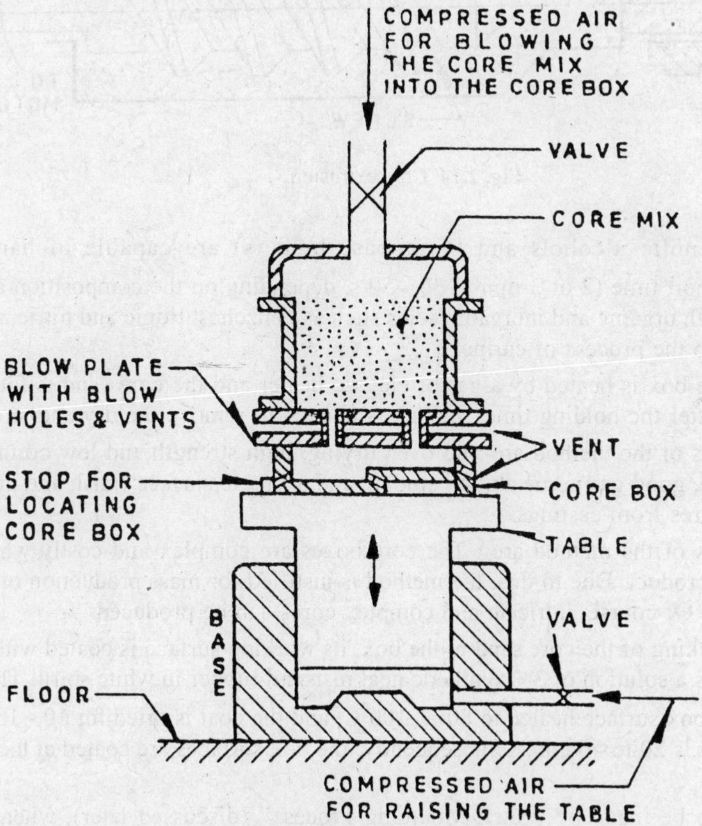

Fig. 2.13. Core blowing machine.

In Fig. 2.13, the core box is placed on the table, located and pressed up against the blow plate. Then the compressed air is introduced above the sand in the reservoir. The core box is filled and rammed in less than a second, and it is then lowered and removed from the machine.

(*b*). **Extrusion Machines.** These machines, (Fig. 2.14), produce small cores of any uniform cross-section. The core sand is extruded by a screw or piston through a replaceable die (the section of the die will depend on the section of the core) onto a plate. The sand goes out in the form of a continuous core which is cut to the required length after baking. The core prints are then ground if necessary. The machines are simple in design, reliable in operation, and highly productive.

The machine produced cores are then cured as usual.

2. Hot Core Box Method. The main drawback of the conventional method of coremaking is that baking or curing of the cores is needed to impart strength to them. This lengthens the production process and lowers the operating efficiency of the foundry.

The procedure of core drying in ovens is eliminated by the process of production of cores from quick-set synthetic resins-bonded core sands. These binders (phenol formaldehyde, Urea formaldehyde, Phenolic alcohols and Furan-base binders) are capable of hardening at 230 – 250°C in a short time (2 or 3 min to 30 - 50 s, depending on the composition and size of cores). Catalysts (both organic and inorganic acids such as benzenesulfonic and nitric acids), may be added to speed up the process of curing.

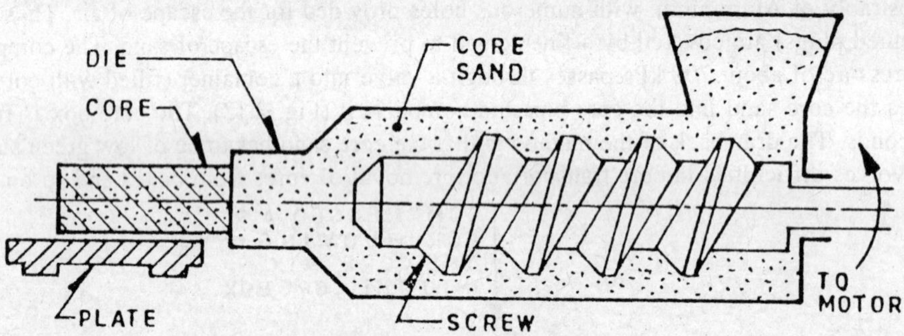

Fig. 2.14. Core extrusion.

The metal core box is heated by a gas or electric heater and the core sand is forced into it by a core blower. After the holding time, the box is opened to remove the dry and strong core.

The advantages of the method are : no oven drying, high strength and low crumbleness of cores in the dry state, good gas permeability, smooth and accurate surface finish and provides for easy shake out of cores from castings.

The limitations of the method are : The core boxes are complex and costly which add to the cost of finished product. Due to this, the method is justified for mass production of small and medium-sized cores. Of course, intricate and complex cores can be produced.

To prevent sticking of the core sand to the box, its working surface is coated with a parting composition which is a solution of 3% synthetic heat resistant rubber in white spirit. The solution is sprayed on to the box surface heated to 80 – 100°C, and the coat is dried for 10 - 15 min. The durability of the coat is 20 to 50 cores. Subsequently, the box surfaces are coated at the operating temperature.

Cores can also be made by "Shell Moulding Process" (discussed later), where a heated match plate is used.

3. Synthetic Resin - Bonded Cold Curing Core Sands Method. These sands do not call for heat treatment to get the required strength of the core. The same binders are used as in the hot box method. Catalysts are added to speed up the hardening of binders. The ready mix is poured into the core boxes where it is held for sufficient time until the required compressive strength is reached (300 to 500 kPa). The core is then removed from the box and allowed to stand in the air for 30 to 120 min. As a result, the core acquires a strength of about 800 to 1200 kPa.

Cold curing sands have a lower strength than sands curable in hot boxes, and therefore, they are largely used for the production of cores of moderately complex and simple configuration. The total hardening time may reach a few hours.

The advantages of the method are :

Core boxes of wood, plastic and metal can be used ; core production process can be mechanized, improves the quality of castings and increases the output. The method has found the widest application in the batch production of moderate sized and large-sized castings from iron and steel.

No Bake Sand :

It is a special type of core making sand. The cores made from these sands do not require heat for baking or curing. It is based on the fact that when the sand mixture containing part A (binder) and part B (Catalyst) is mixed with sand mixture containing part C (hardener), it gets hardened. The hardening time depends upon the composition of part C and the room temperature. Greater is the percentage of part C, shorter will be the hardening time. The commonly used hardener is Tetra (or Tri) ethyl Amine gas (TEA gas). It acts as hardener due to its addesive properties.

For making a core, the sand mix (Part A and Part B) is blown into the core-box with air pressure of 0.3 to 0.6 N/mm^2. Then, the TEA gas is passed to harden the core. The amount of TEA gas passed depends upon the size of the core. This no bake sand is called the ABC sand.

In Furan no-bake core sand

Furan no-bake resin = 2%

Activator (phosphoric acid) = 40%

4. Cold Curing CO_2 process. The CO_2 process has already been discussed for the drying and hardening of moulding boxes. The sand mixture includes : pure silica sand (without clay) and sodium silicate liquid base binder (water glass). These two are mixed in a muller for about 3 to 4 min. The moisture content should not exceed 3%. Sawdust (about 1.5%) and asbestos powder (upto 5%) make the cores more deformable and collapsible. The prepared mixture is packed in the core box for core making manually or by machines. After this, CO_2 gas is forced through the core sand under a pressure of about 140 kPa. This CO_2 reacts with sodium silicate to form 'silica gel' which binds the sand grains together to provide strength and hardness to the core. The chemical reaction is as follows :

$$Na_2O \,.\, x\, SiO_2 + n\, H_2O + CO_2 \rightarrow Na_2CO_3 + x \,.\, \underbrace{SiO_2 \,.\, n\,(H_2O)}_{\text{Stiff gel.}}$$

x is usually 2.

The gassing process lasts for about 15 to 30 s only. Immediately after gassing, the core is ready for use.

This method is used in the piece and batch production of steel and iron castings. To improve the strength of these cores, they are dried at 200 – 250°C.

2.5.4. Finishing and Coating of Cores. Before the cores are transported to the stores, they are subjected to additional treatment : burr trimming, sealing for correcting small defects, and so on. The cores are trimmed off with an emery stone, file, wire, hard rubber and emery paper. The grinding process is done to give smooth surface finish and accurate dimension to the core.

The coatings on the core surface prevent metal penetration, increase the surface strength, decrease the crumbleness of core and provide clean and smooth appearance. The coatings consists of refractory materials along with binding materials. The choice of coating depends on the kind of metal cast, mass of the casting and the moulding method. For large iron castings, the coating may consist of black lead, bentonite and binders. For small and medium sized castings, the coating may be of graphite or of silica flour mixed with coal and ground coke. For steel castings, the coating consists of silica flour or Zirconium silicate mixed with some binders.

2.5.5. Types of Cores. The cores can be classified in many ways :

(*a*) **According to state of Core**

(*i*) Green sand core

(*ii*) Dry sand core

A green sand core is formed by the pattern itself when it is being rammed [Fig. 2.15 (*a*)]. the core is made of the same sand as the moulding sand. Such cores are weak and can be used only for light castings.

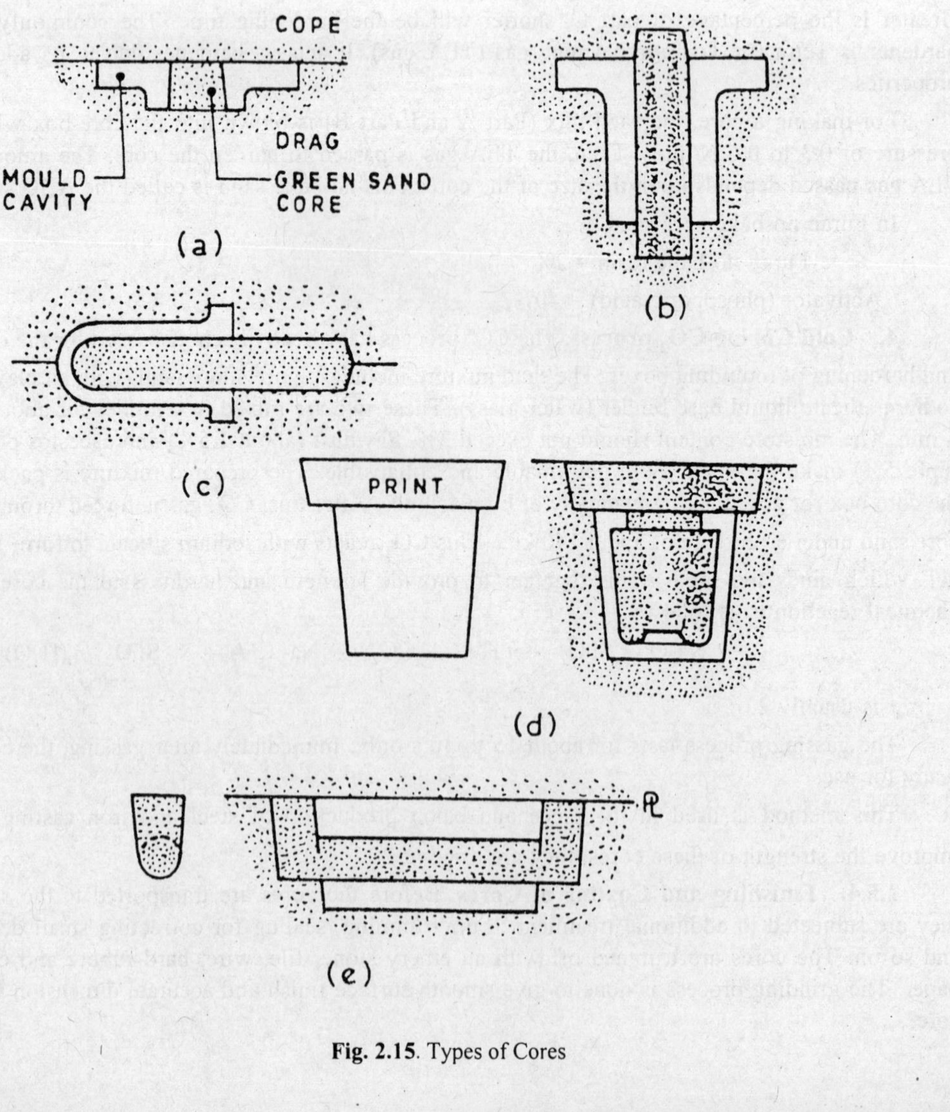

Fig. 2.15. Types of Cores

Dry sand cores are made separately (as explained above) and then positioned in the mould, after the pattern is taken out and before the mould is closed. These are the cores most commonly used.

(*b*) **According to the position of the Core in the Mould.**

(*i*) **Horizontal Core.** This core is placed horizontally in the mould and is very common in foundaries. It is usually positioned along the parting line of the mould.

(*ii*) **Vertical Core.** These cores are positioned vertically in the mould, [see Fig. 2.15 (*b*)]. The two ends of the core rest on core seats in cope and drag. The maximum portion of the core is supported in the drag.

(*iii*) **Balanced Core.** [Fig. 2.15 (*c*).] These cores are used when blind holes along a horizontal axis are to be produced. Since the core is supported only at one end, the core seat should be of sufficient length to prevent its falling into the mould.

(*iv*) **Hanging Core.** [Fig. 2.15 (*d*).] These cores are used when a cored casting is to be completely moulded in the drag, with the help of a single piece solid pattern. The core is supported above and hangs into the mould. A hole in the upper part of the core is provided for the metal to reach the mould cavity.

(*v*) **Drop Core.** [Fig. 2.15 (*e*).] This core is used when a hole which is not in line with the parting surface is to be produced at a lower level.

2.5.6. Chaplets. Sometimes it is not possible to provide sufficient support for a core in the mould being poured, particularly if the cores are very big in size. In such cases, the core is supported with rigid metal pieces, called "Chaplets", [Fig. 2.16 (*a*)], placed between the core and the mould face. The material of the chaplet must conform to the molten metal or alloy being cast in order to effect proper fusion. For iron or steel castings, the chaplets are made of tin plate and low-carbon steel.

The surface of the chaplets must be clean, without any trace of corrosion, moisture or oil. Chaplets are shot peened (to increase their compressive strength) and then tinned before use. Clean chaplets fuse well with the metal and do not cause the formation of holes and discontinuities. It is, however, undesirable and even, sometimes, impermissible to use chaplets for castings to be used as pressure vessels.

Fig. 2.16 (*b*) shows the various forms of chaplets used in the foundary.

2.6. GATES AND RISERS

2.6.1. General. After the mould has been prepared, the molten metal is to be poured or injected into the mould cavity. The problem of introducing the molten metal into the mould cavity so as to obtain a sound casting, is one that exists in connection with all casting processes. Unfortunately, because of the large number of variables present, the problem has not been solved completely by scientific analysis. Therefore, differences exist as to the correct procedures to use and the solution in a particular case usually depends on the application of previous knowledge and trial and error, guided by an understanding of certain basic factors.

The problem basically is one of introducing the molten metal into the mould cavity in a manner which does not erode the mould, fills the mould completely without entrapping gases and permits cooling and solidification to progress in such a way that the accompanying shrinkage may take place without resulting in voids, cracks or porosity within the casting.

As the molten metal cools, solidifies and further cools, shrinkage occurs in connection with each step. Normally, the metal adjacent to the walls of the mould solidifies first, resulting in a shell of solid metal. The centre of the section remains liquid while there is a zone between the liquid interior and solid exterior wherein the metal is in a semisolid or mushy state. As

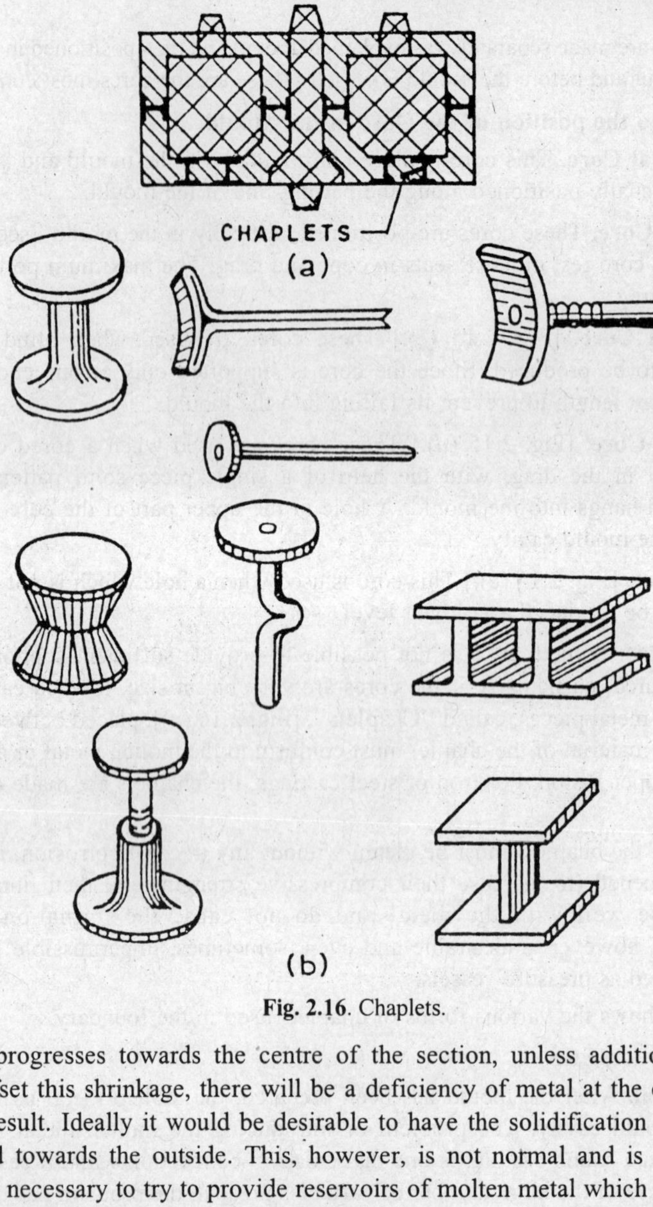

Fig. 2.16. Chaplets.

solidification progresses towards the centre of the section, unless additional liquid metal is supplied to offset this shrinkage, there will be a deficiency of metal at the centre and a void or porosity will result. Ideally it would be desirable to have the solidification started at the centre and progressed towards the outside. This, however, is not normal and is difficult to achieve. Therefore, it is necessary to try to provide reservoirs of molten metal which can feed the various sections of the casting as solidification and shrinkage occurs.

2.6.2. Gating System. The molten metal from the ladle is not introduced directly into the mould cavity, because it will strike the bottom of the mould cavity with a great velocity and can cause considerable erosion of the bottom of the mould cavity. Due to this, the molten metal is introduced into the mould cavity from the ladle, through a gating system. The gating system for a casting is a series of channels which lead molten metal from the ladle into the mould cavity. It may include any or all of the following, [Fig. 2.17 and 2.17A] :

(*i*) Pouring Basin

(*ii*) Sprue

(*iii*) Sprue base or well

(*iv*) Runner

(v) Choke

(vi) Skim bob

(vii) Gates and / or ingates.

(viii) Riser

1. Pouring Basin. A pouring basin or cup is a reservoir at the top of the sprue that receives the stream of molten metal poured from the ladle. Sometimes, the metal is directly poured into the top of the sprue, which is made with a funnel shaped opening. However, better results are usually obtained with the use of a pouring basin.

The pouring basin may be moulded into the cope at the top of the sprue, or it may be made of core sand and placed on the cope above the sprue. It may also be made from C.I. for non-ferrous alloys, if chilling is not a problem, or it may be lined with core sand to reduce the chilling effect. Such pouring basins are also placed on the cope above the sprue.

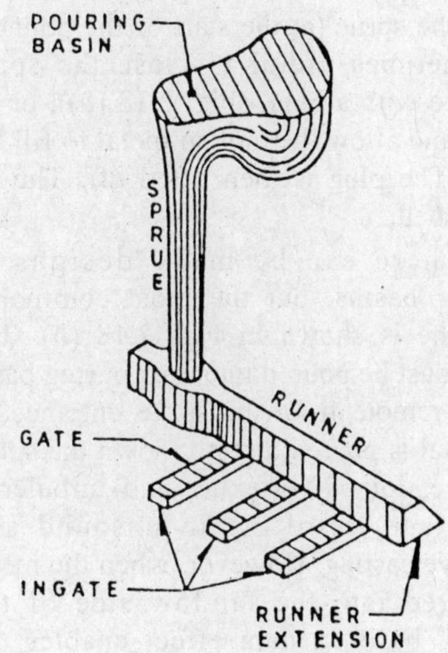

Fig. 2.17. Gating System.

The pouring basin is filled quickly at the start of the pour and it should remain full of molten metal during pouring. Thus, dross consisting of oxides and slags which float, may be kept

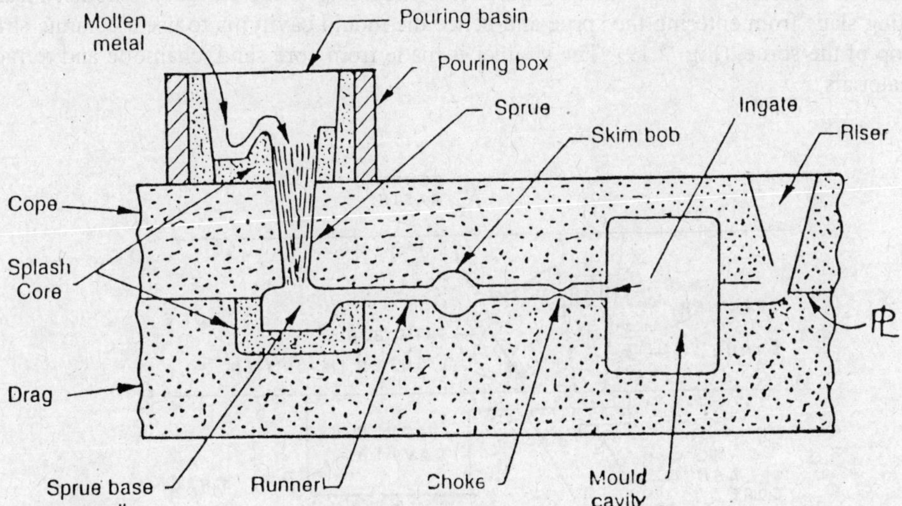

Fig. 2.17A. Gating System

from entering the sprue. If the depth of metal in the cup is insufficient, a funnel is likely to form above the sprue entrance, through which air and slag may get into the sprue and then into the mould cavity. The depth of the pouring basin is a function of sprue entrance diameter. This will prevent the formation of a vortex at the top of the sprue and hence the drawing in of air and dross into the sprue.

Experience has shown that liquid metal depth above the sprue entrance should be 2.5 times the spruce entrance diameter to prevent the formation of a vortex.

To prevent the slag from flowing down the sprue (at the start of the pour), it is sometimes useful to close the sprue entrance with a plug, (Fig. 2.18 (*a*)), or tin plates and allow the molten metal to fill the basin. The plug is then lifted off. The tin plates melt.

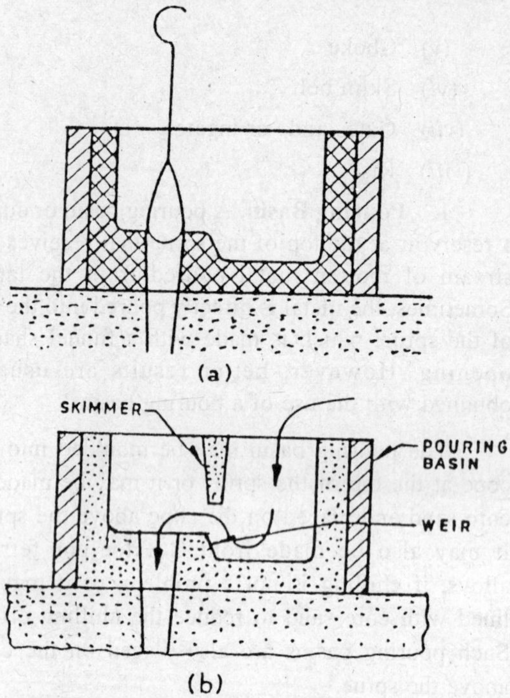

There can be many designs of pouring basins, but the most commonly used one is shown in Fig. 2.18 (*b*). The metal must be poured into the pouring basin that is remote from the sprue entrance. If the metal is poured directly down the sprue entrance, lot of vortexing and turbulence take place, resulting in unsound and defective casting. However, when the metal is poured into the far low side of the pouring basin, a dam effect enables the operator to reach and optimum pouring speed before any metal enters the sprue. This condition must be maintained, to prevent dross and slag from entering the sprue.The weir will hold back the heavier inclusions and the skimmer or dam will hold back the dross (oxides and other lighter inclusions rising to the surface). Another means of preventing slags from entering the sprue and hence the mould cavity, is to use a ceramic strainer at the top of the sprue, (Fig. 2.19). The strainer is made from core sand, chamotte and refractory fibre materials.

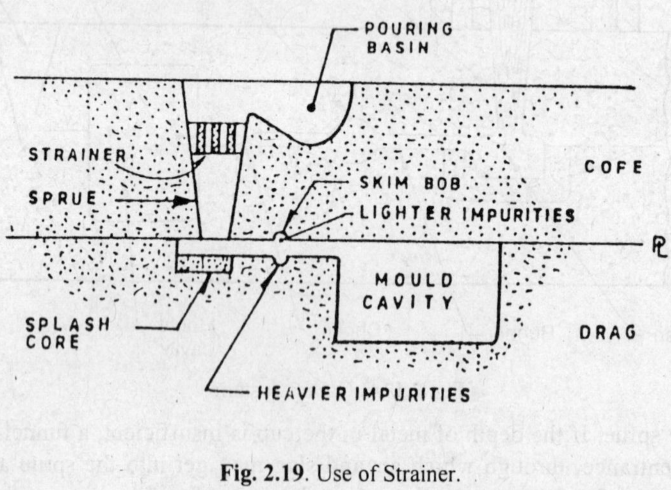

Fig. 2.19. Use of Strainer.

2. Sprue. From the pouring basin, the molten metal is transported down into the mould cavity by means of the sprue or downgate. It is a vertical channel that connects the pouring basin with runners and gates.

As the metal flows down the sprue, its velocity increases. Hence the section of the sprue should decrease, otherwise the sprue will not remain full of metal with the metal leaving the walls of the sprue. This will result in aspiration of gases through the surrounding moulding sand. To prevent this, the sprues are made tapered downward. The taper ranges from 2° to 4°. The taper also helps in convenient moulding of the sprue. For moulding large castings, the sprue and other channels in the mould are constructed of standardised chamotte tubes.

3. Sprue Base. Where a sprue joins a runner, usually an enlargement in the runner is made. This enlargement which is called as "Sprue Base" or "sprue well" serves a dual function. A molten metal pool is an excellent device for preventing excessive sand erosion where the molten metal impinges on the runner at the sprue base. Also, there is sudden slowing of flow which dissipates kinetic energy and helps to drop out inclusions, scum and various other refractory materials that may have been washed in with the molten metal. Ceramic or wiremesh filters are sometimes placed at the sprue base to filter out dross and other large inclusions.

The well dimensions have been empirically recommended by Kura, as follows : (*a*) For a narrow deep runner:—

(*i*) Its diameter should be 2.5 times the width of the runner in a two way runner, that is, when the metal flows laterally from the well in two opposite directions.

(*ii*) Base diameter = 2 × width of runner in a one runner system.

(*b*) For square, round or wide shallow runners, the well should have a cross-sectional area

= 5 × area of the sprue bottom

and, depth = 2 × that of the runner.

Splash Core. This is a piece of ceramic or baked sand core inserted in the mould directly beneath the sprue. Its function, (Fig. 2.19), is to prevent the erosion of the mould sand where the molten metal strikes it at the base of the sprue.

4. Runners. A runner is commonly a horizontal channel which connects the sprue with gates, thus allowing the molten metal to enter the mould cavity. The runners are of larger cross-section and often streamlined to slow down and smooth out the flow, and are designed to provide approximately uniform flow rates to the various parts of the mould cavity. Runners are commonly made trapezoidal in cross-section.

Skim - bob. A skim bob is an enlargement along the runner, whose function is to trap heavier and lighter impurities such as dross or eroded sand, (Fig. 2.19). It, thus, prevents these impurities from going into the mould cavity.

5. Gates. Gates are channels which connect the runners to the mould cavity and through which the incoming metal directly enters into the mould cavity. The gates should break off easily from the casting after solidification. For this, at the junction to the cavity, the gates are much reduced in thickness. This will also choke the flow of metal and ensure its quiet entrance into the mould cavity. From experience, the best cross-section for gates is a trapezoidal one that smoothly passes into a rectangular section at the junction of the cavity.

Ingate. This is the end of the gate where it joins the mould cavity and through which the molten metal will be introduced into the mould cavity.

Runner extension. The leading edge of the molten metal flowing in a stream follows the path of least resistance and continues to build up kinetic energy until it reaches the end of the runner. If a runner extension is used, the kinetic energy may be absorbed, thus causing a smoother flow of metal in the runners and into the mould cavity. It also serves as a trap for inclusions carried into the runner by the first metal.

Gate Ratio. It is defined as the ratio of sprue base area, followed by the total runner area and the total ingate area. The sprue base area is taken as unity.

Choke. The choke is that part of the gating system which has the smallest cross-sectional area. Its function is to control the rate of metal flow to help lower the flow velocity in the runner, to hold back slag and foreign material and float these in the cope side of the runner and to minimize sand erosion in the runner. Depending upon its location, the gating system may be classified as :

(*i*) Pressurised or choked system.

(*ii*) Unpressurised or free system.

Pressurised gating system. In this system, the ingates serve as the choke. This system maintains a back pressure and causes the entire gating system to become pressurised. It is usual to cut runners in the cope and gates in the drag. And since in the pressurised system, the runner flows full, the slag will remain in the runner while floating up to its upper section. The system will flow full even if a straight sprue is used once the flow has been established. The full flow system will minimize aspiration and oxidation in the gating system. However, the molten metal will enter the mould cavity with relatively high velocity, causing turbulence. This will cause erosion and oxidation in the mould cavity. But, in this system the molten metal will enter the mould cavity uniformly through all the gates. A typical gating ratio in this system can be : 1 : 0.75 : 0.50.

Gating ratio is defined as : Sprue area : Runner area : Ingate area..

Unpressurised gating system. In this system, the sprue base serves as the choke. Typical gating ratios in this system can be :

1 : 2 : 4 or 1 : 4 : 4.

This system reduces velocity and turbulence and hence minimizes erosion and oxidation in the mould cavity. But, since the system will not flow full, it may favour absorption of gases and oxidation of metal in the gating system. Also, the slag will not float up and will get access to the mould cavity. This system must have tapered sprues, sprue base wells and pouring basins. This system can deliver metal uniformly to multiple gates only if the runner is reduced after each gate.

For casting alloys such as light metals and Manganese bronze, usually the unpressurised system is used. In contrast, for casting heavy and ferrous metals, pressurised system has been used. However, according to Briggs, for high quality steel castings, a gating ratio of 1 : 2 : 2 or 1 : 2 : 1.5 will produce castings nearly free from erosion, will minimize oxidation, and will produce uniform flow. But a gating ratio of 1 : 4 : 4 might favour the formation of oxidation defects. Heine suggests a typical gating system for Malleable iron using a four hole strainer and tapered sprue as : 1 : 2 : 9.5, that is, an unpressurised system.

6. Risers. Risers or feed heads are a part of the feeding system. These are reservoirs of molten metal that feed the melt in the casting proper as it solidifies. The riser thus provides the feed metal which flows from the riser to the casting to make up for the shrinkage which takes place in the casting metal as it changes from liquid to solid. The riser is a passage of sand made in the mould during ramming. The molten metal rises in the riser after the mould cavity is filled up.

Risers find use in castings of heavy sections or of high shrinkage alloys.

To understand the functions of risers, the mechanism of solidification of the casting has to be understood. When the casting solidifies, it shrinks or contracts in size. This shrinkage occurs in three stages :

(*i*) **Liquid Contraction or Shrinkage.** This shrinkage occurs when the metal is liquid. It occurs when the molten metal cools from the temperature at which it is poured to the temperature at which solidification occurs.

(*ii*) **Solidification Shrinkage.** This occurs when the metal looses its latent heat, that is, during the time the metal changes from the liquid state to the solid state.

(*iii*) **Solid Shrinkage.** This occurs when the metal cools from the freezing temperature to the room temperature.

Only the first two shrinkages are considered for risering purposes, since the third is accounted for by the Pattern maker's contraction allowance. Liquid shrinkage is generally negligible, but solidification shrinkage can be substantial (12 to 15%).

Gray C.I. with a carbon equivalent of 4.3% has negative shrinkage, that is, it actually expands upto 2.5% because of graphite precipitation. So, for this, no risering is needed.

Solidification shrinkage for other ferrous metals varies from 2.5 to 4%.

Solidification shrinkage for non-ferrous metals is even more, for example,

For pure aluminium it is 6.6% and for copper, it is 4.9%. Solidification shrinkage in case of their alloys is somewhat less.

As already discussed, when the mould cavity is filled with molten metal, the metal adjacent to the walls of the mould cools and solidifies first. This results in a shell of solid metal, with the centre of the section remaining liquid, while there is a zone between the liquid interior and solid exterior wherein the metal is in a semisolid or mushy state. The solidification then proceeds inwards towards the centre of the section. This solidification is called as "Lateral or Progressive solidification". Longitudinal solidification occurs at right angles to the lateral solidification at the centre line, (Fig. 2.20). The gating should be designed in such a manner that it permits solidification and cooling to progress in such a way that the accompanying shrinkage may take

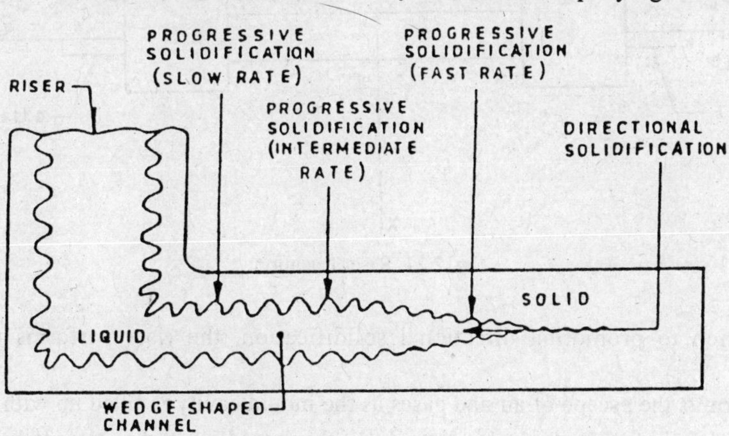

Fig. 2.20. Directional Solidification.

place without resulting in voids, cracks, or porosity within the casting, so as to obtain a sound casting. This type of solidification is called "Directional Solidification". For this to occur, the following two conditions should be satisfied :

(*a*) The longitudinal solidification must progress from the thinnest faster cooling sections to the heavier hotter sections.

(*b*) The temperature gradient, in addition to being properly directed, must be sufficiently steep so that the liquid metal can pass through the wedge shaped channel to compensate for shrinkage as it occurs at the centre line. It implies that the progressive solidification is controlled in such a way that no portion of the casting is isolated from liquid metal feeding channels, during the complete solidification cycle.

If the depth of a section is quite large as compared to its cross-section, then the progressive solidification rate will exceed the longitudinal solidification and result in fine center line porosity or even a larger or series of large cavities. To prevent such defects, the cross-section of the casting should increase towards the hotter sections.

When it is discovered in the foundary that progressive solidification of a casting is uncertain or impossible because of unique sections or isolated masses, three remedies may be available :

(*i*) The foundary-man can add a riser to feed the remote thick sections.

(*ii*) The designer can thicken (pad) the thin sections that serve as feed paths to the heavier remote sections.

(*iii*) The designer can reduce the mass of the remote thick sections or increase their rate of cooling (with the help of chills).

In riser system, the effective directional solidification is one which starts in those parts of the mould which are farthest from the risers and proceeds through the casting towards the risers.

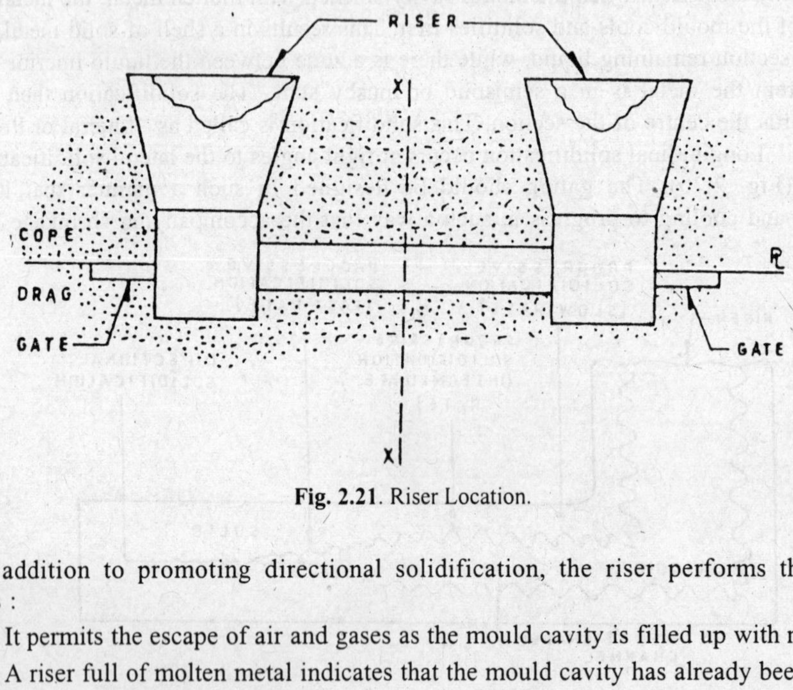

Fig. 2.21. Riser Location.

In addition to promoting directional solidification, the riser performs the following functions :

(*a*) It permits the escape of air and gases as the mould cavity is filled up with molten metal.

(*b*) A riser full of molten metal indicates that the mould cavity has already been completely filled up.

(*A*) **Requirements of a Riser**

(*i*) To be effective, a riser must be the last part of the casting to solidify, such that all the shrinkages that are likely to occur should be in the riser.

(*ii*) The volume of the riser must be sufficient to compensate for metal shrinkage within the casting.

(*iii*) The riser must cover completely the casting section that requires feeding.

(*iv*) The fluidity of the metal inside the riser must be maintained so that the metal can flow from it and penetrate to the last contraction cavity.

(*v*) The riser should be effective in establishing a pronounced temperature gradient with the casting so that the casting will solidify directionally towards the riser.

(*B*) **Riser Location.** Before the shape and size of the riser is determined, its location must be specified. Any casting, no matter how complex, can be subdivided into several geometrical shapes, consisting of two heavier sections joined by a thinner section, (Fig. 2.21). A riser should be located close to each heavier section. Also, it should be located in such a manner that it is the last portion of the casting to solidify.

(*C*) **Types of Risers.** Depending upon its location, the riser is described as 'Top riser' or 'Side riser'. If the riser is located between runners and casting, (Fig. 2.22 (*a*)), it is known as 'side riser'. It is also called a 'live or hot riser' since it is filled last and contains the hottest metal. If a riser must be placed at the top of the casting or at the end of the mould cavity, (Fig. 2.22 (*b*)), then it is called as 'Top riser' or 'dead or cold riser'. These types of risers fill up with the coldest metal and are likely to solidify before casting. The side riser receives the molten metal directly from the runner before it enters the mould cavity and is more effective than the top riser. A riser may either be an 'Open riser' or a 'Blind riser'.

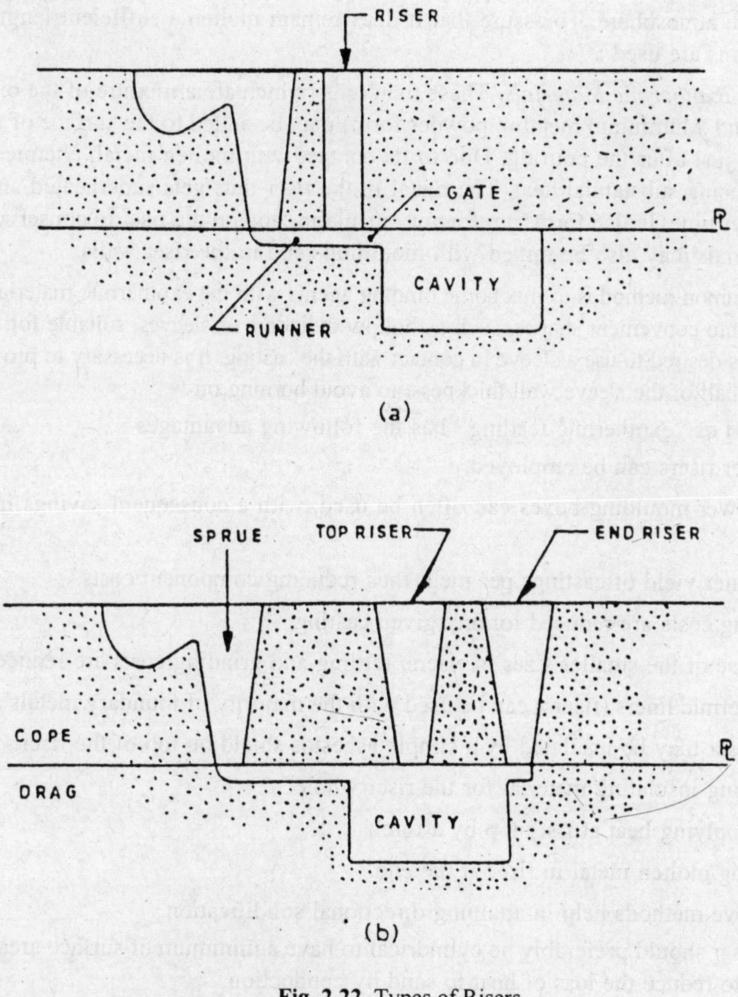

Fig. 2.22. Types of Risers.

(*I*) **Open Risers.** These risers open to the atmosphere at the top surface of the mould, (Fig. 2.22). Such risers have the following advantages :

(*i*) It can be easily moulded.

(*ii*) As it is open to atmosphere, it will not draw metal from the casting as a result of partial vacuum in the riser.

(*iii*) These risers serve as collectors of nonmetallic inclusions floating up to the surface.

The limitations to the usefulness of open risers are as follows :

(*a*) They can be moulded only in the cope (either at the top of the mould cavity or on the side at the parting line).

(*b*) They should have the height commensurate with the height of the cope. This reduces the 'yield' of the casting. The 'yield' is the percentage of total metal poured into a mould which actually ends up in the casting.

(*c*) Open risers are holes through which foreign matter may get into the mould cavity.

(*d*) Since the top of the riser is open to the atmosphere, the molten metal in it starts loosing heat to the atmosphere. To assure that it does remain molten a sufficient length of time, the following means are used :

(*i*) *Use of Exothermic Materials.* These materials (which are a mixture of the oxide of the metal to be cast and Aluminium metal in powder form) may be added to the surface of the molten metal in the riser just after the pouring. Due to its contact with molten metal, chemical reaction takes place, producing substantial heat. The metal in the riser thus gets superheated and remains molten for a longer time. It also forms a refractory insulating top on the riser to conserve this heat. Exothermic materials may also be mixed with moulding sand in the riser walls.

Another common method is to mix some binding agents with the exothermic material, enabling it to be moulded into convenient shapes, such as hollow cylinders or sleeves, suitable for lining riser cavities. When it is desired to use a sleeve in contact with the casting, it is necessary to provide a base of core sand, one half of the sleeve wall thickness, to avoid burning on.

This method of "exothermic feeding" has the following advantages :

1. Smaller risers can be employed.
2. Shallower moulding boxes can often be used, with a consequent savings in sand and moulding costs.
3. A greater yield of castings per melt, thus reducing component costs.
4. Melting costs are reduced for any given casting.
5. Because of the smaller sizes of risers, fettling and grinding costs are reduced.
6. Exothermic liners (risers) can be used with the majority of foundary metals and alloys.

(*ii*) The riser may be insulated by a simple asbestos shield on top of the riser.

(*iii*) By using insulating material for the riser whole.

(*iv*) By supplying heat at riser top by a torch.

(*v*) Pouring molten metal in riser at the end.

All the above methods help in attaining directional solidification.

An open riser should preferably be cylindrical to have a minimum of surface area in contact with the mould, to reduce the loss of heat to sand by conduction.

(*II*) **Blind Risers.** A riser which does not break to the top of the cope and is entirely surrounded by moulding sand is known as 'Blind Riser' (Fig. 2.23). Blind risers are set up in high moulding boxes, where the use of open risers would entail a large consumption of molten metal. The advantages of blind risers are :

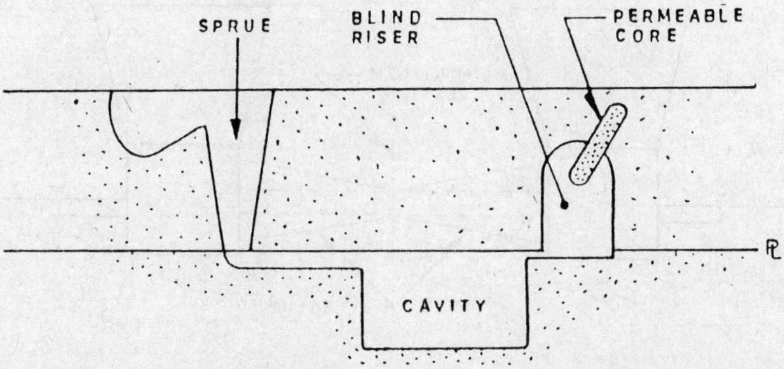

Fig. 2.23. Blind Riser.

(*i*) Considerable latitude or flexibility is allowed for positioning the riser, either in a cope or in drag.

(*ii*) It is surrounded on all sides by moulding sand. Thus it looses heat slowly which helps in better directional solidification of the casting.

(*iii*) A blind riser can be smaller than a comparable open riser, therefore, more yield is obtained.

(*iv*) A blind riser can be removed more easily from the casting than an open riser.

The main drawback of a blind riser is that when the metal in it cools, metal skins may quickly form on its walls. This will result in a vacuum in the riser and the riser will not feed and may actually draw metal from the casting. To avoid this, a permeable dry sand core is inserted into the riser cavity, connecting it to the mould sand layers. Through these sand layers air passes into the riser interior. The riser thus operates under atmospheric pressure.

Blind risers are often made rounded at the top to effect metal savings, because a hemi-spherical shape has the smallest surface area to volume ratio.

(*III*) **Shape and Size of Riser.** The risers are designed to solidify last so as to feed enough metal to heavy sections of the casting to make up for the shrinkage before and during solidification. For this, they should loose heat at a slower rate. Amount of heat content is proportional to the volume and the rate of heat dissipation depends upon the surface area of the riser. Thus, the risers should be designed with a high V/A (Volume/Surface Area) ratio, for a given size. This will minimize the loss of heat, so that the riser will remain hot and the metal in the molten state as long as possible. This condition can be met when the riser is spherical in shape so that its surface area is minimum. The next best practical shape is a cylinder. Rectangular sections are very inefficient and must be avoided as far as possible. As risers of ideal spherical shape are difficult to mould, consequently, the cylinder is probably the best shape to employ for general run of casting.

The riser size for a given application depends primarily on the metal poured, but numerous other factors are also important, for example, gray C.I. requires little or no risering to compensate for solidification contraction because the precipitation of flake graphites causes compensating expansion. Steel, white iron and many non-ferrous alloys which have long freezing ranges, may produce sound castings only with the best application of foundary engineering and experience.

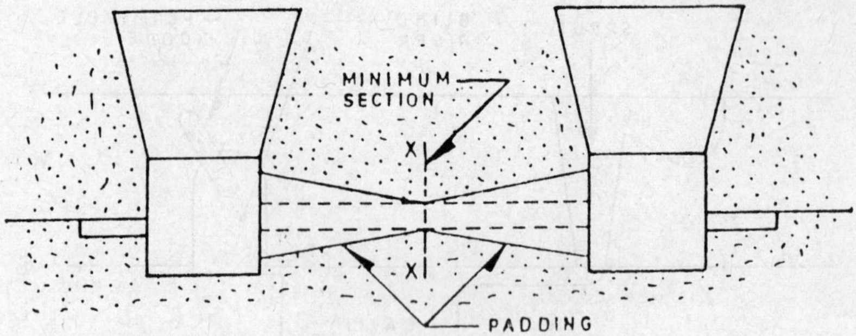

Fig. 2.24. Use of Padding.

As regards the correct height of the cylindrical risers, there is as yet no one opinion. The general points to be considered are that they must be tall enough so that any shrinkage cavity in the riser (pipe formation) does not penetrate into the castings. The shrinkage cavity must lie above the neck (a portion that connects the riser with the casting). The neck should be as short as possible and also solidify longer than the casting. In general,

[Height of Cylindrical Riser = 1.5 × Diameterof Riser]

Riser Size : As has been noted above, the freezing time or solidification time of a casting (Riser is also a casting) depends upon the amount of heat in a casting (directly) and depends (inversely) upon the surface area of the casting. Based on these facts, many relations have been suggested for determining the riser size

1. Chvorinov's rule :–

Solidification time or Freezing time

$$t = C \cdot \left(\frac{V}{SA}\right)^2$$

where V = volume of Casting

SA = Surface area of casting

C = Constant of proportionately that depends upon composition/properties of cast metal (including its latent heat). pouring temperature, mould material etc,

Since, the metal in the riser must be the last to solidify, to achieve directional solidification.

$$\left(\frac{V}{SA}\right)_{Riser} > \left(\frac{V}{SA}\right)_{Casting}$$

In practice,

$$\left(\frac{V}{SA}\right)_{Riser} = 1.05 \text{ to } 1.075 \left(\frac{V}{SA}\right)_{Casting}$$

Since V and SA for the casting are known, $(V/SA)_{riser}$ can be determined, Assuming height to diameter ratio for the cylindrical riser, the riser size can be determined.

Chvorinov's formula is not very accurate, because, it does not take into account the solidification contraction or shinkage. The method is valid for calculating the proper riser size for short freezing-range alloys such as steel and pure metals. There is no satisfactory relationship for determining riser size for non-ferrous alloys.

2. Caine's Formula : caine's method of determining riser size is based on an experimentally determined hyperbolic relationship between relative freezing times and volumes of the casting and the riser.

Relative freezing time or Freezing ratio is defined as

$$X = \frac{(SA/V)_{Casting}}{(SA/V)_{Riser}}$$

Volume ratio, Y, is given as

$$Y = \frac{\text{Volume of riser}}{\text{Volume of Casting}}$$

Then, Caine's formula is given as,

$$X = \frac{a}{Y - b} + c$$

where a = Freezing characteristics constant

b = Liquid – solid solidification contraction

c = Relative freezing rate of riser and casting

Typical values of a, b and c for commonly used cast metals are given below :-

Cast Metals	*a*	*b*	*c*
Steel	0.12	0.05	1.00
Aluminium	0.10	0.06	1.08
Gray Cast Iron	0.33	0.03	1.00
Cast Iron, Brass	0.04	0.017	1.00
Aluminium Bronze	0.24	0.017	1.00
Silicon Bronze	0.24	0.017	1.00

A typical Caines hyperbolic curve is shown in Fig.

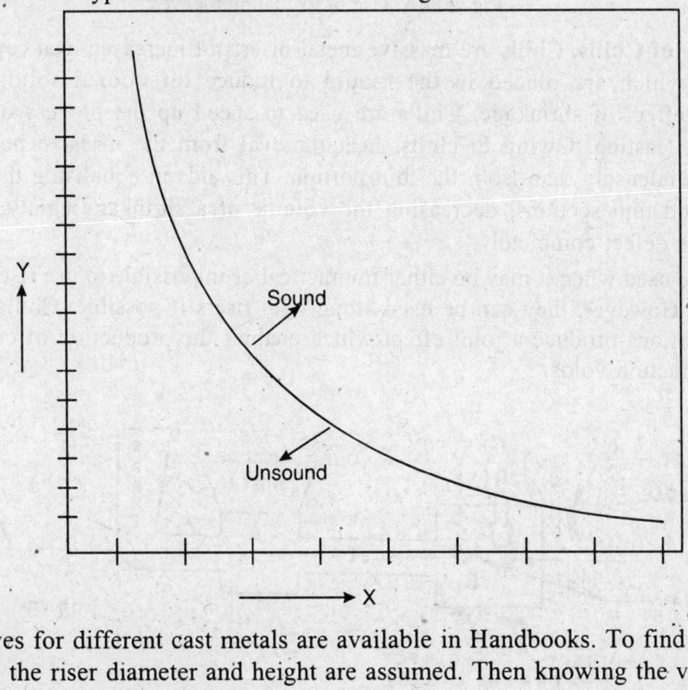

Such curves for different cast metals are available in Handbooks. To find the riser size for a given casting, the riser diameter and height are assumed. Then knowing the values of a, b and c, the values of X and Y are Calculated. These values of X and Y are plotted on the hyperbolic curve figure. If the values of X and Y meet above the curve, the assumed riser size is satisfactory. Otherwise, a new assumption is made.

(*IV*) **Use of Padding.** The casting in Fig. 2.21 is symmetrical about axis X - X. The two thick portions of the casting have been provided with risers. The cooling of this casting will initiate from the section X - X and move towards the heavier sections. As already discussed, for the production of a sound casting, the temperature gradients should be such that no part of the casting remains without metal during the full solidification cycle. For this, the thinner section should be tapered towards heavier sections to change the progressive solidification to directional solidification, (Fig. 2.24). This tapering of thinner section towards thicker section is known as 'padding'. However, this will require extra material. If padding is not provided, centre line shrinkage or porosity will result in the thinner section.

'Insulating Sleeves' may also be used around the thinner sections to retard their rate of cooling, so that the metal in them remains in the liquid state for a longer period, Fig. 2.24 A.

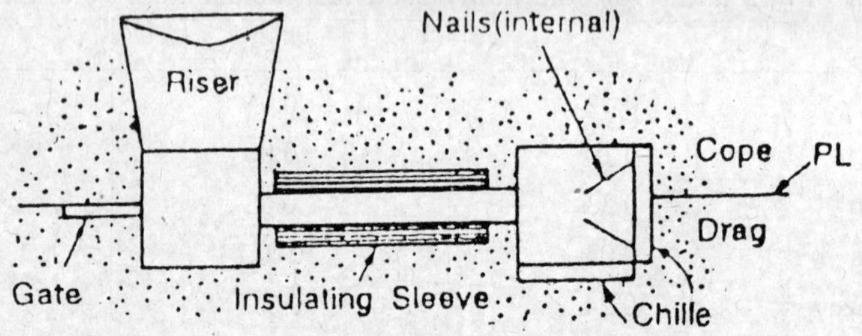

Fig. 2.24A. Use of Insulating Sleeves.

(*V*) **Use of Chills.** Chills are massive metal inserts of increased heat capacity and thermal conductivity, which are placed in the mould to induce directional solidification and help overcome the effect of shrinkage. Chills are used to speed up the process of cooling of thick sections of the casting. Owing to chills, heat removal from the massive portion of a casting proceeds more intensely than from the thin portion. This aids in equalising the rates of freezing of the thick and thin sections, decreasing the volume of a shrinkage cavity, or in some case, eliminating this defect completely.

Chills are used when it may be either impractical or impossible to use riser on thick sections of the casting. However, they can be used along with risers if possible. Chills and risers placed on massive portions produce a joint effect which enables the production of castings completely free from contraction voids.

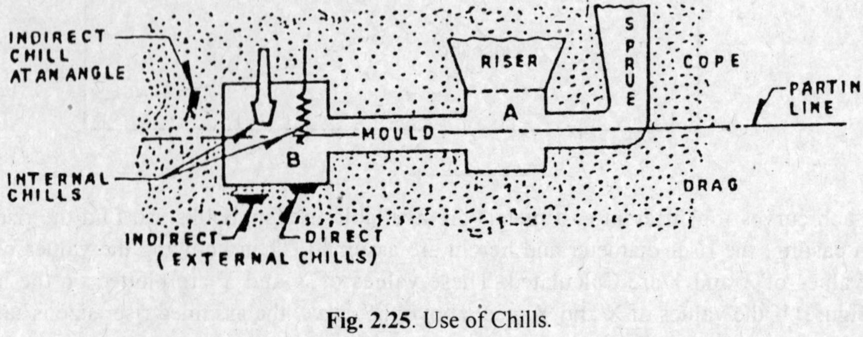

Fig. 2.25. Use of Chills.

The use of chills is illustrated in Fig. 2.25. The casting is gated at A in order to put the coldest metal in portion B of the casting. If the casting were gated at B, the chills will absorb heat from the hot metal and become saturated with heat, thereby, loosing much of their chilling ability. Also, cooled metal will flow into the thin section of the casting, which is not desirable from the point of view of directional solidification.

The chills can be either external or internal chills. The external chills are rammed up in the mould walls, whereas the internal chills are inserted in the mould cavity, (Fig. 2.25). The external chills may be direct or indirect chills. External chills are made of iron or steel. Of course, the most effective chills will be made of copper, because of its high heat conductivity. External chills are coated with red lead or moulding sand to keep them from welding upto the casting. External chills can be used again and again.

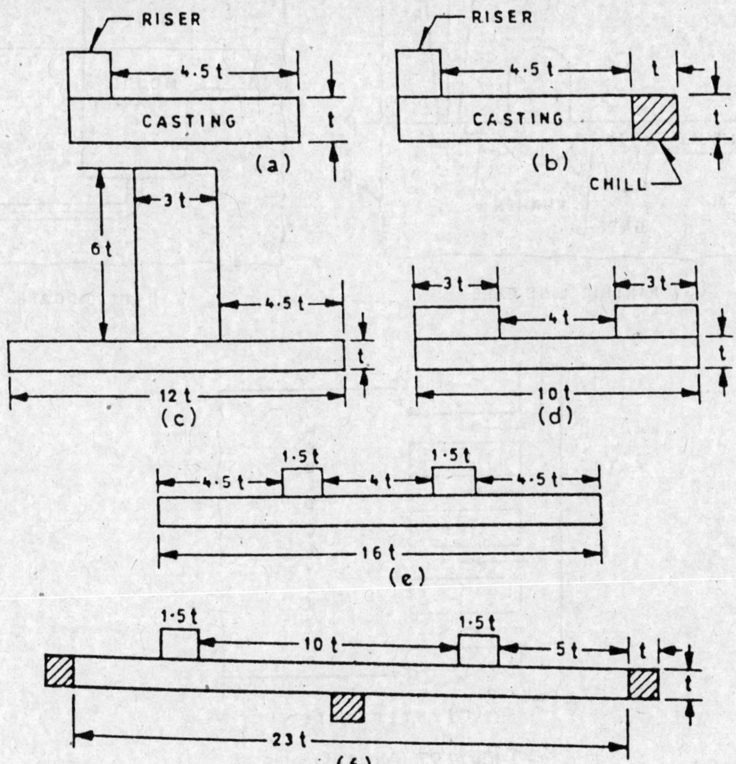

Fig. 2.26. Feeding Ranges of Risers.

Because the internal chills will fuse into the casting, these must be made from alloys whose chemical composition is the same as that of alloys of castings. In pouring the mould, the metal washes over the internal chills, so that they partially melt and fuse with the metal. Internal chills must weld up well with the metal of the casting since they will become a part of it. But sometimes, internal chills poorly weld up with the base metal, therefore, it is not advisable to use them in important casting.

Before being placed in the mould, the chills should be thoroughly cleaned of dirt, oil, grease, moisture or other gas forming materials.

Chills are metal objects in the form of rods, wire and nails, either cast or made from rolled stock.

(*VI*) **Risering versus Padding and Chilling.** Because risers add to the cost of a casting in several ways (time is consumed in providing them, they reduce the yield of the metal poured and time is required for their removal), every effort should be made to design castings so that the number of risers is minimised. Padding is one method of doing this.

(*VII*) **Flow-offs.** Flow-offs are vertical channels, just like sprue, made in the mould Fig. 2.27 (a). They serve many purposes in a casting process. In the case of castings, where risers are not needed or where blind risers are used, they permit the escape of gases from the mould cavity.

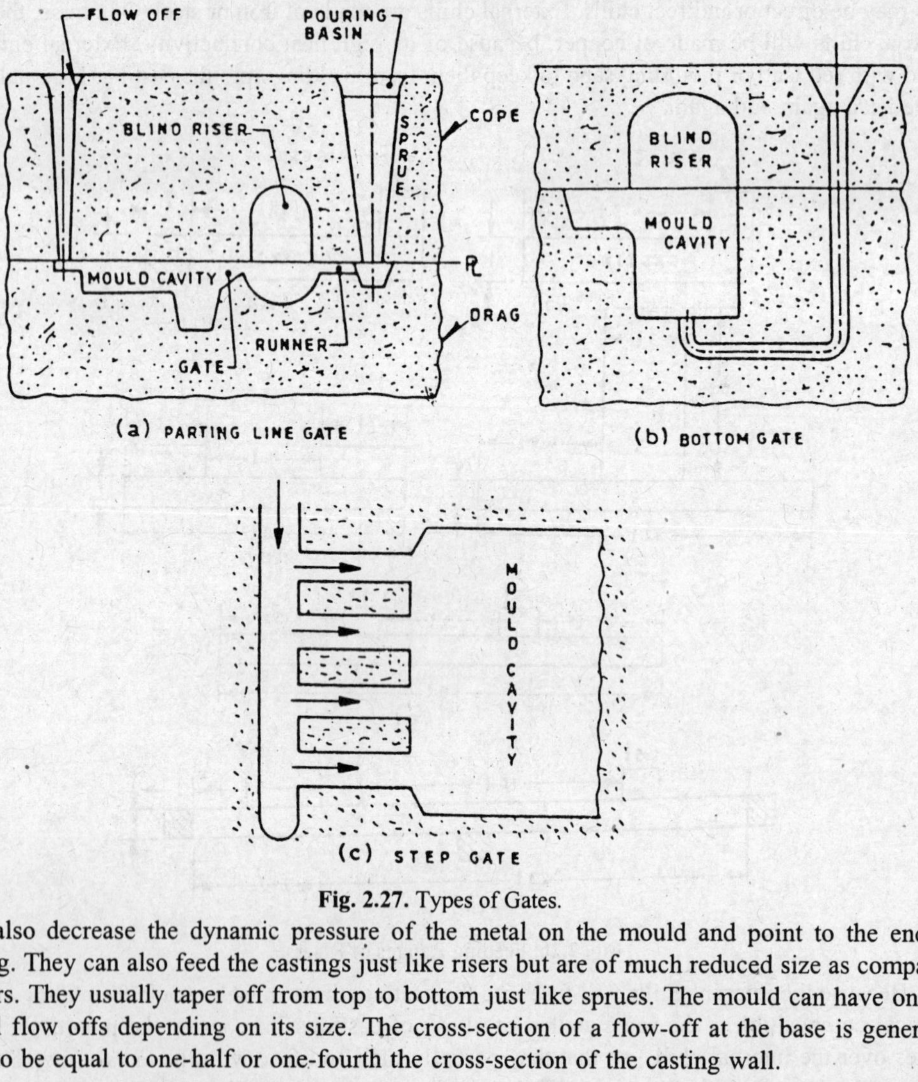

Fig. 2.27. Types of Gates.

They also decrease the dynamic pressure of the metal on the mould and point to the end of pouring. They can also feed the castings just like risers but are of much reduced size as compared to risers. They usually taper off from top to bottom just like sprues. The mould can have one or several flow offs depending on its size. The cross-section of a flow-off at the base is generally taken to be equal to one-half or one-fourth the cross-section of the casting wall.

(*VIII*) **Feeding Range of Risers.** It has been seen that for steel castings, an adequate riser will provide complete sounding of the casting upto a length equal to 4.5 times the minimum thickness of the casting. A metal chill of minimum section equal to the cross-section of the casting will, when placed at the end of the casting, merely adds a distance equal to the thickness of the casting. However, the same chill placed in-between two risers will permit them to be 2t further apart without impairing soundness. It has also been seen that a riser in the centre of a casting has a much greater feeding range that at the end of a casting. Fig. 2.26 illustrates the feeding ranges of risers.

(*IX*) **Requirements of Gating System.** From the above discussions, it is clear that the following design considerations should be given a weightage while designing a gating system :

1. The metal should flow with a minimum of turbulence and aspiration of mould gases.

2. The metal must be rapidly poured into the mould cavity or else it will solidify before completely filling the mould.

3. The use of higher superheat of metal is undesirable since the amount of gas dissolved in the liquid metal increases with an increase in superheat.

4. Loose sand, slag or oxides should be prevented from entering the mould cavity.

5. Erosion of the mould surface should be avoided. For this, the metal must not impinge against the mould with enough energy to erode it, particularly in a green sand mould.

6. Provide adequate gates and runners. It is better to have a lower yield than much rejections.

7. The lengths of gates and runners should be kept to a minimum to reduce the amount of heat loss from the liquid before it enters into a mould cavity.

8. Always calculate the cross-sectional areas of the gates, runners and sprues to get good results.

9. A round gate or runner has the best surface area. Use this shape to have lower surface friction.

10. As far as possible, provide gating system in the pattern to have smooth surface.

11. A gating should provide equal distribution of metal to several mould cavities in a mould.

12. If several gates branch off from one runner, the cross-sectional area of the runner should be decreased after each branching gate.

13. If the gating system is properly positioned, strainer core and other traps are unnecessary.

14. Do no tilt the mould white pouring.

15. For small castings of uneven thickness, gating into a heavy section with short gate will prove satisfactory in most of the cases.

16. To fill high castings, the end of the mould where the gate is to be located should be lowered so that the metal is compelled to run in a slight incline.

17. The metal should be introduced into the mould cavity in such a way that the temperature gradient produced on the mould surface and within the mould facilitates the production of directional solidification towards the riser.

(*X*) **Types of Gates.** Basically, there are three types of gates, (1) Parting line gate (2) Bottom gate, and (3) Top gate depending upon the level at which the molten metal enters the mould cavity relative to the level of parting line.

(1) **Parting Line Gate.** Here, the molten metal enters the mould cavity at the parting line, [Fig. 2.27 (*a*) and Fig. 2.27 (*b*)]. These gates are used most because they are easy to form. They have the advantage of speed and simplicity. Their chief drawback is that the metal drops into the mould cavity and for deep cavities, it may cause erosion or washing of the mould. In case of non-ferrous metals, this drop aggravates the dross and entraps air in the metal, which make for inferior castings.

(*i*) It is a general practice to cut runners in the cope and ingates in the drag. The main reason for this is to trap the slag and dross, which are lighter and thus trapped in the upper portions of the runners.

(*ii*) Whenever possible, surfaces to be machined should be cast in the drag side of the mould. This helps to provide a cleaner casting surface on the bottom, while dross, oxides and other impurities float to the top. Any shrinkage would also be on the top surface and not on the surface to be machined. If finished surfaces must be cast in the cope side of the mould, an extra allowance should be made.

(2) **Bottom Gate.** For deep cavities, bottom gating may be employed. Here, the metal enters the mould cavity at or near the bottom of the mould cavity, (Fig. 2.27 (*b*)). This offers a smooth flow of metal into the cavity with a minimum of mould core erosion and the turbulence of metal. However, these gates are difficult to make and require extra skill to mould. Many times, these gates use a core gating system. Also, their main drawback is that in castings which are bottom gated and fed by top risers, an unfavourable temperature gradient is created in the casting. Hottest metal enters the cavity and the coldest metal will be in the riser. This drawback sometimes can be partially or completely corrected by special casting techniques discussed earlier, that is,

(*a*) Use of exothermic materials in the making of the riser or spreading these at the top of the riser.

(*b*) Radiation shielding on top risers reduces radiant heat loss.

(*c*) Insulating riser sleeves help maintain riser heat.

(*d*) Chills can promote directional solidification or deduce undesirable heat concentration.

(*e*) Pouring the risers last with fresh metal places the hottest metal there.

Another technique to minimise the drawbacks of bottom gating is to use a 'Step gate', (Fig. 2.27 (*c*)). The sizes of ingates are normally increased from top to bottom such that the metal enters the mould cavity from the bottom most gate and then progressively moves to the higher gates. This

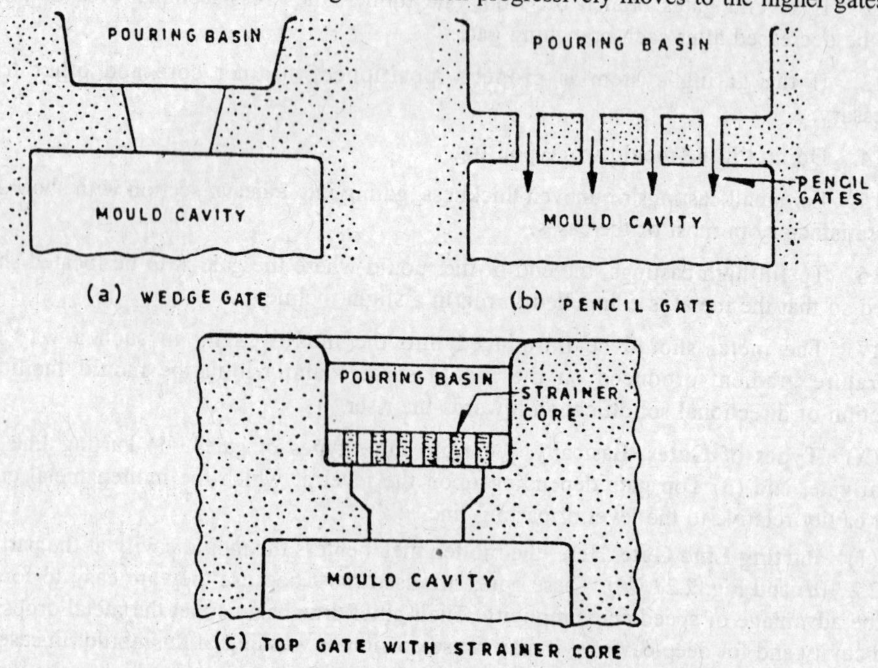

Fig. 2.28. Top Gates.

ensures a gradual filling of the mould without any mould erosion and produce a sound casting. However, a difficulty with step gating is that most of the metal will continue to enter through the bottom most gate.

(3) **Top Gate.** In this type of gating, the metal enters the mould cavity above the level of the parting line. Its main advantage is that it is conducive to a favourable temperature gradient. The pouring basin and gates themselves may serve as risers, (Fig. 2.28). However, its big disadvantage is mould erosion, lot of turbulence and excessive aspiration of air and gases. It is not recommended for metals which oxidise easily, for example, Aluminium and Magnesium, since such metals have a tendency to form excessive dross when agitated and produce unsound castings. Top gates are at times used for gray castings of simple design, for example thin, flat plates. The different designs of top gates are shown in Fig. 2.28.

Pencil Gate. (Fig. 2.28 (*b*)). A series of small round gates enter the mould cavity from above and come from a common pouring basin. This gate is employed to reduce the rate of flow of metal and to control the inclusions of sand, slag or dross. However, excessive turbulence is still present.

Use of a strainer (to trap sand, slag or dross) in a top gate is shown in Fig. 2.28 (*c*)

Other types of Gates. The three main types of gates are sometimes modified to meet certain specific conditions. Some of these gates are discussed below :

(4) **Branch Gate.** (Fig. 2.29 (*a*)). Two or more gates leading into the casting cavity. It is designed either to feed a single casting at several points or a number of individual castings. Also, see Fig. 2.17.

(5) **Ring Gate.** (Fig. 2.29 (*b*)). A gate so formed that the number of small gates conduct the metal from a circular runner to a mould in the centre.

(6) **Whirl Gate.** (Fig. 2.29 (*c*)). A gate and sprue arrangement which tangentially introduces metal into a riser so that the centrifugal action forces dirt or slag to the centre of the riser and away from the riser connection as the metal enters the mould cavity. It thus prevents dirt and slag from entering the gate and the mould cavity.

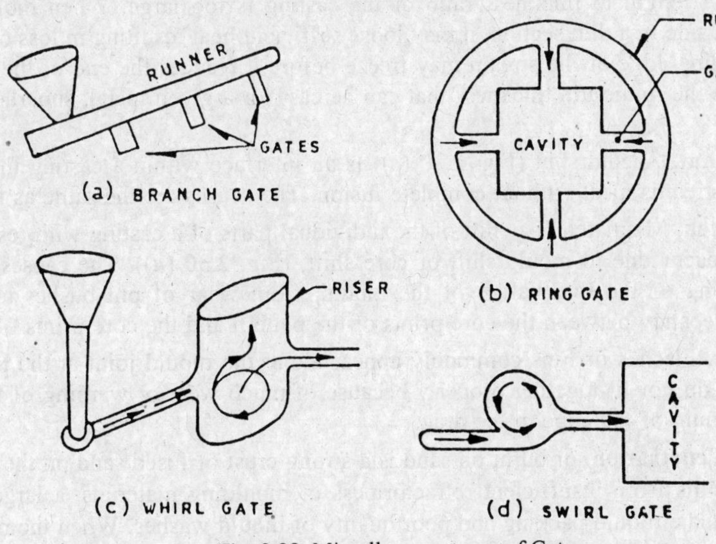

Fig. 2.29. Miscellaneous types of Gates.

(7) **Swirl Gate.** A gate used with a feeder and runner designed to swirl the metal in the feeder to remove the impurities, Fig. 2.29d.

2.7. SAND CASTING DEFECTS.

A properly designed casting, a properly prepared mould and correctly melted metal should result in a defect free casting. However, if proper control is not exercised in the foundary–sometimes it is too expensive – a variety of defects may result in a casting. These defects may be the result of : improper pattern design, improper mould construction, improper melting practice, improper pouring practice and also because of moulding and core making materials. Defective castings, even at advanced foundaries, account for 2 to 5% and sometimes from 10 to 25% of the number of produced castings. Some of the common casting defects are discussed below :

1. **Blow Holes.** (Fig. 2.30 (*a*)). Blow holes, gas holes or gas cavities are well rounded cavities having a clean and smooth surface. They appear either on the casting surface (open cavities) or in the body of a casting (closed cavities). These defects occur when an excessive evolved gas is not able to flow through the mould. So, it collects into a bubble at the high points of a mould cavity ad prevents the liquid metal from filling that space. This will result in open blows. Closed cavities or gas holes are formed when the evolved gases or the dissolved gases in the molten metal are not able to leave the mass of the molten metal as it solidifies and get trapped within the casting. The solubility of the gases in a molten metal decreases as the temperature decreases. So, the dissolved gases will try to leave the metal as it cools and solidifies.

These defects are caused by : excessive moisture content (in the case of green sand moulds) or organic content of the sand, moisture on chills, chaplets or metal inserts, inadequate gas permeability of the moulding sand (due to fine grain size of sand, high clay content, hard ramming), poor venting of mould, insufficient drying of mould and cores, cores not properly vented, high gas content of the molten metal, low pouring temperature and incorrect feeding of the casting etc.

As the gas follows the path of least resistance, these defects usually occur in the cope portion of the casting. Internal gas holes are revealed when the casting is machined or cut into sections.

2. **Misrun or Short run.** This defect is incomplete cavity filling, (Fig. 2.30 (*b*)). The reasons can be :- inadequate metal supply, too low mould or melt temperature, improperly designed gates, or length to thickness ratio of the casting is too large. When molten metal is flowing from one side in a thin section, it may loose sufficient heat resulting in loss of its fluidity, such that the leading edge of the stream may freeze before it reaches the end of the cavity. This defect determines the minimum thickness that can be cast for a given metal, superheat, and type of mould.

3. **Cold Shut.** A cold shut (Fig. 2.30 (*c*)) is an interface within a casting that is formed when two metal streams meet without complete fusion. The causes are the same as for misrun.

4. **Mismatch.** Mismatch is a shift of the individual parts of a casting with respect to each other. This may occur due to mould shift or core shift, (Fig. 2.30 (*d*)). The causes can be—an inexpert assembling of the two halves of the mould, from wear of pin bushes and pins and dimensional discrepancy between the core prints of the pattern and the core prints of the core.

5. **Flashes.** Flashes or Fins commonly appear along the mould joint at the places where the mould halves do not fit together properly because of much wear or warping of flask halves, or improper fastening of the cope to the drag.

6. **Metal Penetration.** or burnt on sand is a strong crust of fused sand on the surface of a casting which results from insufficient refractoriness of moulding materials, a large content of impurities, inadequate mould packing and poor quality of mould washes. When the molten metal is poured into the mould cavity, at those places when the sand packing is inadequate, some metal will flow between the sand particles for a distance into the mould wall and get solidified. When

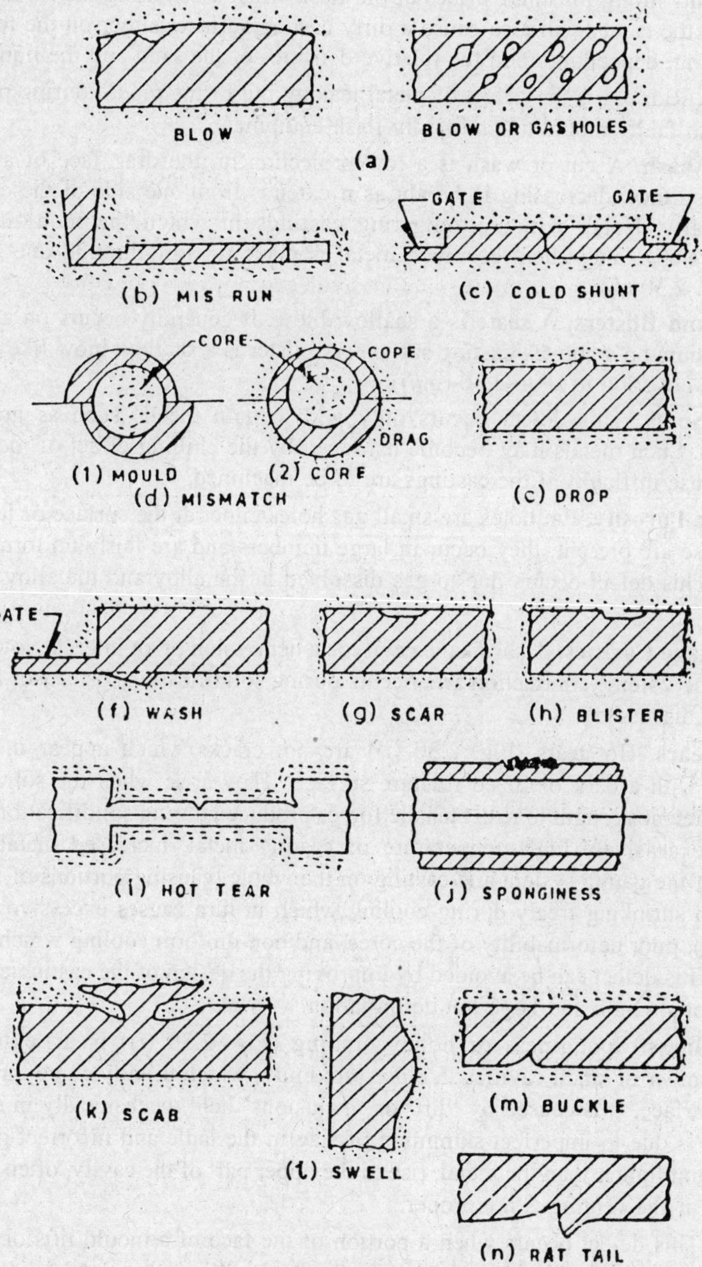

Fig. 2.30. Casting Defects.

he casting is removed, this lump of metal remains attached to the casting. Of course, it can be removed afterwards by chipping or grinding.

7. **Drop.** Drop or crush in a mould is an irregularly shaped projections on the cope surface of a casting (Fig. 2.30 (*e*)). This defect is caused by the break-away of a part of mould sand as a result of weak packing of the mould, low strength of the moulding sand, malfunctioning of

moulding equipment, strong jolts and strikes at the flask when assembling the mould. The loose sand that falls into the cavity will also cause a dirty casting surface, either on the top or bottom surface of the casting, depending upon the relative densities of the sand and the liquid.

8. **Run out.** Run out is the defect of metal leaking out of the mould during pouring. This defect occurs due to faulty moulding and faulty flask equipment.

9. **Cut or Wash.** A cut or wash is a low projection on the drag face of a casting that extends along the surface, decreasing in height as it extends from one side of the casting to the other end. It usually occurs with bottomgating castings in which the moulding sand has insufficient hot strength, and when too much metal is made to flow through one gate into the mould cavity, (Fig. 2.30 (*f*)).

10. **Scars and Blisters.** A scare is a shallow blow. It generally occurs on a flat surface, whereas a blow occurs on a convex casting surface. A blister is a shallow blow like a scar with a thin layer of metal covering it, (Fig. 2.30 (*g*,*h*)).

11. **Hard Spots.** This defect occurs only with certain metals such as gray C.I. with insufficient silicon. Such metals may become hardened by the chilling effect of moulding sand. Hard spots will cause difficulty if the castings are to be machined.

12. **Pinhole Porosity.** Pin holes are small gas holes either at the surface or just below the surface. When these are present, they occur in large numbers and are fairly uniformly dispersed over the surface. This defect occurs due to gas dissolved in the alloy and the alloy not properly degassed.

13. **Shrinkage Cavities.** A shrinkage cavity is a depression or an internal void in a casting that results from the volume contraction that occurs during solidification. Its causes and remedies have already been discussed.

14. **Hot Tears.** Hot tears (Fig. 2.30 (*i*)) are hot cracks which appear in the form of irregular crevices with a dark oxidized fracture surface. They arise when the solidifying metal does not have sufficient strength to resist tensile forces produced during solidification. They occur chiefly from an excessively high temperature of casting metal, increased metal contraction, incorrect design of the gating system and casting on the whole (causing portions of the casting to be restrained from shrinking freely during cooling which in turn causes excessive high internal resistance stresses), poor deformability of the cores, and non-uniform cooling which gives rise to internal stresses. This defect can be avoided by improving the design of the casting and by having a mould of low hot strength and large hot deformation.

15. **Sponginess.** Sponginess or honeycombing (Fig. 2.30 (*j*)) is an external defect, consisting of a number of small cavities in close proximity, which usually come through and are apparent on the surface. It is caused by 'dirt' or 'inclusions' held mechanically in suspension in molten metal and is due to imperfect skimming of slag in the ladle and incorrect gating design. The impurities being lighter than the metal, rise to the upper part of the cavity, often accompanied by bubbles of gas if the venting is not proper.

16. **Scab.** This defect occurs when a portion of the face of a mould lifts or breaks down and the recess thus made is filled by metal (Fig. 2.30) (*k*). When the metal is poured into the cavity, gas may be disengaged with such violence as to break up the sand (sand upheaving) which is then washed away and the resulting cavity filled with metal. The reasons can be :- too fine a sand, low permeability of sand, high moisture content of sand, and uneven mould ramming.

17. **Swell.** A swell is a slight, smooth bulge usually found on vertical faces of castings resulting from liquid metal pressure Fig. 2.30 (*l*). It may be due to low strength of mould because of too high a water content or when the mould is not rammed sufficiently.

18. **Buckle.** A buckle (Fig. 2.30 (*m*)) is a long, fairly shallow, broad, vee depression that occurs in the surface of flat castings. It extends in a fairly straight line across the entire fla

surface. It results due to the sand expansion caused by the heat of the metal, when the sand has insufficient hot deformation. It also results from poor casting design providing too large a flat surface in the mould cavity.

Buckling is prevented by mixing cereal or wood flour to sand. When the molten metal is poured into the mould cavity, these organic particles are oxidised and burnt. They leave the mould as a gas. This results in small void spaces for the expanding sand particles to move into without inducing severe compressive stresses in the mould.

19. **Rat-tail.** A rat-tail is a long, shallow, angular depression in the surface of a flat casting and resembles a buckle except that it is not shaped like a broad vee. The difference between a buckle and a rat-tail are shown in Fig. 2.30 (*n*). The reasons for this defect are the same as for buckle.

2.8. DESIGN OF CASTINGS

A cast part must have such a design as to ensure a high level of its working characteristics (strength, rigidity, stiffness, tightness, and corrosion resistance) at a given mass and shape of the part. Proper attention to design details can minimize casting problems and lower costs. For this, a close collaboration between the designer and the foundary engineer is important.

The main feature of the casting process is that the molten metal poured into the mould contracts as it cools and solidifies. The main consideration is that the shape of the casting should allow for directional solidification. Some of the important design considerations are discussed below:

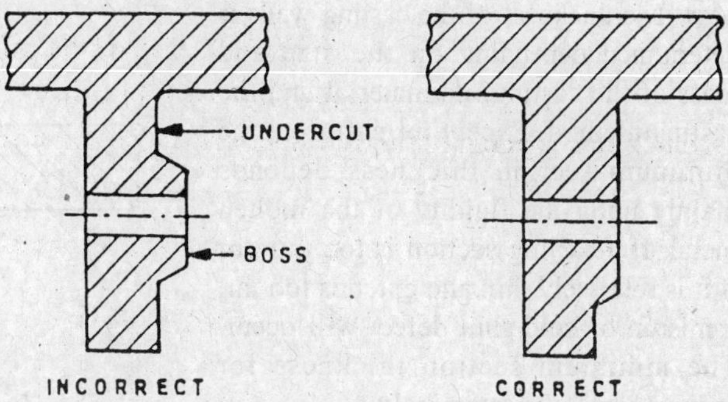

Fig. 2.31

1. The shape of the castings should be as simple as possible. That helps to reduce the cost of patterns, cores and moulds.

2. Castings should be made as compact as possible. Large steel castings of complex shape are divided into two or more castings, which can be easily cast, and then joined by welding to produce a 'cast-weld' construction.

3. Projecting details (bosses, lugs etc.) or undercuts should be avoided. Or, the pattern elements for them should be made so that they do not hinder the removal of pattern from the mould, (Fig. 2.31).

4. To facilitate removal, provision should be made for draft $\left(\frac{1^\circ}{2} \text{ to } 3^\circ\right)$ on the castings vertical surfaces. The draft is greater for the inside surfaces than for the exterior surfaces.

5. Wherever possible, avoid complex parting lines on the pattern, because these increase the cost of moulding operations. Parting lines should be in a single plane, if practicable, (Fig. 2.32). The design (*b*) is better, not only because the lines are simpler and the pattern is less costly but also because a plane parting can be used in the process of moulding.

6. Avoid concentration of metal so that no shrinkage cavities are formed. For this reason, bosses, lugs, pads should be avoided unless absolutely necessary. Metal section is too heavy at bosses which is difficult to feed solid, (Fig. 2.33).

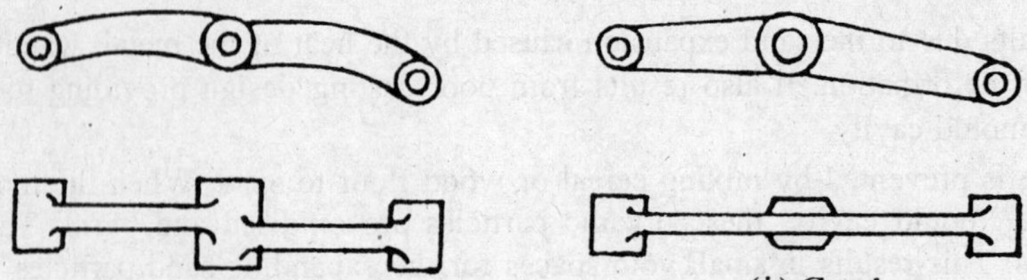

(a) Bad (b) Good

Fig. 2.32

7. The position of the casting's surfaces during metal pouring must be taken into account, since gas blow holes may form on the casting's upper horizontal surfaces. Critical surfaces of castings should lie at the bottom part of the mould.

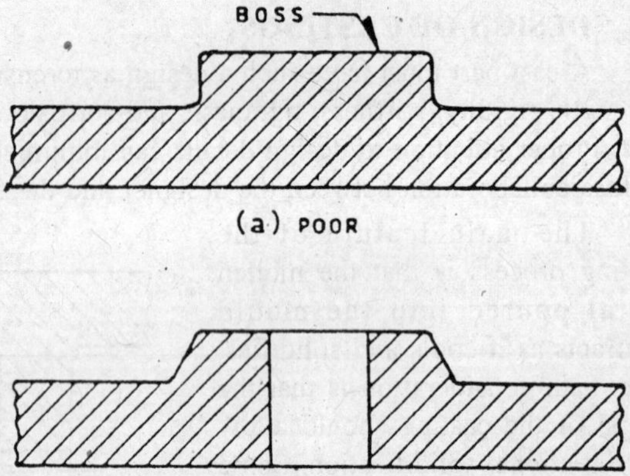

Fig. 2.33

8. **Minimum Section Thickness.** The thickness of the casting walls is determined depending on the size and mass of the casting, its material and the casting method. Except for cast iron, the minimum section thickness depends mainly upon the fluidity of the molten metal. If a casting section is too thin, or if it is relatively thin and extends too far, a misrun or cold shut defect will occur. The minimum section thickness for various metals is given below :

Al = 4.75 mm ; Cu = 2.38 mm ; Fe = 2.38 to 3.2 mm

Mg = 3.95 mm ; Steel = 6.35 to 12.7 mm.

For gray cast iron parts made in sand moulds, the minimum wall thickness can be obtained as follows :

$$t = \frac{L}{200} + 4 \text{ mm}$$

where L = greatest dimension of the casting.

Inner sections of the castings, resulting from complex cores, cool much slower than outer sections and cause variations in strength properties. Due to this, the internal walls should be 10 to 20% thinner than the external walls.

9. Whenever feasible, the castings should be designed with uniform section thickness because shrinkage defects (porosities, cracks) may arise in the thickened portions. Where it is not possible, the difference in thickness of adjoining sections should not exceed 2 to 1 and the two sections must be joined gradually with the help of large fillets, (Fig. 2.34 (*a*)). An abrupt change in section and sharp corners act as stress raisers in the finished casting, create turbulence during pouring and hinder proper feeding of the casting.

10. Sharp corners, (Fig. 2.34 (*b*)), in addition to creating troubles discussed under point 9

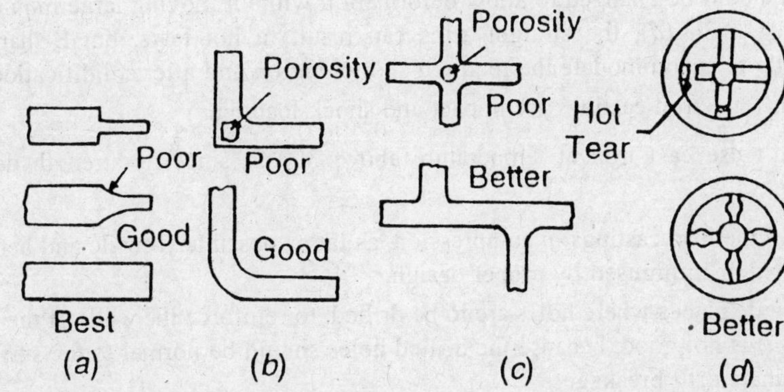

Fig. 2.34

cause local heavy sections. The sharp corners are eliminated with a radius at the corner from one-half to one-third of the section thickness.

11. Ribs are used for two purposes : (*i*) to increase stiffness (*ii*) to reduce weight

Too shallow in depth and too widely spaced ribs prove in effectual. Thickness of rib should be 0.8 times the casting thickness. Ribs should be rounded at edges and correctly filleted. Avoid complex ribs. Wide and low ribs are safer than thin and high ones.

When two ribs cross each other, (Fig. 2.34 (*c*)), localized heavy cross-sections result. This creates hot spot where the melt solidifies only after adjacent zones have solidified, resulting in shrinkage cavities. This can be avoided by offsetting the ribs.

The staggered ribs, (Fig. 2.35 (*b*)) cause less distortion than the regularly spaced ribs, (Fig. 2.35 (*a*)).

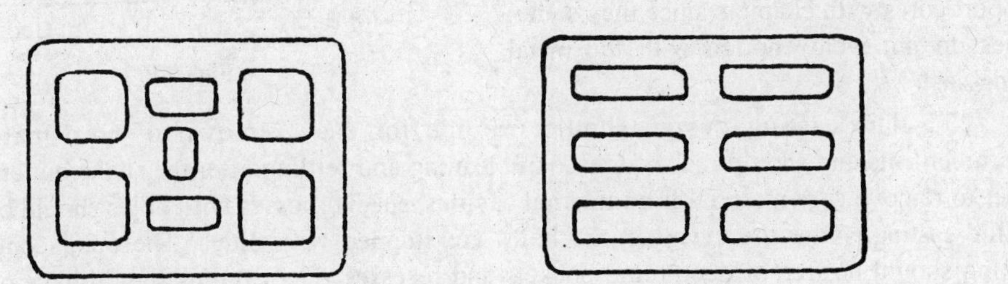

Fig. 2.35

Ribs should be used chiefly for static loads. These should be avoided where impact loads are expected since these increase the rigidity of the parts.

12. Due to the same reasons as for cross ribs, bring or join minimum number of sections together.

13. Inside diameter of a cylinder or bushing should be greater than the wall thickness of the casting. If it is less, it is better to cast solid. Holes can be produced by cheaper and safer methods than by coring.

14. A material that has a large solidification shrinkage will result in hot-shortness (hot tears) if the moulding material does not collapse sufficiently to allow shrinkage or the moulds

should be of simple shape so as not to develop tensile stress during solidification. If possible, the casting design should be changed to allow deformation without moving large mould masses. For example, in Fig. 2.34 (*d*), the straight arms can result in hot tears, but S-shaped arms can straighten a little to accommodate the required shortening on and after solidification.

15. Don't use iron castings for impact and shock loading.

16. Don't use cast iron at temperature above 300°C, since its strength decreases after 300°C.

17. Load the iron castings in compression, as far as possible. Tensile and bending stresses can be eliminated or minimised by proper design.

18. Provide places where holes are to be drilled, to reinforce the walls of the casting, (Fig. 2.36). Design (*a*) is not good, because the drilled holes should be normal to the surfaces, top and bottom to eliminate drill breakage.

Design (*b*) is much more satisfactory because it eliminates drill breakage, reduces drilling time and saves lot of material.

19. The casting shape should allow easy cut-off of the gating system elements and removal of cores.

20. It is preferable to dispose the entire casting in the drag if the casting construction permits doing so. This helps rule out mismatch.

21. The mould should have a minimum number of cores or no cores at all, if possible. It is advisable to use projection cods.

22. The cavities in castings should have extensions roomy enough to receive the core prints of cores. It is undesirable to support cores with chaplets since they sometimes do not weld enough with the metal being cast.

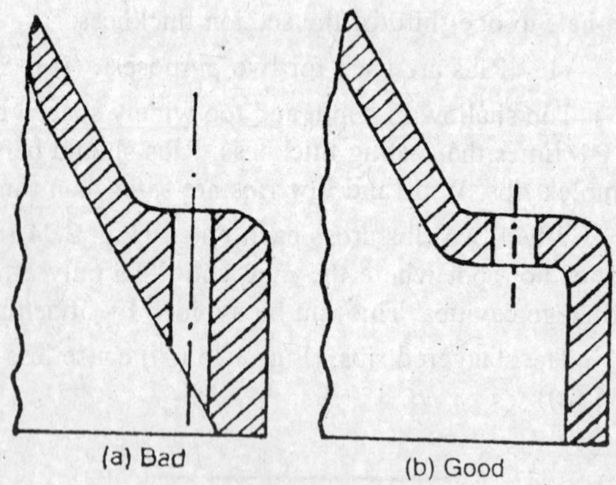

Fig. 2.36

23. The casting design should provide for easy removal of core materials and reinforcements and should make for ease of cleaning and fettling after the shake out operation. In order to remove core material from internal cavities, special bosses with holes should be provided on the casting. After the cleaning, the holes are stopped with plugs. The outer contour of the casting should be free of deep blind pockets and recesses. The cavities should have openings of sufficient size to facilitate stripping.

24. Casting drawings should indicate locating surfaces which are to be used in the machining of the castings and also in the checking of the castings. The locating surfaces should be formed by the pattern and should lie in the same mould half, so that relative displacements of mould parts and the cores do not affect accuracy of these surfaces. These locating surfaces are not needed for part functioning and can be removed after machining when necessary.

Examples of locating surfaces are :

(*i*) Centre holes on shafts.

(*ii*) Centring recess 1 and end face 2 on the skirt of an automotive engine piston, (Fig. 2.37 (*a*)).

(*iii*) Flats 1 of bosses 2 on cast blanks, provided for their proper loading and clamping., (Fig. 2.37 (*b*)).

(*iv*) Bosses 1 on turbine blade blanks, (Fig. 2.37 (*c*)).

(*v*) Two locating holes 1 on housing-type castings (Fig. 2.37 (*d*)).

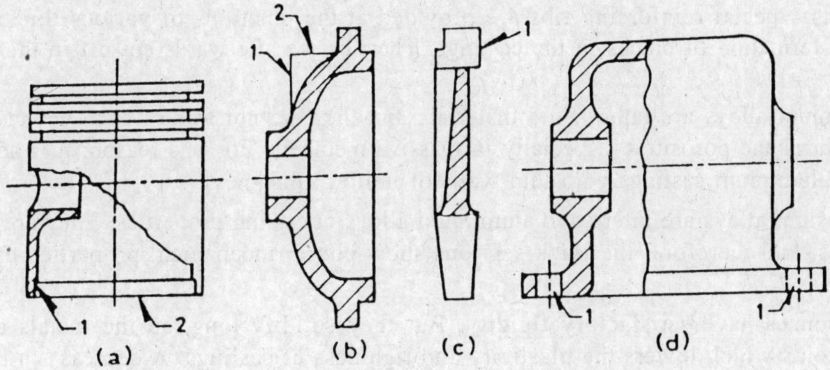

Fig. 2.37. Locating Surfaces types.

25. Cooling and solidification of molten metal start at the surface of the mould cavity and crystals form there first. Then the crystal growth takes place inwards, normal to all surfaces. At the central plane of the casting, the formations from different sides intersect. The metal in this plane is weak because the crystals do not join perfectly or even porosity may result there. In a casting with sharp corners, (Fig. 2.38 (*a*)), a plane of weakness extends from corner to corner, exactly where the stresses are apt to be maximum. In castings with corners rounded off, (Fig. 2.38 (*b*)), the plane of weakness runs uniformly throughout the centre, where the stresses are extremely low. Also, the heat flows propagating normally to the corner walls intersect and develop a 'hot zone' within the corner inner area. The corner walls here are made 20 to 25% thinner than those in the areas farther from the corner.

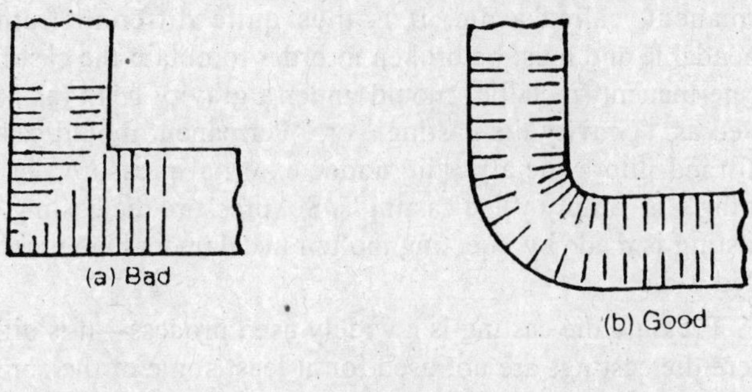

Fig. 2.38

26. The design of a cast part must take into account the casting properties of the alloy being cast.

Gray iron has good casting properties. The structure of gray iron is more affected by the rate of cooling than other grades. A change in the iron structure affects the mechanical properties of iron, its machinability, durability, corrosion resistance, and other working properties. If the metal cools fast, all the carbon in the thin walls of castings, proves to be in the bound state, forming the cementite, Fe_3C. This structure is of high hardness and is not amenable to machining. This defect in gray-iron castings is called hard spots. In the thick walls of the same casting, metal solidifies slowly resulting in a structure of low strength. So, the need for the uniform wall thickness in gray-iron castings acquires a particular importance.

High strength grades of iron show decreased fluidity and enhanced shrinkage. These factors should be properly considered while designing castings.

The low fluidity of steel causes difficulties in producing thin-walled long castings. High shrinkage of steel may result in blows, cracks and porosities in the thickened sections of castings, at the junction of walls and in bosses. Risers placed at these locations help eliminate these defects. Special reinforcing ribs are provided at the junctions of various thickness, to exclude the formation of cracks in the castings. The ribs are afterwards cut off in the fettling operation.

Aluminium alloys are rather fluid materials; but they are not free of such defects as gas holes and shrinkage porosities, especially in massive members. For this reason, it is advisable to produce aluminium castings with thin walls of uniform thickness.

Magnesium alloys are inferior to aluminium alloys in casting properties. They are subject to porosities, and therefore, the thick sections show poorer mechanical properties than thin members.

Tin bronzes have satisfactory fluidity. But they solidify long, so the metals develop shrinkage voids which lowers the plasticity and tightness of castings. A decrease in the wall thickness of the castings increases the rate of solidification and thus aids in improving the properties of castings.

Tin free bronzes show enhanced shrinkage. Due to this, the casting design should provide for risering of castings.

2.9. MISCELLANEOUS CASTING PROCESSES

2.9.1. Die Casting. In the die casting process, the mould used for making a casting is permanent, called a die. It is thus quite different from sand casting where the mould is expendable and must be broken in order to obtain the casting. If the molten metal is poured into the permanent (metallic) mould under a gravity head (as is done in sand casting) the process is called as ''Gravity die casting'' or ''Permanent mould casting''. However, if the molten metal is forced into a metallic die under external pressure, the process is known as ''Pressure die casting'' or simply ''Die casting''. So, pressure die casting can be defined as a process by which a casting is made by injecting molten metal under high pressure into a permanent (metallic) mould, called a die.

Pressure die casting is a widely used process—it is difficult to name a product or appliance where die castings are not used for at least some of the component parts. In every industry there is a steadily rising demand for high quality die castings. Pressure die cast parts are used in Automobile, Autoancilliary, Electrical equipments, Electrical motors. Business machines, Telecommunication equipments, Building hardwares, Toys and Home appliances and so on. The main reasons for the wide spread use of pressure die casting as a production process are listed below:—

1. Dimensional tolerances can be held to remarkably close limits which reduces or sometimes completely eliminates subsequent machining operations. For small castings, the tolerance range can be ± 0.03 to 0.25 mm. Closest tolerances are obtained, when Zinc alloys are die Cast.

2. Surfaces are smooth and clean and require a minimum of preparation for chrome plating, anodising, painting or other finishes.

3. A casting with walls of varying thickness can be produced.

4. Intricate castings (with thin walls) with high tensile strength can be btained, resulting in reduced raw material input. They are superior to stampings in this regard.

5. The production rate is rapid which lowers the conversion cost.

6. Cored holes down to 0.75 mm diameter at accurate locations are possible.

7. Inserts of any metal can be successfully embedded into a die casting, which largely eliminates secondary operations such as drilling and certain types of threading.

8. Die casting dies retain their usefulness and accuracy over a very long time of production.

9. The sprue, runners and gates can be remelted, resulting in low scrap loss.

The metals most widely used for die casting are non-ferrous and include: Lead, Tin, Zinc alloys, Aluminium alloys, Magnesium alloys and copper base alloys (Brass). Because of its many desirable properties (high dimensional accuracy, high castability and long die life), Zinc accounts for over $\frac{2}{3}$ rd of all die castings. Because of its light weight, the use of Aluminium as a die casting material is increasing. Magnesium die castings have superior vibration absorbing properties. Other main characteristics of Magnesium are :— lightness $\left(\frac{2}{3}\right.$ weight of Aluminium$\left.\right)$, excellent machineability, dimensional stability, shorter solidification time, less lubrication problems and little tendency to solder to the die. Although, Brass has high strength and resistance to corrosion, the high melting temperature has a serious effect in reducing the die life.

2.9.1.1. Product Applications :—

1. Household Equipment:— Decorative parts and mechanical parts for mixers, fans, dishwashers, vacuum cleaners, washing machines, refrigerators, meat slicers, can openers and stoves.

2. Business Equipment:— Components for offset duplicating machines, pencil sharpeners, staplers, typewriters, data processing machines and so on.

3. Hard ware:— Cabinet handles, knobs, catches, latches, bathroom fixtures, electrical fixtures; Taps, valves, fittings, burners (of brass) etc.

4. Industrial Equipment:— Motor housings, parts for hoisting equipment, motors, switches, rotor fan, Impeller wheel (Pump industry), electrical terminal sockets and so on.

5. Automotive. Windshield frames, window channels, bodies of fuel pumps and carburettor (of Zinc), instrument panel housings, handles, rear view mirror parts, switch housings, stearing wheel hubs, brake shoe (Al), clutch housing (Al), handle bar top, crank case cover etc.

6. Music and Communications. Components for microphones, fire alarm systems, Telephones, Television sets, speakers, drums, clarinets, record players and so on.

7. Toys. Pistols, electric trains, model aircraft, automobiles, game equipment and so on.

2.9.1.2. Limitations of Die Castings

1. Only small parts can be made. The maximum practical weight for a die casting is about 200 newton for zinc, 100 newton for Aluminium and 80 newton for Magnesium. The minimum weight can be as low as 0.05 newtons.

2. Only non-ferrous metals and alloys can be commercially cast. For ferrous metals (C.I. and steel), the chief drawback to their use is the lack of suitable and strong die materials for withstanding the high pouring temperatures.

3. Because of the high cost of equipment and dies, the process is economical only for mass production where the number of parts produced is 1000 or more. The dies will retain their size and shape over long periods of production during which many thousands of identical castings have been made. In general, the dies are heat treated and water cooled when in operation. The die life can be:

600,000 pieces of zinc

150,000 to 200,000 pieces for Aluminium or Magnesium,

and 40,000 to 50,000 pieces for Brass.

4. Because of some entrapped air (due to intensive stirring of metal with air under a rapid filling of the die), the die castings are usually porous, resulting in reduced mechanical properties. However, this drawback can be overcome by some methods discussed later.

2.9.1.3. Die-Casting Machines

A die-casting machine consists of four main parts:—

1. The frame
2. The die
3. The mechanism for opening and closing the dies.
4. Apparatus for injecting the molten metal into the die, along with source of molten metal.

The frame must be rigid and strong to support the weight of the dies, since often the weight of an assembled die may exceed several tonnes. The machine frame incorporates a stationary platen and a movable platen to which the die halves are attached (Die is made in 2 halves to facilitate removal of the casting). Die casting machine frames generally are four bar *presses*, although the solid one piece frame has gained wide acceptance for small machines.

The basic function of a casting machine is to open and close the die and to hold the two die halves together against the pressure of the molten metal developed by the injection system. The locking force must ensure leak proof clamping at the die parting. In some modern die-casting machines, locking forces may approach 10 MN, depending on the die size and the molten-metal pressure employed. The maximum force tending to open a die will be equal to the maximum molten metal pressure times the total projected area of the mould cavity and gating.

The opening and closing mechanism, which actuates the movable platen can be — Pneumatic, Hydraulic, Mechanical or a combination of these. In straight hydraulic method, live pressure backs up the die at all times, when it is closed. The most common method is the combination of hydraulic with mechanical system, where compound toggles are used with the force supplied by the hydraulic cylinder. Such machines have a fixed stroke, beyond which there is no force. In a toggle system, the full locking force is generated only when the toggle system

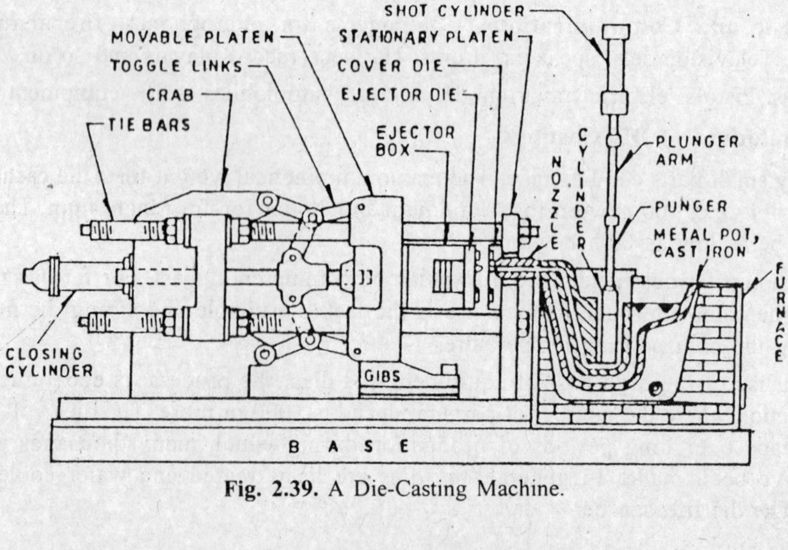

Fig. 2.39. A Die-Casting Machine.

is fully extended. Hence to accommodate dies of various thicknesses, the locking system is capable of sliding along the machine base. The die opening stroke is therefore independent of the die thickness. Fig. 2.39 shows the important features of such a system, (minus the molten metal injection system). The die is closed when the piston in the hydraulic cylinder moves to the right. This causes the toggle links to be straightened so that they close the die with great force. Due to the inherent kinematic properties of the toggle system, the movable platen moves slow in the beginning of the stroke, moves fast subsequently and finally slows down towards the end of the stroke. This serves as a built-in safety even if no other safety is provided. Double toggle system ensures parallel motion of all parts of the die throughout the stroke even with heavy dies.

The toggle system is self-locking and will not snap back even if there is a power failure. The system is,flexible and the closing speed can be infinitely variable and the locking force can be adjusted to suit the requirements by adjusting the toggle nuts. The other closing systems can be : wedge-lock hydraulic, cam lock hydraulic. Die casting machines are normally rated by the magnitude of the clamping force. Another method specifies the shot-weight capacity of the *injection* system. Machine capacities range from 50 kN to 25 MN.

The steps involved in the production of a die-casting are :—

1. Closing and locking the two halves of the die.
2. Forcing the molten metal into the die cavity under pressure.
3. Maintaining the pressure for a short time, and permitting the metal to solidify.
4. Opening the die halves.
5. Ejecting the casting with its assembly of sprue, runners and gates, by pins, and the above cycle is repeated.

After the casting has been ejected, the parting lines and gates are removed by trimming the excess metal from them. The casting is then ready for the finishing operations.

To obtain uniformity of die castings and maximum speed of operation, a predetermined and automatically controlled time cycle must be employed. A machine cycle is started by pushing a button and from then onwards the cycle continues automatically, stopping at the end of one complete cycle.

2.9.1.4. Molten Metal Injection Systems. Die-casting machines can either be hot chamber machines or cold chamber machines. In hot chamber machines, the metal pot and furnace are an integral part of the machine. The injection ram (plunger) and cylinder are submerged in molten metal. In cold chamber system, the molten metal is contained in a separate holding furnace. It is ladled into the cold chamber or shot sleeve and then forced into the die by a ram or plunger.

In hot chamber machine, the melting pot is usually made of steel. So, this method is not suitable for casting high melting point metals and alloys such as Aluminium, Magnesium and Copper-base alloys, to avoid wear and erosion of the melting pot, injection ram and cylinder. Molten aluminium readily reacts or alloys with the solid steel. Thus, it takes in iron from any steel with which it comes in contact. So, hot chamber system is used for casting Zinc, Lead, Tin and low melting point alloys. Cold chamber method is used for higher melting point non-ferrous metals and alloys such as Aluminium, Magnesium and Copper-base alloys. Since molten aluminium contact with the cold chamber and plunger is only momentary, pick up of iron by aluminium is held to a minimum. The lower melting point alloys (Zinc, Lead, Tin etc.) can be cast in cold chamber machines, but they are more economically cast in the faster operating hot chamber machines. Cold chamber system is slower than the hot chamber system, since hand ladling is used. However, hand ladling can be eliminated by employing electromagnetic pumping to bring the correct amount of metal into the cold chamber. With this, the production rate of a cold chamber machine can be increased by 25 per cent. Also, better castings are obtained since cleaner metal is pumped from below the surface of the metal in the furnace.

The metal injection pressure in hot chamber system is generally below 14 N/mm^2, whereas in cold chamber system, it may range from 35 to 175 N/mm^2 depending upon the size and type of castings to be made. However in some hot chamber machines, the injection pressure may be upto 40 MPa, that is, 40 N/mm^2.

Note. Aluminium alloys can be cast with hot chamber system by using ceramic pumps, but there are still some problems with the durability of the pumps.

Small to medium size castings can be made at a cycle rate of 100 to 800 (even upto 1000) die fillings or shots per hour.

2.9.1.5. Hot Chambers Machines. There are two designs of such machines :

(*i*) Submerged Plunger Machine

(*ii*) Air-blown system (employing "gooseneck" machine).

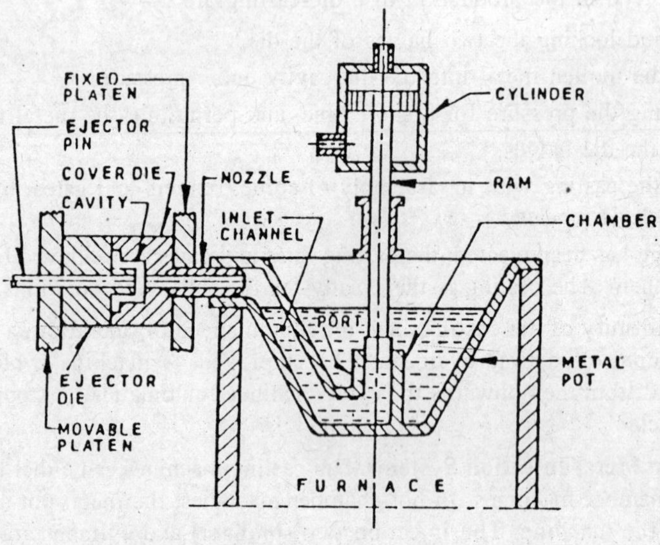

Fig. 2.40. Submerged Plunger Machine.

1. Submerged Plunger Machine. Fig. 2.40 shows the submerged-plunger type hot chamber machine. The ram or plunger is submerged below the surface of the molten metal, and operates with in a cylinder. The working medium can be pneumatic or hydraulic. The casting cycle consists of the steps already explained. The dies are locked together. When the ram is raised, it uncovers a part in the chamber wall, through which molten metal enters and fills the chamber. In operation, the plunger is forced downward closing the port, and then forcing the confined metal up through an inlet channel and nozzle into the die cavity. After a present time interval, the ram (plunger) is again raised, allowing the molten metal in the channel and nozzle to drop pack into the metal pot. Then, the die is opened and the solidified casting is ejected by ejector pins. Metal injection speeds and pressures can be controlled to suit different metals and castings.

2. Air-blown Machine:— In this machine, also known as the "direct air-injection machine", there is a closed pot or chamber often called a "gooseneck", Fig. 2.41. This container is supported above a larger metal pot from which the container is filled by dipping below the metal

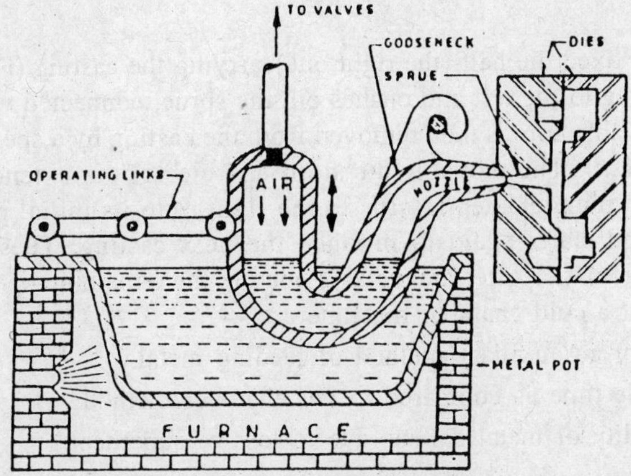

Fig. 2.41. Air Blown Machine.

level (by lowering the container with the help of operating links). The container is then raised above and its nozzle is locked against the die. Air under pressure (usually 2 to 4 MPa) is admitted through a valve and forces metal through the nozzle (spout) into the die. Next, the die is closed, air above the metal is released and exhausted, and the gooseneck is lowered to be refilled, breaking away at the end of the sprue where the spout is kept hot by a flame. After the casting formed in the die solidifies, the die is unlocked and opened and the casting is ejected. After, relocking the die, the cycle is repeated.

The metal enters the die in a spray of finely divided particles. Because of this spraying action, air is entrapped in the cavity resulting in porosity in the casting. Metals which have a dissolving effect on iron and steel (material of gooseneck) should not be used. The machine is rather slow and expensive to operate. Due to the above factors, the machine has been gradually replaced by the cold-chamber die-casting machine.

2.9.1.6. Cold Chamber Machines. There are two types of cold chamber machines depending upon the direction of movement of the injection plunger: Horizontal and Vertical.

1. Horizontal Cold Chamber Machine. The injection plunger or shot sleeve is in a horizontal position and the injection plunger movement is in a horizontal direction. Fig. 2.42 shows such a machine in the various stages for producing a casting. When the working cycle begins (Fig. 2.42*a*), the two die halves are closed with the core being in the working position. The injection plunger is in the extreme right hand position. Melt of a quantity sufficient for one shot is transferred (often mechanically) to the injection cylinder or shot sleeve. The plunger moves to the left and forces the molten metal into the cavity of the die (Fig. 2.42*b*). After a sufficient time for solidification the die automatically opens and the movable die half (left one)

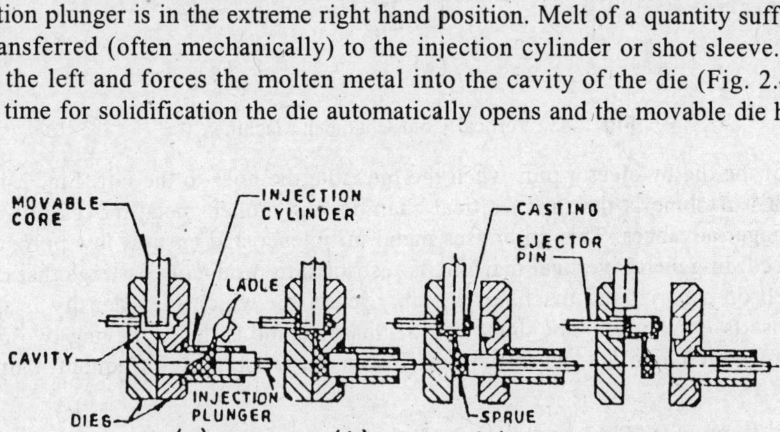

Fig. 2.42. Horizontal Cold Chamber Machine.

moves away from the fixed die half (the right one)carrying the casting (Fig. 2.42*c*) with it. The plunger goes on moving to the left and pushes out the sprue, connected with the casting, out of the fixed die. The movable core is then removed from the casting by a special device (not shown in the figure (Fig. 2.42*d*): The movable die stops and ejector pin extends from it to eject the casting. The injection plunger then returns to the right, into its initial position. The working cycle is now over and starts again to produce the next casting. These machines are fully hydraulic and semi-automatic. After the metal is poured, the remaining operations are automatic. Main disadvantages of a cold chamber machine are:—

1. The need for an auxiliary method of feeding metal.
2. Longer cycle time as compared to hot chamber method.
3. The possibility of metal defects due to loss of superheat.

2. Vertical Cold Chamber Machine. In this machine, the injection cylinder (shot sleeve) is in a vertical position and so the injecting plunger moves in a vertical direction. Fig. 2.43 shows a vertical cold chamber machine for producing a casting. At the start of the working cycle, the injecting plunger is up and the molten metal is poured into the chamber from a ladle. The return piston is held up by the spring and shuts the opening of sprue adapter. This prevents the metal from getting into the die cavity by gravity flow, under no external pressure. When descending, the injection plunger presses against the metal and moves the return piston downward, forcing a shot of metal into the die cavity through the sprue channels (Fig. 2.43*b*). After the holding period, the injection plunger moves upward into its initial position, the return piston actuated by the pusher removes excessive metal (biscuit) from the chamber. The casting

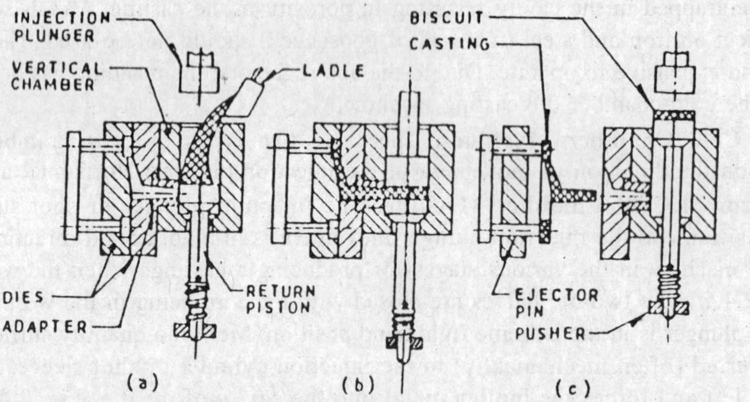

Fig. 2.43. Vertical Cold Chamber Machine.

is ejected out of the die by ejector pins when the movable die goes to the left (Fig. 2.43*c*). The advantage of this machine is that in a vertical chamber the molten metal moves in a compact mass as the plunger advances. This minimises metal turbulence and ensures low porosity in the castings produced. In general, vertical machine is restricted to producing castings that could not be made as well on a horizontal machine - notably, parts for which high density is specified, or the use of inserts is required. One drawback of this machine is that in having two operating plungers, one of which also acts as a trimmer (lower one), it requires frequent maintenance work.

It is clear from Figs. 2.42 and Fig. 2.43 that in the case of horizontal cold chamber machine, the molten metal is injected at an offset position (below the centre line of the die cavity) into the die cavity, whereas in vertical machine, the metal is injected into the centre of the die cavity. Again, in the horizontal machine, the slug of excess metal comes along with casting, but in vertical machine it is rejected at the chamber.

Horizontal Chamber Machine Vs Vertical Machine

Machines with horizontal cold chambers have some advantages as compared to the vertical chamber machines in the following respects:—

1. Shorter working cycle since there is no need to remove excess metal, the latter is pushed out along with the casting when the die opens. So, production rate on horizontal machines is higher compared to vertical machines.
2. Short travel of the metal from the chamber into the die cavity with no turn at right angle, resulting in less pressure drop and velocity drop in horizontal machines than in vertical machines. This allows the metal to fill up the die cavity under higher pressure resulting in better overall strength of the castings.
3. More complex articles can be cast in horizontal machines.
4. Metal consumption per shot is lesser in horizontal machines.

Offset Injection Vs. Central Injection. The die design in the case of offset injection (as in hot chamber system and horizontal cold chamber machine) is simpler and therefore, manufacture and maintenance of dies are easier and less time consuming. However, there are certain parts that can best be made with centre gating (vertical machine), that is, parts having thick centre hubs and thin outer sections, for example, wheels and blower impellers etc.

2.9.1.7. Die Casting Design. Some of the factors to be kept in mind while designing a casting to be made by casting method are discussed below:—

1. **Minimum Wall Thickness.** Keep wall thickness of the casting as uniform as possible or one that tapers slightly from the thinnest section farthest from the gate to the heaviest section at the gate. The transition should be as gradual as possible. The minimum wall thickness depends upon the fluidity of the metal to be cast. The wall thickness should be sufficient to permit proper filling but sufficiently small for rapid chilling of the metal to obtain maximum physical properties (maximum density). The commonly used wall thicknesses for the die-castable metals are given below :

Minimum Wall Thickness

(*a*) Cu, Mg = 1.27 to 2.29 mm
(*b*) Al = 0.76 to 2.03 mm
(*c*) Zn = 0.38 to 1.29 mm
(*d*) Pb, Sn = 0.76 to 1.52 mm

Maximum Wall Thickness

Preferably 7.93 mm, Usually $<$ 12.7 mm

The section thickness increases as the surface area of the casting increases.

2. **Undercuts.** Undercuts can be provided on a casting, but wherever possible, a part should be redesigned to eliminate undercuts, since these result in large added cost and reduced production rate.

3. **Ribs.** Ribs may be employed to increase strength and stiffness of the casting. These reduce the weight of the casting and makes for better distribution of metal within the die.

$$\text{Height of ribs} \ntriangleright 5 \times \text{ wall thickness}$$

Draft and fillets should be ample to obtains a smooth transition into the thinner section.

4. **Inserts.** Inserts such as bearings, wear plates, bushings, shafts and screws can be incorporated in die castings. They must be easily and precisely located in the die. The

surrounding material must shrink on to the insert so that it does not come loose. Inserts, however, result in reduced production rate.

5. **Cored Holes.** Holes vertical to the die parting line can be readily cast with stationary core pins. Holes in planes not perpendicular to the die parting plane require special core pulls. The cored holes may be round or of any other desired shape. The following data may be used for cored holes:—

(*a*) *For Al and Mg :*

Minimum: 3mm diameter × 5 mm deep, advancing proportionately to: 3 diameter deep at 6 mm diameter and above.

(*b*) *Zn :*

Minimum: 1.5 mm diameter × 2.5 mm deep, increasing proportionately to: 6 diameter deep at 6 mm diameter and above.

6. **Corner radii and fillets.** Avoid sharp corners in all castings. Corners must be given proper radii. Diecasting permits the smallest corner radii as compared with other casting processes, because the metal is injected into the die under high pressure. For low melting point alloys,

$$\text{Fillet radius} = \frac{1}{2} \times \text{ wall thickness}$$

To avoid an undesirable concentration of metal at a corner, the fillet radius should not be greater than wall thickness.

7. **Draft.** The amount and location of draft on a casting will depend upon its location in the die and whether it is an external surface or a cored hole. Draft on the die surfaces normal to the parting line permits the casting to be ejected without excessive wear on the die cavity. Draft allowance may be taken as follows:—

(*a*) On Al and Mg: 1 in 100

(*b*) On Zn: 1 in 200

(*c*) On Cu alloys: 1 in 80

Taper in cored holes:—

(*a*) For Al and Mg: 1°upto 10 mm diameter increasing to 2° on diameters over 25 mm.

(*b*) For Zn: Half the above values

8. **Other characteristics :**

(*a*) *Dimensional Tolerance:—*

0.025 to 0.178 mm upto first 25 mm size.

and 0.025 to 0.050 mm for additional 25 mm size.

(*b*) *Machine finish allowance:—* 0.79 to 0.40 mm

(*c*) *Surface Roughness:—* 1.02 to 2.54 μ m

(*d*) *Overall size:—*

Maximum:—

Zn = 34 kg, Al = 45kg, Mg = 20 kg

Cu = 2.3 kg, Sn = 4.5 kg, Pb = 6.8 kg

Minimum:— < 0.028 kg.

(*e*) *Optimum Lot Size:— 1000 to 100,000.*

(*f*) *Cost:—*

(*i*) Tooling Cost : High

(*ii*) Direct labour cost, finishing cost, Scrap loss : Low

2.9.1.8 Types of Die-Casting Dies. Depending upon the number of cavities in the die and their function, the die-casting dies are of the following types:—

1. Single-cavity Dies. These dies contain only one die cavity and produce only one casting at a time. Such dies are used for large or complex castings.

2. Multiple-cavity Dies. If the quantity of castings to be produced is large and they are relatively small in size, a multiple-cavity die can be used. In such a die all the cavities are identical.

3. Combination Die. It is again a multiple-cavity die but the cavities are not alike. Such dies are often used to produce several parts of the same assembly. They are frequently made of insert blocks that can be removed so that other die blocks can be substituted.

4. Unit Dies. Results achieved by multi-cavity and combination dies are sometimes also achieved by using a group of unit dies. Unit dies are separate small dies having usually one or more cavities, that are inserted in a common master holding die, and are connected by runners to a common opening or sprue hole, so that all dies in the group are filled simultaneously. When this is done, units can be arranged in various combinations and different unit dies may be interchanged in the master die. This results in the production of a variety of castings with one master die. Unit dies are commonly low in cost and often are employed with high economy.

2.9.1.9. Die-Materials. Since the die casting is accomplished at elevated temperatures and alternating heating and cooling, the die material is subject to heavy working conditions. The die material should possess the following properties :

1. Sufficiently high melting point to withstand erosion by the hot liquid metal.
2. High enough strength so as not to deform in repeated use.
3. A high thermal fatigue strength to resist premature crazing, that is, the formation of thermal fatigue cracks.
4. Resistance to softening, that is, material should retain hardness at elevated temperature.
5. Low thermal expansion co-efficient.
6. Low adhesion to melt to prevent welding of part to the die.

The performance of die materials is directly related to the injection temperature of the molten alloy, the thermal gradients within the die, the production cycle and the production quantities. The die material may be close grained C.I., although alloy steels are the most widely used. Alloy steels of increasing alloy content are required as the injection temperature of the molten metal and the thermal gradients within the die increases and the production cycle becomes shorter. For casting higher melting point alloys (brasses and ferrous metals), the die steel must contain large proportions of stable carbides so that strength is retained at higher temperatures. Composition of die-casting die-steels is given in Table 2.4.

Table 2.4 Die-casting Die-Steels

Type of Steel	*Composition, %*							*Casting Metal*
	C	*Cr*	*Mo*	*W*	*V*	*Co*	*Ni*	
Hot work Tool Steel:								
H11	0.35	5.00	1.50	—	0.50	—	—	Zn
H12	0.35	5.00	1.50	1.50	0.40	—	—	Al
H13	0.35	5.00	1.50	—	1.00	—	—	Al

H19	0.40	4.25	—	4.25	2.00	4.25	—	Cu alloys
H20	0.35	2.00	—	9.00	—	—	—	,,
H21	0.35	3.50	—	9.00	—	—	—	,,
Mould Steel P2	0.30	1.70	0.40	—	—	—	—	Zn
Maraging Steel:								
250 grade	—	—	5.00	—	—	8.00	18.00	Cu alloys
300 grade	—	—	5.00	—	—	9.00	18.00	,,

The die material is heat treated to minimum hardness of 35 Rc for Zinc. It will be higher for Mg, Al and Cu alloys and lower for lead and tin alloys.

Recent work with refractory materials-particularly Mo, W and their alloys-has greatly improved die life for casting copper alloys.

Materials for Cores and Slides :

(*i*) For Zn alloys :— H11, H12, H13

(*ii*) *For Al or Mg alloys* : H11, H12, H13

(*iii*) Copper alloys :— H20, H21

Materials for Pins :

(*i*) For Zn alloys :— H11, H12 nitriding steel, H13

(*ii*) For Al or Mg alloys :— H11, H12 nitriding steel, H13

(*iii*) *For Cu alloys* :— H20, H21

Note. Sliding components, except those used with dies for casting copper alloys are nitrided for improved wear resistance.

2.9.1.10. Die-Manufacture. The dies used for die-casting may be either wrought or cast. The cavities in the wrought die blocks are made by the conventional machining methods including die sinkers (special purpose milling machines). The unconventional machining methods such as ECM, EDM, and USM have gained wide applications in making complex dies.

Advantages of Wrought Dies :

1. More economical whenever a single cavity or core is required with relatively shallow machining.
2. More economical whenever cavities can be *hobbed* with simple hobs and where production requirements are low enough to make the hobbing of steels economical.
3. More economical to machine for low production requirements and for dies with simple configurations that can be sunk in low cost steels.
4. Are usually more economical with conventional machining and engraving whenever engraving is to be raised on the casting.
5. They are generally more economical whenever EDM can be used in shallow impressions and whenever EDM electrode configuration is simple and few in number. Whenever, extremely close tolerance is required, this method is superior, especially on prehardened blocks.
6. They are necessary for the production of refractory material dies, which cannot be precision cast at present.
7. Have the advantage in applications where cores are long and fragile.

Advantages of Cast Dies :

1. Give improved service life.

2. Their production requires no additional costs to form complex die parting line.
3. This method cheaply, rapidly and accurately produces cavities and cores after patterns are made.
4. Casting of dies permits these to cycle at higher speeds, because cores, waterways etc. may be cast directly into the dies
5. Eliminate difficult metal removal problems when cores are deep and complex.
6. Better and more economical when textures must be die cast or when engravings must be sunk into the die castings.
7. With the use of reversed patterns, casting allows simplified production of many dies which would be difficult to machine.

2.9.1.11. Die Lubrication and Coatings. In all die casting processes, it is essential to apply a die lubricant to the die surfaces in between each shot. Lubrication prevents a casting from adhering to the die and provides the casting with a better finish. The lubricant is usually graphite or MoS_2 in an oily carrier, which is then dispersed in water. Evaporation of water aids cooling.

A coating or dressing is applied to the die and core surfaces to serve as a barrier between the molten metal and the surfaces while a skin of solidified metal is formed. A coating is applied for the following purposes :

1. To protect the die and reduce heat transfer, thereby minimising thermal shock to the die material.
2. To prevent soldering of molten metal to the die, reduce adhesion and facilitate ejection of the casting.
3. To prevent premature freezing of the molten metal.
4. To control the rate and direction of solidification of the casting and thus its soundness and structure.

Coatings or dressings are composed of refractory powder in a suspending material. Refractory coatings (graphite, silicone) are sometimes built upto thicker layers for the purpose of reducing temperature fluctuations on the die surface. Coatings are an important element in the system and their uniform application is most important.

2.9.1.12. Preheating of Dies. The die casting dies work as heat exchangers. At the start of the production cycle, the dies must be preheated to the desired temperature. This is done to produce sound castings. The molten metal can fill the thinnest and remotest sections of the casting. Also, it facilitates the application of mould coatings/ lubricant to the die surfaces. During steady state production of castings, heat given off by the molten metal while solidifying, is removed by means of radiating pins or fins or circulating water. The die faces are also cooled by the evaporation of water from lubricants and coatings. The typical temperatures to which the dies must be preheated are:—

(*1*) For Zn: 150 to 200°C

(*2*) For Al: 225 to 330°C

(*3*) For Mg: 250 to 275°C

(*4*) For Cu alloys:300 to 700°C

2.9.1.13. Vacuum Die Castings. In this method, a vacuum is applied to remove all air from the die cavity before a shot of molten metal is forced into the die cavity. This completely removes porosity from the casting. However, application of vacuum is not a suitable substitute for good die practices.

2.9.1.14. Defects in die castings

Die-casting defects, basically can be divided into three categories:

1. *Defects due to the cooling of the die :*

The possible defects under this category are:

(*a*) Non-filling of die cavities

(*b*) Cold shut

(*c*) Severe chill

(*d*) Chill

(*e*) Flow line

2. *Defects due to the heating up of the die or the metal: These defects include :*

(*a*) Soldering

(*b*) Crack

(*c*) Bent part

(*d*) Broken parts

3. *Defects due to one or the other reasons :*

These defects include:

(*a*) Scaling

(*b*) Blister

(*c*) Porosity

(*d*) Mechanical defects.

Below, we discuss the possible causes and the remedies for these defects:

(i) *Non-filling of die-cavities:* The causes are :

— Quantity of metal is less than needed

— Cooling of the die or the metal

— Less period of the metal shot

The remedies to avoid the occurrence of this defect are:

— Required quantity of metal should be used. Do not allow the metal or the die to be cooled, that is, avoid any delay in completing the operation. Heat the metal to the required temperature and do pre-heating of the die if needed.

— Apply the shot with the proper speed

(*ii*) *Cold shut/Mis-run :*

This defect is the same as in sand mould casting. The reasons and remedies are also similar. The reasons are:

— Cooling of die or loss of plasticity of the metal

— Shot speed less

— Air-vent or overflow is closed

The remedies are:

— The metal should be heated to the required temperature. Die should be preheated if needed. The thickness of the casting should not be less than the minimum recommended design thickness.

— Shot speed should be proper.

— Air vent should be opened.

(*iii*) *Severe chill :* This defect is similar to the defect of cold shut/mis-run, with the difference that whereas, mis-run/cold shut is a localised phenomenon, severe chill covers a large portion of the casting. The reasons and remedies are the same as mentioned under the cold shut/mis-run defect.

(*iv*) *Flow-Line :* This defect is similar to cold shut/mis-run and chill defects. The causes, thus, are:

— Low temperature of the metal or die

— Excess of die lubricant

The remedies are:

— Keep the optimum temperature of the metal and the die

— Do not use excess of lubricant.

(*v*) *Soldering :* This defect pertains to the sticking of the cast metal to the die. This results in certain other defects like cracks, porosity etc.

The causes are :

— high temperature of the metal or the die.

— excess of die-lubricant

— roughness of the die surface

— content of iron is less in the metal

The remedies are:

— Keep the temperature of the metal and die, optimum.

— Do not use excess of lubricant.

— Smoothen the die surface

— Use the metal with proper iron content.

(*vi*) *Broken part :* Sometimes, during ejection of the casting from the die cavity, the total casting does not come out of the die-cavity. Some part of the casting remains sticking with the die.

The causes are :

— Soldering

— Die-lubricant.

The remedies are :

Avoid soldering as discussed under point (*v*)

Use proper die-lubricant.

(*vii*) *Heat mark :* Heat mark is a prominant line on the surface of the metal. Sometimes, due to the closure of the air vent or overflow, some gas is entrapped inside the die. Due to this, the size of the heat mark gets increased.

The causes of this defect are:

— improper die steel

— improper die cooling

— production done with the cooled die

The remedies are:

— Use proper die steel

— Proper cooling time should be given

— Production should not be done with the cooled die.

(*viii*) *Scaling* : Due to chemical reaction and oxide scale formation, an irregular surface appears on the surface of the die.

The remedies are :

— Use Caustic solution

— Polish the Casting

— only adequate use of lubricant.

2.9.1.15. Recent trends in Die-Castings process :

(1) *Die Casting of Ferrous metals.* Because steel cannot withstand the temperatures of molten C.I. and steel, it has not been feasible to use die casting method for Ferrous metals. However, advances in the use of W and Mo as die materials has increased interest in the die casting of C.I. and steel. More recently, refractory metal alloys, particularly the precipitation-hardenable molybdenum alloy TZM (0.015% C, 0.5% Ti, 0.08%Zr), have found increasing application. Graphite dies can also be used for steel, but only for relatively simple shapes. These materials have got the following advantages for ferrous die casting :

(*i*) High melting point

(*ii*) Low co-efficient of thermal expansion.

(*iii*) High thermal conductivity. This causes rapid solidification of the castings, in as little as tenths of a second in several cases, resulting in fine grained structure and rapid cycle time. Fine grained steels and C.I. exhibit strength values that approach the properties of forged steel.

(*iv*) Unusual resistance to thermal fatigue.

Characteristics of Ferrous Die-Castings :

(*a*) Good dimensional accuracy.

(*b*) Tolerance = ± 0.05 mm across parting line of 19 mm thick casting. Closer tolerances are possible within die cavities.

(*c*) Wall thickness upto 3.2 mm.

(*d*) Surface roughness = 0.8 to 1.6 μm Ra.

(*e*) Draft can be held to a minimum if the casting is removed from the die while it is hot. No draft is necessary in internal die cavities and 3° to 5°draft can be for long diameter, short cores.

(*f*) Certain refractory dies last as long as 15000 cycles.

(2) *Reduction of porosity in Casting :*

(*i*) A recent trend is to keep the injection pressure at medium level, followed by much higher pressure after the casting has started to solidify. This tends to improve the density and reduce porosity.

(*ii*) In the method known as " Squeeze Casting" or " Melt forging", also known as "Liquid forging", typical forging dies are used. A pre-measured amount of molten metal is poured into a die. When the casting has cooled to the forging temperature, the dies are closed and solidification is completed. This method which is a transition between die casting and conventional hot forging has been discussed in Chapter 4. The method produces a highly refined grain structure and is also used for producing Al alloy parts such as diesel engine pistons reinforced with Al_2O_3 fibres.

(3) *Use of Robots.* Programmable robots can perform the repetitive tasks involved in die casting, for example, opening and closing the die, removing and quenching the casting, insertion into a trim press and also application of coatings. Some of the newer die casting machines use indexing turrets to move the casting through trimming and minor machining operations.

2.9.2. Centrifugal Casting. Centrifugal casting is the method of producing castings by pouring the molten metal into a rapidly rotating mould. The metal is thrown out towards the mould face by the centrifugal force under considerable pressure. This results in better mould filling and a casting with a denser grain structure, which is virtually free of porosity. The mechanical properties of most metals are improved by 10 to 20%. The method is chiefly used for casting parts having the shape of bodies of revolutions. According to the shape of the mould, the centrifugal casting method can be classified as :

1. True - Centrifugal Casting
2. Semi - Centrifugal Casting
3. Centrifuge Casting.

2.9.2.1. True - Centrifugal Casting. True centrifugal castings are produced by pouring molten metal into the cavity of a rapidly rotating metal mould to whose walls the metal is thrown by centrifugal force and where it solidifies in the form of a hollow casting. Thus the distinguishing feature of true-centrifugal casting is the production of hollow casting by the centrifugal force alone and without the aid of a central core.

The centrifugal casting machines used to spin the mould may have either a horizontal or a vertical axis of rotation. For short castings, that is, when diameter to length ratio is rather large (bronze bushings, worm wheels and piston ring blanks), the rotational axis is vertical. Long parts, that is, castings with relatively long length in relation to their diameters (C.I. water supply and sewerage pipes, steel gun barrels etc.) are made in horizontal axis machines.

The principle of the vertical axis machine is shown in Fig. 2.44 (*a*). As is clear, the central hole will not be completely cylindrical, but will be slightly paraboloidal, which will need machining after the casting is made. However, nearly cylindrical holes can be made by employing higher spinning speeds. The moulds may be permanent moulds made of metal or graphite or may be sand lined.

There are two types of horizontal axis centrifugal casting machines which differ in the way the metal is distributed along the length of the mould during pouring : In one, the pouring trough (ladle or spout) travels horizontally, while the spinning mould is stationary. In the other type, the ladle is stationary and the mould travels. The mould or flask is made of steel or graphite. The metal mould can be used as it is or its inner surface may be lined with green sand or dry-sand. The choice depends upon the geometrical dimensions of a casting, its material and the production volume. Metal moulds are preferred for the production of a large number of single-type parts which show no retarded shrinkage and have an outer shape allowing easy extraction of the casting from the mould. Sand-lined moulds are preferred for castings in which retarded shrinkage takes place and the outer surface of these castings causes difficulties in removing them from the moulds. Such moulds are used to advantage for the production of a small lot of cast iron pipes without chilling. These pipes will not require annealing heat treatment, as will be needed for similar pipes made in metal moulds, since the cooling rate will be slow because of sand lining. The sand lining also protects the metal mould from the intense heat of the molten metal and, increases its durability.

The method of casting pipes in which the ladle is stationary and the mould travels is called as "The deLavaud Process". This is explained with the help of the following steps, (see Fig. 2.44 (*b*)),:

1. The machine consists of an accurately machined metal mould surrounded by water. Dry heat-insulating coats (refractory mould wash of materials such as quartz sand and phosphate flour) are applied to the inner surface of the mould to secure stable thermal conditions, increase its durability and produce iron castings free of chill. The machine is mounted on wheel so that it can be moved lengthwise on a track.

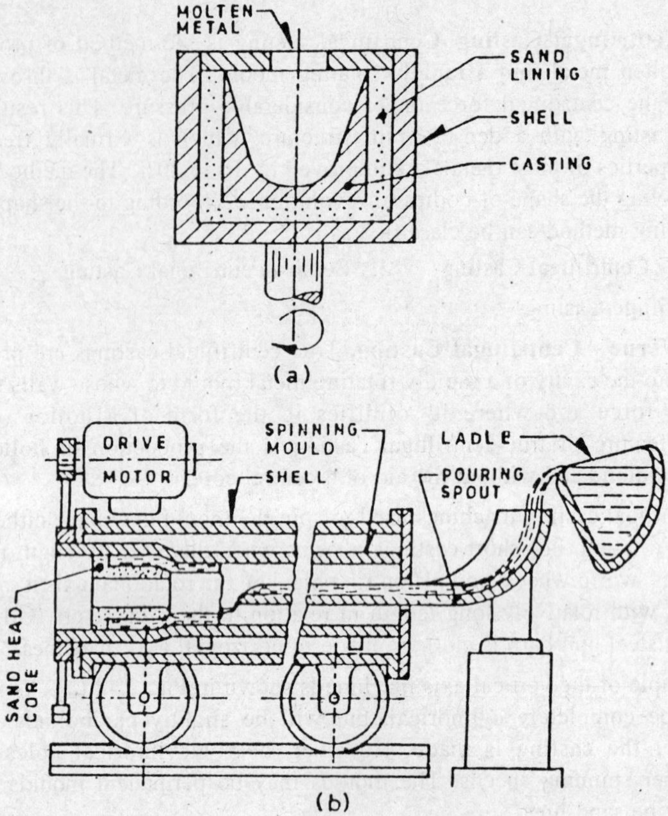

Fig. 2.44. True Centrifugal Casting.

2. The metal mould is heated to 150 – 200°C before pouring. After pouring, the mould is cooled to by circulating water through the surrounding shell.

3. The long pouring spout is inserted to the far extremity of the mould.

4. As pouring proceeds, the rotating mould, that is, the machine is moved slowly on the track, leftwards.

5. At the end of the process, the machine will be at the lower end of its track, with the mould spinning continuously till the pipe has solidified.

6. The cast pipe is then extracted from the mould by inserting a piper puller.

The use of the sand head core is to form the inside contour of the bell end of the pipe. If desired, the outer contour of the casting can be varied, while the inside remains cylindrical. The spinning speed of the mould will depend upon the diameter of the pipe, the composition of the pouring metal and a number of other factors.

Advantages :

1. Castings acquire high density and are distinguished for their fine-grained structure and high mechanical strength.

2. Inclusions and impurities such as oxide, slag and gas etc., being lighter than the molten metal, will segregate toward the centre and cling to the inside surface of the casting, where they may be removed by subsequent machining operations.

So, the castings will be of mainly clean metal.

3. Gates and risers are not needed, which results in saving in material and increasing the yield.

4. High output.

5. Formation of hollow interiors without cores.

6. The castings are less subject to directional variations than static castings.

Disadvantages :

1. Contaminations of inner surface of the casting with segregates and non-metallic inclusions, which make it necessary to increase finishing allowance for subsequent machining of inner surfaces.

2. An inaccurate diameter of the inner surface of a casting.

3. Not all the alloys can be cast by this method due to the increased segregation of alloy constituents during pouring under centrifugal forces.

Product applications. Parts with rotational symmetry : C.I. water supply and sewerage pipes, steel gun barrels, pump components, rolls, liners, hollow shaftings, bearings, sleeves, hydraulic jack cylinders and air cylinders, large propeller shaft sleeves and stern tubes, chemical reactor vessels, heat exchanger bodies, refinery and. petrochemical applications ; pressure vessel bodies, reactor tubes and pressure piping for nuclear power plants ; paper mill rolls, textile rolls, steel mill rolls and machinery drive rolls etc.

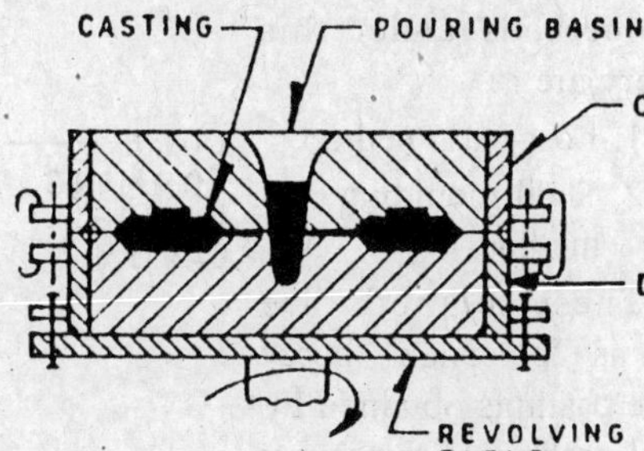

Fig. 2.45. Semi - centrifugal Casting.

2.9.2.2. Semi - centrifugal Casting. When the moulds prepared by any of the techniques under expendable or permanent moulds are rotated about the central vertical axis and the casting is symmetrical about the axis of rotating (that is the axis of the casting coincides with the axis of rotation), the process is called as "Semi-centrifugal casting". The centrifugal force aids the flow of the molten metal from a central feeding sprue and can thereby produce a somewhat more dense structure as compared to conventional sand casting, (Fig. 2.45). Casting shapes can be more complicated than true centrifugal casting. Cores are used if central hole is needed, the spinning speed being considerably less than that for true centrifugal casting. An advantage of the process is that stack moulding can be used to cast many parts at the same time.

Product applications. Disk-shaped parts : wheels, rings, rollers, sheaves, pulleys, flywheels, gear blanks, Turbo-Supercharger diaphragm disks and steel railroad wheels etc.

2.9.2.3. Centrifuge Casting. When a group of small moulds are arranged in a circle (to balance each other) around the central vertical axis of the flask and the flask is rotated about the vertical axis, the process is called as "centrifuge casting", [Fig. 2.46]. It is clear that the moulds are not symmetrical about the axis of rotation, that is, the axes of the castings and the axis of rotation do not coincide with one another. Here again, the centrifugal force is used to obtain higher pressure on the metal and get more dense castings. The molten metal will flow to all the moulds under centrifugal force from a central feeding sprue. The process is again adaptable to stack moulding.

Product applications. Products may be irregular or non-symmetrical: valve bodies, plugs, valve bonnets, pillow blocks and yokes etc. Jewelery is centrifugally cast by spinning the investment moulding during pouring. Centrifugal force transfers the metal and ensures good mould filling.

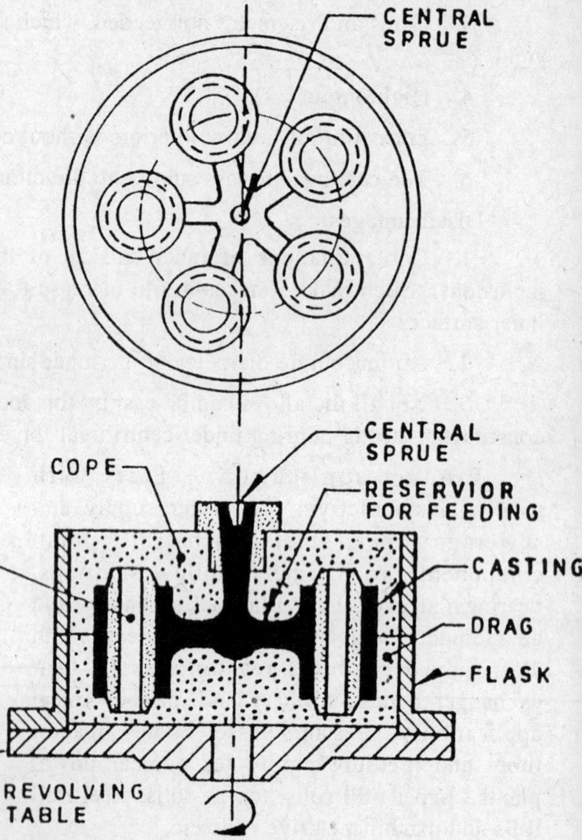

Fig. 2.46. Centrifuge Casting.

2.9.3. Precision Investment Casting. The two common and widely used methods under this category are :

1. Lost-wax method
2. Shell-moulding method.

These methods are called as "precision" because the castings obtained by these methods have very smooth surfaces and possess high dimensional accuracy. The term "investment" refers to the layer of refractory material with which the pattern is covered to make the mould. Like sand casting method, the mould is destroyed everytime a casting is made.

The difference between lost-wax method and shell moulding method is that in the former, the pattern is made of either wax or a plastic material, whereas in the latter method, it is made of a metal.

2.9.3.1. Lost-wax method. The Lost-wax method, sometimes, also called simply as 'precision-investment casting' has been used for many years by jewelers and dentists. Since world-war-II, the method has been adapted to the production of small and precise industrial castings. Basically, the method involves the use of expendable (heat disposable) pattern surrounded with a shell of refractory material to form the casting mould. Castings are formed by pouring molten metal in the mould cavities created by melting out the pattern. Since the pattern made of wax is melted out and gets destroyed, that is, why the name "lost - wax method".

The steps involved in this method are explained below :

1. Making a master pattern of the part to be cast. The pattern is usually made of a metal that can be easily machined, such as, brass, aluminium alloy or steel, or a fusible alloy (alloy of tin, lead and bismuth). It is made oversize to compensate for wax and metal shrinkage.

2. Making a composite die to the master pattern (by casting) for casting the wax/plastic patterns. The die-material is a low melting-point alloy like bismuth alloys or even aluminium and C.I. The master dies can be directly made from die-blocks of steel by machining the cavities.

3. Making of wax / plastic patterns. For this, the two halves of the master die are clamped together and molten wax/plastic is injected into the die cavity under pressure, (Fig. 2.47 (*a*)). The molten wax is slightly above its melting point and the injection pressure is about 4 bars with the dies preheated to about 65°C. For plastic materials (polystyrene, polythenes etc.), the injection pressure is of the order of 35 bars with higher temperature.

4. Assembling the wax patterns to a wax gating system. Several small wax patterns are assembled together to a wax gating system connected to a central sprue, by wax welding with the aid of heated tools. In this manner, a cluster or "tree" of wax patterns is formed, (Fig. 2.47 (*b*)).

5. Investing the wax patterns. The wax patterns are invested in two stages. Firstly, a thin coating (about 1 mm) of primary investment slurry is made around the wax patterns by dipping these in the slurry. This slurry is made by mixing extremely fine silica sand with a water/ethyl silicate or gypsum solution. This slurry coating being in direct contact with the surfaces of the wax patterns, will determine the surface quality of the castings. After this primary coating has dried sufficiently, the final investment layer consists of coarser and less expensive slurry. The final investment moulds may be either solid type or shell type. A solid type mould is formed by placing a metal container type flask over the cluster of patterns and then pouring a hard-setting moulding material into the flask, (Fig. 2.47 (*c*)). Shell type investment moulds are made by dipping the cluster of patterns in a ceramic slurry and the procedure is repeated until the required thickness of the mould or shell (about 6.5 mm) is obtained. The common refractory used for investment material is silica. The binder is gypsum or water-based sodium silicate.

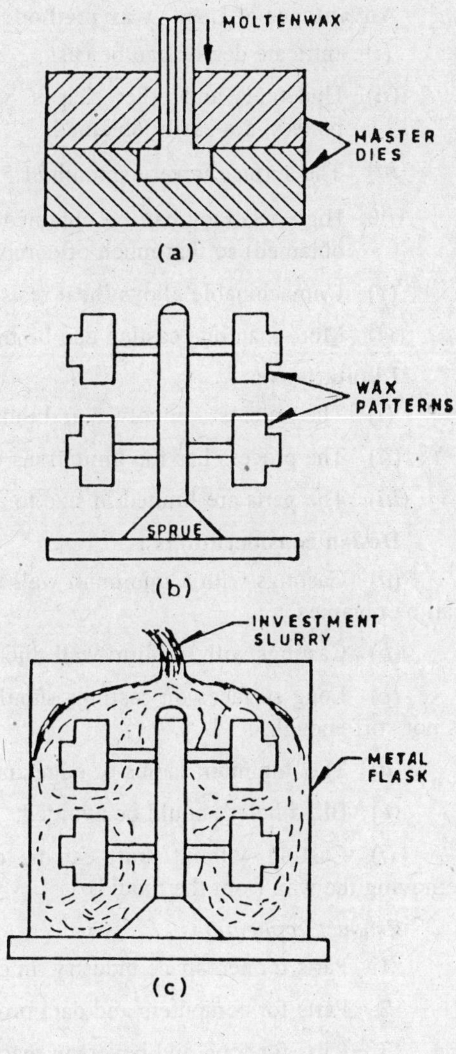

Fig. 2.47. Lost-wax Method.

6. Melting out the wax patterns and baking the mould. The finished mould is dried in air for 2 to 3 hours and then baked in an oven for 2 hours to melt out the wax. At a temperature of 100 to 120°C the wax melts and the moulds are inverted so that all or most of the wax will run out of the sprue.

7. Melting the metal and pouring the mould. The mould or flask is transferred to a drying furnace where it is first held at 150°C and then gradually heated to 800 to 900°C. This will vaporize any remaining wax in the mould. The preheated mould also ensures that the molten metal will completely fill the mould cavity. The preheated mould is poured with molten metal which fills the cavity and is allowed to cool and solidify. Another way to fill the cavity is to place the mould on a vacuum table and drawing a vacuum through the slurry which will help suck the metal right into the mould or by placing the mould in a centrifugal casting machine and pouring the metal in this manner.

8. Shaking out the castings, removing gates and feeders. After cooling, the fragile mould material is broken away freeing the castings. The gates and runners are removed in the normal way.

9. Cleaning and inspecting the casting

Advantages of Lost - wax method:

(*i*) Intricate details can be cast.

(*ii*) Undercuts and other shapes, which would not allow the withdrawal of a normal pattern, are easily obtained.

(*iii*) The surface is very smooth (1.5 to 2.25 μm Ra) and there is no parting line.

(*iv*) High accuracy can be obtained (tolerances of the order of ± 0.003 mm/mm can be obtained) so that much of complicated and costly machining can be eliminated.

(*v*) Unmachinable alloys (heat resisting steels or nimonic alloys) can be cast.

(*vi*) More than one casting can be made at a time.

Limitations :

(*i*) The process is involved and thus expensive.

(*ii*) The process has the limitations in use of and location of holes.

(*iii*) The parts are limited in size to a few kg.

Design considerations :

(*a*) Castings with a minimum wall thickness of 1 to 2 mm and a hole diameter of 2 mm can be obtained.

(*b*) Castings with uniform wall thickness are desirable.

(*c*) Long surfaces on castings should be avoided because ceramic material of the mould is not stiff enough.

(*d*) The minimum radius of edge and corner rounding should be 1 to 3 mm.

(*e*) Blind holes should be avoided.

(*f*) Casting without draft can be obtained (a draft angle of 0.5° is needed only for removing the wax from the mould).

Product applications.

1. Parts for aerospace industry, in aircraft engines, frames, fuel systems and instruments.
2. Parts for computers and data processing equipment.
3. Parts for food and beverage machinery.
4. Parts for machine tools and accessories, scientific instruments, sewing machines.
5. Nozzles, buckets, vanes and blades for gas turbines.
6. Costume jewellery.
7. Rock drill thread chaser holder blocks.
8. Dentures and special metal implants for orthopedic surgery.
9. Parts for movie cameras and projectors.
10. Radar wave guides.

11. Parts such as reciprocating slides for cloth cutting machines.

Precision investment casting is used, when :

1. Forging of the component material is very difficult.
2. Pressure die-castings are too expensive.
3. Non-machinable materials are preferred for the component.
4. Frequent design changes are required.
5. The volume of production is too small to justify much tooling.
6. The machining operations call for expensive tools.
7. The design of the component calls for intricate machining operations.
8. Assembly costs can be eliminated by redesigning and making a single casting.

2.9.3.2. Shell Moulding. The process involves the use of match plate patterns similar to cope and drag patterns which are used for green sand casting. The process is based on the capability of a thermosetting resin and sand mixture to assume the shape of a preheated metal pattern and to form a dense, quickly hardened shell mould. The two matching shell halves (made from cope and drag parts of the pattern) are clamped or glued together to form the mould cavity. Castings are obtained by pouring molten metal into this mould cavity. The shell thickness varies from 5 to 8 mm. As in sand casting process, the mould is expendable and must be broken away to facilitate the removal of the solidified casting.

Shell moulding process involves the following steps.

1. Preparation of the metal match plate cope and drag type patterns. The patterns are machined from copper alloys, C.I., Al or Steel depending upon the life of the pattern. They are given the usual shrinkage and draft allowances, are polished and then attached to the metal match plates.

2. Mix the investment material. The investment material consists of dry, fine and thoroughly washed silica and 5 to 10% of thermosetting phenolic resin (phenol formaldehyde) which acts as a binder. The two are thoroughly mixed together.

3. Heat the pattern. The pattern plate is heated to about 200 to 300°C for investing it. To prevent/minimize the mould adhesion of the investment material and to make the removal of the shell convenient, a release agent containing a silicone is next sprayed over the hot pattern plate surface.

4. Invest the pattern. For this, the investing material is placed in a container called "dump box" (about one-third full) and the heated pattern plate is clamped to it, with the pattern surface inwards. The dump box is then quickly inverted on its trunnions. The investing material falls over the surface of the heated pattern. The heat penetrates the mixture, softens the binder making it effective. This will result in the formation of a uniform shell around the pattern surface after the binder thermosets. The shell thickness will depend upon the time of contact of the mixture with the heated pattern. In about 30 to 45 s, we can get a normal shell thickness of 5 to 8 mm. The dump box is then rotated back to its normal position. The loose and unaffected investing mixture will fall down, leaving the shell adhering closely to the pattern.

5. Curing the Shell. For this, that is, to cure or harden the shell, the pattern plate is removed from the dump box and transferred to an oven and cured for about 1 to 3 minutes at a temperature of about 250 to 450°C.

6. Remove the Shell. The shell is then stripped from the pattern plate with the help of a number of ejector pins which pass through the plate. When these ejector pins are pressed simultaneously, the tops of these pins at the pattern plate surface, more upward and strip the shell from the pattern plate.

7. Repeat the steps 2 to 6 for the other half of the metal match plate pattern to produce the other half of the shell.

8. **Assemble the Shells.** The two halves of the cured shells are securely assembled by clamping or bonding them together at the parting line with a quick-acting thermosetting resin adhesive such as phenol-formaldehyde.

9. **Pour the Mould.** It is done in the same manner as for Lost-wax method, that is static pouring or vacuum pouring or centrifugal pouring.

10. **Remove the Casting.** After cooling and solidification, the shells are broken or shaken away from the castings in the usual way. The steps are clear in Fig. 2.48.

Advantages :

1. A high precision/accuracy of castings with tolerances of ± 0.002 to 0.005 mm/mm, is possible.
2. Smooth surface finish and reduced machining allowances possible. Many shell mould castings require almost no subsequent machining. The common value of surface roughness is in the range of about 3.2 μm Ra.
3. Complex parts can be made by this method, even of difficult to machine metals and alloys.
4. The moulds can be stored until required, because the used resins are not hygroscopic.
5. Less sand is used as compared to sand casting.
6. Permeability of the thin shell moulds is higher compared with other types of moulds, so the gases escape easily through the walls. Also, thin shell does not have as great a chilling effect as a sand mould. Due to the above two factors, a better quality of the casting is assured.
7. The process enables foundary automation to be introduced.
8. Cleaning is considerably reduced and in some cases eliminated.
9. Saving of metal through use of smaller gates, sprues and risers, resulting in increased yield.

Disadvantages :

1. Due to the higher cost of match plate patterns and the cost of binder, the process is economical for mass/batch production.
2. Size of the casting is limited - 10 to 13.5 kg.
3. Serious dust and fume problems during sand and resin mixing.
4. Carbon pick up in the case of steels.

Design considerations :

1. A single parting line should be provided for the mould.
2. Detachable pattern parts and cores should be avoided.
3. The casting walls of uniform thickness are desirable, the minimum wall thickness being 2 to 2.5 mm.
4. Draft angles of no less than 1° and rounding radii of 2.5 to 3 mm should be used.

Product applications. For making cores, small pulleys, motor housing, mixers, fan blades and valve bodies etc. Other parts include : Bomb casings, brake drums and bushings, cams, camshafts, pistons and piston rings, permanent-magnet pole pieces, pinions, impellers, pipe bends, air-compressor crank cases and cylinders, conveyor rollers and so on.

Core-making by Shell Moulding Method. Hollow cores to very close limits are made by this process. Fine sand and upto 6% resin are thoroughly mixed. The halves of the metal core box are heated upto 160°C, and are coated with stripping or parting solution and are then assembled. The sand-resin mixture is poured into the heated core-box. The box is then either vibrated or the sand is joggled to fill all the core-box cavities. After about 10 to 20 s, the loose sand is poured out of the box. The assembly is then placed in an oven for curing, for about 20 s.or intricate cores, multiple-piece core boxes can be used. Cores for steel castings have an oil bond and in some cases, Zircon flour, alcohol and other agents are added to the mixture.

Shell moulded cores are used to increase dimensional accuracy of the normal green-sand mouldings for such items as manifolds and cylinder blocks.

2.9.3.3. The Mercast Process. This process is a variant of "Lost-wax method" where mercury is used in place of wax, as the pattern material. The steps are similar. The mercury patterns are made by injecting it into a die and are then frozen (at – 40°C). The frozen mercury patterns are invested and the mercury is then reclaimed by allowing the mould/flask to come to the room temperature so that mercury flows out of the mould cavity. Then the flask undergoes the heating, curing and pouring steps. The use of mercury has the advantage that superior accuracy (surface/ dimensional) is obtained as compared to 'Lost-wax process'. Also, mercury changes size little on melting and is not as likely to crack a thin/weak would as is wax or plastic. Again, recovery of mercury is more than that of wax. The co-efficient of expansion of mercury is about $\frac{1}{3}$ of that for wax. So, thinner walls may be used for the shell. Mercury has low creep properties so that mould dimensions are more consistently produced. Frozen mercury has a high state of self-diffusion, so that two surfaces will adhere immediately they contact each other. This is known technically as 'booking' and facilitates the assembly of intricate patterns.

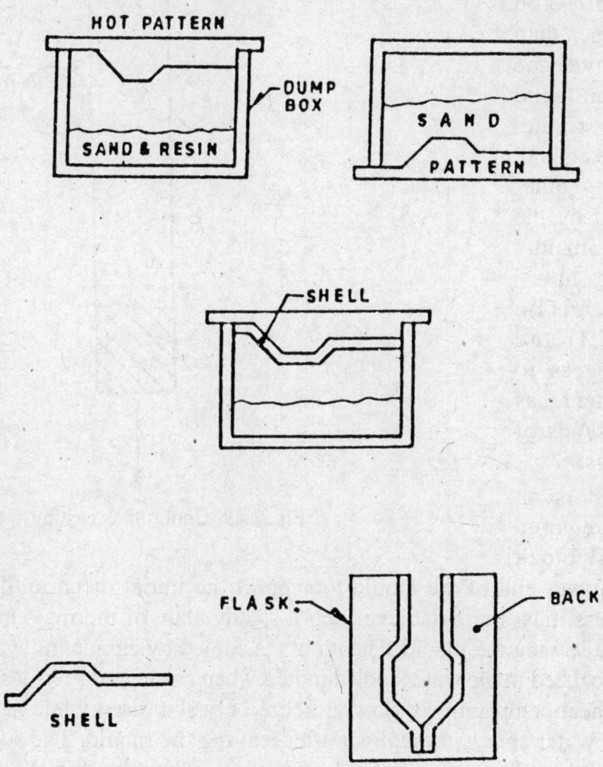

Fig. 2.48. Shell Moulding Process.

However, initial cost of mercury is high and the vapours of both mercury and acetone (needed for obtaining low temperature) are poisonous and call for special means of extraction.

2.9.3.4. Ceramic - mould Casting. In this process, the slurry is made up of fine-grained refractory powders of zircon ($Zr\,Si\,O_4$), alumina ($Al_2\,O_3$), or fused silica ($Si\,O_2$) together with patented bonding agents. The slurry is applied as a thin coating to the pattern and is backed by a less expensive fire clay. The mould is baked at 1000°C and the melt is poured while the mould is still hot. This process is suitable for all materials due to the use of better ingredients for the slurry.

2.9.4. Continuous Casting. In essential, the process consists of pouring molten metal into a short vertical metal die or mould (at a controlled rate), which is open at both ends, cooling the melt rapidly and withdrawing the solidified product in a continuous length from the bottom of the mould at a rate consistent with that of pouring. Fig. 2.49. The process is used for producing blooms, billets and slabs for rolling structural shapes. The process (a German invention perfected in the early 1960's) allows steel producers to bypass the costly process of pouring molten metal (steel) into ingots and then reheating them for milling (roll) and finishing. This is cheaper than rolling from ingots. Any shape of uniform cross-section (solid or hollow) such as : round, rectangular, square, hexagonal, gear toothed and many other forms can be produced by the process. The process is mainly employed for copper, brass, bronze and aluminium and also increasingly with C.I. and steel. Again, the process to quite a considerable extent, has replaced the batch methods of producing ingot castings.

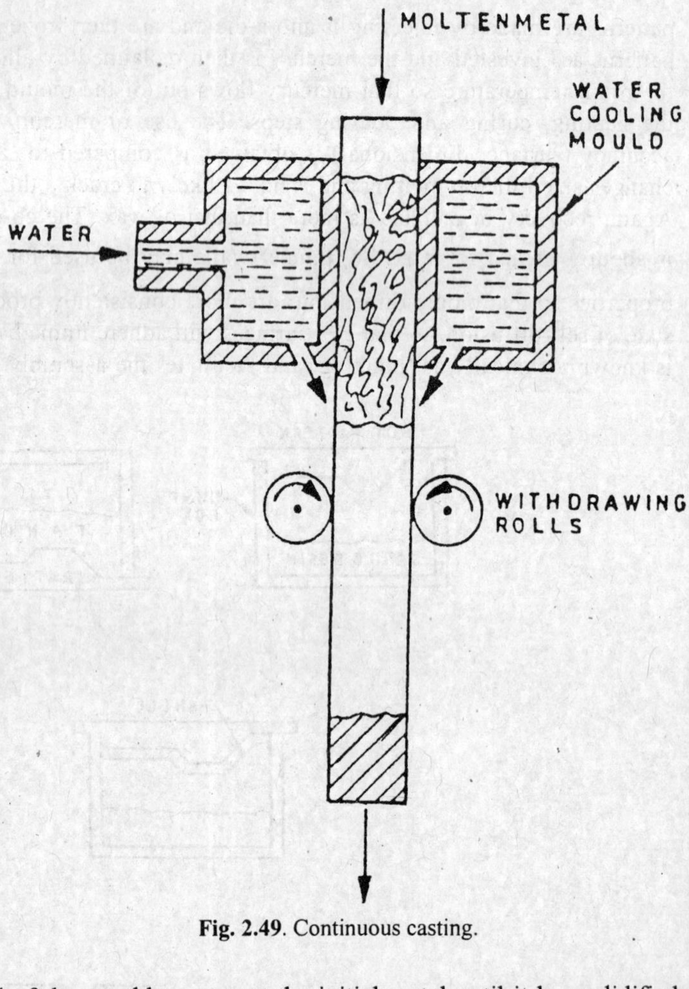

Fig. 2.49. Continuous casting.

Since the mould is an open cylinder of the required cross-section, a metal block must be placed at the lower end of the mould to support the initial metal until it has solidified. The object of the process must be to achieve such a steady state of thermal equilibrium that the casting solidifies before leaving the mould. The mould is cooled by circulating water around it. The solidi-fication zone is localized in the water cooled mould when casting non-ferrous metals. However, ferrous metals conduct heat comparatively slowly, freeze only skin deep while in the mould. They are further cooled with water sprays throughout after leaving the mould. The solidified casting is withdrawn gradually from the bottom end of the mould at a controlled speed by the withdrawing rolls, typically at speeds of 25 mm/s.

Theoretically, the process may go on almost indefinitely. In that case, the required lengths of billet, bloom or slab are cut up during their movement with a flying saw or torch. In the "semi continuous casting process", the required lengths of billets, blooms and slabs or ingots are obtained by successively pouring sufficient molten metal for each length and the process is interrupted periodically to remove the casting. Large diameter and long iron water pipes with flanged ends are also cast by the semi-continuous process with a separate charge of molten metal for each pipe.

Moulds are made of graphite or copper (particularly for casting of steels due to its high-heat conductivity). To prevent the adhesion and welding of the melt to the mould, its surface is given a coating of ceramic powder for steel castings or of a lubricating type parting compound often containing graphite or molybdenum disulphide. Some systems employ vibrating or reciprocating mould to keep the casting from sticking.

Continuous casting process has many advantages : High quality metal can be obtained because it is protected from contamination while melting and being poured. Its yield in rolled shapes is about 10% more as compared to that from ingots. The physical properties and surface qualities are comparable to those obtained in other permanent mould processes. Ease of automation, Reduced energy consumption, Fueleconomy, Less installation and operating costs and safety. Also, increased production capacity, inproved product structure and reduced waste metal.

Fig. 2.49 shows the principle of continuous casting of steel. The system contains water cooling mould as well as secondary cooling Zone where the sprays of water brings the ingot to complete solidification. As it passes through the withdrawal-roll set, the ingot enters the carriage of a cutting device moving together with the hot ingot. A gas cutter cuts off the ingots into billets. Roller conveyers then carry the billets for rolling.

2.10. CLEANING AND FINISHING OF CASTINGS

After a casting has solidified and cooled down sufficiently in an expendable mould, the first step is freeing the casting from the mould. This operation is called as the "Shake out operation". Since a great deal of heat and dust are involved in this operation, the operation is usually mechanised. Shake out is usually done by means of vibratory knockouts, jolting grids and vibraters. The mould is intensively jolted and broken up.

After shaking the casting out of the mould, it is conveyed to the fettling (dressing) shop for cleaning and finishing. The fettling process (cleaning and finishing) consists of the following operations : Decoring or core removal, cleaning of surfaces, removal of gates, risers and fins etc., repair of defective castings if possible and heat treatment.

Before starting the fettling process, the castings are examined for obvious defects such as misrun, drop, cold laps and coldshuts etc. Defective castings are set aside and are not cleaned.

1. **Core removal or Core knockout.** Due to the reasons mentioned under "shake out operation", this operation is also done mechanically. Hammering and vibrating will loosen and break up cores. Stationary or portable vibrators are employed for this purpose. To knockout cores from heavy castings, it is advantageous to use air drills. Removal of cores by hydro-blasting is more sanitary process keeping in view the dust problems. The operation consists of breaking up and washing out the cores with a jet of water delivered at a pressure of 25 to 100 atm. A recent method for core removal is hydro-sand blasting in which sand is mixed with the water jet. This method results in a sufficiently clean surface.

2. **Cleaning of Surfaces.** This operation involves the removal of all adhering sand and oxide scale and produce a uniformly smooth surface. Mechanical methods are employed for this purpose, since cleaning by hand with wire brush is tedious and costly. These methods include :

(*i*) **Tumbling.** This method is used for cleaning small and light castings. The castings are loaded into a tumbler or a barrel along with white iron picks (jack stars) (in an amount of 20 to 35% of the mass of castings). Rotation of the barrel causes the castings and jack stars to tumble. Jack stars abrade the surfaces of the castings and also the castings abrade one another. This operation removes the adhered sand and oxide scale from the surface of the castings. The rotational speed of the barrel is 30 rev/min.

(*ii*) **Sand blasting and Shot-blasting.** These methods are widely used to clean surfaces of light, medium and heavy castings. In these machines, dry sand or shots (white C.I. shot, steel shot) or grit of white C.I. or steel (grit is made by crushing shot) is blown by a stream of compressed air against the surfaces of the casting. The impact of the abrasive particles traveling at a high speed, on the surface removes the adhering sand and oxide scale. Velocity of the abrasive particles leaving the nozzle of the machine is in the range of 35 to 75 m/s and the air pressure is in the range of 0.7 MPa.

(*iii*) **Airless shot blasting.** In airless shot blasting or mechanical impact cleaning, shots are hurled on the surface of a casting by a fast rotating paddle wheel. For harder castings, the shots are made of white iron, malleable iron or steel, whereas for softer non-ferrous castings, these are made of copper, bronze, glass or mild steel. The wheel rotates at 1800 to 2500 rev/min. The velocity of shots striking the casting surface is about 60 to 72 m/s.

This method offers the following advantages : a high output (10 times the output of a pneumatic shot blasting machine), low power consumption, shot jet speed regulation by changing the speed of the rotor, and better working conditions. The drawbacks of the method are : rapid wear of rotor blades and poor cleaning of shaped castings with intricate cavities.

(*iv*) **Hydroblasting.** This is the most effective surface cleaning method. Here two operations are accomplished simultaneously : Core knockout and surface cleaning. Castings are placed on a rotary or stationary table and high velocity jets containing about 15% sand and 85% water, under a pressure of 10 to 20 MPa, are directed at the casting surface. The jet velocity can be upto 100 m/s. The method produces no dust but consumes lot of water.

3. Removal of Gates and Risers. Gates, runners, risers and sprue can be removed before or after cleaning operations. In brittle materials, these are simply broken off from the castings. In more ductile materials, the following methods are used to remove them : power hacksaws, band saws, disk type cutting benches, Abrasive cut off wheels, flame cutting with an oxyacetylene cutting torch and arc cutting for heat-resistant and acid-resistant steels which are not amenable to gas cutting. The surface of cut becomes rough and needs additional treatment.

Fins or flash (that forms when melt flows into gaps between two mould halves or at cores), ends of nails and other unwanted projections are removed by : chipping, sawing, flame cutting, flame gouging and grinding.

Powder Cutting. Powder cutting is a process by which large risers and gates can be rapidly removed from castings made of oxidation resistant alloys. Preheated iron powder is introduced into an oxygen stream. This burning iron then attacks the metal riser or gate by a process of fluxing and oxidation.

4. Some minor defects detected may sometimes be repaired by welding without affecting the function of the finished casting.

5. The finished castings are sometimes subjected to various heat treatments to modify mechanical properties or reduce residual stresses.

2.11. INSPECTION AND TESTING OF CASTINGS

The aims of inspection and testing of castings (that is quality control) are to prevent defective castings being supplied from the foundary and to reduce the pe:centage of inevitable processing rejects

The various inspection and testing procedures may be classed as follows :

1. Visual Inspection. Visual inspection of castings can reveal many of the common surface defects such as misrun, cracks and warping etc. This method is very common and is applicable both in piece and mass production of castings. The inspection is carried out in two steps : prior to cleaning and annealing and then after the final finishing operation.

2. Dimensional Inspection. Geometric dimensions of castings are checked by means of measuring tools such as plug and snap gauges, template gauges, marked-out plates and special alliances, to establish whether the dimensions of the casting conform to the drawing or not and to make certain that the pattern and core boxes are correct. The deviations of dimensions should not exceed the permissible limits.

3. Metallurgical Control. Under this the chemical composition and the mechanical and other properties are determined in a laboratory.

The chemical composition of castings is checked by the methods of chemical and spectral analysis. For this, the test pieces are commonly cast-on test bars, that is, cast integral with the casting, or separately cast test specimens prepared for checking strength properties.

The strength or mechanical tests include test in : bending, tension, hardness, compression, shear and creep.

4. Pressure Testing. This test is carried out on those castings to be used for conveying liquids or gases. The castings are checked for pressure tightness or impermeability and leakage. The tests include : water or air-pressure tests.

In water pressure test, the casting is held under a certain pressure of water, the test pressure depends on the conditions under which the casting has to function. The outer surface of the casting must be dry, otherwise it will not be possible to detect the traces of leakage, if any.

In the air-pressure test, a soap solution is applied to the surface of the casting. When the casting is subjected to air-pressure testing, bubbles will appear on the surface showing the place of leakage, if any.

5. Radio-graphical Testing. Internal defects in a casting such as cracks, voids, cavities and porosity etc., as well as surface cracks can be revealed by radiographic inspection using x-rays and γ-rays.

In x-ray testing, (Fig. 2.50) short wave length rays from an x-ray tube are passed through a casting and recorded on a special film held against the opposite face of the casting. If the casting has an internal defect, the

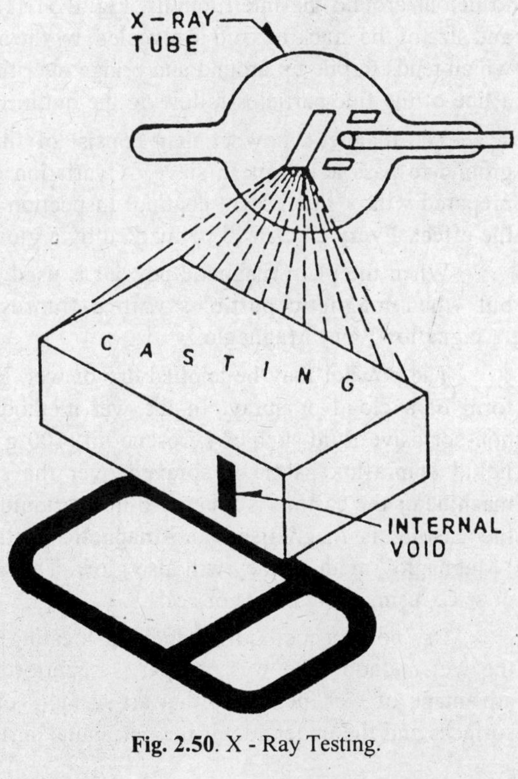

Fig. 2.50. X - Ray Testing.

density of the material at that spot will be less as compared to the surrounding material. This area will allow more penetration of the rays, that is, the sections of the casting with cracks cavities will absorb a smaller amount of x- rays as compared to fully dense material. This will result in the appearance of a dark shadow on the x-ray film reproducing the contour of the defect. The power source used for x-ray tube is a high-voltage source : 200 kV for casting thickness upto 50 mm and 1 million volts for thickness form 50 to 180 mm.

γ-ray testing is used for checking heavy-walled castings since these rays are more penetrating and less scattering as compared to x-rays. γ-ray radiate from Radium or its salts contained in a capsule. Another source can be radio-active element Co^{60}.

6. Magnetic Testing. In this method, the casting to be tested is magnetised and then placed between the poles of an electromagnet or in the magnetic field of a solenoid coil. The energized coil is now moved along the casting. If the coil comes across a defect on its way, the magnetic flux changes its direction and induces an emf in the coil turns, the value of which shows up on the galvanometer.

The method can detect defects (cracks) on the surface or slightly below the surface of a casting. Thus, it supplements the radiographical methods which ordinarily can not detect small cracks. However, the method can be applied to castings made from ferromagnetic metals.

7. Magnetic Particle testing: This method of inspection is a procedure used to determine the presence of defects at or near the surface of ferromagnetic castings.

The method is based on the principle that, if an object is magnetised, surface cracks and voids in the material, which are at an angle to the magnetic lines of force, interrupts the magnetic field which gets distorted. That is, there is an abrupt change in the path of a magnetic flux flowing through the casting normal to the surface defect. This results in a local flux leakage field and hence interference with the magnetic lines of force. The magnetic lines spread out in order to detour around the interruptions, Fig. 2.51. This interference is detected and hence the shape and size of the crack or void is revealed, by the application of a fine powder of magnetic material, which tends to pile up around and bridge over the discontinuities. A surface crack is indicated by a line of the fine particles following the outline of the crack.

The magnetic powder may consist of fine iron filings, but Fe_2O_3 is preferred which is ground to pass a 100-mesh sieve. A variation of the method is that the magnetic particles are prepared with a fluorescent coating. Inspection will be carried out under U.V. light to intensify the effect. Every crack will be marked by a glowing indication.

When the plain magnetic powder is used, the trade name of the method is "Magna-flux", but when magnetic particles with a fluorescent coating are used, the method is called "Magnaflow" or "Magnaglo".

The powder may be applied dry or wet. For the dry method, the powder is applied in the form of a cloud or spray. In the wet method, the powder is suspended in a low viscosity, non-corrosive fluid such as kerosene oil (100 g of magnetic powder in about 5 l of K.oil). This liquid (supraflux paste) is sprayed over the surface to be tested, by hydraulically operated machine or the casting is immersed in the liquid. Then the casting is allowed to dry. Now, when the casting is magnetised, the magnetic particles will gather around the crack and in the "Magnaglo" method, they will also glow. The magnetic fields can be generated either with D.C. or A.C., using yokes, bars or coils.

Dry powder method is better for locating near surface defects and is also less messy than the wet method. The wet method is superior for detecting fine surface defects. Another big advantage of wet method is that all surfaces of the casting can be reached, including vertical surfaces and the underside of the horizontal surfaces, by housing or by immersion in the liquid.

From the above discussion, it is apparent that cracks that are in a direction parallel to the magnetic field would not be detected. The cracks which are perpendicular to the direction of magnetic field are the easiest to detect.

8. Eddy current Inspection. In this method, the material of the casting need not be ferromagnetic. The test includes a probe which is supplied with a high frequency current. It induces an electric field in the casting. The field changes in the presence of surface or near-surface defects. These changes show up on instruments.

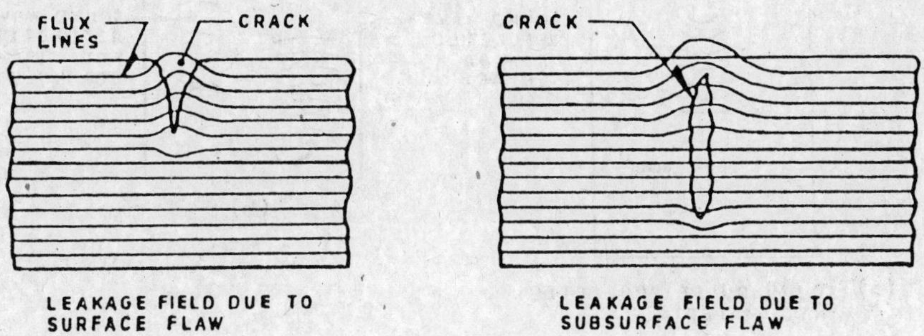

Fig. 2.51. Magnetic Particle Testing.

9. Liquid - penetrant Inspection. This method can reveal surface defects only but can be used for any material. The surface of the casting is thoroughly cleaned and dried. Then the liquid penetrants are applied as sprays or by immersion. The penetrant liquid contains either a material which will fluoresce under black light or a dye that can be visually detected. The liquid penetrant will be readily drawn into extremely small surface cracks. The surface is cleaned and dried. Then, a powder material called a "developer" is sprayed on the surface. The penetrant trapped in defects bleeds out due to blotting action and delineate defects during development. The extent of the discontinuity in the casting surface will be proportional to the amount of penetrant bleeding out. If a fluorescent penetrant is used, defects show up as glowing yellow green dots or lines against a dark back ground. In dye penetrants, defects are revealed as red dots or lines against a white background.

10. Ultrasonic Testing. This test is based on the fact that a beam of ultrasonic waves (frequency 20,000 Hz) passes through a solid (dense) material with little loss but is partially reflected from surfaces. Therefore, this method can detect voids, cracks and porosity within a casting.

The ultrasonic waves are produced by the application of reverse Piezo-electric effect. That is, if an electric potential is applied across the flat ends of a crystal (quartz crystal), it will either contract or elongate in the normal direction. The crystal is held against a smooth surface of the casting with the help of a coupling fluid. A high frequency A.C. (1 million c/s) is impressed across the faces of the crystal with the help of an oscillator. The sound waves produced travel through the casting. These will get reflected from the other end of the casting and the signals are measured with a C.R.O. If the casting has same flaws within it, some of the sound waves will be reflected back and will return to the instrument earlier. The location of the defect from the testing surface may be readily obtained by measuring the relative position of the flaw "pip" between the two "pips" representing the metal thickness, (Fig. 2.52).

The method is not very suitable for a material with high damping capacity, e.g. C.I., because in such a case, the signal gets considerably weakened over some distance.

The tests for determining the mechanical properties of the castings are called as "Destructive testing" since the castings or cast test specimens which undergo these tests, become unserviceable. All other tests discussed above are called as "Non-Destructive Testing", that is, "NDT", since the casting after the test can be used in the usual way, if found suitable.

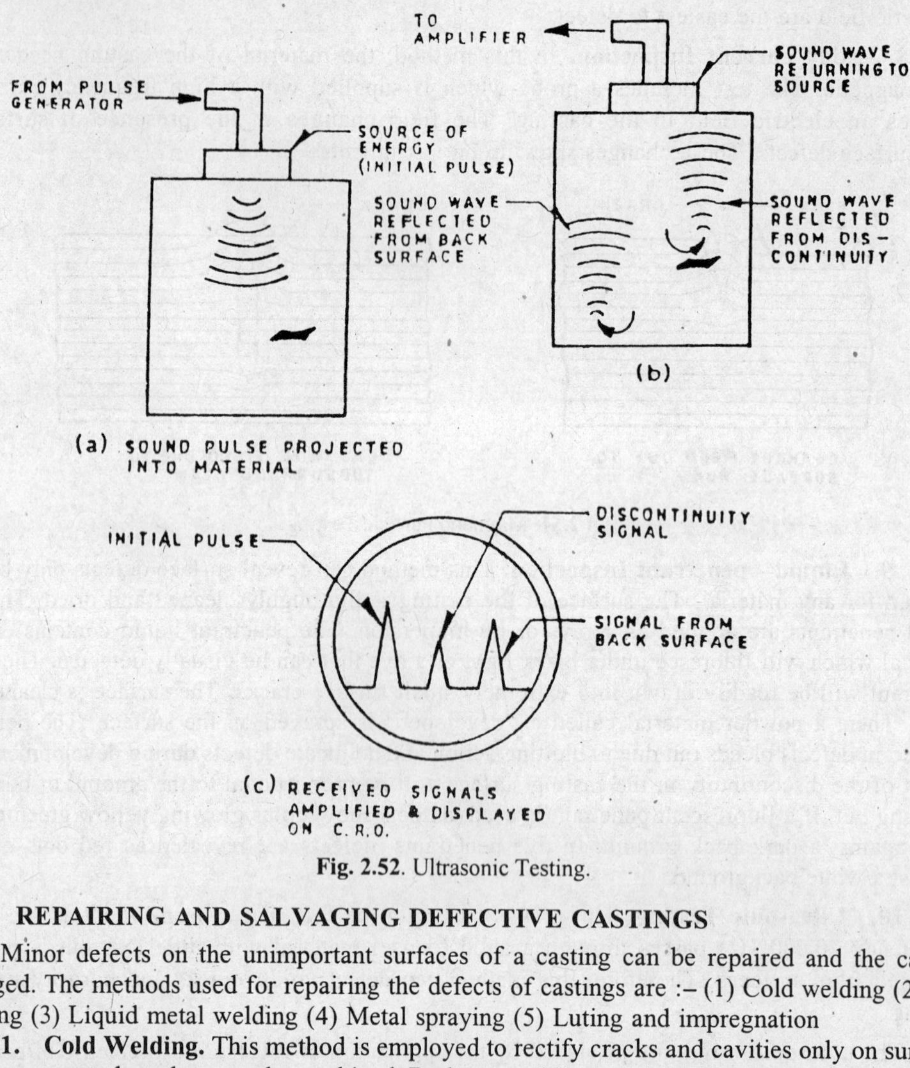

Fig. 2.52. Ultrasonic Testing.

2.12. REPAIRING AND SALVAGING DEFECTIVE CASTINGS

Minor defects on the unimportant surfaces of a casting can be repaired and the casting salvaged. The methods used for repairing the defects of castings are :– (1) Cold welding (2) Hot welding (3) Liquid metal welding (4) Metal spraying (5) Luting and impregnation

1. Cold Welding. This method is employed to rectify cracks and cavities only on surfaces which are not to be subsequently machined. Both gas welding or electric arc welding can be used for this. Various filler materials used are : Steel, copper, steel-sheathed copper, copper nickel. These are used in the form of long rods 5 to 6 mm in diameter. Defective areas to be repaired are grooved with pneumatic chippers or drilled out. Cracks should be cut out to the entire depth.

2. Hot Welding. Hot welding is used to repair large cavities, holes and cracks. Before welding, a casting is preheated to 500° to 600°C to preclude the appearance of cracks, stresses and chilling of the casting metal. The casting is held at this temperature for 45 to 60 minutes. After this, the defects can be repaired either with an OAW flame or electric arc welding. OAW with preheating is a suitable method for rectifying the defects in gray iron castings of complex configuration, whose sections sharply vary in thickness. The filler rods used are of the following composition :- C : 3.2 to 3.5%, Si : 3.5 to 4%, Mn : 0.5 to 0.6%. The flux to be used is a mixture of : Borax : 50%, soda : 47% and silica sand : 3%. After welding, the casting is annealed at 500° to 600°C and is removed from the furnace at 50° to 60°C and left to cool slowly to exclude chilling in the weld.

3. Liquid Metal Welding. This method for repairing the casting defects is resorted to when the cavity or crack on the surface of the casting is large. The process consists in first cleaning the surfaces of cavity or crack and then moulding a cup or basin of clay above it. After this, the whole casting is preheated uniformly and molten metal is then poured into the cup. Finally, the casting is cooled slowly. For "Braze welding" and "Bronze welding" see P 363.

4. Metal Spraying. Metal spraying or metallising consists in coating the surface to be repaired with a layer of atomised metal. The method is used to fill small cavities and shrinkage pores on unimportant surfaces of the casting, which are not to be machined. The method is also used to decrease porosity in gray iron castings after the welding operation. The surface to be repaired or the weld is cleaned and then coated with 0.3 to 0.8 mm in thickness by a metal spray gun, which gives a coat of metal 0.03 mm thick in one pass. The small drops of molten metal are blown out by the compressed air on to the casting portion requiring repair.

Metal Spray

The metal is melted by an electric arc or by an oxy-acetylene flame, and the compressed air pressure used in the metal spray gun is of the order of 0.06 MPa.

The method is also commonly used to restore shafts running in ball/roller bearings, for which the method provides a fast and economical repairs.

The process of metal spraying or metallization of cylindrical parts being carried out on a lathe is shown in Fig. 2.52A.

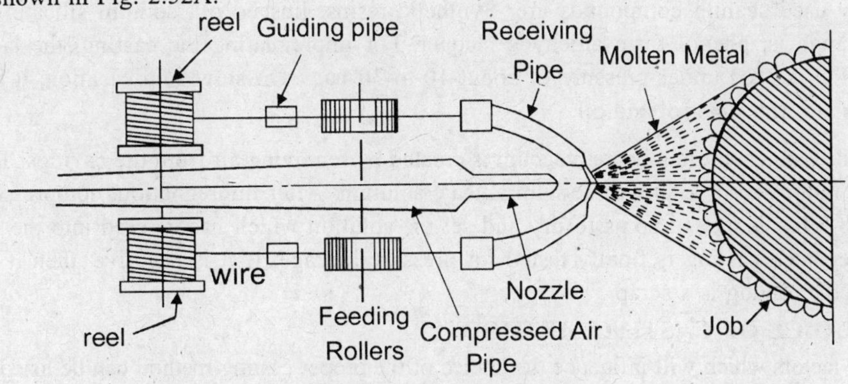

Fig. 2.52A : Metallization

The method involves the following steps :–

1. The surface to be metallized is first prepared (made rough) by knurling or by sand blasting or by cutting rough threads.

2. The surface is now cleaned of oil and dirt.

3. The job is then held in the chuck or between centres and the metallizing device is mounted on the lathe carriage.

4. It is preferable to preheat the surface to be metallized to about 200°C by rotating it while mounted in the lathe, locally by oxy-acetylene flame. Preheated surface takes better deposition.

5. The welding wires from the reels are fed by the rollers through the guiding pipes and the receiving pipes. The arc is struck at the meeting ends of the wires, which melts the metal. The molten metal is sprayed on the surface to be metallized by means of compressed air entering through the compressed air pipe and the nozzle.

6. Welding wires of different materials can be fed from the two sides and the composition

6. Welding wires of different materials can be fed from the two sides and the composition can be varied by adjusting different rates of feed of the two wires.

7. The sections to be protected against metal spraying (keyways etc.) are plugged with a wooden key.

Advantage : – In this process, the parts get heated upto a much lower temperature than in welding. Thus, there is a less chance of distortion of the part.

5. Luting and Impregnation. In these methods, the defects in the castings are repaired with the help of sealing agents.

Lutes or putties are plastic substances prepared from epoxy resins. These are used for smoothing out surface imperfections on castings and improving the appearance of the finish. After the defective surface is sealed with a lute and its solidification, the repaired surface is trimmed and the casting is made ready for machining. The lute in the solidified state possesses high strength and readily lends itself to machining and grinding.

If the casting is found to be positive for pressure leakage, then the salvaging operation called "impregnation" is carried out. The pressure leakage is due to some "micro-defects" in the casting. Air gets entrapped in these minute cavities. The operation consists in sealing the microcavities with liquid sealing agents. The operation is carried out by forcing the sealing compound into the microcavities under pressure alone or by a combination of pressure and vacuum treatment. The sealing compound is finally cured either by air drying or by baking. The commonly used sealing compounds are: Synthetic resins, linseed oil, sodium silicate etc. The most common is: phenol form aldehyde lacquer. For impregnating the casting, the lacquer at about 200°C is used under pressure of about 10 to 30 bar. On slow solidification, it becomes impervious to water, petrol and oil.

While using sodium silicate, vaccum is created by removing air from the cavities. Then, the casting is immersed in a tank of sodium silicate solution. After impregnation, compressed air at about 4 to 5 bar is passed so as to dry and set the solution which has entered into the cavities. The impregnated casting is finally tested for pressure leakage. If it is negative, then it is O.K., otherwise the casting is a scrap.

2.13. CHOICE OF CASTING METHOD

The factors which will influence the choice of the proper casting method can be listed below :

1. Complexity of the shape, that is, external and internal shape, minimum wall thickness and types of cores needed.
2. Surface quality (Tolerance and surface finish) required.
3. Volume of Castings required.
4. Cost of the pattern or die.
5. Weight
6. Working characteristics (strength, rigidity, stiffness, tightness, corrosion resistance etc.) required.

Because of the many factors involved, it is difficult to say just which casting method will be best for making a given part. However, the qualitative information of Table 2.5 can often be used in deciding on a suitable casting method in a particular instance.

2.14. COMPARISON OF CASTING PROCESS WITH OTHER PROCESSES

The casting process has the following plus points, as compared to the other manufacturing processes :

1. The process is particularly suited to the parts which contain internal details that are too complex, too large or inaccessible, to be easily produced by the machining process.

Casting Methods

Table 2.5. Design Considerations

S. No.	*Method of Casting*	Material choice	Complexity of Part	Casting mass, tonnes, t	Minimum wall Thickness, mm	Precision and tolerance	Surface Smoothness'	Surface Details	Mechanical Properties	Remarks
1.	Sand Casting	Wide	Conside-rable	Unlimited (100 t and over)	3	± 0.6 to 0.3 mm/mm	Poor	Poor	Fair to High	Usually require machining before use
2.	Shell Mould Casting	Wide	Moderate	Best for Smaller Parts	1.5	± 0.003 to 0.005 mm/mm	Good	Good	Good	Best for low costing methods
3.	Permanent Mould Casting	Restricted : Al, Brass, Bronze, Some gray iron	Limited	Moderate, 250 newtons For Al.	2.5	± 0.06 mm/mm	Good	Good	Fair	Economical for Mass production only
4.	Machine Moulding	Wide	Conside-rable	3 to 5	3	± 0.125 to 0.125 mm	Good	Good	Fair to High	Well mechanized
5.	Casting in Cored Moulds	— do —	— do —	Unlimited	3	— do —	– do –	– do –	Good	Same
6.	Plaster Mould Casting	Narrow : Brass, Bronze, Al	Conside-rable	Moderate, upto 80 newtons	0.75	± 0.25 το 0.125 μμ	Good	Good	Fair	Little finishing required
7.	Investment Casting	Wide : includes materials hard to forge and Machine	Conside-rable	Moderate, Best for parts under 10 newtons	0.75	± 0.125 μμ/μμ	Excellent	Excellent	Good	Best for too complicated parts
8.	Diecasting	Narrow : Nonferrous	Conside-rable	300 newtons for Al; 1000 newtons for Zinc	0.06	0.025 mm/mm	Good	Good	Fair to Good	Most economical where application
9.	Centrifugal Casting	Wide	Conside-rable	0.01 to 0.10 t	5 to 8	0.6 mm/mm	Good	Good	Good	Economical for Mass production
10.	Full mould Casting	Steel, C.I.	Considerable	upto 15 t	6 to 8 for steel	± 0.003 mm/mm	Excellent	Excellent	Good	Casting cycle is shortened

Casting Method
Cost Considerations

S.No.	*Method of Casting*	Raw Materials Cost	*Tool and Die Costs*	*Direct Labour Cost*	*Finishing Cost*	*Scrap Loss*	*Optimum Lot Size*
1.	Sand Casting	Low to Medium, depending on metal	Low	High	High	Moderate	Side range, From a few pieces to large quantities
2.	Shell Mould Casting	Low to Medium	Low to Moderate	Relatively Low	Low	Low	From few to Quantity depending upon complexity
3.	Permanent Mould Casting	Medium	Medium	Moderate	Low to Medium	Low	Large-in thousands
4.	Plaster Mould Casting	Medium	Medium	High-Skilled operators necessary	Low	Low	100-2000 best range
5.	Investment Casting	High	Low to Moderate	High	Low	Low	Wide-although best for Small Quantities
6.	Die Casting	Medium	High	Low to Medium	Low	Low	Large-1000 to hundreds of thousands
7.	Centrifugal Casting	Medium	Medium	Moderate	Lot to Moderate	Low	Large-in thousands

2. The process (die-casting) is advantageous to cast complex parts in large numbers, especially of non-ferrous metals and their alloys.

3. For producing one of a kind part, in a variety of materials.

4. For processing precious metals, since there is little or no loss of material (Lost Wax Method).

5. Parts made by casting process are isotropic, which can be an important characteristic in some applications.

However, the casting process is not viable for producing parts of metals, with high melting points, such as tungsten etc. Casting is not economical, when parts can be made by sheet metal working processes. Also for many parts, particularly of lower melting point, non-ferrous alloys, the extrusion process may be preferrable to casting process.

2.15 FOUNDRY FURNACES

2.15.1 Cupola

The cupola is the most widely used furnace in the foundary for melting ferrous and non-ferrous metals and alloys. A cross-section of a cupola is shown in Fig. 2.53. A cupola is a shaft furnace of cylindrical shape erected on legs or columns. The cupola shell is made of steel plate 8 or 10 mm thick. The interior is lined with refractory bricks to protect the shell from getting over-heated. The charge for the cupola consists of metallic materials, fuel and fluxes. The metallic part of the charge is made up of definite quantities of pig iron of various grades, cast iron and steel scrap, foundary scrap (gating, splashes and spills, rejects and chips) and a small amount of ferroalloys. Foundary coke and, sometimes, anthracite are used as fuel for melting the metallic charge. Fluxes used in melting both grey iron and steel may be limestone, dolomite ($CaCO_3$ and $MgCO_3$), fluorspar (CaF_2) and CaC_2 etc. The fluxes melt and react with contaminants and non-metallic elements and inclusions, the resulting slag (Calcium silicate) floats to the surface of the melt. By definition, a flux is a substance which reacts with a slag (accumulation of oxides and other unwanted materials) to lower its melting point. Fluxes used in melting non-ferrous metals are usually chlorous and fluoric salts of alkali and alkaline-earth metals.

The solid materials (metal, coke, flux) are charged into the cupola from the charging floor through charging hole (door). Charging is mechanised in large foundaries where drop bottom buckets are employed. Air for combustion of the fuel is delivered from a blower and enters a chamber called the wind box. The wind box completely encircles the cupola. Its purpose is to supply air evenly to all the tuyers, which are the openings which extend through the steel shell and refractory wall to the combustion zone. The tuyers are arranged above the cupola well (the lower part of the cupola from sand bottom to tuyers is called the well). The products of combustion (flue gases) pass out through stack and spark arrester. The function of spark arrester is to catch the incandescent dust carried with the flue gases and thereby prevent fires. The bottom of the cupola is rammed of weak moulding sand possessing high refractoriness. After each heat the bottom is dropped. A taphole is provided at the lower part of the sand bottom to let the molten metal into a forehearth or a ladle. A forehearth is a metal receiver in front of a cupola. It may perform the following functions :

1. Mix the molten metal collected over a period of time thus ensuring a more uniform metal.

2. Hold the molten metal in temporary storage.

3. Permit an intermediate metallurgical treatment of the metal such as desulphurising or adding of inocculants (Ferrosilicon, ferromanganese, ferro-chromium, ferro-silicon-chromium, ferro silicon-cadmium and others). Slagging is done, as required, through a slag spout.

Dimensions of a Cupola. The cross-sectional area of a cupola depends upon the designed hourly output and is determined from the following formula :

$$A_C = \frac{\pi}{4} d^2 = \frac{Q}{Q_1} m^2$$

where d = Clear diameter of the cupola, *m*

Q = designed cupola output, *t/h*.

Q_1 = specific output of the cupola in tonnes per sq m of cross-sectional area per hour

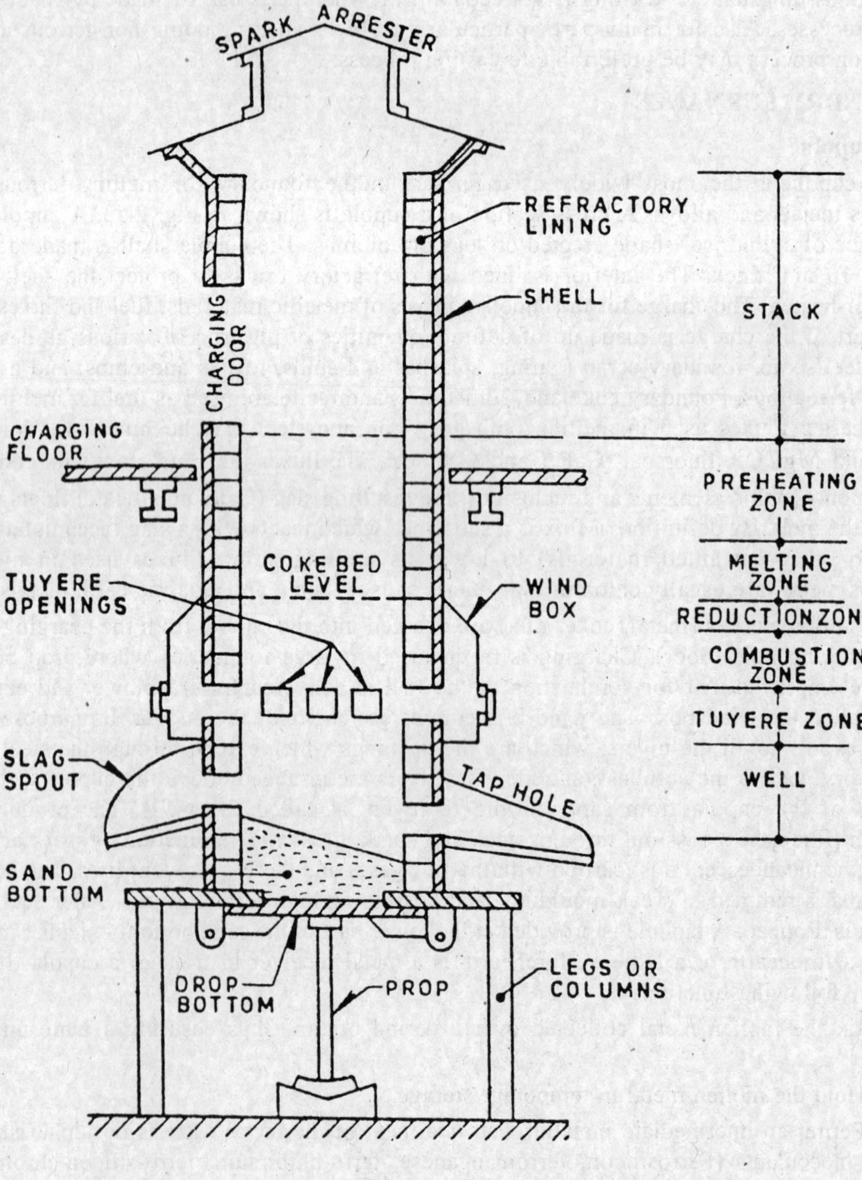

Fig. 2.53. Cross-section of a Cupola.

As a rule Q_1 = 6 to 8.

The useful height of a cupola (distance from the axis of the *tuyers* to the lower edge of the charging door), is taken as 3 to 5 times the diameter of the cupola. The cupola height directly affects the melting rate, fuel consumption and the temperature and the quality of the molten. If the cupola height is too much, the coke may be crushed as the charge drops. If it is too low, the metal is not heated to the desired level, the drought is reduced and the cupola output is decreased. The common dimensions of a cupola are :

(*i*) Outside diameter : 900 to 2700 mm

(*ii*) Inside diameter : 500 to 2100 mm

(*iii*) Height : upto 12 m.

The cupola *tuyers* should deliver a uniform blast sufficient for complete combustion of fuel. This requires that their total cross-sectional area should be in a definite relation to the cupola cross-section. Cupola output and the temperature and quality of the metal depends upon the tuyer design.

Total cross-sectional area of the tuyers

$$= \left(\frac{1}{5} \text{ to } \frac{1}{6}\right) \text{ of the area of the cupola in the clear}$$

Usually, the tuyer size is : 50×150 or 100×300 mm.

The tuyers of the receiver cupola (cupola with a forehearth) should be as near as possible to the sand bottom. In cupolas without receivers, the position of the tuyers above the bottom is determined by the amount of molten metal that it is necessary to accumulate in the well before tapping.

Height of well = 150 to 200 mm for a receiver cupola

= upto 450 mm for a cupola without receiver

Cupola Operation. A newly built cupola (or one after inspection and repairs if any) should be thoroughly dried before firing. The bottom doors are then closed and held shut by means of a vertical prop. A layer of sand about 150 mm thick is placed over the doors and sloped towards the tap hole. In firing a cupola, fire of kindling wood is started on the sand bottom. Coke is then added in several portions to a level slightly above the tuyers and the air blast is turned on at a lower than normal blowing rate. This intensifies coke combustion. Then new portions of coke are charged into the cupola to reach a height of 700 to 800 mm above the tuyers. This layer of coke is called "Coke bed". The height of the coke bed is very important to the cupola process. It affects the temperature, melting rate and chemical composition of the grey iron tapped from the cupola. As soon as the coke bed is thoroughly ignited, alternate charges of limestone, iron and coke are added in weighted portions until level with the charging door.

Limestone = 2 to 4% by weight of metal charge

Coke = 8 to 12% of the metal charge.

Full blast, delivered to the tuyers from the blower is turned on after completely charging the cupola. The first molten metal would appear at the tap hole within 5 to 10 minutes. When the well or the crucible becomes full, the slag is first drained off through the slag spout. For intermittent tapping (cupola without a receiver), the tap hole is closed with a suitable lump of sand and clay called a "bot". For tapping, the bot is punctured with a long bar and the cast iron flows out the tap spout into a holding ladle. When the crucible is emptied of cast iron, another bot is rammed into the tap hole to accumulate another melt. As the cupola is operated, additional charges of limestone iron and coke are charged through the charging door when there is sufficient room for them.

Various zones in a cupola are :

1. Well or crucible.
2. Tuyer Zone.
3. **Combustion Zone.** This zone extends from the top of the tuyers to a surface boundary below which all oxygen of air blast is consumed by combustion of coke. The reaction is exothermic and the highest temperatures are developed in this Zone, which may reach 1600° to 1700°C.

$$C + O_2 \rightarrow CO_2 + \text{Heat}$$

4. **Reducing Zone.** This zone is above the combustion zone upto a height of initial coke bed charge. The CO_2 flowing upward through this zone reacts with hot coke and the reaction is endothermic :

$$CO_2 + C\ (\text{Coke}) \rightarrow 2CO - \text{Heat}$$

Because of this reaction, the temperature in the reducing zone gets reduced to about 1200°C. However, due to the reducing atmosphere, this zone protects the metal charge above from getting oxidised.

5. **Melting Zone.** This zone includes the first layer of iron above the initial coke bed. In this zone, the charge starts melting and trickling through the coke to the bottom of the cupola. A significant part of carbon pick up by metal also occurs in this Zone according to the following reaction :

$$3\,Fe + 2\,CO \rightarrow Fe_3\,C + CO_2$$

6. **Preheating Zone.** This zone includes all the layers of cupola charges above the melting zone to the top of the last charge. In this zone, moisture and volatile matter are evaporated and the layers of charge are heated by the outgoing gases and the temperature here is upto about 1090°C.

Lining of the Cupola. The cupola lining may be basic or acidic.

In basic cupola, the refractory material lining is magnesia bricks or dolomite plaster. A basic cupola is used primarily to eliminate sulphur from iron as,

$$FeS + CaO + C \rightarrow Fe + CaS + CO$$

Sulphur content may be reduced to upto 0.005%. The main use of basic cupola is for the making of ductile cast iron. In acidic cupola, the lining is of fireclay bricks, containing silicon. Sulphur content of the charge cannot be reduced while melting. On the contrary, the metal absorbs a certain amount of sulphur from fuel.

Note. Lining of a cupola with a refractory is being abandoned in favour of water cooled steel jackets.

Output of a Cupola. The output of a cupola is defined as the tonnes of liquid metal obtained per hour of heat. It depends upon the dimensions of the cupola and the intensity of coke combustion which is defined as the tonnes of coke burned per m^2 of the cross-sectional area of the cupola in unit time.

The output of a cupola can be increased by :

1. Oxygen enrichment of the air blast (upto 30 - 35% O_2)

2. Utilisation of the heat of the outgoing gases to preheat the blast upto 180°C – 260°C or higher. Such a cupola is called a "Hot blast cupola".

Advantages of a Cupola :

1. Its continuous operation
2. Low cost of operation.
3. Easy to operate.
4. Composition of the melt can be controlled.
5. Temperature of the melt can be controlled

Inspite of its many advantages, the cupola has the following drawbacks :

(*i*) It is difficult to maintain close temperature control.

(*ii*) Since the fuel (Coke) and the molten metal come into contact with each other, some of the elements like *Si* are lost and others like S are picked up. This changes the final analysis of the molten metal.

2.15.2 Other common foundry furnaces are discussed below :

1. Crucible or Pot furnaces : These are the simplest of all the furnaces used in afoundry. They are used for melting many ferrous and non-ferrous metals. For melting copper-based alloys,

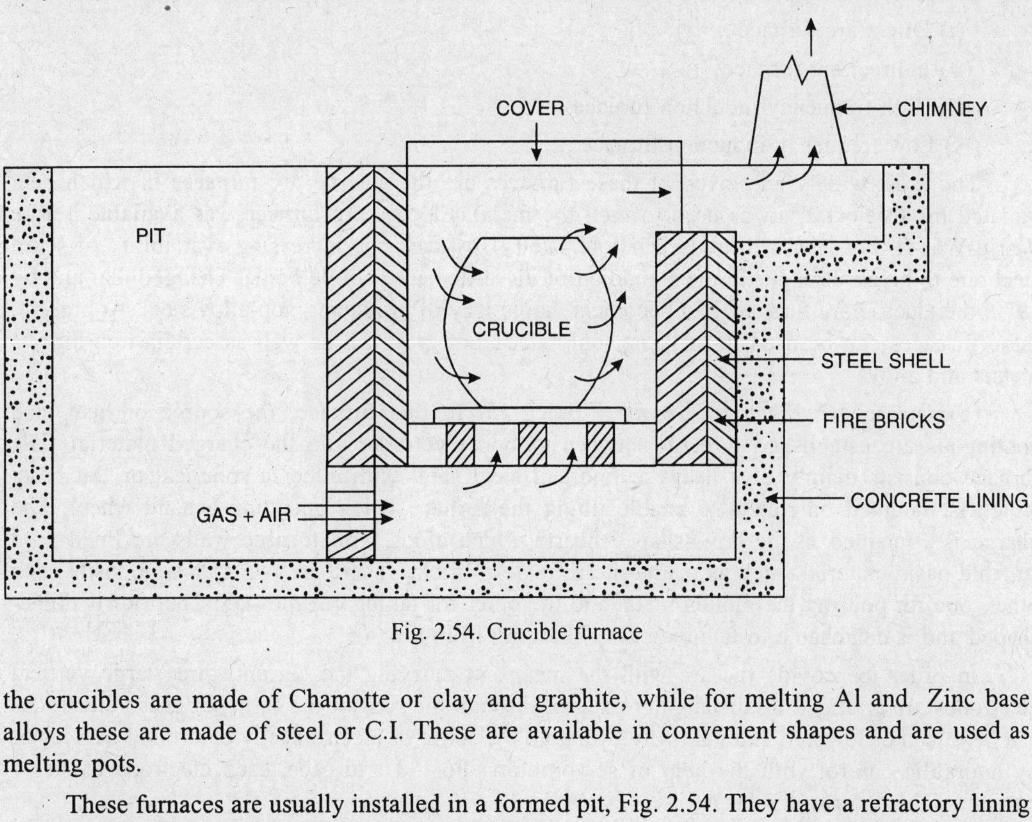

Fig. 2.54. Crucible furnace

the crucibles are made of Chamotte or clay and graphite, while for melting Al and Zinc base alloys these are made of steel or C.I. These are available in convenient shapes and are used as melting pots.

These furnaces are usually installed in a formed pit, Fig. 2.54. They have a refractory lining inside. These furnaces are fired by solid fuel (Coke), pulverised fuel, liquid or gaseous fuel (or natural gas). Both natural or artificial draughts can be used. In coke fired furnaces, the crucible containing metal is placed over the bed of coke. The products of fuel combustion heat the walls of the crucible or pot which transmit the heat to the metal inside. Thus, in most cases, the products of combustion do not come into direct contact with the metal, the metal does not become saturated with flue gases and high quality castings can be made. These furnaces may be stationary, tilting or movable. Tilting-pot furnaces are mounted above the floor level on supports. The furnace is tilted on trunnions to pour the molten metal. The capacity of these furnaces ranges from 30 to 150 kg. The melting time depends upon the metal and the type of fuel used. A disadvantage of these furnaces is the high fuel consumption.

2. **Electric Furnaces :** Electric furnaces are one of the major methods of melting in iron and steel factories. These furnaces are thus finding increasing use in foundries. Since there are no products of combustion in these furnaces, the loss of constituents elements is minimum. This results in the production of high quality metal for foundary purposes. Also, there is higher and

accurate metallurgical control in getting proper compositions and temperature for casting intricate high duty alloys having various metallurgical and physical properties. Unlike cupola or air furnaces, electric furnaces possess greater adaptability and flexibility and provide precise control over the temperature of molten metal. The high cost of electric power is a limitation, but this is outweighed by several overwhelming advantages listed above.

Electric furnaces are now being used for melting and refining all kinds of steel, including stainless steel and other alloy steels. At times, they are also used for melting plain and alloy cast irons.

Electric furnaces are of four types:

(*i*) Direct arc furnace

(*ii*) Indirect arc furnace

(*iii*) High frequency induction furnace

(*iv*) Low frequency induction furnace.

The most widely employed of these furnaces are the electric arc furnaces in which heat radiated by an electric arc is used to melt the metal. Electric arc furnace was available before World-War-II, but it was little used before 1960. However, the increasing availability of scrap steel due to break throughs in the shredding of discarded automobile bodies changed that and by 1970, the electric arc furnace, had become a viable way of producing non-alloy steel. At present, these furnaces can be used to melt, not only steel, but grey C.I. as well as certain non-ferrous metals and alloys.

(a) *Direct Arc Electric Furnace* : Fig 2.55. In this furnace, the source of heat is a continuous arc established directly between carbon electrodes and the charged material. The furnace consists mainly of a heavy cylindrical steel shell with either a spherical or flat base, which is mounted on rollers to enable tilting the furnace when operating a hand wheel. The charge is contained in the bowl-shaped hearth which along with furnace walls arc lined with suitable basic material, such as magnesite refractory bricks. There are two spouts opposite each other, one for pouring the molten metal and the other for taking out the slag. The roof is dome-shaped and is detachable to facilitate easy charging from top.

In order to provide the arc with the means of carrying the current, three large vertical electrodes are arranged in a triangular pattern, through the roof of the furnace. These electrodes can be raised or lowered automatically by suitable electric or electronically controlled devices or by hydraulic control with the help of servomotors. For 50 t furnace, each electrode carries a current of the order of 25000 A.

The furnace works on the principle that heat is generated when resistance is offered to the flow of electricity. In this case, it is the metal in the charge that provides the resistance to the flow of current. When the metal is molten, the slag offers the resistance to the flow of current. Thus to maintain proper heating even when the metal is molten, the electrodes must be raised so that they just touch the slag layer.

The electrodes should have high electrical and low thermal conductivity, good refractoriness and resistance to oxidation or chemical reaction. They should also possess good mechanical strength at elevated temperature. Out of graphite and amorphous carbon, graphite is preferred as it has higher electrical conductivity and is lighter. The charge consists of scrap plus a small amount of carbon and limestone. The scrap consists of 40% heavy scrap such as heads, risers and bloom heads, 40% medium scrap and 20% light scrap. The charge should be so distributed as to facilitate the formation of a pool under the electrodes. It is sound practice to place on the furnace base a small amount of light scrap topped by heavier scrap. Light scrap or turnings form another layer to reduce the electrode breakage and allow fast melting. The heaviest pieces of scrap are then placed directly below the electrodes, where the heat is the greatest.

The charge is placed in the furnace through the roof which is lifted up for charging. The roof is then closed and the electrodes are lowered. The power supply is switched on. The three carbon electrodes carry the current to contact the metal charge. Arcs are established between the electrodes and the metal charge. Within about 2 hours, the charge melts. The power supply is switched off, the electrodes are raised and the furnace is tilted to pour the molten metal into a

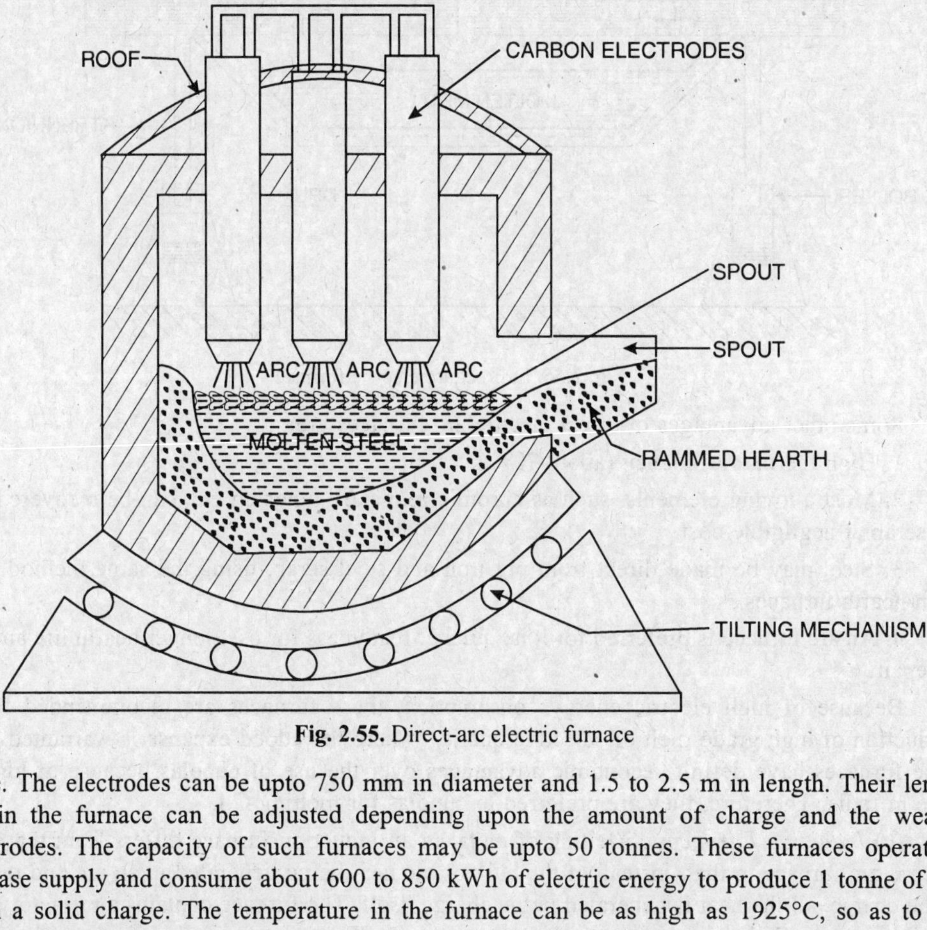

Fig. 2.55. Direct-arc electric furnace

ladle. The electrodes can be upto 750 mm in diameter and 1.5 to 2.5 m in length. Their lengths within the furnace can be adjusted depending upon the amount of charge and the wear of electrodes. The capacity of such furnaces may be upto 50 tonnes. These furnaces operate on 3-phase supply and consume about 600 to 850 kWh of electric energy to produce 1 tonne of steel from a solid charge. The temperature in the furnace can be as high as 1925°C, so as to melt metals with a high concentration of components with high melting points (Chromium, Tungsten, Molybdenum and other admixtures).

(*b*) *Indirect arc electric furnace* : Here, the arc is struck between two electrodes, Fig. 2.56, instead of between electrode and the metal charge. Such furnaces are commonly used for melting copper base alloys. The furnace consists of a horizontal refractory lined drum. Two graphite or carbon electrodes between which the arc is struck, are inserted in the centres of the end walls. The drum is mounted on trunnions and rollers due to which rocking motion can be given to the furnace. When needed, the furnace is tilted on trunnions and the molten metal is poured into a ladle. The capacity of such a furnace is upto 1000 kg.

As noted above, this furnace is specially designed for non-ferrous metals. However, these furnaces are excellent for melting and refining plain carbon and low alloy and high alloy steels, in addition to non-ferrous metals.

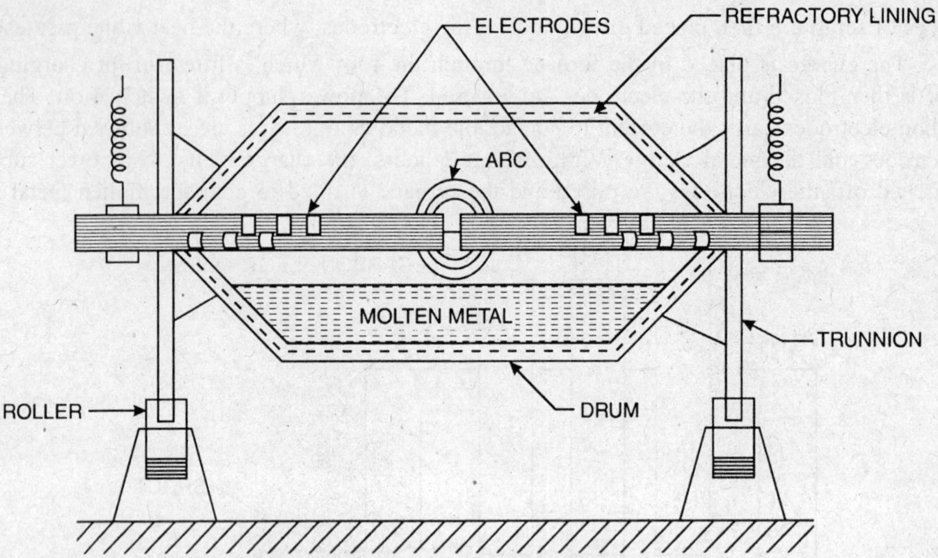

Fig. 2.56. Indirect arc electric furnace

Some other advantages of direct electric arc furnaces are:

1. High thermal efficiency (about 70%).

2. Most alloying elements, such as chromium, nickel, tungsten etc. can be recovered from the scrap at negligible cost.

3. Steel may be made direct from pig iron and steel scrap, using the same method as for open-hearth furnaces.

4. An arc furnace is preferred for it is: quicker readyness for use, longer hearth life and ease of repair.

Because of high electric energy consumption, these furnaces are recommended for the production of high grade melt of selective quality, where the added expense is warranted. Also, these furnaces have definite economic advantages over the use of cupolas because of high-ash coke in India. Therefore, they are preferred to cupolas for melting C.I.

(*c*) *Induction Furnaces :* Melting of metal in an induction furnace differs from that in the electric arc furnace in that, instead of the bulk of the heat being generated in an arc and radiated to the charge, all the heat is generated in the charge itself. The furnace contains a refractory lined crucible surrounded by a water cooled copper coil. The furnace works on the principle of transformer. Water cooled copper coil is the primary and the secondary is the metal charge. When A.C. is passed through the copper tubing, a magnetic field is set up. This magnetic field induces eddy currents in the crucible charge, which melts the metal. If ferrous metals show magnetic property, the loss due to hysterises produces extra heat. Very high temperatures can be obtained by this method of melting, the only limitation being the ability of the furnace lining to withstand the temperature developed.

The charge ratio is:

Steel : 40 to 60%

Pig iron : 4 to 7%

Returns : Rest

In case, Pig iron is not available, then, charge ratio is.

Steel : 40 to 65%

Returns : Rest

The induction furnaces are suitable especially for smaller quantities, 50 kg to 10 tonnes. Since, these furnaces do not require electrodes, this prevents carburisation of the metal and simplifies control of the process.

These are two designs of induction furnaces:

(*i*) *Core-less or high frequency induction furnace* : Fig. 2.57, is a cross-section of a high frequency induction furnace. Here, a water-cooled copper coil constituting the primary coil of a transformer completely surrounds the crucible. A high frequency current (10000 cps to 500,000 cps) is passed through the coil. A heavier, secondary current is induced in the charge constituting the secondary coil. The resistance of metal charge will cause it to heat up to the desired temperature.

(*ii*) *Core or Channel furnace or Low frequency induction furnace:*

Here the coil surrounds only a small portion of the crucible and the A.C. passed through the coil is of low frequency (50 to 60 cps).

The induction furnaces are employed to melt steel, iron, bronze, brass and Al base alloys. They are especially widely used to melt precision alloys, that is, alloys of exact composition based on iron and other metals. These furnaces have capacities ranging from few kgs to 4 or 6 tonnes.

Advantages:

(*i*) *Simple controls.* These furnaces can be readily started or stopped. High rate of melting and can deliver metal at regular intervals.

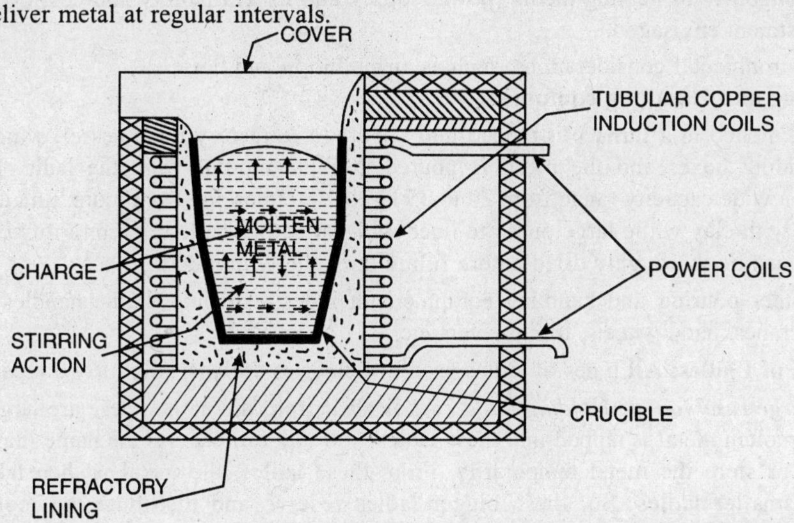

Fig. 2.57. High frequency induction furnace

(*ii*) Metal can be heated to very high temperatures without the danger of localised overheating that is sometimes observed in electric arc furnaces.

(*iii*) Gas atmosphere within the furnace can be controlled or a vacuum may be set up if necessary.

(*iv*) The motor effect of the strong electro-magnetic forces in the bath produces a circulatory motion, that is, automatic stirring or mixing. This facilitates better mixing between metal and slag. This mixing characteristic is excellent for alloying and adding new charge of metal. All this results in more uniform composition of the metal.

(*v*) Automatic mixing and stirring of the metal in the bath also has a favourable effect on its degassification and upon the removal of non-metallic inclusions. Thus high quality metals and alloys free of H_2 and N_2 can be easily produced.

The main drawback of high frequency induction furnace is its high initial cost. The low frequency furnace has better efficiency. Low frequency furnace is commonly used in non-ferrous foundries. It is particularly suitable for super heating (which improves fluidity), holding (keeping the metal temperature constant for a longer time which makes it suitable for die-casting) and duplexing (using two furnaces, melting in one furnace and then transferring to the second).

Choice of Furnace : There are a number of furnaces available which can meet the requirements of melting and casting metals and alloys in foundries. The quality of castings and economics of casting operations will depend upon the furnace selection. Therefore, the following factors should be given careful consideration while selecting a furnace for a job:

1. Capacity of molten metal/alloy required.
2. Melting rate and temperature control desired.
3. Quality of melt required.
4. Economics of melting, that is, the initial cost and operation and maintenance charges.
5. Forms and types of charge material.
6. Method of pouring and types of products contemplated.
7. Control of furnace atmosphere.
8. Availability of heating media (power supply and its availability and cost of fuels), and capital investment envisaged.
9. Environmental considerations, such as air pollution and noise.

2.16. Pouring and Handling equipment:

Metal, melted in a furnace is tapped into a ladle (a refractory lined vessel) which is taken to the moulding boxes and the metal is poured into the moulds from the ladle. Ladles are available in a wide capacity range from 20 to 35 kg to 10 tonnes and even more. Small ladles are luted inside with clay while large ones are lined with refractory ladle brick or a refractory mass. The lining must be thoroughly dried before filling the ladle with metal.

The other pouring and handling equipment includes:– Shanks (Ladle handles), trolleys, monorails, cranes, hand wheels, tilting levers etc.

Types of Ladles: All types of ladles used in foundries are basically of two types:

(*a*) *Large reservoir or Holding ladle* : As is clear from the name, these are large capacity ladles. The molten metal is tapped into these ladles from the furnace. As the name implies, these ladles hold or store the metal temporarily. From these ladles, the metal is then taken to the moulds in smaller ladles. So, these bigger ladles receive and distribute the molten metal simultaneously. These ladles are steel shells lined with firebricks. The bottom and sides of the ladle can also be lined with fire sand and clay, which is hardened by baking. These are also employed when there may be a mismatch between the rates of melting and using material.

(*b*) *Crane or Monorail ladles* : These are also large capacity ladles but smaller in size as compared to holding ladles. These are conveyed with the help of a crane or monorail to the place where the moulds are kept ready for pouring. The capacity of these ladles can range from 350 kg to 100 tonnes (for very big castings).

These ladles are of the following types :

(*i*) *Lip pouring ladle* : The action of this ladle is just like that of pouring water out of a jug. To avoid going of slag and other impurities into the mould, the molten metal in the ladle is skimmed with the help of a metal skimming bar before it is poured into the mould. Or a refractory dam can be built in front of the lip to achieve the same results. To pour the metal into the mould, the ladle is tilted by hand wheel or lever.

(*ii*) *Teapot ladle :* This ladle resembles a tea pot except that the snout is inside the body of the ladle. Since the metal leaves the ladle near the bottom, it is clean and free of any slag etc. For pouring, the ladle is tilted with the help of a hand wheel.

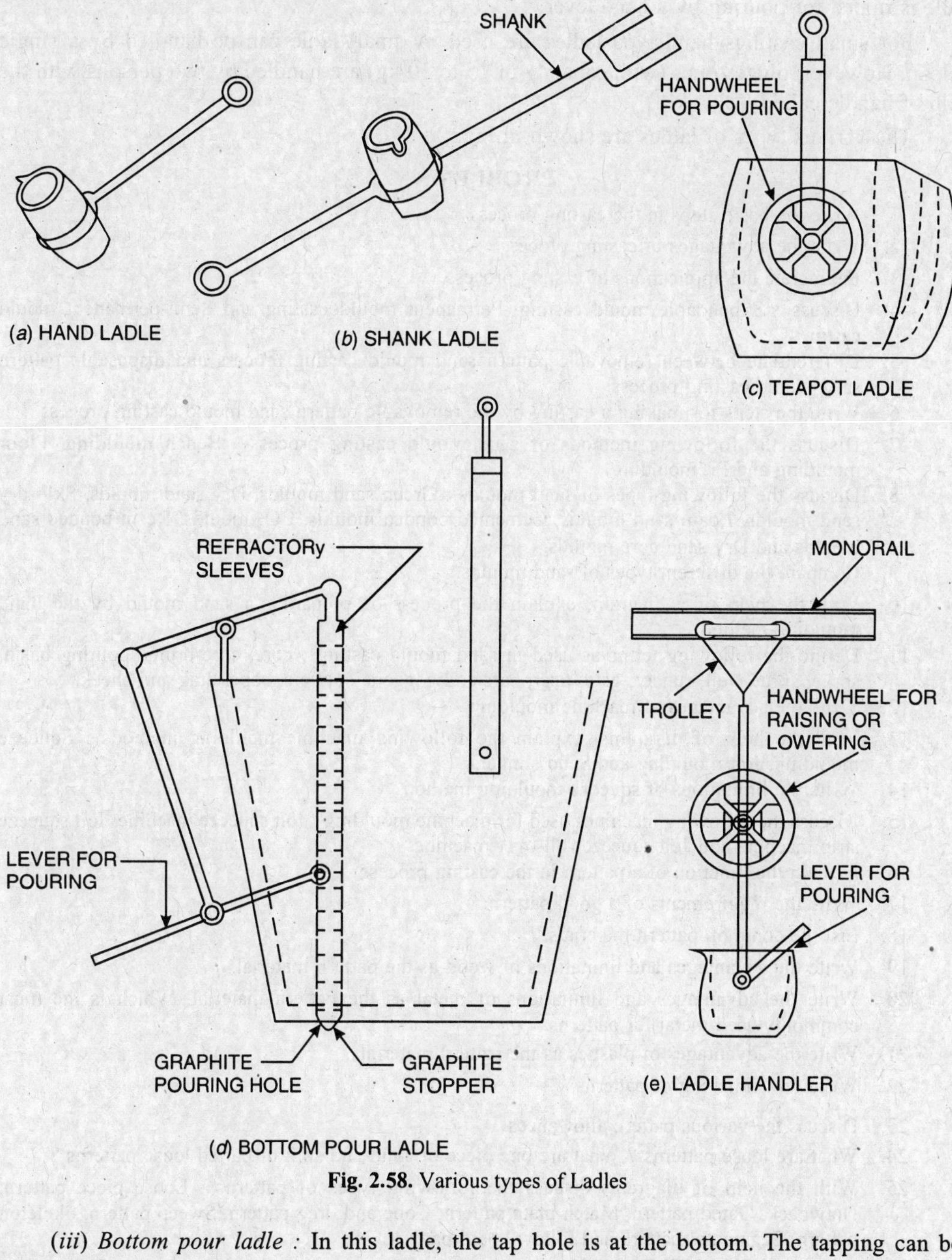

Fig. 2.58. Various types of Ladles

(*iii*) *Bottom pour ladle :* In this ladle, the tap hole is at the bottom. The tapping can be stopped with a refractory covered vertical rod stopper. Clean metal goes into the mould.

However, in this ladle, the momentum of the pouring metal is greater as compared to lip and tea pot ladles.

Crane or monorail ladles are employed in case of very heavy castings. A mechanical device called "ladle handler" which hangs from the overhead monorail, enables one man to pour the metal from the ladle. A large handwheel is used for readily raising or lowering the ladle, and the ladle is tipped for pouring by a hand lever.

For small castings hand type ladles are used. A small ladle can be handled by a single person. However, bigger ones (with capacity of 15 to 30 kg) are handled by two persons with the help of handle called "Shank".

The various types of ladles are shown in Fig. 2.58.

PROBLEMS

1. Write the basic steps in the casting process.
2. Write the advantages of casting process.
3. Enumerate the applications of casting process.
4. Discuss : Expendable mould casting, Permanent mould casting and Semi-permanent mould casting.
5. Differentiate between removable pattern sand mould casting process and disposable pattern sand mould casting process.
6. Write the steps for making a casting by the removable pattern sand mould casting process.
7. Discuss the following methods of sand mould casting process : Bench moulding, Floor moulding and Pit moulding.
8. Discuss the following types of sand moulds : Green sand moulds, Dry sand moulds, Skin-dry sand moulds, Loam sand moulds, Cemented bonded moulds, CO_2 moulds, Resin bonded sand moulds and Dry sand core moulds.
9. Compare the different types of sand moulds.
10. With the help of a diagram, explain the process of preparing a sand mould by the hand moulding method.
11. Define the following terms as used in sand mould casting : core, core-prints, pouring basin, sprue, sprue well, runner, gate, riser, bars and gaggers, soldiers, cope, drag and cheek.
12. Write the advantages of machine moulding.
13. With the help of diagrams, explain the following machine moulding methods : Squeeze moulding, Jolt moulding and Sand slingers.
14. Write the limitations of squeeze moulding method.
15. Discuss the following machines used for machine moulding : Jolt squeeze machine, Jolt squeeze strip machine and Jolt squeeze roll-over machine.
16. Explain the function of a pattern in the casting process.
17. Write the requirements of a good pattern.
18. List the common pattern materials.
19. Write the advantages and limitations of wood as the pattern material.
20. Write the advantages and limitations of metal as the pattern material. Which is the most commonly used metal for patterns.
21. Write the advantages of plastics as the pattern material.
22. Write on : finishing of patterns.
23. Discuss the various pattern allowances.
24. What are loose patterns ? What are one piece or solid and split or parted loose patterns ?
25. With the help of diagrams discuss the following types of patterns : Loose piece pattern, Drawback, Gated pattern, Match plate pattern, Cope and drag pattern, Sweep pattern, Skeleton pattern, Segmental pattern and follow board pattern.
26. What is a "Stop-off" ?
27. Why a colour scheme for patterns is needed ? Illustrate a common colour scheme.

28. What is "Stack moulding" ?
29. Name the various moulding materials used in foundary.
30. Name the essential constituents of moulding sand.
31. Write the advantages of silica sand as a moulding material.
32. What are the functions of a binder in moulding sand ?
33. What is meant by sand "at temper" ?
34. What are the functions of additives in moulding sand ?
35. How the moulding sand is classified on the basis of clayey matter it contains ?
36. Discuss: natural sand, synthetic sand and chemically coated sand.
37. Give the composition of a typical green moulding sand.
38. Discuss the various binders used in moulding sand.
39. Discuss the functions of the following additives in moulding sand : Coal dust, cereals, saw dust, wood flour, silica flour, fuel oil, iron oxide and dextrin.
40. What is the function of facing materials in moulding sand ? Name the common facing materials.
41. What is the function of cushion materials in moulding sand ? Name the common cushion materials.
42. Write about the following types of sands : Facing sand, Backing sand, System sand, parting sand.
43. Write on "parting materials" used in sand moulding.
44. Discuss the "mould surface coatings" used in sand mould casting.
45. Write about the moulding sand used for non-ferrous castings.
46. Discuss the various properties of moulding sand.
47. Discuss the essential qualities of a core. What is core sand ?
48. Explain the manual method of core making.
49. What is a core dryer ?
50. Why the cores are reinforced ?
51. What is core venting ?
52. Discuss synthetic resin core binders.
53. With the help of a diagram explain the core-blowing machine method for core making.
54. With the help of a diagram explain the extrusion method of core making.
55. Discuss the "Hot core box method" of core making.
56. Discuss the "Synthetic resin-bonded cold curing core sands method" of core making.
57. Discuss the CO_2 cold curing process of core making.
58. Write on "finishing and coating of cores".
59. With the help of diagrams discuss the various types of cores used in sand mould casting.
60. Explain the function of chaplets. Sketch the various forms of chaplets used in the foundary.
61. Sketch a common gating system. Label it and explain the function of its various elements.
62. Discuss the various methods of preventing the slag from flowing into the mould cavity.
63. Why the section of a Sprue reduces downwards ?
64. Explain the functions of : splash core, skim bob, runner and runner extension.
65. What is "gate ratio" ?
66. What is "choke" ? Discuss the "pressurised gating system" and "unpressurised gating system".
67. What is the function of a riser ? Write the requirements of a good riser.
68. What is "directional solidification" ? Explain it with the help of a diagram.
69. Write about "riser location".
70. Compare top riser and side riser.
71. What are open and blind risers ? Give their advantages and drawbacks.
72. Write the methods of keeping the metal hot and in the molten state in the top riser ?

73. Why the risers should be made with a high volume / surface area ratio.
74. Discuss the various shapes of a riser.
75. Write about the height of risers.
76. With the help of diagrams discuss the use of : (*a*) padding (*b*) chills and (*c*) Sleeves, in achieving directional solidification and thereby obtaining sound castings.
77. What are "flow offs" ?
78. With the help of a diagram, discuss the feeding ranges of risers.
79. Discuss the requirements of a good gating system.
80. Sketch and compare : Parting line gate, top gate and bottom gate.
81. What are : Pencil gate, Wedge gate, Whirl gate, Ring gate, Branch gate and Swirl gate.
82. Sketch the various sand mould casting defects. Give their causes and remedies.
83. With the help of sketches, discuss the various design considerations for designing the sand mould castings.
84. Differentiate between Pressure die Casting and Permanent mould casting.
85. List the main reasons behind the widespread use of pressure die castings process.
86. What are the limitations of Pressure die casting method ?
87. List the product applications of Pressure die casting method.
88. How are the dies opened and locked in die casting.
89. What is meant by the capacity of a die casting machine. Write its usual values.
90. Write the steps for making a casting by die casting process.
91. Why die-casting method is mainly used for non-ferrous metals.
92. Compare Cold-chamber and Hot-chamber methods of die casting.
93. List the metals normally cast by the cold-chamber method and hot chamber method.
94. Compare horizontal die casting machines and vertical die casting machines.
95. Compare off-set and central metal injection techniques.
96. Discuss the various factors for die casting design.
97. Name the various types of die-casting dies.
98. What are the various types of cores used in die-casting ?
99. Discuss in detail the design of gating system for die-casting.
100. Discuss the runner design of a die casting process.
101. Why venting is necessary in die-casting and how it is achieved ?
102. What is the difference between a multiple cavity die and a combination die ?
103. What are unit dies ?
104. Why are ejector pins required in a die-casting ?
105. What factors are important while deciding about the parting line in a die-casting ?
106. What is the use of 'Slides' in die-casting ?
107. Why cooling of dies is necessary during their operation ?
108. List the factors affecting the selection of die material.
109. Discuss some common die materials.
110. How the dies for die casting are manufactured ?
111. List the advantages of wrought dies and cast dies.
112. What is vacuum die-casting ?
113. Can the ferrous metals be cast by die-casting method ? if yes, then how ?
114. What is "Squeeze casting" ?
115. What is a "book mould" ?
116. List the materials commonly used to make permanent moulds.
117. What are the major types of cores used in permanent mould casting ?

118. Explain permanent mould venting.
119. Explain permanent mould gating system.
120. Why are risers usually required in permanent mould casting method.
121. Why permanent moulds are preheated before operations ?
122. What metals are cast by permanent mould method.
123. Discuss the mould coatings.
124. List the steps needed for permanent mould casting operation.
125. List the advantages and limitations of permanent mould casting method.
126. Discuss the casting design for a permanent mould casting method.
127. How the permanent moulds are manufactured ?
128. Explain a Turntable moulding machine.
129. Discuss "Slush Casting" and "Low pressure permanent mould casting".
130. Name the products made by Slush casting process.
131. Define the method of centrifugal casting.
132. With the help of diagrams discuss the following casting methods : (*a*) True-centrifugal casting (*b*) Semi-centrifugal casting, and (*c*) Centrifuge casting.
133. Write the advantages, disadvantages and product applications of True-centrifugal casting method.
134. Write the product applications of Semi-centrifugal and centrifuge casting methods.
135. What is meant by "Precision in vestment casting" ?
136. With the help of diagrams, discuss the "Lost wax method".
137. Write the advantages, limitations and product applications of Lost -wax method.
138. With the help of diagrams, discuss the "Shell moulding" method.
139. Write the advantages, disadvantages and product applications of Shell moulding method.
140. Discuss the following casting methods :

     (*a*) The Mercast process (*b*) Plaster - mould casting (*c*) Ceramic mould Casting (*d*) Full - mould casting, and (*e*) Continuous casting.
141. Write the advantages and disadvantages of Plaster - mould Casting.
142. What is "Shake-out operation" ?
143. How the cores are removed from Castings ?
144. Discuss the various methods of cleaning the surfaces of Castings.
145. How the gates and risers are removed from a Casting ?
146. What is the aim of inspection and testing of Castings ?
147. Discuss the following methods of inspection and testing of castings : (*a*) Radio-graphical testing (*b*) magnetic testing (*c*) Magnetic particle testing (*d*) Eddy current inspection (*e*) Liquid penetrant inspection (*f*) Ultrasonic testing.
148. Discuss the various methods of repairing and salvaging defective castings.
149. With the help of a diagram, explain the working of a "Cupola".
150. Write the advantages of a Cupola.
151. What is the necessity of sand testing ?
152. Write brief notes on :-

     (*a*) Moisture content test (*b*) Clay Content test. (*c*) Permeability test.
     (*d*) Fineness test (*e*) Hardness test (*f*) Strength tests
153. Discuss the "Chvorinov's rule" and "Caines rule" for determining the Riser size.
154. Discuss the various defects in die-castings :their causes and remedies.
155. List the drawbacks of Cupola as the foundry furnace.
156. Sketch and discuss a Crucible or Pot furnace.
157. List the advantages of Electric furnances.
158. List the types of Electric furnance.

159. Sketch and write on Direct arc electric furnace.
160. Sketch and write on Indirect arc electric furnace.
161. List the advantages of Direct arc electric furnaces.
162. What is the principle of Induction furnances.
163. Write on the two types of Induction furnaces.
164. Sketch and write on High frequency induction furnance."
165. List the advantages of Induction furnaces."
166. List the main draw back of high frequency induction furnace.
167.. List the advantages of Low frequency induction furnance and its applications.
168. Write a brief note on : Choice of furnace.
169. What is a Ladle ?
170. Sketch and discuss the various types of Ladles used in a foundry.
171. What are Jobbing foundries and Captive foundries ?
172. Compare the solidification times for castings of three different shapes of same volume : cubic, cylindrical (with height equal to its diameter) and spherical.

Solution. : – From Chvorinov's equation,

$$\text{Solidification time} = C.\left(\frac{\text{Volume of Casting}}{\text{Surface Area of Casting}}\right)^2$$

Since all the three castings have the sme volume (Let it be equal to unity),

$$\text{Solidification time, } t = C.\left(\frac{1}{\text{Surface Area}}\right)^2$$

(a) Cubic Casting :– Let 'a' be the side of the cube

$\therefore\ V = a^3 = 1 \quad \therefore a = 1$

Surface Area = $6a^2$ = 6 units

$$\therefore \qquad t = C.\left(\frac{1}{6}\right)^2 = 0.0278\ C$$

(b) Cylindrical Casting : Let r be the radius and h be the height of the cylinder. $\therefore\ h = 2r$

$$V = \pi r^2.h = 2\pi r^3$$

$$\therefore \qquad r = \left(\frac{1}{2\pi}\right)^{\frac{1}{3}}$$

$$\therefore \text{ Surface Area} = 2\pi r^2 + 2\pi rh = 6\pi r^2$$

$$= 6\pi\left[\frac{1}{2\pi}\right]^{2/3}$$

$$= 5.53 \text{ units}$$

$$\therefore \qquad t = C\left(\frac{1}{5.53}\right)^2 = 0.0327\ C$$

(*C*) Spherical Casting : – Let r be its radius,

$$\therefore \qquad V = \frac{4}{3}\pi r^3 \qquad \therefore\ r = \left(\frac{3}{4\pi}\right)^{\frac{1}{3}}$$

$$\therefore \qquad \text{Surface Area} = 4\pi r^2 = 4\pi.\left(\frac{3}{4\pi}\right)^{2/3} = 4.83 \text{ units}$$

$$\therefore \qquad t = C.\left(\frac{1}{4.83}\right)^2 = 0.0429\ C$$

∴ spherical casting has the maximum solidification time (∴ it will solidify the slowest) and the cubic casting has the least solidification time (∴ it will solidify fastest)

173. The first type of crystal that is formed upon the solidification of a metal is classified as :

(*a*) A dendrite crystal (*b*) A columnar crystal

(*c*) A chilled crystal (*d*) An equi-axed crystal. (*D.U., L.U*) (**Ans.** : *a*)

174. Draw an approximate sketch of a dendritic crystal, what are columnar crystals ?

175. Define : Nucleation, Grain growth, Grain boundaries.

176. Derive an expression for the critical radius of nucleous.

177. Define the term "Segregation". Discuss the various types of segregations during cooling and solidification of metals.

178. Write a brief description of sequence of events taking place in solidification of a pure metal (*UPSC, AMIE, DU*).

179. Which one of the following metals is not suitable for testing with ultrasonics ?

(a) Cu (b) Al (c) C.I. (d) Stainless stell

(*BTE, UPSC, I.Mech-E*) (**Ans.** : c)

180. Which one of the following tests is of a destruction nature ?

(*a*) Ultrasonic testing (*b*) Radiographic testing

(*c*) Macroscopic testing (*d*) Electro-magnetic testing (*BTE, UPSC*)

(**Ans.** : C)

181. Describe the details of magnetic particle test. State one of its limitations

182. State briefly the ultrasonic test and describe its specialities.

183. Radiography plays a vital role in the body of methods of non-destructive testing. Justify the statement. (*L.U.*)

184. How the calculations of Cupola Charge are done ?

185. A steel rectangular section 100 × 75 × 200 mm is to be cast. Calculate the size of a teakwood pattern if an Al pattern is to be obtained first to be subsequently used for making steel castings.

Shrinkage allowance for Al = 16 mm/m

Shrinkage allowance for steel = 21 mm/m

186. Figure shows finished dimensions of a brass bushing in which the central hole is to be cored. Calculate the dimensions of a teakwood pattern showing shrinkage. Take shrinkage allowance for Brass = 16 mm/m

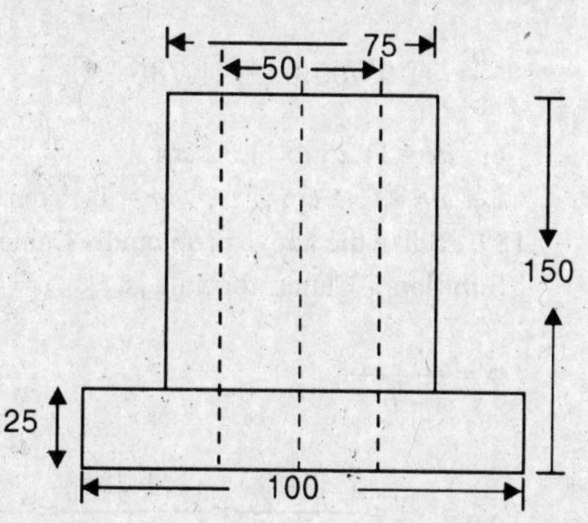

187. Calculate the dimensions of the pattern in problem 186, by considering. Draft allowance also. Take draft = 5 mm/m.

188. Calculate the size of a cylinderical riser (height and diameter equal) necessary to feed a steel slab casting 25 × 25 × 5 cm with a side riser, casting pouring horizontally into the mould. (*PTU. Dec. 2004*).

Solution : –

Volume of casting = 25 × 25 × 5 = 3125 cm^3.

Shrinkage volume = Depending upon metal, this srinkage volume varies from 2.5 % to 7.5% of volume of casting

For steel, let us assume it to be 3%.

∴ Shrinkage volume = 0.03 × 3125 = 93.75 cm^3

A riser should be designed with the minimum possible volume, while having a longer solidification time than the casting.

Now from practice, minimum volume of riser is approximately three times the shrinkage volume.

$\therefore$ Minimum volume of riser, $V_r = 3 \times 93.75 = 281.25$ cm^3

$$\therefore \quad \frac{\pi}{4} dr^2 \times hr = 281.25$$

Now, the optimum ratio diameter and height of a cylindrical side riser is,

$$hr = dr$$

$$\therefore \quad dr = 7.1 cm$$

$$\left(\frac{A}{V}\right)_r = \frac{6}{dr} = \frac{6}{7.1}$$

Now, for sound casting, the metal in the riser should be the last to cool, that is, the riser should have a longer solidification time than the casting. So,

$$\left(\frac{V}{A}\right)_r \geq \left(\frac{V}{A}\right)_C$$

$$\text{Now} \quad \left(\frac{V}{A}\right)_C = \frac{3125}{2(25 \times 25 + 5 \times 25 + 5 \times 25)} = \frac{3125}{1750} = 1.786 cm^{-1}$$

$$\text{and} \quad \left(\frac{V}{A}\right)_r = \frac{7.1}{6} = 1.1833$$

As is clear, $\left(\frac{V}{A}\right)_r$ is $< \left(\frac{V}{A}\right)_C$, which is not desirable.

For the condition.

$$\left(\frac{V}{A}\right)_r \geq \left(\frac{V}{A}\right)_C$$

In practice, $\left(\frac{V}{A}\right)_r = (1.05 \text{ to } 1.075) \left(\frac{V}{A}\right)_C$

$$\frac{dr}{6} = (1.05 \, to \, 1.075) \times 1.786$$

or $dr = 11.25$ to 11.52 cm

Let $dr = 11.5$ cm ; $\therefore$ $hr = 11.5$ cm. $\therefore$ $V_r = 1194.55$ cm^3

189. Solve the above problem by Caines formula.

Solution :- Caines formula is,

$$x = \frac{a}{y - b} + C$$

$$\text{Now} \quad X = \frac{(A/V)_C}{(A/V)_r} = \frac{(1/1.786)}{(6/dr)} = \frac{dr}{6 \times 1.786}$$

$$Y = \frac{V_r}{V_c} = \frac{\pi}{4} dr^3 \times \frac{1}{3125}$$

For steel, $a = 0.12,\ b = 0.05,\ c = 1$

$$\therefore \quad \frac{dr}{6\times 1.786} = \frac{0.12}{\dfrac{\pi dr^3}{12500} - 0.05} + 1$$

$$= \frac{1500}{\pi dr^3 - 625} + 1$$

$$\therefore \quad \pi d_r^4 - 625\, dr = 6\times 1.786\left(\pi d_r{}^3 + 875\right)$$

$$= 33.67\, d_r{}^3 + 9376.5$$

$$\therefore \quad \pi d_r^4 - 33.67\, d_r{}^3 - 625\, dr - 9376.5 = 0$$

By trial and error

dr ≅ 13.2 cm.

3

Fabrication Processes

Introduction. Any machine may consist of hundreds and thousands of individual components. These individual components are manufactured by an individual or a combination of manufacturing processes listed in chapter 1.

1. These individual components are then joined / fastened / bonded / assembled together to get the machine. These joints may either be permanent (welded, brazed, soldered and adhesive bonded, riveted or temporary (nut & bolt, pin, key etc.). The later ones are called 'Mechanical fastenings'. All these joining processes are usually termed as fabrication processes. The following are the fabrication processes dealt with in this chapter :

(a) Welding (b) Brazing (c) Soldering (d) Adhesive bonds

3.1. Welding :–

3.1.1. Definition. There are many definitions of a welding process. But the most comprehensive definition is given below :

Welding is defined as "a localized coalescence of metals, wherein coalescence is obtained by heating to suitable temperature, with or without the application of pressure and with or without the use of filler metal. The filler metal has a melting point approximately the same as the base metals".

The welding process is used to metallurgically join together two metal pieces, to produce essentially a single piece of metal. The process results in what is known as a 'permanent joint'. A good welded joint is as strong as the parent metal. The product is known as 'Weldment'.

3.1.2. Applications. The welding process finds widespread applica-tions in almost all branches of industry and construction. Welding is extensively employed in the fabrication and erection of steel structures in industrial construction and civil engineering, for example, structural members of bridges and buildings etc. ; vessels of welded-plate construction (steel reservoirs, boilers, pressure vessel tanks and pipelines etc.) ; and concrete reinforcement. It is the chief means of fastening panels and members together into automobile bodies and also in aviation industry. It has taken the place of castings for a large proportion of machine, jig and fixture bases, bodies and frames. The process is also extensively used for the repair of broken parts, in the building of worn surfaces and in the repair of defective castings. In fact, the future of any metal may depend upon how far it would lend itself to fabrication by welding.

3.1.3. Advantages. The widespread use of welding at the present time is due to its following advantages :

1. Welding results in a good saving of material and reduced labour content of production.
2. Low manufacturing costs.
3. Dependability of the medium, that is, the weldments are safer.

4. It gives the designer great latitude in planning and designing.
5. Welding is also useful as a method for repairing broken, worn or defective metal parts. Due to this, the cost of re-investment can be avoided.
6. Without welding techniques, the light weight methods of fabrication, so vital to the automotive and aircraft industries, would be unthinkable.

The welding process has the plus points that it is readily adaptable to streamline structure and the welded joints are very tight. Welded joints are strong, especially under static loading. However, they have poor fatigue resistance due to stress concentration, residual stresses and various weld defects, such as cracks, incomplete fusion, slag inclusions and the like. But, all these drawbacks can be overcome to a large extent.

Advantages of Welded Joints over Riveted Joints :

1. Economy of material and lighter weight of structure owing to :

(*a*) Better utilization of metal elements (plates, angles) since their working sections are not weakened by the rivet holes, consequently the sections of the welded pieces can be made smaller than the sections of riveted elements, for the same acting forces.

(*b*) Possibility of a wide use of butt-jointed seams, requiring no additional elements, such as cover straps.

(*c*) Lighter weight of the joining elements (rivets weight more than the welds). The weight of welds comprises about 1 to 1.5 per cent of work weight, while the weight of rivets is about 3.5 to 4 per cent. The use of welding instead of riveting, saves on an average 10 to 20 per cent in weight.

2. Greater strength of the joints, due to absence of holes needed for riveting.

3. Less labour is required, since there is no need for marking out and drilling or punching holes. Riveting consumes much more labour and is a much more complicated and less productive job, than welding, which can be often largely automated.

4. Possibility of joining curvilinear parts.

5. Tightness and impermeability of the joint.

6. Noiselessness (riveting is inevitably accompanied by noise).

Advantages over Casting Process :

1. Lighter weight and saving of material due to :

(*a*) lesser machining allowances and

(*b*) the possibility of utilizing smaller sections, since the wall thickness of cast parts, determined in many cases by the casting process, is as a rule 2 to 3 times greater, and sometimes even more than that of welded parts.

The saving of metal in welded machine components, as compared with cast ones, may amount to 40%.

2. Lower Cost.

3. Greater strength.

4. Maximum homogenity.

Whether machine parts should be welded, is decided in each particular case, by design and economy considerations.

The drawbacks of welding can be : Not all the metals are satisfactori-ly weldable and the weldments are less readily machinable, as compared to castings.

3.1.4. Classification. Welding processes may be classified accord-ing to the source of energy employed for heating the metals and the state of metal at the place being welded. Depending upon the source of heat, the welding processes are classified as :-

1. Chemical (oxygen + combustible fuel gas, e.g., acetylene, propane, butane, natural gas, hydrogen etc.).
2. Chemico – Mechanical (Pressure gas welding, Thermit welding)
3. Electro – chemical (Atomic hydrogen welding)
4. Electro-Mechanical (Electric resistance welding)
5. Electric – Arc welding.

These may be divided into two groups as follows :

(*a*) **Pressure Processes.** In these processes, the parts to be joined are heated to a plastic state (fusion may occur to a limited extent) and forced together with external pressure to make the joint. Some of the more common processes in this group are mentioned below :

1. Forge welding
2. Thermit Pressure welding
3. Pressure Gas welding
4. Electric Resistance welding

(*b*) **Fusion Processes.** In these processes, the material at the joint is heated to the molten state and allowed to solidify to make the joint, without the application of pressure. Here, some joints may be made without the addition of a filler metal, but, in general, a filler metal must be added to the weld to fill the space between the parts being welded. The filler metal deposited should ordinarily be of the same composition as the base metal.

Some of the common welding processes in this group are listed below :

1. Gas welding
2. Electric Arc welding
3. Thermit Fusion welding

The welding processes can also be classified as :

Autogeneous, Homogeneous and Hetrogeneous

In 'Autogeneous' processes, no filler metal is added to the joint interface, for example, cold and hot pressure welding processes and electric resistance welding.

In 'Homogeneous' processes, filler metal is added and is of the same type as the parent metal, for example, welding of plain low- C steel with a low- C welding rod and welding of 70 - 30 brass with a 70 - 30 brass welding rod etc.

In 'Hetrogeneous' processes, a filler metal is used but is of a different type from the parent metal, for example, brazing and soldering processes. Brazing and Soldering are not strictly the welding processes in view of the definition of welding process given above. However, these processes also belong to the family of welding processes.

The two most widely used welding methods are : Gas welding and Arc welding. The temperature of an electric arc is much higher than that of a gas flame, so the joint zone melts practically instantaneously in arc welding. Gas welding involves long preheating period which raises the metal adjacent to the joint to a high temperature. This exerts an unfavour- able effect on the crystalline structure, which results in considerable stresses being set up. So, gas welding is unsuitable for relatively large cross - sections due to this trouble and the time involved in preheating. Plates above 20 mm thick are, therefore, best welded by arc welding.

3.1.5. Types of Weld Joints. The relative positions of the two pieces being welded determine the type of joint. Welded structures are assembled by five basic types of joints : Butt, Lap, Corner, T and Edge joints, shown in Fig. 3.1. All other possible joints are variations of these basic joints. To make a weld, the two pieces to be joined are located and clamped or held in the correct position.

Butt joints are formed by welding the end surfaces or edges of the members.

In Lap joints, the two members being welded should overlap by an amount from 3 to 5 times their thickness. Edge of each piece is welded to the surface of the other.

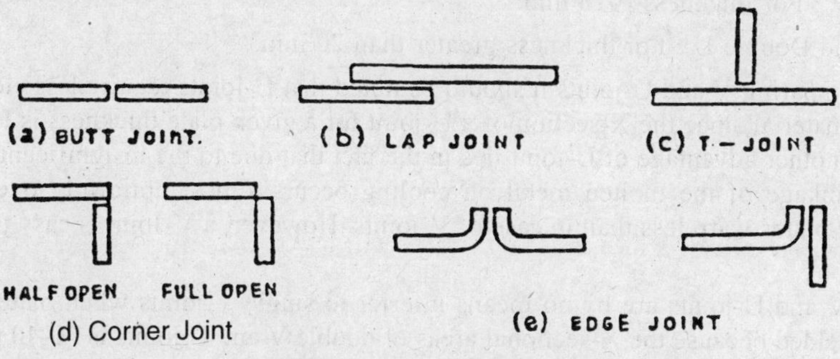

Fig. 3.1. Basic Welded Joints.

T-joint is used to join two pieces whose surfaces are approximately at right angles to each other.

Corner joints are used to join the edges of two pieces whose surfaces are approximately at right angles to each other.

Edge joint is used to join the edges of two pieces in which a part of the surface remains parallel to each other.

3.1.6. Types of Welds. Fig. 3.2 shows the various types of welds used in making a joint. A 'bead' weld is one in which the filler metal is deposited at a joint where the two surfaces adjoining the joint are in the same plane. A 'bead' is defined as a single run of weld metal. A 'fillet' weld is one in which the filler metal is deposited at the corner of two intersecting surfaces, such as a T or Lap joint. A 'groove' weld is one in which the filler material is deposited in a groove formed by edge preparation of one member or of both the members. A 'plug' or 'slot' weld is one in which a hole is formed through one of the pieces to be welded and the filler material is then deposited into this hole and fused with the mating part.

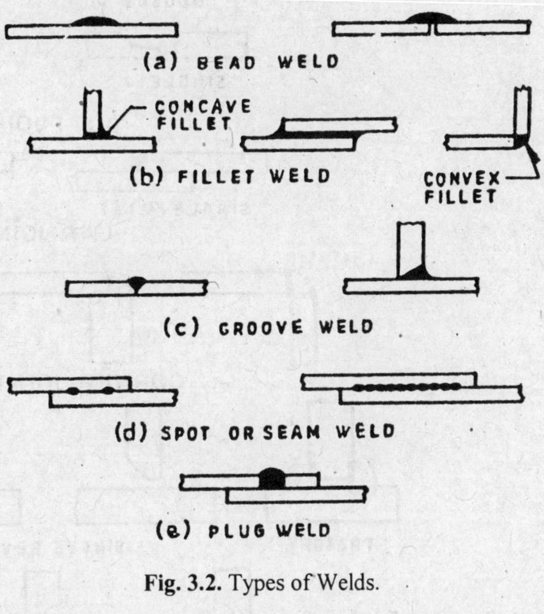

Fig. 3.2. Types of Welds.

3.1.7. Edge Preparation. The preparation of the edges of the pieces to be welded depends upon the thickness of metal being welded. Edge preparation is necessary when thickness increases so that heat would be able to penetrate the entire depth. This ensures formation of sound welds. The edge preparation is done by bevelling the edges of the pieces after the rust, grease, oil or paint are completely removed from their surfaces. The various edge preparations are illustrated in Fig. 3.3. There are five basic types of chamfers put on the mating edges prior to welding : Square, V, Bevel, U and J.

Butt Joints. The straight square butt joints with no special edge preparation is used when the thickness of the two plates to be welded is small so that heat of welding penetrates the full depth of joint. These joints are suitable for thicknesses from 3 to 8 mm. However, if the plate thickness is more than 4.5 mm, edge preparation is recommended.

Single V : For thickness upto 16 mm

Double V : For thickness > 16 mm

Single and Double U : For thickness greater than 20 mm.

When comparing V and U-joints it should be noted that U-joints require less electric power and electrode material since the X-section of a U-joint for a given plate thickness is less than that of a V-joint. Another advantage of U-joint lies in the fact that due to the insignificant level of the edges, the shrinkage of the molten metal on cooling occurs almost uniformly over the entire section and the plates warp less than in case of V-joints. However, a V-Joint is easy to make than a U-joint.

Double V and U-joints are by no means inferior to single V-joints when plates of 12 mm and over are welded because the X-sectional areas of double V and U joints is 30-40 per cent less than that of single V-joints at the same bevel angle. The short comings of double V and U-joints accrue from poor penetration liable to accur inside the butt and the greater cost of edge preparation especially in the case of U-joints.

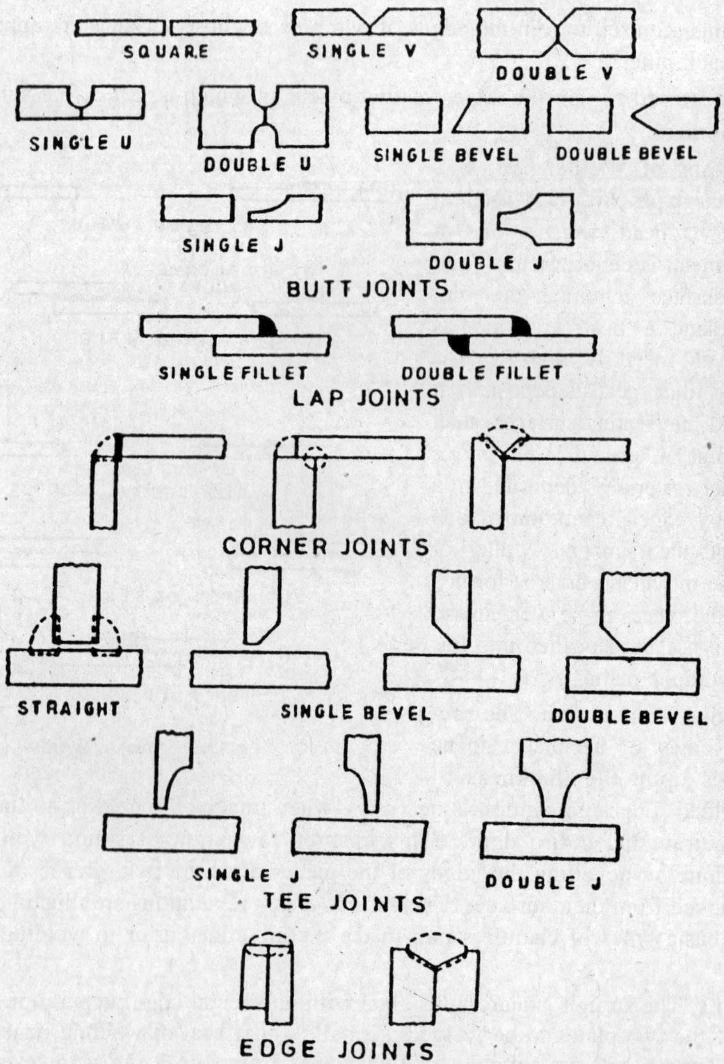

Fig. 3.3. Edge Preparation.

Other edge preparations for a butt joint are : Single bevel, Double bevel, Single J, Double J. Butt joints are made by Bead or Groove welds.

Lap Joints are used to join thin sheets, usually less than 3 mm thick. These joints do not need any special edge preparation. The joint is produced by fillet welds.

Corner Joints are used to join sheets upto 5 mm thick. These joints are welded with or without edge preparation, with the help of fillet and/or groove welds.

Tee Joints. Only structures subjected to low static loads can be welded without edge preparation. Single bevel joints are employed for critical structures in which the members are from 10 to 20 mm thick and Double bevel designs are used for thicker metals. Single J and Double J joints can also be used for thicker metals. Tee joints are made by using fillet and/or groove welds.

Edge Joints. Edge joints are used for metals upto 3 mm thick. The height of flange should be twice the thickness of the sheet. These joints are made by Bead or Groove welds.

• Fig. 3.4 gives the various definitions concerning welding joints and welds.

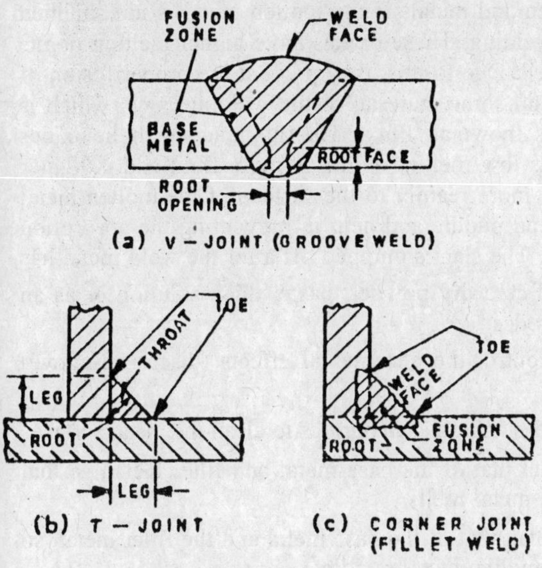

Fig. 3.4. Definitions.

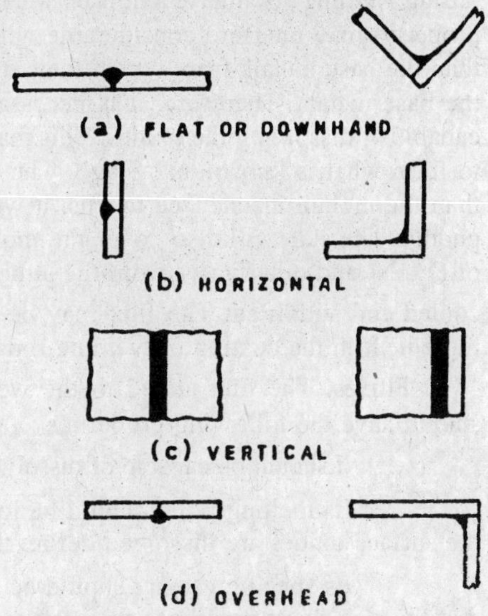

Fig. 3.5. Welding positions.

3.1.8. Welding positions. The basic welding positions are shown in Fig. 3.5. The flat position is the easiest and quickest for welding. In this position, welding benefits from the force of gravity and maximum deposition rates are obtained resulting in strongest welded joints. Next in ease of welding is the horizontal fillet position in which the force of gravity helps to some extent. However, the major defects that can occur while welding in horizontal position are : undercutting and overlapping of the weld zone.

For welding in these positions, the joint should be level or nearly so when possible. Welding in positions other than flat requires the use of manipulative techniques and welding rods/electrodes that result in faster freezing of the molten metal and slag, to counteract the effect of gravity. In vertical position, the welder can deposit the bead in the uphill or downhill direction. Uphill welding is preferred for thick metals because it produces stronger welds. Downhill welding is faster than the uphill welding and is suited for thin metals.

3.1.9. Cleaning the Joint and Fluxing. The weld area (joint) must be free of oil, dirt, grease, paint or moisture etc. Firstly, these will interfere with the proper fusion on the metal and result in a weak joint. Secondly, gases may evolve from these (under the action of heat) which may be absorbed by molten metal.

For removing oily substances, the cleaning is done by organic solvents like acetone, CCl_4 and trichloro ethylene. For removing foreign substances, cleaning is done with a rag soaked in the solvent. Heavier oxide films may be removed by acid pickling, wire brushing or emery. When organic solvents are used for cleaning, these should be completely evaporated from the interfaces before any welding is attempted. Otherwise, highly poisonous gases such as phosgene may form under intense heat of welding.

In addition to many gases which may evolve if the joint is unclean, gases such as O_2 , N_2 and H_2 are present in the atmosphere. Also, water vapours in air, breakdown under intense heat of welding to form O_2 and H_2. All these gases may be absorbed by the molten metal. As the metal cools after welding, its solubility for these gases decreases. The gases try to leave the metal as it solidifies. If they are not able to leave, they will form porosities and gas holes in the weld. To avoid this, the joint must be properly cleaned and the molten metal properly shielded from surrounding air.

Most oxides present on the metal surface (if the joint is not cleaned properly) or formed during welding (by the reaction between oxygen and metals if the molten metal is not shielded properly) may interfere considerable with the welding. These oxides have higher melting points than the base metals from which they are formed. Due to this, they prevent the proper fusion of the base metals. Therefore, it is necessary to add some material to the welding zone, which is capable of dissolving the oxides. This material is known as 'Flux'. The flux reacts with the oxides to form what is known as "Slag". The slag has low melting point, is more fluid and is lighter than the molten metal. Due to this, it will float more readily to the surface of the molten metal puddle. The slag will also cover the molten metal puddle and help in preventing the absorption of O_2, N_2 and other gases from the atmosphere. The slag is chipped off after the weld metal has cooled and solidified. The flux may be applied as a dry powder, paste, thick solution or as an ingredient in the coating of welding rods /electrodes.

Fluxes. The flux added to the weld to control the detrimental effects (discussed above) should have the following properties :

1. It should be capable of dissolving oxides and other impurities to clean the metal surface.
2. Its melting point should be lower than that of the base metal and filler metal, so that the surface oxides are dissolved before the base metal melts.
3. Its specific gravity should be lower than that of the base metal and the filler metal, so that the slag floats readily to the surface of the molten metal puddle.
4. It should control the flowability of the molten metal.
5. It should not affect the base metal adversely.

The type of flux used will depend upon the type of the metals being welded. The common fluxes used are listed below :

(*i*) **For Aluminium and Aluminium alloys.** Mixture of alkaline fluorides, chlorides and bisulphates, for example, mixtures of chlorides of sodium, potassium, Lithium and Barium, fluorides of calcium (Fluorspar, CaF_2), sodium and potassium and sodium bisulphate.

(*ii*) **Copper and Copper alloys.** Mixtures of sodium and potassium borates, carbonates, chlorides, sulphates; Borax ($Na_2B_4O_7.10H_2O$) Boric acid (H_3BO_3) and Di-sodium hydrogen phosphate (Na_2HPO_4) are suitable for dissolving oxides called cuprous oxides.

(*iii*) **Ferrous metals.** For welding C.I., Mixtures of Borax, Sodium carbonate and Potassium carbonate ; Sodium carbonate and Sodium bicarbonate ; Borax, Sodium carbonate and Sodium nitrate or Borax alone are suitable fluxes.

For Carbon steels : Dehydrated borax and calcium oxide dissolved in liquid are common fluxes.

For Alloy steels : Mixture of Boric acid, dehydrated borax and Calcium fluoride.

Borax forms compounds with iron oxide and the carbonates cleans and promotes fluidity.

3.2. GAS WELDING

Gas welding is "a group of welding processes wherein coalescence is produced by heating with a gas flame or flames with or without the application of pressure and with or without the use of filler material".

It may be noted here that out of the various types of gas welding processes, pressure is used only in Pressure Gas Welding (PGW). In all other types of gas welding processes, joint is made without the application of any pressure. The edges or surfaces to be joined are melted (alongwith the filler material when used) by a suitable gas flame and the molten metal is allowed to flow together to form a permanent joint on cooling.

The welding heat is obtained from burning a mixture of oxygen and a combustible gas. The gases are mixed in the proper proportion within a welding torch (also called Blowpipe) which provides control of the welding flame. The mixture is burnt at the blowpipe nozzle or tip. The ge4neral name of the method is : Oxy fuel Gas welding (OFW)

The common commercial gases used in gas welding include : acetylene, hydrogen, propane and butane. The most common form of gas welding is oxygen-acetylene (oxy-acetylene) welding, OAW. Acetylene (C_2H_2) produces higher temperature (in the range of 3200°C) than other gases, (which produce a flame temperature in the range of 2500°C) because it contains more available carbon and releases heat when its components (C and H) dissociate to combine with O_2 and burn. The cost of production of acetylene is low and the gases (O_2 and C_2H_2) can be stored at high pressure in separate steel cylinders. The main drawback of acetylene is that it is dangerous if not handled carefully.

3.2.1. Oxy- Acetylene Welding (OAW). There are two systems of OAW depending upon the manner in which acetylene is supplied for welding : High pressure system and Low pressure system. Acetylene is supplied either from generators or it may be purchased in metal cylinders. O_2 is always supplied from metal cylinders.

1. High Pressure System. In this system, both O_2 and C_2H_2 are supplied from high pressure cylinders. Oxygen cylinders are charged to a pressure of 120 atm. gauge. Due to the danger of explosion, pure acetylene cannot be compressed to a pressure more than 0.1 Pa above atmosphere. Therefore, acetylene is supplied in cylinders in the form known as "Dissolved acetylene". It is stored in cylinders in which it is dissolved in acetone under a pressure of from 16 to 22 atm gauge. At normal pressure, one litre of acetone dissolves about 25 litres of acetylene. For every additional atmosphere of pressure another 25 volumes of acetylene will be dissolved. As a safety measure, the cylinder of acetylene is filled with a porous filler (usually charcoal) forming a system of capillary vessels. It should not be with drawn from cylinder too rapidly, since some acetone may then be with drawn along with acetylene. The maximum recommended pressure when taking acetylene from a cylinder through a rubber hose is 1 bar (1×10^5 Pa). In H.P. system, the pressure of acetylene at the welding torch is from 0.06 to 1.0 bar.

2. Low Pressure System. Here, acetylene is produced at the place of welding by the interaction of calcium carbide and water in an acetylene generator, according to the reaction :

$$CaC_2 + 2H_2O \rightarrow Ca(OH)_2 + C_2H_2 + 127.3 \text{ kJ per mol.}$$

As is clear from above, a great deal of heat is evolved in this reaction.

The produced acetylene is supplied to the blowpipe at a low pressure from a gas holder incorporated in the generator. Acetylene is cleaned by passing it through a purifier. To prevent the possibility of an explosion by oxygen or air blowing back and entering the generating plant, a back pressure valve is arranged between the blowpipe and the gas holder.

The pressure of acetylene at the torch is upto 0.06 bar

For oxygen, the desired pressure at the welding torch is :

(*i*) High Pressure System (Welding and Cutting) = 0.1 to 3.5 bar

(*ii*) Low Pressure System (Welding) = 0.5 to 3.5 bar

3.2.2. Gas Welding Equipment. Equipment required in gas welding includes : cylinders for compressed gases (acetylene generator in place of acetylene cylinder in low pressure system), regulators, blowpipes, Nozzles, Hose and Hose fittings. The assembled basic equipment required for high pressure system is shown in Fig. 3.6.

1. Cylinders for Compressed Gases. The Oxygen cylinder is painted Black and is made of steel. Acetylene cylinder is painted Maroon and is made of steel.

2. Pressure Regulators. A pressure regulator or pressure reducing valve, located on the top of both O_2 and C_2H_2 cylinders, serves to reduce the high cylinder pressure of the gas to a suitable working value at the blow pipe and to maintain a constant pressure. The pressure is

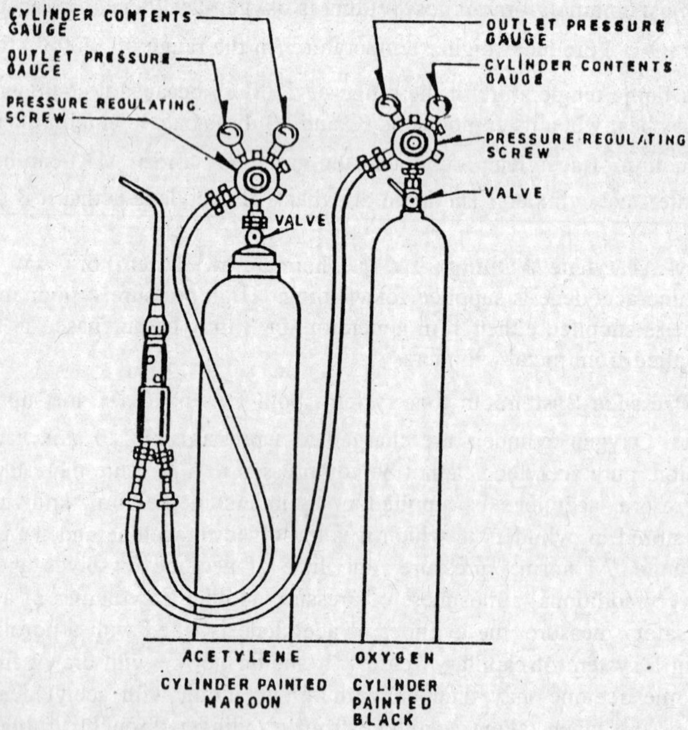

Fig. 3.6. High Pressure Gas Welding Equipment.

regulated with the help of a spring loaded diaphragm. Variations of pressure is necessary for different sized nozzles (inside diameter) and the pressure is controlled by a graduated adjusting screw which serves to vary the compression of the spring.

3. Pressure Gauges. Each gas cylinder is provided with two pressure gauges. One gauge indicates the pressure of the gas inside the cylinder and the other indicates the pressure of the gas supplied to the blow pipe.

4. Blow Pipe. The blow pipe or weld-ing torch serves to mix the gases in proper proportion and to deliver the mixture to the nozzle or tip where it is burned. The gases from the cylinders are taken to the blow pipe through reducing valves and with the help of rubber tubes (hoses). On the shank of the blowpipe, two control values (needle type) are provided, one for controlling the flow of acetylene and the other of oxygen, entering a chamber called mixing chamber where the two gases are mixed in a correct proportion. The control knobs of the control valves are usually coloured, red for acetylene and blue for oxygen.

Depending upon the system of OAW, there are two designs of blow pipes : Injector type and High pressure type.

The injector type blow pipe is used in the low pressure system. The low pressure of acetylene is not sufficient to force it through the small passage, into the mixing chamber. To overcome this difficulty, high pressure oxygen at a very high velocity is led through an injector nozzle inside the body of the blow pipe. It produces a vacuum in the acetylene channel, drawing acetylene into the mixing chamber, (Fig. 3.7 (*a*)). From the mixer, the gas mixture flows to the tip. In the high pressure system, (Fig. 3.7 (*b*)), both the gases are at high pressure. The gases flow on their own to the mixing chamber and no injector is needed. Under no circumstance, must a high pressure torch be used on a low pressure system, because the absence of injector makes it unsafe. Low pressure blow pipe is expensive than the high pressure blow pipe, because the whole head containing both the nozzle (tip) and injector is to be replaced with every nozzle change.

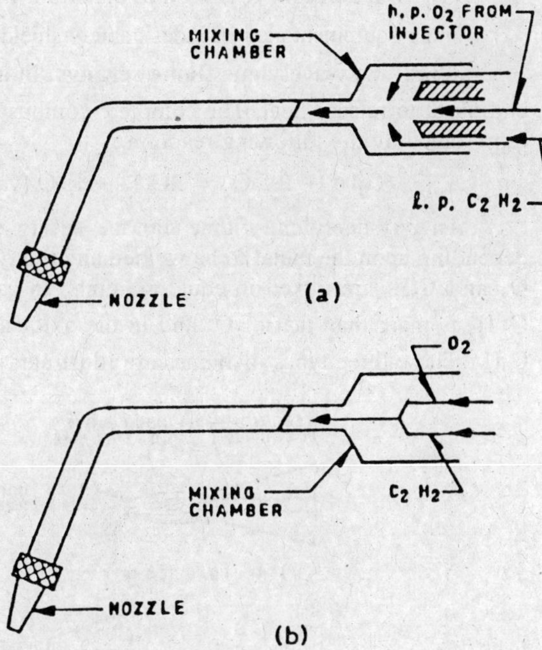

Fig. 3.7. Blow Pipes.

5. Nozzle or Tip. The nozzle is a device screwed to the end of the blow pipe. It is used to permit the flow of oxyacetylene gas mixture from the mixing chamber of blow pipe to the tip of the nozzle to facilitate burning. In order to vary the size of flame (and heat supply) necessary to weld varying thicknesses of metal, a selection of tips is available for the blow pipe. For this, the nozzles are interchangeable so that the correct nozzle is fitted at the end of the blow pipe. Each nozzle is marked showing its gas consumption in litres/hour and a table supplied with the blowpipe shows which tip should be used to weld any required thickness of metal. The delivery pressure from the regulator must be varied according to the size of the tip used, and instructions are supplied to obtain the correct conditions.

6. Hose and Hose fittings. The hose connects the outlet of the pressure reducing valve and the blow pipe. Rubber tubing is necessary for flexibility but it should be of the highest quality, specifically manufactured for this purpose. In accordance with International standards, hose is manufactured to a colour code : Blue for Oxygen and Red for acetylene. Hose fittings are provided at the ends of the hoses for attachment to the blow pipe and the outlet of the pressure reducing valves. To prevent the interchange of fittings, the oxygen hose connection nut has a right handed thread and the acetylene fuel gas fittings have a left handed thread.

The other equipment used during gas welding are :

7. Goggles. Welding goggles must be worn to protect the eyes from the heat and light radiated from the flame and molten metal in the weld pool.

8. **Welding gloves.** They protect the hands from the heat and metal splashes.

9. **Spark lighter.** It is used to conveniently and instantaneously light the blow pipe.

10. **Chipping hammer.** It is made of steel and is used to remove metal oxides from the welded bead.

11. **Wire-brush.** It is used to clean the weld joint before and after welding.

Other equipment also includes : safety shields and protective clothing.

3.2.3. Oxy-acetylene flame settings. In an oxyacetylene flame, O_2 and C_2H_2 are mixed and burnt to release heat. The complete combustion of acetylene in an atmosphere of oxygen is represented by the following reaction :

$$C_2H_2 + 2.5\,O_2 = 2CO_2 + H_2O\,(\text{vapour}) + 1284.57\ \text{k J / mol}$$

An oxy-acetylene flame can be set to : neutral, carburising or oxidising condition depending upon the metal to be welded and the type of filler metal to be used. In a neutral flame, O_2 and C_2H_2 are mixed in equal amounts. In carburising or reducing flame, the percentage of C_2H_2 is more than that of O_2 and in the oxidising flame, the percentage of O_2 is more than of C_2H_2. These three types of flames are illustrated in Fig. 3.8.

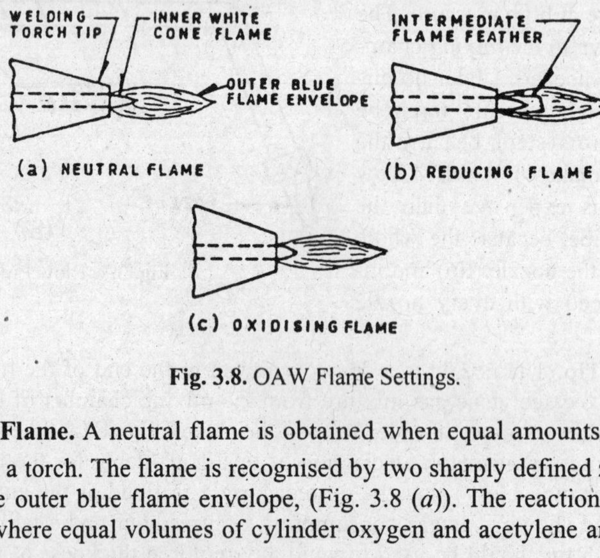

Fig. 3.8. OAW Flame Settings.

1. **Neutral Flame.** A neutral flame is obtained when equal amounts of O_2 and C_2H_2 are mixed and burnt in a torch. The flame is recognised by two sharply defined zones, the inner white cone flame and the outer blue flame envelope, (Fig. 3.8 (*a*)). The reaction at the inner cone for the neutral flame where equal volumes of cylinder oxygen and acetylene are used is,

$$C_2H_2 + O_2 \longrightarrow 2\,CO + H_2$$

This provides the most concentrated heat with the highest tempera-ture for welding at a distance of 3 to 5 mm from the end of the inner cone. It is also apparent that the environment within the outer envelope consists of carbon monoxide and hydrogen and is relatively inert to materials that oxidise readily.

The reactions at the outer envelope are :

$$2\,CO + O_2 \longrightarrow 2\,CO_2$$

$$H_2 + \frac{1}{2}O_2 \longrightarrow H_2O\,(\text{vapour})$$

For these reactions, the oxygen is supplied from the surrounding air. During actual welding, the outer envelope spreads over the surface of the work material and serves as a protective shield from the ordinary atmosphere. Also, since the heat developed is not as concentrated, this outer envelope of flame contributes only to preheat the work material for welding.

2. Carburising Flame. A carburising or reducing flame is obtained when an excess of acetylene is supplied than which is theoretically required (O_2 : C_2H_2 = 0.85 to 0.95). A reducing flame is recognised by three distinct sections : the inner cone (which is not sharply defined) and an outer envelope as for the neutral flame. The third zone surrounds the inner cone and extends into the outer enveloping zone. It is whitish in colour and is called "excess acetylene feather". Its length is an indication of the amount of excess acetylene. (Fig. 3.8 (*b*)).

To obtain a neutral flame, first the reducing flame is obtained. Then the supply of oxygen is gradually increased until the intermediate feather disappears. The resulting flame will be a neutral flame.

3. Oxidising Flame. This flame has an excess of oxygen over that required for a neutral flame (O_2 : C_2H_2 = 1.15 to 1.50). To obtain an oxidising flame, the flame is first set to the neutral condition. Then the acetylene valve is turned down gradually to reduce the amount of acetylene, giving an excess of oxygen. The flame resembles the neutral flame except that it acquires a light blue tint and the inner cone is slightly shorter and more pointed than in a neutral flame (Fig. 3.8 (*c*)). An oxidis-ing flame burns with a harsh sound and the outer envelope is short and narrow.

Applications of three types of flames :

(*a*) **Neutral Flame.** In most welding situations, it is theoretically desirable to use a neutral flame, but in practice it is very difficult to discern whether the flame is neutral or oxidising, either a slightly reducing or a slightly Oxidising flame is used.

If possible, most welding should be done with a neutral flame, since such a flame has a minimum chemical effect upon most heated metals. The flame is widely used for the welding of steel, stainless steel, cast iron, Copper and Aluminium.

(*b*) **Reducing Flame.** The acetylene being excess in this flame, the available carbon is not completely consumed. With iron and steel, it will form iron carbide (hard and brittle). Therefore, metals that tend to absorb carbon should not be welded with a reducing flame. This flame is used for materials that oxidise rapidly like steel and aluminium. It is suitable for welding low alloy steels, low c-steels and for welding those metals (for example, non-ferrous) that do not tend to absorb carbon. Such flames are also used for welding Monel metal and Nickel and in hard surfacing with high speed steel and cemented carbides.

(*c*) **Oxidising Flame.** This flame has limited use because the excess oxygen tends to combine with many metals to form hard, brittle, low strength oxides. Also, excess of oxygen causes the weld bead and the surrounding area to have a skummy or dirty appearance. So, this flame is not used for welding steel. It is mainly used when welding materials which are not oxidised readily and which have a high solubility of hydrogen in the molten state and low solubility in the solid state, for example, brasses, bronzes and gold. For these metals the oxidising atmosphere creates a base metal oxide that protects the base metal. For example, in welding brass, Zinc has a tendency to separate and fume away. The formation of a covering copper oxide prevents Zinc from dissipating.

Lighting up the Flame. For lighting up the torch for welding, the following steps should be followed :

1. Open both cylinder valves slowly.
2. Adjust the pressure regulators to the required working pressures.
3. Open the acetylene gas valve on the blow pipe and adjust the pressure regulating screw until the gauge reads correctly.
4. Close the acetylene gas valve on the blow pipe.
5. Repeat steps 3 and 4 for oxygen.
6. Turn on the acetylene gas valve on the blowpipe and allow the gas to flush the system.

7. Using a friction spark-lighter, ignite the gas. This will produce an acetylene flame.
8. Adjust the blowpipe valve until the flame just stops smoking and releasing soot.
9. Gradually open the oxygen control valve on blowpipe and adjust it until the required flame is obtained.

Closing down procedure. When closing down :

1. First turn off the acetylene blowpipe control valve.
2. Then turn off the oxygen blowpipe control valve.
3. Close the cylinder supply valves.
4. Purge each hose in turn by opening the control valves on the blow pipe, first for oxygen and then for acetylene.
5. Release the pressure on the two regulators.
6. Finally check that the blow pipe valves are closed.

Power of Blowpipe and Nozzles.

The power of a blowpipe is denoted in various ways :

(*i*) By the bore diameter of the nozzle. The gas consumption and hence the size of the gas flame will depend upon this dimension.

(*ii*) By the amount of gas flow in litres / hour. Acetylene consump-tion varies from 50 to 2800 litres per hour. Oxygen consumption varies from 50 to 3100 litres/hour.

(*iii*) By a number which represents the thickness of the plate that can be welded with this blowpipe.

It is thus clear that the power of the blow pipe determines the size of the flame produced and hence the heat available for welding. The power will be influenced by the velocity and the quantity of the gas mixture leaving the nozzle. Both these factors will depend upon the supply pressure of the gases and the size of the orifice of the nozzle.

The velocity of the gas mixture leaving the nozzle orifice ranges from 60 to 200 m/s. The lower velocity flame will burn quietly and is called 'soft flame'. The higher velocity flame burns with a hissing sound. In general, a flame of average velocity will give best results.

3.2.4. Oxy-Acetylene Welding Techniques. A sound welded joint can be obtained by the proper selection of the torch size, filler material, method of moving the torch along the weld and the angle at which the torch is held, as well as proper regulation of the welding flame.

As discussed above, the size of torch required will depend upon the thickness and heat conductivity of the metal being welded. Metals with higher conductivity will require a torch head with a larger gas consumption.

The joint is first heated with the torch until the welding pool is formed. A welding rod (if used) should be held in the flame so that its end melts at about the same time as the base metal and the filler metal is added to the pool.

In OAW, there are two techniques commonly used, called as : Leftward and Rightward techniques. The choice of either technique will depend upon the metal to be welded, its thickness, any test requirements, and total cost. To compare the two techniques, the blow pipe is held in the right hand and the filler rod in the left hand (with a right handed person). The filler rod is carried to the left of the blowpipe.

1. Leftward Welding Technique. In left ward (also called as forward or forehand) welding technique, the torch flame progresses from right to left, (Fig. 3.9 (*a*)). The angles of blow pipe and filter rod are shown in the figure. The method allows preheating of the plate edges immediately ahead of the molten pool and this is the method more commonly used. The blowpipe is given a very slight side to side movement and with the filler rod is moved progressively along the joint. With the blowpipe movement, the flame moves away from the just welded portion of the joint, which starts loosing heat and cooling starts soon. Due to this, this technique is restricted to welding of mild steel plates upto 5 mm thick, cast iron and non-ferrous metals.

2. Rightward Welding Technique. In the rightward (or back hand or backward) technique, welding commences at the left hand side of the plates and proceeds towards the right, (Fig. 3.9 (*b*)), The blow- pipe points in the direction of the completed weld with the inner cone of the flame directed towards the bottom of the joint, concentrating the maximum amount of heat into the plates. Due to this, the weld puddle is kept hot for a longer time and a narrow and deeper weld results. Hence, this technique is principally used for welding thick sections (over 5 mm thick).

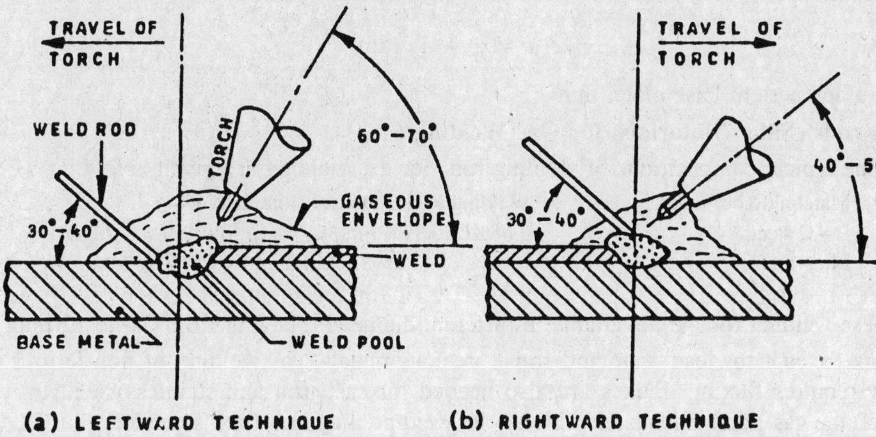

Fig. 3.9. OAW Techniques.

During welding, the blowpipe moves regularly along the weld seam without any lateral movement. On the other hand, the end of the welding wire describes a series of loops instead of moving steadily.

Chief advantages of Rightward Technique :

1. The rightward technique is faster by comparison with the leftward technique. This is because, in the left ward method, the view of the joint edge is interrupted and it is necessary to remove the end of the filler rod to inspect the progress. This action slows down the process. Also, the end of the filler rod becomes oxidised resulting in an unfavourable weld structure.
2. Plates upto 9 mm thick can be welded with square edge preparation, whereas with the leftward technique, plates over 3mm thick will have to have edges bevelled. For this reason, this technique is often limited to materials upto 5 mm thick.
3. The technique consumes less gas and filler material in comparison to left ward technique. On larger thicknesses, this technique requires no or little edge bevelling and therefore, less filler metal is required resulting in corresponding savings in gas and time.
4. The mechanical properties of the weld are better due to the annealing effect of the flame which is directed on the completed weld.
5. The included angle of edges is smaller in right ward technique as compared to leftward technique. The heat remains confined to the weld seam and there is less spread of flame. Due to this, the amount of distortion in the work is minimum.

Note:

(*i*) Back hand welding is faster by 20 to 25% and from 15 to 25% less acetylene is need in comparison to forehand welding.

(*ii*) Horizontal and overhead welding are usually done by the backhand technique.

(*iii*) Vertical joints are welded by the forehand technique.

(*iv*) The angle at which the torch is inclined to the surface being welded depends upon the thickness of the metal. Thicker metals require a higher concentration of heat and consequently a larger torch angle.

3.2.5. Filler Material and Fluxes. The welding wire or rod used as filler material in gas welding should have a chemical composition similar to that of the base metal. The diameter of the welding rod is selected to suit the thickness of the base metal. The welding rod diameter d can be approximately determined by the following empirical relation :

$$d = \frac{t}{2} + 1, \text{ mm}$$

where t = thickness of base plate, mm

Welding rods (Filler Materials) for Gas Welding :

Some typical compositions of welding rods for gas welding are given below :

	Material to be Welded	Welding Rod Chemical Composition
1.	L – C steels	0.08%C, 0.36% Mn, 0.13% Cr, 0.013% Ni, 0.20%P
2.	Mn – Steels	0.14% C, 0.12% Si, 0.81% Mn, 0.25% Ni
3.	Cr – Steel	0.24%C, 0.21% Si, 0.42% Mn, 0.96% Cr, 0.17% Ni, 0.35% S

These welding rods are available in different diameters ranging from 0.3 to 12 mm.

Flux. Except for lead, zinc and some precious metals, gas welding of non-ferrous metals generally requires fluxing. Fluxes are also needed for cast iron and stainless steel. In welding mild steel, the gas flame adequately shields the weld pool and no flux is needed. For C.I., fluxes are required to increase the fluidity of the fusible iron silicate slag, as well as to aid in the removal of slag.

Fluxes compose of borates or boric acid, soda ash and small amounts of other compounds such as sodium chloride, ammonium sulphate and iron oxide. Mixture of equal amounts of boric acid and soda ash, 2% aluminium sulphate and 15% powdered iron makes a satisfactory flux.

Gas welding fluxes must melt at a lower temperature than the metals being welded so that surface oxides will be dissolved before the metal melts.

3.2.6. Advantages and Disadvantages of OAW.

Advantages :

1. The equipment is low cost, versatile, self-sufficient and usually portable. It requires little maintenance, and can be used with equal facility in the field and in the factory. The oxy-acetylene can be used for welding, brazing, soldering, preheating, postheating and metal cutting etc.
2. It can weld most common materials.
3. The gas flame temperature is lower and easily controllable which is necessary for delicate work. Therefore OAW is extensively used for sheet metal fabrication and repairs.
4. The process is well adapted for short production runs.

For the above reasons, gas welding is used in :

Automotive and aircraft industry

Sheet metal fabrication plants, and in fabrication of industrial pipes.

OAW is best suited for : joining thin sheet metal, thin small tube, small pipe and assemblies with poor fit up and for repairing rough arc welds.

Disadvantages :

1. Oxygen and acetylene gases are expensive.
2. There are safety problems involved in their handling and storing.
3. The flame takes considerably longer for the metal to heat up. Due to this, OAW is not suitable for thick sections.
4. Because the flame is not concentrated, considerable areas of the metal are heated and distortion is likely to occur.
5. Flux applications and shielding provided by OA flame are not so effective as in inert gas arc welding.

Metals unsuited for welding with OAW torch are : refractory metals (Columbium, Tantalum, Molybdenum, Tungsten) and the reacting metals such as Titanium and Zirconium.

3.2.7. Other Fuel Gases. As already noted, the other fuel gases which can be used in gas welding are : propane, butane, natural gas or hydrogen. But because of the lower temperature of the flame (about 2500°C) obtained with these gases, these are used particularly for welding lower melting point alloys such as aluminium, zinc, lead and some precious metals.

MAPP Gas. Methyl acetylene propadiene gas is replacing acetylene gas particularly when portability is important, because :

1. It is more dense, thus providing more energy for a given volume.
2. It can be stored safely in ordinary pressure tanks.

3.2.8. Pressure Gas Welding (PGW). In this process, the abutting surfaces of the parts being welded are heated by an Oxy- acetylene flame to a state of fusion or plasticity and then coalescence is produced by the application of pressure and without the use of a filler material. The method is widely employed in butt welding bars, pipes, tubes, railroad rails, tools and rings of low and medium carbon steels and of low and medium alloy steels.

Two methods are used commercially for PGW :

1. Closed - joint method. In closed joint pressure gas welding, clean square surfaces are butted together under moderate pressure. The surfaces are then heated by a water cooled oxyacetylene torch, Fig. 3.10, until the correct temperature is attained. Then an additional

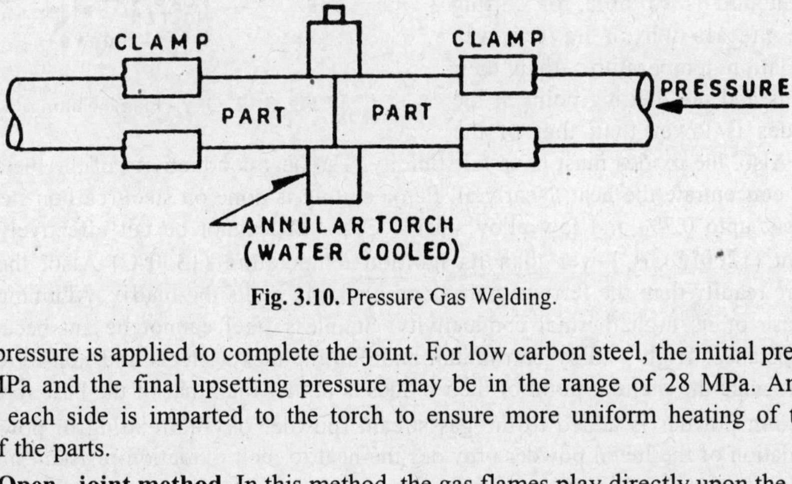

Fig. 3.10. Pressure Gas Welding.

upsetting pressure is applied to complete the joint. For low carbon steel, the initial pressure is less than 10 MPa and the final upsetting pressure may be in the range of 28 MPa. An oscillating motion to each side is imparted to the torch to ensure more uniform heating of the abutting surfaces of the parts.

2. Open - joint method. In this method, the gas flames play directly upon the square weld joint faces, which have been spaced a short distance apart. When the ends of the joint have reached the fusion temperature, they are brought rapidly in contact under pressure to effect welding under upsetting.

Both methods are generally used in partially or fully mechanised set ups.

3.2.9. Oxy-Acetylene Flame Cutting. Oxy gas cutting is based upon the ability of certain metals to burn in oxygen with the evolution of a great deal of heat thereby melting the metal and forming oxides. The torch for flame cutting is similar to the welding torch, with two exceptions. First, the welding tip contains only one hole in the centre of the tip through which the mixture of C_2H_2 and O_2 gases flow, whereas the flame cutting tip contains a centre hole through which pure oxygen, which does the actual "cutting" flows. There are also several concentric holes around the centre hole through which mixture of C_2H_2 and O_2 flows and the flame produced by its burning preheats the metal, (Fig. 3.11). Second, the cutting torch has an additional, or third, valve for controlling the flow of pure oxygen.

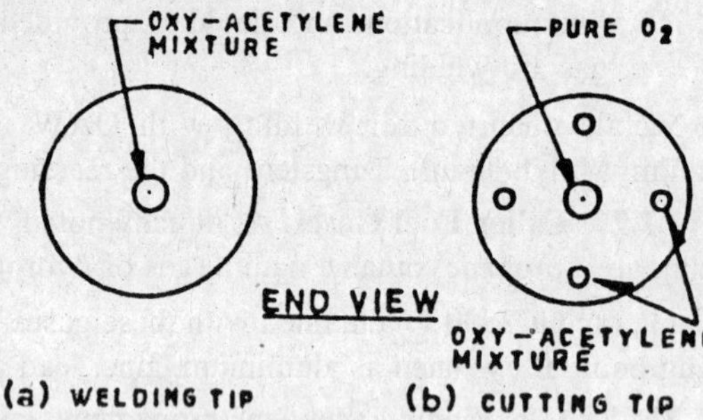

Fig. 3.11. Welding and Cutting tip.

Flame cutting is done both manually and with motor driven heads. As the metal is burnt and eroded away, the torch is moved steadily, (Fig. 3.12), along the path of cut. A uniformly wide slot called a 'kerf' is cut by the jet of oxygen. The faster the rate of traverse, the more the bottom lags behind the top of the cut. This is known as 'drag' and must be kept small. Thicknesses upto 1.5 m can be cut.

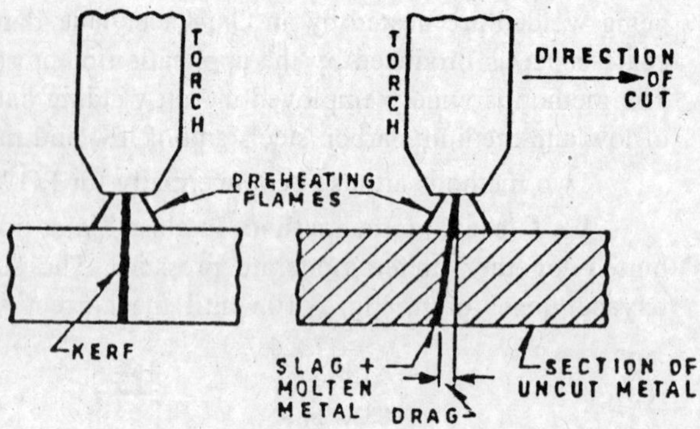

Fig. 3.12. Oxy - Flame Cutting.

This method is suitable for cutting only those metals which have lower ignition/oxidation temperatures than their melting points and the melting point of the formed oxides is lower than that of the metal itself. Also, the oxides must have fair fluidity. The heat conductivity of the metal must be low so as to concentrate the heat. Nearly all flame cutting is done on steel (carbon steels with a carbon content upto 0.7% and low alloy steels). Cast iron cannot be cut effectively since its melting point (1200°C) is lower than its ignition temperature (1350°C) Also, the graphite oxidizes more readily than the ferrous matrix and it simply melts the matrix. Aluminium cannot be cut because of its high thermal conductivity. Stainless steel cannot be cut because of its oxidation resistance. High – alloy chro-mium and chrome-nickel steels and non-ferrous alloys cannot be cut since the melting point of their oxides is higher than that of the base metals.

When iron powder is added to the gas stream (powder oxyfuel cutting or powder metal cutting), oxidation of the metal powder provides the heat to melt oxidation-resistant materials.

Underwater Cutting. Techniques have been developed for cutting metal underwater in ship building and repair work, construction and repair requirements associated with offshore exploration, drilling and recovery of oil and natural gas.

A specially designed torch is employed for this purpose. An auxiliary skirt surrounds the main tip of the torch. Compressed air is supplied through the passages in the skirt. The compressed air performs two functions : it expels the water away from the tip area and it provides secondary oxygen and thereby stablises the flame.

For depths upto 7.5 m, oxy-acetylene torch is used but for greater depths, oxy-hydrogen torch is employed, because at such depths, C_2H_2 will have to be used at higher pressure to neutralise the high surrounding pressure created by the depth of water. The use of C_2H_2 at high pressure is very unsafe. But H_2 can be compressed to a higher pressure without any danger.

The torch is either ignited in the conventional manner before it is taken underwater or is ignited by an electric spark device after it is submerged.

3.3. ELECTRIC ARC WELDING

Electric are welding is the most extensively used method of joining components of metallic parts, the source of heat being an electric arc. An electric arc is a continuous stream of electrons flowing through some sort of medium between two conductors of an electric circuit and accompanied by intense heat generation and radiation.

An electric arc for welding is obtained in the following ways :

1. Between a consumable electrode (which also supplies filler metal) and the workpiece.
2. Between a non consumable electrode (carbon, graphite or tungsten etc.) and the workpiece.
3. Between two nonconsumable electrodes.

The most common electric arc welding method is the one in which the arc is struck between an electrode and the work. This is called as 'Direct arc'. The arc struck between two non consumable electrodes adjacent to the parts being welded is called as "Independent or Indirect arc". The metal is heated by the indirect action (by radiation) of the arc. Due to this, the thermal efficiency of the method is poor. Thus it is very rarely used these days. The method is called as "Double or Twin arc method". Atomic hydrogen welding process belongs to this category and is used in a few special applications.

To strike an arc, the electrode is brought in contact with the work at the point where the welding is to be started, after connecting the work to the welding circuit. After a light contact, the electrode is immediately withdrawn to a distance of from 2 to 4 mm from the work. Only a comparatively low potential difference is required between the electrode and the work to strike an arc. From 40 to 45 V is usually sufficient for D.C. and from 50 to 60V for A.C. This voltage available at the output terminals of a welding set, before the arc is struck, is known as open circuit voltage (OCV). The voltage falls after the arc is established which is normally less than half the OCV. A stable arc can be maintained between a metal electrode and the work metal with a voltage of 15 to 30V while from 30 to 35V is needed to strike an arc between non consumable electrode and the work.

The stable arc required for high quality welding can be achieved with an arc length equal to 0.6 to 0.8 of the electrode diameter. The arc length is defined as the distance between the end of the electrode and the surface of the molten metal on the work.

When the electrode first makes contact with the job, a large short circuit current flows. When the electrode later is immediately withdrawn, the current continues to flow in the form of spark across the air gap so formed. Due to this, the air gap gets ionized, that is, splits into electrons and positive ions. The lighter electrons flow from cathode to anode and the heavier positive ions flow from anode to cathode. Thus, the air gap becomes conducting and current is able to flow across the gap in the form of an arc.

When the lighter, high-velocity electrons strike the anode at great velocity, intense heat is generated at the anode. Heat generated at the cathode is much less, because of the low velocity of impinging positive ions.

Thermal and luminous energy is not uniformly evolved in the welding arc. About 43 per cent of the total amount of heat is evolved on the anode and about 36 per cent on the cathode. The remaining 21 per cent is evolved by the arc.

The temperature of an electric arc depends upon the type of electrodes between which it is struck. It is about 3200°C on the cathode and about 3900°C on the anode for carbon electrodes and 2400°C and 2600°C respectively for metal electrodes. The temperature may reach 6000 to 7000°C in the centre of the arc.

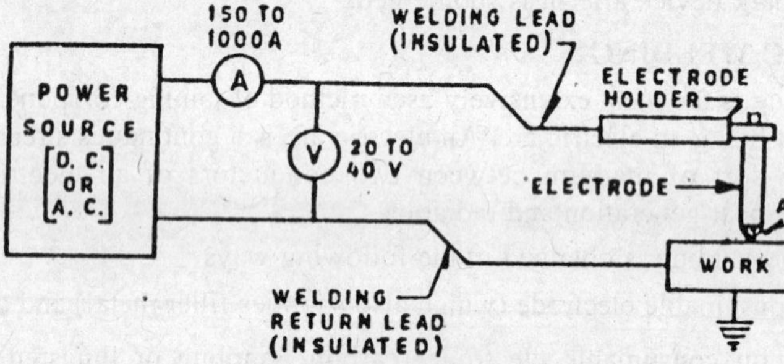

Fig. 3.13. Electric Arc Welding.

Only from 60 to 70% of the heat is utilised in arc welding to heat up and melt the metal. The remaining 30 to 40% is dissipated into the surroundings.

Definition. Electric arc welding is "a welding process wherein coalescence is produced by heating with an arc or arcs, with or without the application of pressure and with or without the use of filler metals."

Electric arc welding is quite versatile and able to weld under many conditions. High quality welds are produced. Metal is deposited rapidly and it is competitive cost wise for many situations.

Fig. 3.13 illustrates an Electric Arc Welding circuit.

3.3.1. Power Sources. Both D.C. and A.C. equipment are employed in arc welding. D.C. arc welding machines are D.C. generators driven by an electric motor or an internal combustion engine. A.C. welding machines are transformers for stepping down the main supply voltage (400/440V) to the voltage (20 to 40V) and current (150 to 1000 A) suitable for arc welding. The two power sources are discussed below :

(*a*) **D.C. Power Source.**

Advantages :

1. It is known to most operators.
2. It can handle all situations and jobs.
3. It is preferred for difficult task like over head welding.
4. Arc stability is much higher with D.C. than with A.C. The reason is that the arc gap cools when an A.C. arc is periodically extinguished and re-established.
5. One advantage of D.C. welding over A.C. welding that offsets its higher cost is the fact that in D.C. supply, the electrons flow in one direction only. Direction of flow of electrons can be changed by simply reversing the cables at the terminals located on

the generator (from electrode to work and vice-versa). This fact can be utilized for obtaining desired penetration of the base metal.

The electrons move from the negative terminal to the positive terminal of an arc, striking it at great speed. Due to this about $\frac{5}{8}$ th of the arc energy goes to the anode surface and about $\frac{3}{8}$ th is liberated at the cathode surface.

DCSP. In D.C. Straight Polarity, work is positive (anode) and elec-trode is cathode, (Fig. 3.14 (*a*)). Therefore, the work material is heated much faster than the electrode. This is of advantage in welding massive pieces because it puts the heat where needed. This system is

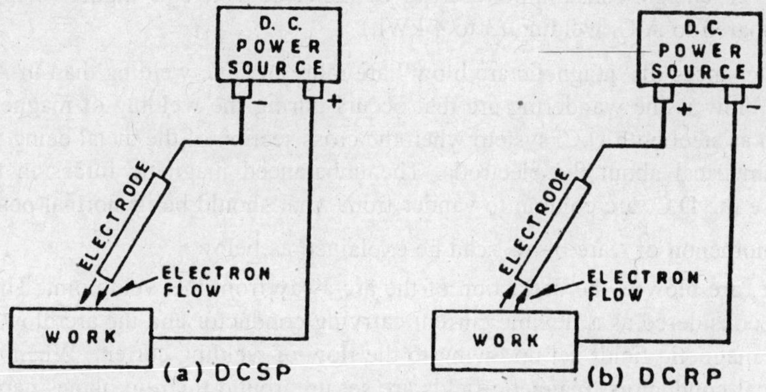

Fig. 3.14. Polarity of D.C. supply.

preferred on all metals except aluminium, magnesium, copper and beryllium. Its advantages are : deeper penetration, Fig. 3.15 (*a*), the narrowest heat-affected zone, the fastest travel speed, less electrode consumption and least distortion because of the faster heat input into the work. Also, we can use bare and medium coated electrodes as they require less amount of heat for melting.

DCRP. In D.C. reverse polarity electrode is positive and the work is negative terminal, (Fig. 3.14 (*b*)). The electrons flow from the workpiece which tends to clean it. But the heat flow

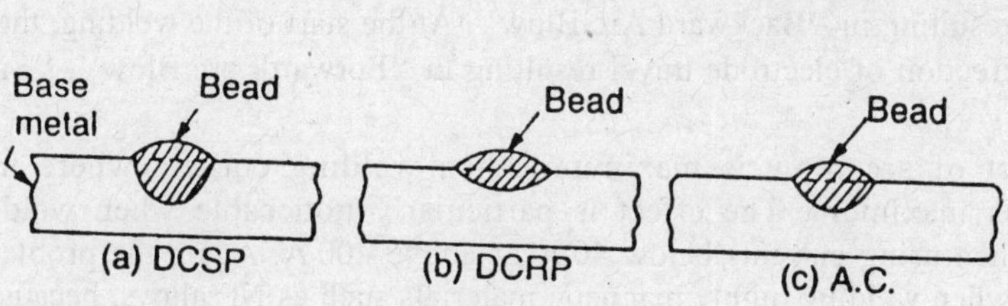

Fig. 3.15. Arc Penetration.

is also towards the electrode making its consumption faster and so limiting the current that can be put through an electrode. This system is preferred in welding thin sections such as automobile bodies and non-ferrous metals to prevent the melting or burning of holes in the metal, since only about 38% of the arc energy is liberated at the work surface in this case. Because this system has an inherent ability to scour the oxide film from the surface of the work, it is very useful for welding aluminium and magnesium whose oxides are very hard and brittle. A wide bead (due to shallow penetration) and cleaner work surface are the characteristics of this system, Fig. 3.15 (*b*).

In A.C. welding, since the polarity is changing at the rate of 50 c/s, half of the arc energy is liberated at the work surface and half at the electrode. Thus, the penetration of the heat zone in the base metal is approximately midway between the two D.C. types, [Fig. 3.15 (*c*)].

6. Higher power factor of the system (0.6 to 0.7) as compared to A.C. welding (0.3 or 0.4).
7. Bare electrodes are also suitable for welding.

Disadvantages :

1. The efficiency of the machine is low (from 0.3 to 0.6).
2. Cost is higher. The equipment is heavier in weight, larger in size and operation and maintenance is difficult because of moving parts.
3. Electric energy consumption per kg of metal deposited is higher (6 to 10 kWh) as compared to A.C. welding (3 to 4 kWh).
4. The chances of "magnetic arc blow" are more in D.C. welding than in A.C. welding. Arc blow is the wandering arc that occurs during the welding of magnetic materials, such as steel with D.C. system when the cross section of the metal being welded is not symmetrical about the electrode. The unbalanced magnetic forces in this situation cause the D.C. arc column to vander from what should be its normal position.

This phenomenon of "arc blow" can be explained as below :

As noted above, arc blow is the deflection of the arc away from the weld point. The electric arc column can be considered as a flexible current carrying conductor and the arc blow is its deflection due to the magnetic fields set up owing to the flow of welding current. When current flows through electrical conductors, magnetic fields are set up around them in planes perpendicular to the direction of flow of current. When the magnetic flux lines surrounding the electrode or the workpiece become unbalanced, they tend to bend the arc away from its intended path resulting in "Arc blow". This effect is more pronounced at the two ends of the workpiece (that is beginning and end of the weld) or when welding corners. When the electrode moves forward along the direction of welding, the magnetic flux lines surrounding it also move with it. When the electrode reaches the end of the workpiece, the flux lines move out of the base metal, which is not possible. This results in concentration of flux lines behind the electrode (opposite to the direction of electrode travel) and hence the unbalanced magnetic field. This causes high magnetic pull on the arc column resulting in "Backward Arc Blow". At the start of the welding, the magnetic pull will be in the direction of electrode travel resulting in "Forward Arc Blow". "Side Arc Blow" occurs rarely.

The effect of arc blow is maximum when welding corners where magnetic field concentration is maximum. The effect is particularly noticeable when welding with bare electrodes or when using currents below 40 A or above 200 A. Again the problem of arc blow gets magnified when welding highly magnetic materials such as Ni alloys, because of the strong magnetic fields set up by these metals.

As noted above, the phenomenon of arc blow occurs mostly with D.C. welding, because of unidirectional flow of current resulting in induced magnetic fields which are constant in direction. With A.C. welding sets, where the current reverses cyclically, the induced magnetic fields cancel each other and hence there are no chances of Arc Blow.

Arc blow results in :

Low heat penetration in the weld zone leading to incomplete fusion and to bead porosity. Also, there is excessive weld spatter, that is, throwing out of tiny droplets of weld metal out of the point on to the base metal resulting in tiny dots on base metal plates.

The effects of arc blow can be minimized with D.C. sets by :

1. Shortening the arc column length.
2. Reducing the current used which will reduce the strength of the induced magnetic field.

3. Reducing the rate of travel of the electrode.

4. Putting steel blocks near the ends of the workpiece so that the flux lines flow through them and not through the air.

5. Placing one ground lead at each end of the workpiece.

6. Wrapping the electrode cable a few turns around the workpiece.

(*b*) **A.C. Power Source**

Advantages :

1. Equipment for A.C. welding is considerably less expensive, light in weight, smaller in size and simpler in operation due to the absence of moving parts. This also results in lower maintenance and operating costs.
2. Little "magnetic arc blow".
3. High efficiency of A.C. welding transformers (0.8 to 0.85).
4. Less electric energy consumption per kg of metal deposited as compared to D.C. welding.

Disadvantages :

1. Low power factor.
2. Unstable arc. The stability of the arc can be increased by raising the voltage or by increasing the frequency of the alternating current.
3. Only coated electrodes are suitable for welding. In case of A.C. supply, the arc cools at zero current position. This tends to deionize the arc path. However, shielding gas keeps the arc space ionized and, therefore, with A.C., coated electrodes (which contain ionizing agents) are must. Bare electrodes without the use of shielding gas should not be used with A.C. supply. However, bare electrodes can be used satisfactorily with D.C. supply only.

3.3.2. Arc - Welding Electrodes. As noted above, the electrodes used for arc welding are of two types : Non-consumable and Consumable.

Non consumable electrodes. These electrodes are made of carbon, Graphite or Tungsten. Carbon and Graphite electrodes are used for D.C. welding only. Tungsten electrodes are used for D.C. as well as A.C. welding. When non-consumable electrodes are used, the filler material is added separately. Since the electrode is virtually not consumed, the arc length remains constant, so that it is stable and easy to maintain.

Consumable electrodes. The consumable electrode provides the filler material also. These may be made of various metals, but should have the same composition as the material to be welded. During welding, they melt and supply the filler material. Therefore the arc length and the resistance of the arc path vary, as a result of the change in arc length and the presence or absence of metal particles in the path. This requires that the electrode be moved toward or away from the work to maintain the arc and satisfactory welding conditions.

The consumable electrodes are of three kinds : (*a*) Bare electrodes (*b*) Fluxed or lightly coated electrodes (*c*) Coated or extruded / shielded electrodes.

Bare electrodes may be used to weld wrought iron and mild steel. They must be used only with straight polarity. Bare electrodes in the form of sticks or rods are used for hand arc welding. For automatic welding, bare electrode is in the form of continuous wire (coil).

Lightly coated electrodes have a coating layer several tenths of a millimetre thick, the weight of the coating is from 1 to 5 per cent of the electrode weight. The primary purpose of a light coating is to increase arc stability and are also called as ionising coatings. This type of coating contains chalk (80 to 85 parts by weight) and water glass (15 to 20 parts by weight).

These electrodes are used in welding only noncritical structures. The welds produced have poor mechanical properties due to the lack of the protection of the molten metal.

Coated electrodes are covered with a relatively high quality covering applied in a layer of from 1 to 3 mm. The weight of such a coating is from 15 to 30% of the electrode rod. The greatest amount of welding is done with coated electrodes. Since the electrode coating is brittle, only straight stick electrodes can be used. This restricts the process to a slow manual operation. If the flux coating is placed inside a long tube, the electrode can be in the form of a bare wire in the form of a coil and then the shielded arc process can be made continuous and automatic.

The stick electrodes are available in the following sizes : 3.2, 4, 5, 6, 8, 9 and 12 mm diameter and lengths of 350 or 450 mm.

Functions of Coatings :

1. Improves arc stability by providing certain chemicals which have this ability, by ionizing the path of arc.

2. Provides a protective gaseous atmosphere to prevent O_2, H_2 and N_2 pick-up by the molten metal.

3. Provides a protective slag over hot metal.

4. Provides flux, which helps to remove oxides and other impurities from the molten weld metal.

5. Reduces spatter of weld metal – when coating burns off slower than core.

6. Acts as deoxidizer.

7. Adds alloying elements.

8. Increases deposition efficiency.

9. Slows down the cooling rate of weld (due to protective layer of slag) to prevent hardening.

10. Coatings are normally insulators of electricity and so permit the use of electrodes in narrow grooves etc.

Electrode Coatings :

1. Slag forming ingredients, asbestos, mica, silica, fluorspar, titanium dioxide, Iron oxide, magnesium carbonate, Calcium carbonate and aluminium oxide.

2. Arc stablising ingredients, or ionizing agents : potassium silicate, TiO_2 + ZrO_2 (Rutile), Mica, Calcium oxide, sodium oxide, magnesium oxide, feldspar ($KAlSi_3O_8$).

The major types of feldspars are :-

(*i*) Potash feldspar, $K_2O \cdot Al_2O_3 \cdot 6SiO_2$

(*ii*) Sodium feldspar, $Na_2O \cdot Al_2O_3 \cdot 6SiO_2$

(*iii*) Lime feldspar, $CaO \cdot Al_2O_3 \cdot 6SiO_2$

3. Deoxidizing ingredients. Cellulose, Calcium carbonate, dolo-mite, starch, dextrin, wood flour, graphite, aluminium, ferromanganese.

4. Binding materials. Sodium silicate, potassium silicate, asbestos.

5. Alloying constituents to improve strength of weld.

6. TiO_2 and potassium compounds increase the melting rate of the base metal for better penetration.

7. Iron powder provides higher deposition rate.

Types of Electrode Coatings :–

There are four main types of coatings which are most commonly used in practice :–

1. Cellulosic coating
2. Rutile coating
3. Iron oxide coating
4. Basic or low H_2 coating.

The first two coatings are almost identical but with one difference. The percentage of Titanium dioxide, TiO_2 (rutile) is somewhat more in rutile coating than in cellulosic coating. When the cellulose burns, it results in the evolution of a mixture of gases. H_2 and CO. These gases provide a protective shield to the molten base metal.

With the same thickness of the work, a cellulosic coating results in deeper penetration as compared to rutile coating. Since the rate of evolution of H_2 gas is high, there is a danger of hydrogen embrittlement of the base metal, with these two coatings.

In iron oxide coating, the evolution of H_2 gas is less as compared to the first two coatings. Therefore, protection to the molten base metal is mainly provided by slag.

In basic coating, the major constituents are calcium carbonate and fluospar (CaF_2). When this coating burns, it results in the evolution of a mixture of CO and CO_2 gases. This mixture provides protection to the molten base metal. This coating has certain adventages over the other coatings : There is no danger of hydrogen embrittlement of the welded joint. The mixture (CO + CO_2) is less reducing than H_2 gas. So, a shorter arc length is maintained for complete shielding of molten base metal.

The thickness of the coating will depend upon the size of the job. Thicker coating results in narrower and deeper penetration.

It should be noted that with A.C. supply, the arc goes off after every half cycle when cycle crosses the zero current line. No such thing happens with D.C. supply. So, with A.C. supply, the arc must reignite after every half cycle. The reignition voltage is higher than the normal arc voltage. However, the process of reignition is facilitated by the presence of ions of low ionization potential, in the gap. Now, pottasium has low ionization potential (4.3V) as compared to sodium (5.1V). Hence, with A.C. supply, the electrode coating has pottasium silicate as the binder, and with D.C. supply, it is sodium silicate.

Coding of Electrodes. There is no standard code designation used to classify the bare electrodes. Although, the composition of the electrode is basically the same as the base metal, small amounts of special additives, such as deoxidizers, are included by many manufacturers to improve the weld quality.

There is a code designation for coated stick electrodes for metal arc welding, for selecting a correct electrode for a specific application IS : 815 -1974 :

The coding for an electrode shall consist of a prefix letter, a code number of six digits in numerals and in some cases one or more suffix letters, each indicating a specific property or characteristic or both of the electrode.

1. Prefix letter indicates the method of manufacturing, of code.

Prefix letter	*Method of manufacturing*
E	Solid extrusion
R	Extruded with reinforcement

2. **First digit.** The first digit of the code indicates the type of covering.

Digit	*Type of covering*
1	Having a high cellulose content
2	Having a high content of titania and producing a fairly viscous slag.
3	contains appreciable amount of titania and producing a fluid slag.
4	contains a high content of oxides and / or silicates or iron and manganese and producing an inflated slag.
5	contains a high content of iron oxides and/or silicates and produce a heavy solid slag.
6	having a high content of calcium carbonate and fluoride.
7	A covering of any other type not classified above.

3. **Second digit.** The second digit of code number indicates the welding position or positions in which the electrode may be used.

Digit	*Welding Positions*
0	F, H, V, D, O
1	F, H, V, O
2	F, H
3	F
4	F, H_f (H_f – horizontal fillet)
9	Any other welding position not classified above.

where F stands for flat

H stands for horizontal-vertical

V stands for vertical-up

D stands for vertical-down

O stands for over-head

4. **Third Digit.** The third digit of the code number indicates the welding current conditions recommended by the manufacturer of the electrode, as follows :

Digit	*Welding Current condition*
0	D +
1	D +, A90
2	D –, A70
3	D –, A50
4	D +, A70
5	D ±, A90
6	D ±, A70
7	D ±, A50
9	Any other current conditions not classified above

The notations are as follows :

D + DC with electrode positive

D – DC with electrode negative

D ± DC with electrode positive or negative

A90 AC with an open circuit voltage not less than 90V

A70 AC with an open circuit voltage not less than 70V

A50 AC with an open circuit voltage not less than 50V

Note 1 : The electrodes may not function satisfactorily at a lower voltage than that for which it is classified but a higher voltage may be used in service with advantage.

2 : The open circuit voltage necessary for striking the arc varies according to the core diameter of the electrode. The reference core wire diameter for coding is 4 or 5 mm. Within the range of sizes 2.5 mm to 8 mm, the open-circuit voltage necessary may be expected to vary approximately as follows :

Code voltage	*Variation in voltage*
90	100 to 70
70	80 to 55
50	60 to 40

The higher voltages are associated with 2.5 mm electrodes and the lower voltages with 8 mm electrodes.

5. Fourth and Fifth Digits. Fourth and fifth digits indicate mechanical properties such as tensile strength and the appropriate yield stress. The range of tensile strength 410 to 510 N/mm^2 is indicated by the digit 41 and the digit 51 indicates the range 510 to 610 N/mm^2. The minimum yield stress for the strength range of 410 to 510 N/mm^2 is 330 N/mm^2 and for the strength range of 510 to 610 N/mm^2, it is 360 N/mm^2.

6. Sixth digit. The sixth digit of code number indicates the percentage elongation in combination with impact value of the deposited metal.

Digit	*Strength code number*	*Minimum elongation in % of gauge value*	*Temp. for min. impact value of 47 J, °C*
0	41	–	–
1	41	20	+ 27
2	41	22	0
3	41	24	– 20
4	41	24	– 30
5	41	24	– 40
0	51	–	0
1	51	18	+ 27
2	51	18	0
3	51	20	– 20
4	51	20	– 30
5	51	20	– 40

7. Suffix Letters. The suffix letters indicate the special properties / characteristics or both of the electrode, as following :

Suffix Letters	*Special properties / characteristics*
H	Hydrogen controlled electrodes
J	Iron powder covering, giving a metal recovery of 110 to 130%.
K	Iron powder covering, giving a metal recovery of 130 to 150%.
L	Iron powder covering, giving a metal recovery of over 150%.
P	Deep penetration

A typical electrode coding will be written as follows :

D 426 - 513 H

Selection of Electrodes :

The selection of a proper electrode for electric are welding is very important to obtain a sound welded joint of proper strength. The various factors affecting the selection of electrodes have been discussed above. It is clear that these factors are :

1. Power Source : A.C. or D.C.
2. Composition of the base metal.
3. Thickness of base metal.
4. Welding position.
5. Welding current conditions recommended by the manufacturer of the electrodes.
6. Mechanical properties desired in the joint.
7. Extent of penetration required in welding and lastly.
8. Economic considerations.

The diameter of the electrode, current and voltage requirements are basically related to the thickness of the metal to be joined. However, the Table 3.1. below is a good guide for selecting a proper electrode (Also see Art 3.3.3 for SMAW) :

Table 3.1 : Selection of Electrode

Metal Thickness mm	*Electrode Diameter, mm*	*Electrode diameter (SWG)*	*Current range, A*	*Voltage across arc, V*
Upto 1.5	2.0	14	50 – 70	15
1.5 – 3	2.5	12	70 – 100	15
3 – 6	3.15	10	85 – 120	20
6 – 10	4.0	8	140 – 180	20
10 – 20	5.0	6	180 – 230	25
20 – 35	6.3	4	230 – 290	30

3.3.3. Types of Electric Arc Welding processes. The common electric arc welding processes are :

1. Carbon Arc Welding
2. Shielded Metal Arc Welding
3. Flux Cored Arc Welding
4. Gas Metal Arc Welding
5. Gas Tungsten -Arc welding
6. Submerged Arc Welding
7. Atomic Hydrogen Welding
8. Plasma Arc welding, and
9. Stud welding
10. Electro Slag welding.

Since most metals are readily oxidized on heating to the temperature required by welding, the molten weld metal, the adjacent heat affected zone in the base metal, and the welding zone must be protected or shielded from the atmosphere. This shielding serves to exclude oxygen and nitrogen from the air and, in this manner, eliminates the formation of metal oxides and nitrides which lower weld metal ductility and toughness and sometimes strength.

Shielding may be accomplished by :

1. a covering applied to the electrode rod or wire.
2. a powdered or granular flux which covers the electric arc, the molten weld metal and the adjacent areas in the base metal.
3. an active gas (generally hydrogen) projected around the arc and over the heated weld-and-base metal area, and
4. an inert gas (generally helium, argon and carbon dioxide) projected around the arc over the heated weld-and-base metal area.

Sometimes a suitable flux solution, emulsion, or paste may also be painted over those parts of the base metal which will be heated by the welding operations, such as the underside of the weld joints.

1. Carbon Arc Welding. In this, Coalescence is produced by heating with an electric arc between a carbon electrode and the work. Shielding is generally not used. Pressure is not used, and filler metal may or may not be used. The electric arc can also be struck by the "twin arc method", that is, between two carbon (graphite) electrodes. Filler metal, when used, is fed into the arc and allows a fairly high rate of weld-metal deposition. Sometimes a filler rod is placed into the joint groove, and the carbon arc is passed slowly along the joint until fusion is completed.

The weld metal is not shielded from contamination of oxygen and nitrogen in the atmosphere. Moreover, very little, if any, carbon is picked up by the weld from the carbon electrode. Thus, this process is generally limited to those materials which are not sufficiently contaminated by these elements, that is, copper alloys, brass, bronze, aluminium alloys etc. It will be better if the filler metal (if used) incorporates deoxidizer (Silicon, phosphorous etc.).

2. Shielded Metal Arc Welding, (SMAW). The method is also known as manual Metal Arc welding, MMAW. This is most widely used arc welding method. It is defined as : "an arc welding process wherein coalescence is produced by heating with an arc between a covered metal electrode and the work. Shielding is obtained from decomposition of the electrode coating. Pressure is not used and filler metal is obtained from the electrode". The chemically coated electrode decomposes and vaporizes with the heat of the electric arc. The ingredients in the vaporised coating creates a protective gas atmosphere over the weld puddle and also, the melting of the coating provides a molten slag cover on the weld. This prevents the molten metal from reoxidizing. The coating performs other functions also as discussed earlier. The coating on the electrode melts at a slower rate than the metal core wire since it melts at a higher temperature than the metallic core, leaving a slight extension of the coating beyond the core wire. This produces a cup shaped end, (Fig. 3.16), which helps to direct and concentrate the heat from the arc.

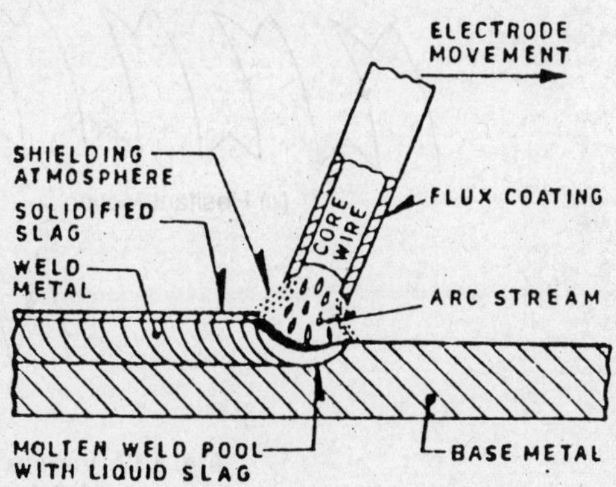

Fig. 3.16. Shielded Metal Arc Welding.

The process is versatile and suitable for field application, but requires considerable skill. Welding in all positions, including overhead welding is possible if the metal and slag solidify fast enough. The method is used almost exclusively for job-shop (nonproduction and general maintenance and repair) work, and also in many high production operations, for example, in the manufacture of machinery, transportation equipment, piping systems and in various structures (on site erection) such as buildings, trusses, machine bases etc.

After welding, the slag or molten flux must be removed to avoid possible flux corrosion.

As already noted, the coated stick electrodes are suitable only for manual arc welding. The method is also known as "Stick Welding".

After the arc is established (as explained in Art. 3.3), the welder feeds the electrode uniformly to the work, maintaining a short-arc. The electrode is held at an angle of 15° to 20° from the vertical, on a flat weld, to produce a weld bead. By changing the angle of inclination of the electrode, the welder can vary the depth to which the parent metal is melted and the cooling rate of the welding pool.

During the process of welding, the electrode is given three movements. The electrode is continuously fed downward along its axis to maintain the arc length. It is progressively fed along the weld and thirdly the electrode tip is given an osillating movement across the weld. The side ways oscillating movement of the electrode tip is given to :-

(*i*) obtain and maintain proper bead width.

(*ii*) Float out slag.

(*iii*) Secure good pentration at the edges of the weld.

(*iv*) Allow gases to escape and thereby avoid porosities.

Transverse oscillating movements of the electrode tip also improve the heating of the joint edges, retard cooling of the molten pool of deposited metal, enable a more homogeneous, sound efficient and strong weld to be obtained and eliminate inadequate joint peretration.

Fig. 3.17. shows the various oscillating motions of the electrode tip.

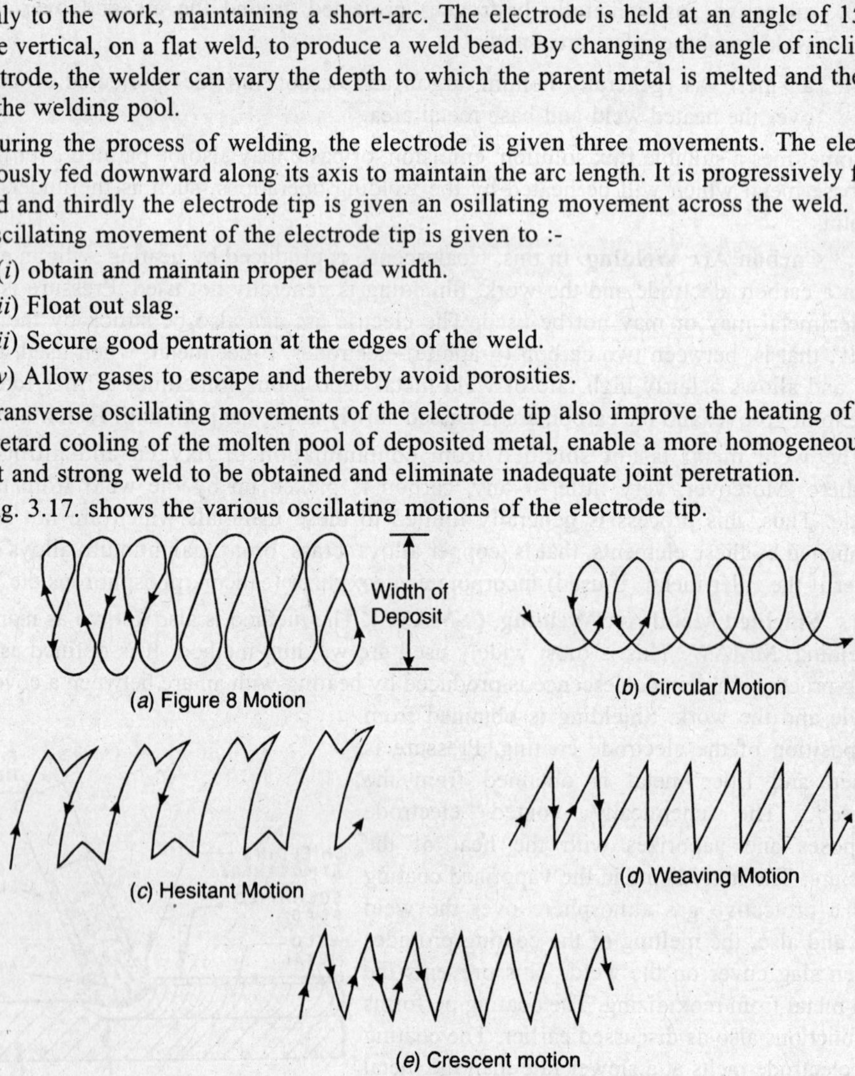

Fig. 3.17. Various oscillating motions of electrode tip

The type of oscillating motion and the rate of oscillations depends upon : thickness of the job, the length of run and the type of electrode and level of current employed. However, these are mainly decided on the basis of experience.

In multiple-pass welding, Fig. 3.18., the brittle slag coating present on the bead after the first pass (root pass), is chipped off and cleaned by wire brush, before the second pass is started and similarly for the remaining passes.

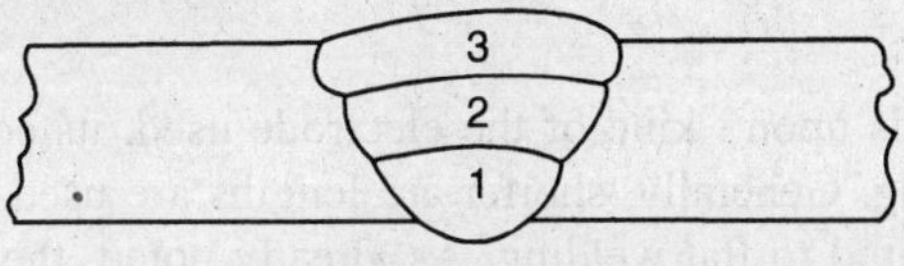

Multiple - pass (3) weld

Fig. 3.18. Multiple-pass Arc Welding

In metal arc welding, the thermal action of the arc melts the end of the electrode and the edges of the work. It has been established that the drops of the molten metal pass along the arc at a velocity of 40 m/s and that the number of drops varies from 10 to 30 per sec.

In hand arc welding, the main factor in high-quality welding, is a proper selection of the welding conditions, which are determined by the electrode diameter, welding current and voltage.

(*a*) **Electrode diameter.** The diameter of the electrode will depend upon the thickness of the metal being welded and the type of the joint. Electrode diameter, according to the thickness of the metal to be welded, is given below :

Metal thickness, mm	:	Upto 2	2 to 4	4 to 6	6 to 8	over 8
Electrode diameter, mm	:	2	3	4	5	5-6

(*b*) **Welding current.** The magnitude of the welding current depends upon : the thickness of the welded metal, type of joint, welding speed, position of the weld, the thickness and type of the coating on the electrode and its working length.

However, for practical purposes, the welding current is determined on the basis of the selected electrode diameter, as given below :

Welding current, $I = k.\ d$, amperes; d is in mm.

where k = a constant

= 45 to 60 for ordinary steel electrodes

= 18 to 22 for graphite = 5 to 8 for carbon electrodes

In general practice, welding is performed with currents over 50 A.

Voltage. The arc voltage depends only upon the arc length. It is given by the relation :

$$V = k_1 + k_2\, l, \text{ volts}$$

where l is the arc length in mm and k_1 and k_2 are constants,

$$k_1 = 10 \text{ to } 12; \text{ and } k_2 = 2 \text{ to } 3$$

The minimum Arc voltage is given by $V_{min} = (20 + 0.04\ I)$, V

Arc Length. As already defined, it is the distance between the end of the electrode and the surface of the molten metal on the work.

For getting good welds, a short arc length is necessary, because :

1. It permits the heat to be concentrated on the joint.
2. Is more stable due to the reduced arc blow effect.
3. The molten weld pool is protected from atmosphere due to the vapours surrounding the electrode metal and molten weld pool.

On the other hand, a long arc results in :

(*i*) Large amounts of heat loss into the atmosphere leading to poor penetration and fusion.

(*ii*) Unstable arc due to increased effect of magnetic arc below.

(*iii*) Weld pool is not protected from the atmosphere. Because of this, the weld gets contaminated due to the absorption of O_2 and N_2.

(*iv*) Weld has : low strength (due to high porosity), less ductility, poor fusion and excessive spatter.

The arc length depends upon : kind of the electrode used, its coating, its diameter, current used and position of welding. Generally, shorter arc lengths are used for vertical, horizontal and overhead welding as compared to flat welding. As already noted, the stable arc needed for high quality welding can be achieved with an arc length of 0.6 to 0.8 times the electrode diameter.

The method is suitable for all metals and alloys except pure copper, Al and some low melting point and reactive metals.

3. Flux - Cored Arc Welding, (FCAW). The difference between SMAW and FCAW is that in the latter case, the consumable electrode is in the form of welding wire which can be coiled. The flux which provides shielding is supplied from the hollow core within the tubular wire electrode. Since the electrode is in the form of a coil, automatic and conti-nuous welding becomes possible with the electrode continuously fed from a spool. There are two designs of this method :

(*a*) **Self shielding.** Here, the shielding is provided as the flux burns and produces protective gas, (Fig. 3.19).

(*b*) **External shielding.** Sometimes, additional shielding is provided (for steel) with a gas (normally CO_2) from an external source.

The method is suitable for quantity production.

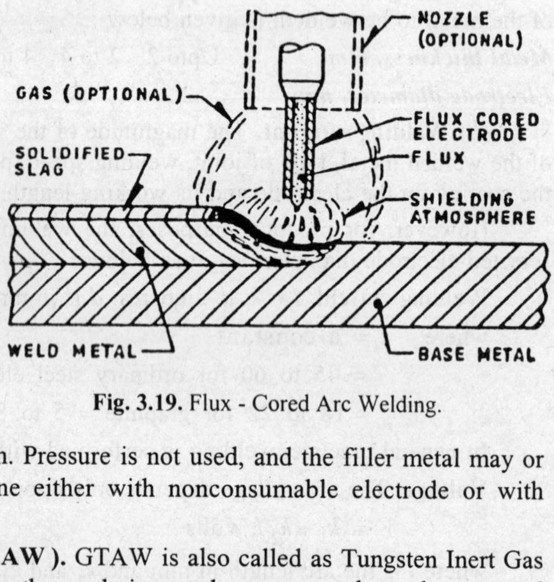

Fig. 3.19. Flux - Cored Arc Welding.

4. Inert Gas Arc Welding. In this, coalescence is produced by heating with an electric arc between a suitable electrode and the work. Shielding is obtained from an inert gas such as carbon dioxide, helium and argon. Pressure is not used, and the filler metal may or may not be used. Inert gas welding is done either with nonconsumable electrode or with consumable metal electrode.

(*i*) **Gas Tungsten Arc Welding, (GTAW).** GTAW is also called as Tungsten Inert Gas Welding (TIG). The arc is maintained between the workpiece and a tungsten electrode protected by an inert gas. The electrode is non consumable since its melting point is about 3400°C. The arc tempe-rature may be in the range of 10,000 K.

The mechanism of TIG is shown in Fig. 3.20. The welding head (gun or torch) consists of a lightweight handle, with provisions for holding a stationary tungsten electrode. In the welding head, the shielding gas flows by or along the electrode through a nozzle at the end of the gun into the arc region. This is done to keep the electrode cooler and permits higher currents to be used. Most guns are also water cooled. A filler metal may or may not be used. In the tungsten electrode, 1 to 2% thorium and zirconium are added to improve electron emission, arc stability, arc striking and current carrying capacity, and to increase the melting point of tungsten. Electrode diameters of 0.8 mm to 5 mm are commonly used with welding speeds ranging from about 8 mm/s to over 50 mm/s. Both hand and automatic operations are possible. The process demands considerable skill. The method was originated in 1940 for welding magnesium which is highly oxidising and

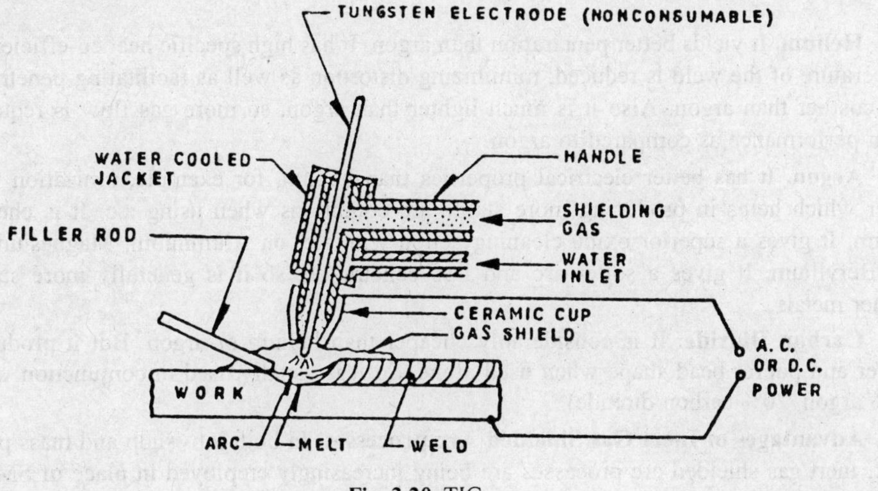

Fig. 3.20. TIG.

whose oxides have a much higher melting point than the parent metal. Now the method has developed as a means for welding a wide range of other materials : aluminium and its alloys, stainless steel, cast iron, silicon bronze, titanium, Nickel, copper and carbon steels. The method is especially suitable for welding thinner metals (below 6 mm). DCRP is rarely used, because it tends to melt tungsten electrode.

DCSP is used when maximum penetration is needed or when welding metals that do not need cleaning of oxide film, like copper and Stainless steel etc. When welding metals that have an oxide film on the surfaces (Aluminium, Magnesuim). A.C. high frequency supply is used to break up the surface oxides.

(*ii*) **Gas Metal Arc Welding, (GMAW).** The method is also known by the acronym, MIG (metal inert gas). The method differs from TIG in that a consumable electrode in wire form is fed into the arc, (Fig. 3.21). The same shielding gases are used. The wire is automatically fed from the reel by a variable speed motor at a preset rate that is determined by the arc voltage. The wire feed increases with an increase in arc voltage caused by an increase in arc length. This process has a greater effi- ciency than TIG and is more rapid. In semi-automatic weld- ing, the welder guides the gun and adjusts process para-meters; in automatic welding all functions are taken care of by the welding machine or robot.

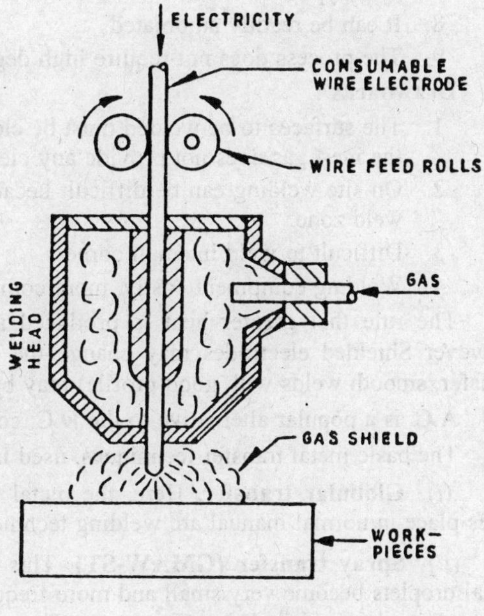

Fig. 3.21. MIG.

Shielding Gases. The shielding gases commonly used are : Helium, Argon and CO_2 or their mixtures. The gases must be pure, clean and-free of moisture.

Helium. It yields better penetration than argon. It has high specific heat co-efficient. So, the temperature of the weld is reduced, minimizing distortion as well as facilitating penetration. But it is costlier than argon. Also it is much lighter than argon, so more gas flow is required for a given performance as compared to argon.

Argon. It has better electrical properties than helium, for example, ionization voltage is lower which helps in producing more stable arc conditions when using a.c. It is cheaper than helium. It gives a superior oxide cleaning action with a.c. on Aluminium, Magnesium, Copper and Beryllium. It gives a softer arc and less penetration, so it is generally more suitable for thinner metals.

Carbon Dioxide. It is considerably cheaper than helium or argon. But it produces more spatter and poorer bead shape when used alone. So it is usually used in conjunction with argon (80% argon, 20% carbon dioxide).

Advantages of Inert Gas Shielded Arc processes. In both job- shop and mass production work, inert gas shielded arc processes are being increasingly employed in place of SMAW, due to the following reasons :

1. Much greater penetration and higher welding speeds are possible, because of the higher currents which can be used.
2. There is no slag to clean off after welding, because no flux is used.
3. Produce very high quality and clean welds on almost any material in any welding position.
4. The weld zone is visible.
5. No possibility of cracks.
6. There is no weld spatter.
7. It is fast and economical, because there are no frequent changes of electrodes, as with stick type electrodes.
8. It can be readily automated.
9. The process does not require high degree of operator skill.

Drawbacks :

1. The surfaces to be welded must be clean and free of oil, grease, paint or rust, because the inert gas does not provide any cleaning or fluxing action.
2. On site welding can be difficult because drafts blow the shielding gas away from the weld zone.
3. Difficult to weld in small corners.
4. Welding equipment costly, more complex and less portable.

The rule that greater heat is produced at anode is valid when using bare electrodes. However Shielded electrodes may change the heat conditions. As such, better results (spray transfer, smooth welds with good profile) may be obtained by using DCRP mode.

A.C. is a popular alternative to the D.C. conditions.

The basic metal transfer techniques, used in GMAW are discussed below :

(*i*) **Globular transfer.** Here, the metal transfer is in blobs, that drop by gravity. This takes place in normal manual arc welding technique, that is, at low current and voltages.

(*ii*) **Spray transfer (GMAW-ST).** This technique requires high current densities. The metal droplets become very small and more frequent. With this, the metal is transferred from the electrode to the weld zone in a fine spray. The method of employed on the heavier butt welds or fillets.

(*iii*) **Dip transfer or Short-circuting technique (GMAW-S).** This technique uses low currents. The electrode wire dips into the weld pool, causing a momentary short-circuit stoppage. The resulting current surge burns back the wire, again forming the arc. This takes place at a very high speed, about 100 dips/s. This frequency may vary with the voltage used and may be as low as 50 dips/s. Drops of liquid metal get transferred by gravity and surface tension.

The technique is very suitable for welding thin sections or sheets (down to 0.7 mm) because of lower heat. Positional welding is quite easily carried out and fit-up is not so critical, since there is greater tolerance of irregularities because of the cooler weld pool. CO_2 gas is used as the shielding gas.

Spray transfer and dip transfer are the techniques used in fine- wire MIG process.

(*iv*) **Pulsed Arc Welding (GMAW-P).** Pulsed arc welding is a new welding technique and has extremely attractive features. By the use of a special dual power unit, pulses or peaks of current are imposed on the welding current at a rate of 50 to 100 per second, which results in rapid acceleration of the droplets in transit across the arc. Both similar and dissimilar metals can be joined in a period of milli-seconds. Because of a short duration of arcing, temperatures are reduced, resulting in a number of benefits :

(*a*) Thinner sections can be welded.

(*b*) Spatter is drastically reduced, because of the smooth and quiet arc produced.

(*c*) Control is improved to alter the shape of the weld pool and vary the penetration.

(*d*) The technique is cheaper, as the energy or power needed to make a weld is considerably less than the other methods.

(*e*) Heat sensitive parts and high conductivity metals can be welded.

(*f*) Distorsion of the weldment is reduced or eliminated.

(*g*) High speed of the process is attractive for productivity.

Types of GMAW process. In addition to the conventional MIG process (bare-wire gas shielded process), other processes have also been developed along similar lines :

(*i*) One variation uses a special magnetic flux (A flux containing magnetic material) in addition to CO_2 shielding. Magnetic powder flux enters the welding head alongwith the shielding gas (Fig. 3.19). Since a magnetic field surrounds a current carrying electrode wire, the flux adheres to it. This results in a flux-coated electrode. The flux is 30 to 50% of the weight of the wire used. The process is more tolerant of variations in base metal composition and of rust, moisture, scale and surface conditions.

(*ii*) Similar results can be obtained by using a specially prepared tabular mould-steel electrode with a flux core, still using the CO_2 gas shield.

The above variations result in higher rates of metal deposition with manual metalarc welding. This makes these processes attractive for general welding, without the need for special preparation or cleaning of weldments.

Table 3.2. Relation between Job thickness and supply current in Inert gas arc welding :

Job thickness, mm	*Current, A*	
	TIG Welding	*MIG Welding*
1 – 1.5	50 – 60	80 – 100
1.5 – 2.0	60 – 80	100 – 120
2 – 2.5	80 – 100	120 – 130
2.5 – 3	100 – 120	130 – 140
3 – 3.5	120 – 140	140 – 150
4 – 4.5	140 – 160	150 – 160

Metal Active Gas Welding : MAG Welding : In MIG arc welding, inert gases are used as shielding gases. These gases (He, Ar) are very costly. To make the method cheaper, many companies use the MAG welding process. The gas used in this method is not inert gas. Mainly CO_2 gas is used for this purpose and it does the job effectively and efficiently.

Welding parameters at Yamaha Motors Escorts Limited (Faridabad) for MAG welding process are :

Voltage = 26 – 30 V
Current = 120 – 150 A
Pressure of gas = 25 – 35 bar
Wire (Cu coated) = 8 mm thick

The Cu coated electrode prevents its oxidation and rusting and also increases electrical conductivity of the electrode wire. For the same reasons, the electrodes used in SAW are coated slightly with Cu mist.

DCRP is used in the MAG process. It does not require high degree of operator skill and deeper penetration is possible.

MAG process or CO_2 welding is being used at Maruti Udyog.

5. Submerged Arc Welding, (SAW). In this, coalescence is produced by heating with an arc between a bare wire electrode and the work. The weld zone is shielded by a blanket of fusible granular flux material supplied directly on the weld seam ahead of the electrode to shield the arc. Pressure is not used and molten filler metal is obtained from the electrode. The flux also acts as a deoxidizer and scavenger and may contain powder metal alloying elements.

The method can be used in fully automatic equipment where the feeds of both the electrode and granular flux are controlled, Fig. 3.22. The method is also adaptable for semiautomatic equipment where the feed of the electrode and granular flux are controlled manually.

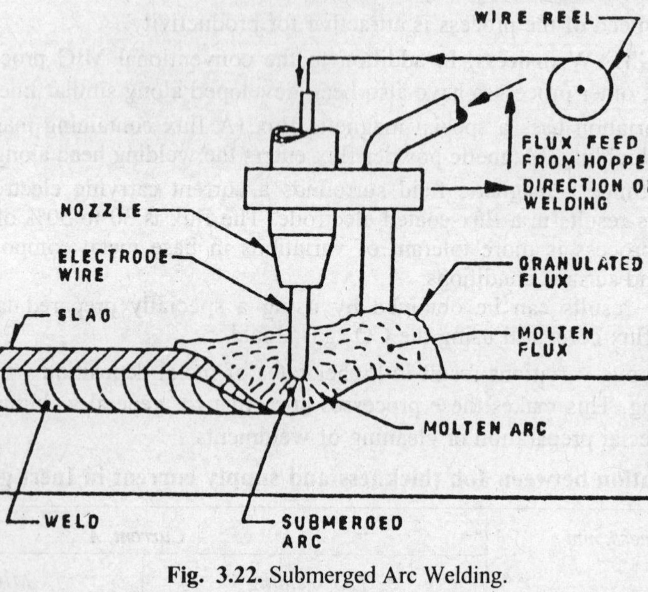

Fig. 3.22. Submerged Arc Welding.

This method is characterised by good appearing welds, high welding speeds and high welding currents upto several thousand amperes. Because of the large diameter electrodes which can be used, rather large welds can be made in one single pass in plates over 25 mm thick. Deep penetration is also obtained.

Since the granular flux must cover the joint to be welded, this method is restricted to making straight welds in the flat position. Thus it is suitable for steel line pipes, boiler pressure vessels, railroad tank cars, structural shapes and cylinders etc. and also for circular welds if the workpiece is rotated under the welding head. Double submerged arc welding (with one weld from the inside, the other from the outside) is used in making spiral- welded pipelines. It can also be used with welding robots, with the workpiece manipulated into appropriate positions.

Submerged arc welding can be used to weld low C- steels, high strength low alloy steels, chromium steels, and austenitic chromium-nickel steels. With special methods, it is also possible to weld high-alloy air hardening steels.

Advantages :

1. Joints can be prepared with a shallow V-groove, resulting in lesser electrode consumption.
2. Wire electrodes are inexpensive.
3. Weld spatters are eliminated.
4. Nearly 100% deposition efficiency is achieved.

Limitations :

1. It can not be used for plates less than 5 mm thick.
2. It can not cut C.I. because of high heat input.
3. Slag has to be removed continuously after it has melted in order to avoid entrapment between passes.

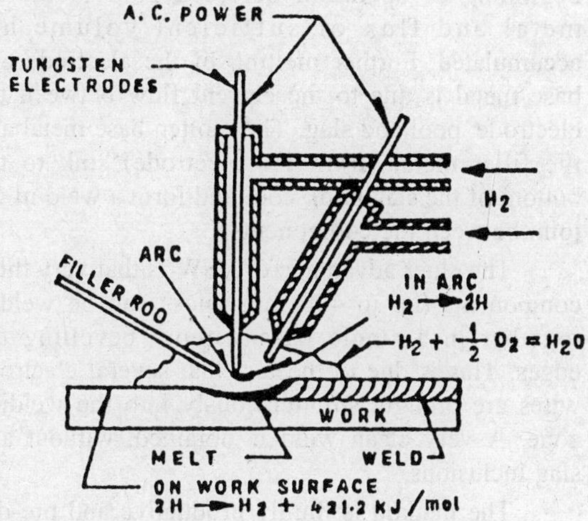

Fig. 3.23. Atomic Hydrogen Welding.

6. Atomic Hydrogen Welding. In this, the coalescence is produced by heating with an electric arc maintained between two tungsten electrodes in an atmosphere of hydrogen. Shielding is obtained from the hydrogen. Pressure is not applied. Hydrogen gas flows by the tungsten electrodes in the holder, keeping them cooler and lengthening the electrode life. Fig. 3.23. Since hydrogen is not inert, it will react chemically with its environment to form water vapours.

In this process an a.c. arc is formed between two tungsten electrodes along which streams of hydrogen are fed to the welding zone. The molecules of hydrogen are dissociated by the high heat of the arc in the gap between the electrodes. The formation of atomic hydrogen proceeds with the absorption of heat :

$$H_2 == 2H - 421.2 \text{ k J / mol.}$$

This atomic hydrogen recombines to form molecular hydrogen outside the arc, particularly on the relatively cold surface of the work being welded, releasing the heat gained previously :

$$2H == H_2 + 421.2 \text{ k J / mol.}$$

The metal is heated by the indirect arc and by the heat evolved above, producing a temperature of about 3700°C of the flame. The hydrogen protects the electrodes and the welding pool from oxidation and from saturation with nitrogen. At the outer surface of this gaseous shield, the hydrogen combined with oxygen of the air to form water vapour, making a bright orange flame.

Atomic hydrogen welding is mainly used in the welding of very thin sheets or small diameter wires (2 to 10 mm thick), particularly those of noble metals and refractory metals. It is so, because of its much lower thermal efficiency as compared to direct arc processes. Since the arc is struck between two electrodes, non-conducting materials such as ceramics may be arc welded. Alternating current is used usually so that both electrodes burn off at the same rate.

7. Electro-Slag Welding, (ESW). In this process, electrode wire (consumable) is fed into a molten slag pool. An arc is drawn initially, but is then snuffed out by the slag, and the heat of fusion is provided by resistance heating in the slag.

The components to be welded are set in a vertical position with a distance of from from 15 to 30 mm between the butted edges depending upon their thickness. The welding wire and the flux are fed automatically into the clearance, (Fig. 3.24). Several electrode wires can be fed simultaneous-ly if the parts to be welded are quite thick. Water cooled copper shoes or slides (dams) travel upwards along the joint. They close off the space between the parts to be welded to prevent the slag and molten metal from spilling out of the pool. They also cool the pool of metal and form the weld.

An arc is established between the steel backing plate and the electrode wire only at the very beginning of operation before a pool of molten metal and flux of sufficient volume has accumulated. Further melting of the electrode and base metal is due to the current flow between the electrode, pool and slag. The molten base metal and the filler metal (from the electrode) sink to the bottom of the slag pool, cool and form a weld in the joint between the components.

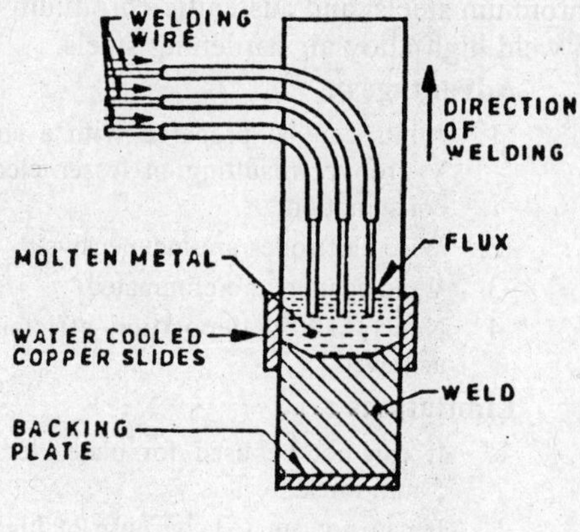

Fig. 3.24. Electro-Slag Welding.

The chief advantage of ESW is that very thick components (up to 400 mm thick) can be welded together in a single pass without bevelling the edges. This is due to the fact that several electrode wires are brought simultaneously into the welding zone. A very clean weld is obtained, without any slag inclusions.

The method is highly productive and pre-dominantly applied in heavy engineering in the manufacture of bed and frames for heavy machinery, drums, boilers etc.

8. Electro-Gas Welding (EGW). This method is a variant of electro-slag welding (ESW). The electric wire is either solid or flux-covered. Protection is provided by an inert gas (typically 80% Ar and 20% CO_2). The molten pool is again retained with copper dams. The major difference between ESW and EGW is that in EGW, an inert gas is used to protect the weld from oxidation and there is continuous arc (such as in SAW) to heat the weld pool.

9. Plasma Arc Welding, (PAW). The term 'plasma' refers to a gas which is sufficiently ionized (by raising it to a high temperature) to conduct current freely. A conventional welding arc is an example of a plasma. A plasma jet is created when the arc is passed through a constrictive nozzle. As a result of this constriction, the plasma jet will take on a narrow, columnar shape with unique properties that make it ideal for welding. Plasma jet welding is an electric-arc welding process which employs a high temperature constricted arc or plasma jet to obtain the melting and coalescence of metal. shielding is obtained from the hot, ionized gas issuing from the nozzle, which may be supplemented by an auxiliary source of shielding gas. Pressure is not applied and filler may or may not be supplied.

The basic circuitry of a plasma arc welding torch is shown in Fig. 3.25. The plasma torch is constructed with an electrode (tungsten) centrally within a metal cup that guides an inert

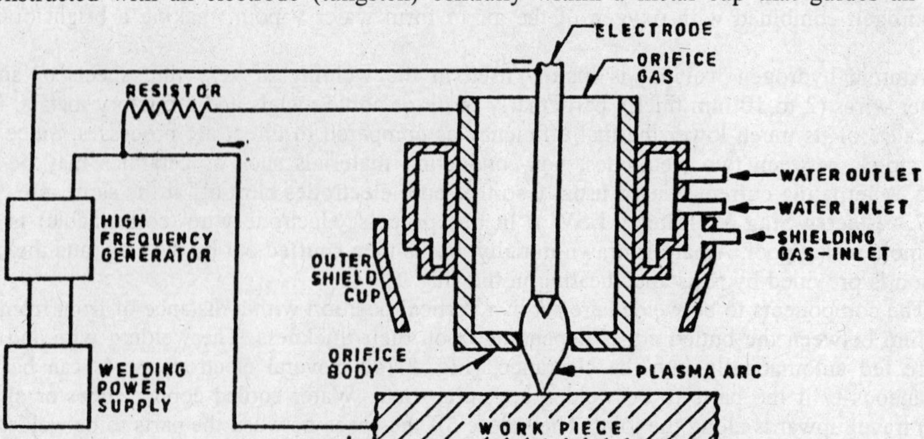

Fig. 3.25. Plasma Arc Welding.

streaming gas past the electrode. The discharge end of the cup is smaller in diameter than the upper diameter so that a discharge nozzle is created. In addition, the inner wall of the nozzle is lined with a ceramic material. The torch has passages for supplying gas and water (to cool it).

There are two arrangements of the plasma arc welding :

(*a*) Transferred plasma arc.

(*b*) Non - transferred plasma arc.

In the transferred plasma arc, the arc is produced between the electrode (– ve) and the work (+ ve). In the non-transferred arc, the arc is produced between the electrode (– ve) and the nozzle (+ ve). The work is heated indirectly by radiation. The heat is carried to the workpiece by the plasma gas.

The gases used should be inert (non-oxidising) and with a higher thermal conductivity, so that they can transfer more heat, making it possible to weld bulky section more easily. Usually, the shielding gas is the same as the orifice (nozzle) gas. Argon, Helium and hydrogen or their mixtures are the gases most frequently used.

Plasma arc provides an intense source of heat (Arc temperature of about 14000 K) and ensures greater arc stability. PAW is used for quality welding and can easily weld 127 mm thick aluminium sections or stainless steel sections upto 100 mm thick. The welded joints have no porosity and display a strong resistance to high stresses and impact loading.

10. Stud Welding. In this coalescence is produced by heating with an electric arc drawn between a metal stud (electrode) or a similarly shaped piece, and the part with which it is to be joined, until the surfaces to be joined are properly heated, after which they are brought together under pressure but without the use of a shielding gas.

The end of the studs have a cap that encases a dry flux and a ceramic ferrule that is used to maintain a gap between the end of the stud and the work, concentrate the heat of the arc, protect against oxidation and confine the melt; making the process an electric arc rather than an electrical resistance one, (Fig. 3.26).

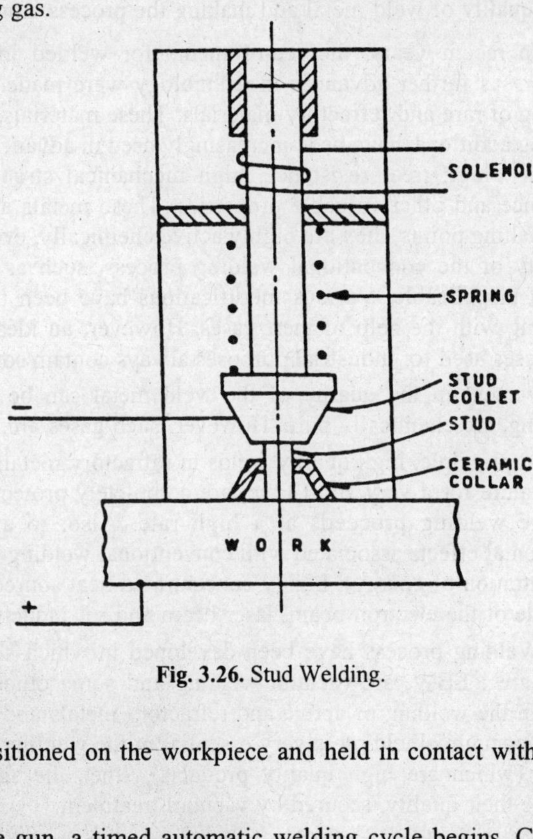

Fig. 3.26. Stud Welding.

The equipment used for stud welding consists of a special welding gun and a timing device. The sequence of operations during stud welding is as follows :

(*i*) The stud, or other projection, to be welded to the workpiece is loaded into the end of the gun, which very often is a portable hand tool. One end of the stud is projecting over which a ceramic. ferrule (collar) is placed.

(*ii*) The stud is then properly positioned on the workpiece and held in contact with it by spring pressure.

(*iii*) By operating a trigger on the gun, a timed automatic welding cycle begins. Current passes through the stud to the work.

(*iv*) Immediately this occurs, a solenoid lifts the stud against the spring pressure, a short distance away from the work to produce an arc.

(*v*) The arc melts the end of the stud and the weld area on the work.

(*vi*) The stud is then released and firmly plunged into the weld pool by the spring.

(*vii*) After a short dwell of about 1⁄25s, the current is switched off and the ceramic collar is removed to allow the end of the stud to freeze in place.

Applications. This process is widely used to join bolts and other projections onto workpieces without the need for drilling and tapping holes and is therefore very economical. It is also used for attachment of handles and feet to appliances and electrical panel construction. The method is extensively employed in the automotive, ship building, railroad and building construction industries.

11. Vacuum Shielded Welding of Metals. As has been discussed before, a very crucial task in fusion-welding is to protect the weld pool against exposure to the atmosphere, to avoid oxidation and contamination of the weld metal, due to the absorption of various gases. The other sources of gases can be : Surface oxides of the metals, rust, paint, grease, coatings, electrode, fluxes etc. Gases present in the weld metal inevitable degrade its physico-mechanical properties. Therefore, cleaning of the weld surfaces and protection of the weld pool are very important. The various shielding mediums in use are : heavy electrode coating, fluxes and inert gases alongwith CO_2 gas. The adoption of these measures have resulted in : increased rate of output, improved and stable quality of weld metal and making the process automated.

In recent years, the requirements for welded joints have become progressively more stringent as further advances in technology were made. Also, the problem arose regarding the welding of rare and refractory materials. These materials, once used on a limited scale, mainly as alloying additions, have been increasingly used in advanced fields of engineering. These materials possess : high treat resistance, high mechanical strength at elevated temperature, corrosion resistance and other valuable properties. These metals are difficult to weld, because apart from high melting points, they are highly active chemically, even at moderate temperatures. As a result, majority of the conventional welding process, such as SMAW, SAW and gas welding, have proved inapplicable. Various modifications have been tried to weld these metals by providing shielding with the help of inert gases. However, an ideal shielding is not possible, because the inert gases used for industrial purposes always contain some amounts of N_2, O_2 and water vapour. Improvement in the quality of the weld metal can be obtained if these inert gases, used for shielding, are chemically pure. However, such gases are very difficult to obtain.

In principle, high quality welds in refractory metals can be made if they are heated to high temperature for a very short time, are completely protected against exposure to the atmosphere, and the welding proceeds at a high rate. Also, to avoid heavy residual strains and other detrimental effects associated with conventional welding methods (where there is a relatively low concentration of energy), highly concentrated heat sources need to be used. Special mention can be made of the electron beam, laser beam and r.f. induction heating.

Welding process have been developed in which shielding has been provided by vacuum. These are : EBW, arc-vacuum welding and some others. Vacuum reliably shields the molten metal in the welding of active and refractory metals and assists in securing a high- quality weld metal. Vacuum-shielding is very essential in the welding of vacuum-melted refractory and active metals (which are high quality products), since the shielding media ordinarily used, usually degrade their quality, secured by vacuum treatment.

Vacuum-shielding offers a number of technical and economic advantages over other shielding methods :

1. It is inexpensive, easy to apply and easy to maintain.

2. It provides shielding, right at the scene of work and makes welding independent of shielding-gas suppliers, cuts down the need for cylinders and reduces the cost of transportation etc.

3. The cost of vacuum-shielding is only a fraction of that of inert-gas-shielding.

4. It secures high-quality welded joints in any metals and non- metallic materials. In welding some materials, it is the only one possible.

Note. Welding of Sheet Metal by Electric Arc Welding :

Normally, the electric arc welding is not recommended for sheet metal welding due to the intense heat generated during the operation and the risk of sheet metal getting burnt. But, sheet metal welding can be made successful by taking the following precautions when welding unequal thickness :

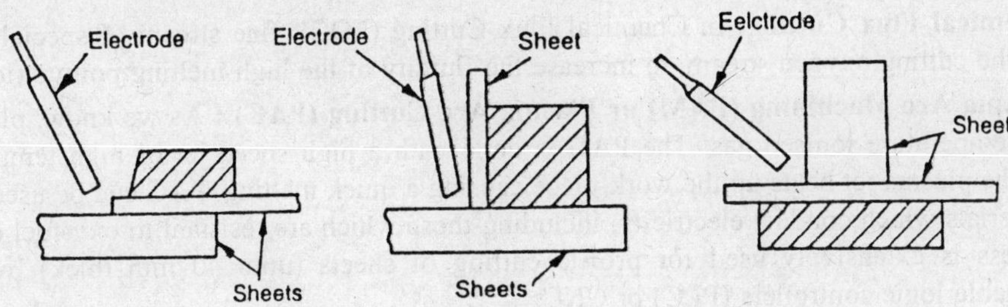

Fig. 3.27. Sheet Metal Electric Arc Welding.

1. Directing the arc towards the thicker sheet.

2. Conducting away heat from the thinner sheet by copper blocks in contact with it, [Fig. 3.27]. The copper blocks are shown by hatched lines. The use of copper blocks prevents melting away of the thinner sheets.

3.3.4. Electric Arc Cutting. Both carbon and metal electrodes are employed in the arc cutting of metals. The cutting is affected by melting the metal with the heat of an electric arc at the cut. Metal arc cutting is done with a coated electrode. Coating being insulator helps the electrode to be inserted deeper into he cut and the coating helps control the arc and form fluid slag. The use of metal electrodes improves the quality of the cut, reduces the width of the kerf and produces more evently cut edges. The application of compressed air just behind the arc, to blow out a molten metal speeds up the cutting process.

O_2 - arc cutting is done with a coated steel tube electrode through which oxygen is blown. The technique is similar to that of oxy-acetylene flame cutting, except that the preheat is supplied by an arc instead of a flame.

It is in effect a combination of flame cutting and metal arc cutting. Because the heat is concentrated into such a small area, the process is very useful for piercing holes in plate upto thickness exceeding 15 cm.

Cuts of better quality are produced by Plasma arc cutting, Electron- beam machining (EBM), and Laser-beam machining (LBM). The last two methods will be discussed in a separate chapter.

Gas-Tungsten Arc Cutting (GTAC). The basic circuit and the protective gases are similar to the used in GTAW. The difference lies in the fact that in GTAC, the jet of shielding gases leaves the nozzle at high velocity to expel the molten metal.

Gas-Metal Arc Cutting (GMAC). To cut the metals by MIG technique, the electrode wire feed rate and other variables are adjusted in such a manner that electrode wire completely penetrates through the workpiece. Thus, cutting this accomplished instead of welding. The width of cut will depend upon the potential difference in the circuit and the quality of cut will depend upon the rate of feed of the electrode wire.

Powder Oxy Fuel Cutting (POC). The method of POFC or Powder Metal Cutting method for cutting oxidation resistant materials or hard- to-cut materials, has been briefly mentioned in Art. 3.2.9. The method consists in introducing into the cutting zone a finely divided iron powder. The powder is carried from a hopper by the stream of oxygen, while the oxy-acetylene cutting torch travels along the reaction zone. The powder burns in the oxygen stream, generating additional heat, which facilitates melting of refractory oxides. The metal is removed from the cutting zone by the stream of oxygen, as usual. This process has proved useful for cutting stainless and heat resistant steels, cast iron and certain non-ferrous metals.

Chemical Flux Cutting. In Chemical Flux Cutting (FOC), fine stream of special flux is added to the cutting oxygen stream, to increase the fluidity of the high melting point oxides.

Plasma Arc Machining (PAM) or Plasma Arc Cutting (PAC) : As we know, plasma is the high temperature ionised gas. The PAC is done with a high speed jet of high temperature plasma. The plasma jet heats up the work-piece causing a quick melting. PAC can be used on all those materials which conduct electricity, including those which are resistant to oxy-fuel cutting. The process is extensively used for profile cutting of sheets (upto 40 mm thick) by using programmable logic controllers (PLC) or CNC.

Plasma is generated by subjecting the flowing gas to the electron bombardment of an arc. For this, the arc is set up between the electrode and anodic nozzle, the gas is forced through the arc. Gases used for plasma arc O_2, N_2 and H_2. The method produces highest temperatures (about 9500°C).

The process has got the following advantages:

1. The process is fast, that is, high cutting speeds.
2. Narrow kerf.
3. Good surface finish.
4. Metal removal rates are much higher than in the EDM and LBM processes.
5. Good reproducibility of parts.
6. Higher productivity than that of oxy-fuel methods.
7. The process is very economical.

Due to the above advantages, it is the preferred way to produce large quantities of cut parts. This attribute makes it the choice for plants in which both economy of operation and high frequency rates are required.

Cutting is done under water to avoid distortion of sheets due to high temperatures generated. Water also helps in:

— accurate cutting.

— cooling the metal.

— protecting eyes from plasma arc.

As noted above, PAC is highly automated process to-day due to the use of PLC and CNC.

EBM and LBM : Electron beams and Laser beams are used for very accurate cutting of a large variety of metals. As compared to other thermal cutting processes, the kerf is narrower and the surface finish is better.

3.4. RESISTANCE WELDING

Resistance Welding is "a group of welding processes wherein coalescence is produced by the heat obtained from resistance of the work to electric current in a circuit of which the work is a part, and by the application of pressure and without the use of a filler metal".

The amount of heat generated at the contacting area of the elements to be welded, is determined from Joule's law,

$$Q = I^2Rt, \text{ joules}$$

where I = current in amperes

R = resistance of the circuit at the contacting area of the elements in ohms.

t = time during which the current flows, in seconds.

The voltage can be low, typically 0.5 to 10V, but currents are very high (thousands of amperes).

Materials of high heat conductivity and specific heat (Al and Cu) call for very high currents to prevent dissipation of heat.

A special feature of resistance welding is the rapid heating of the surface being welded (in hundredths of a second) due to the application of currents of high amperage. The advantages of the resistance welding are : The heat is localised, action is rapid, no filler metal is needed, the operation requires little skill and can be easily mechanised and automated, suitable for large quantity production, all the common metal and dissimilar metals can be resistance welded, the parent metal is normally not harmed and none is lost, many difficult shapes and sections can be processed, and a high degree of reliability and reproductibility can be achieved.

The main disadvantage is high cost of equipment, and there are limitations to the types of joints made.

The resistance welding process finds extensive application in construction and the engineering industries.

There are six types of resistance welding processes :

(1) resistance spot welding (2) resistance seam welding (3) projection welding (4) flash welding (5) upset welding, and (6) percussion welding.

1. Resistance Spot Welding (RSW). In resistance spot welding, the overlapping metal parts are held between two bar type metal electrodes which apply pressure, while an electric current is passed through them, Fig. 2.28 When the current is switched on and applied for a

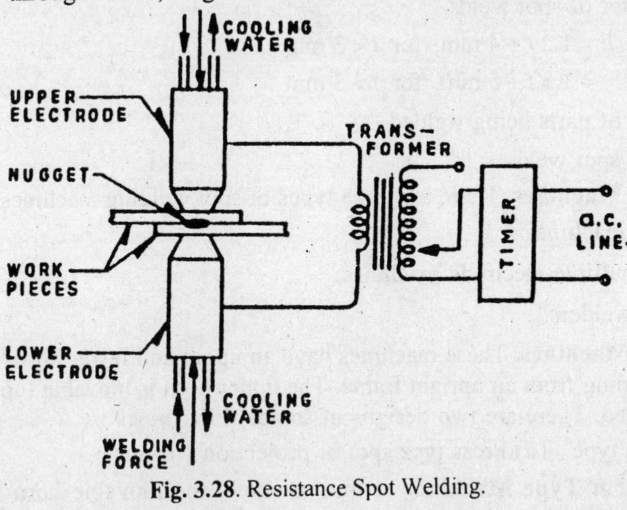

Fig. 3.28. Resistance Spot Welding.

predetermined number of cycles (in the automotive industry 20 - 30 cycles), the lapped pieces of metal are heated in a restricted area. The generated heat melts the surface layer of the metal in the central, more highly heated, area of contact with the electrodes, and the adjacent layers of metal are softened to a plastic state. Then the current is switched off and the electrodes are pressed and the pressure is released only after the weld spot (nugget) has solidified. The spot weld (nugget) is approximately the same size as the electrode tip, usually between 3 or 6 mm in diameter. Electrode tip diameter = Thickness of plate to be welded. Distance between the nearest edge of the plate and centre of weld = 1.5 × electrode tip diameter. The sheet surface shows a slight depression and decolouration.

The welding cycle to produce one spot can be written as :

(*i*) Position the workpieces and squeeze between the electrodes.

(*ii*) Apply a low voltage current to the electrodes.

(*iii*) Hold until the proper temperature is attained.

(*iv*) Release current, continue pressure.

(*v*) Release pressure and remove work.

The electrode pressure can be in the range of upto 2 kN

The electrodes used must possess high electrical and thermal conductivity and retain the required strength at temperatures upto 400°C. Electrodes are made of cold-rolled electrolytic copper with some Cd, Cr, or Be additions, or copper - tungsten or molybdenum alloys. Electrodes are usually of hollow construction and are cooled with water during operation to prevent their overheating.

Because of the widespread application of sheet metal parts, resistance spot welding finds extensive application in the aircraft and automobile industries for manufacturing air frames and bodies (there are some 8000 to 10,000 spot welds per car), attaching handles of cookware, as well as in railway car building and the instrument industry and for making reinforcement in concrete construction etc.

Spot welding is primarily restricted to thin metals (for example 0.025 to 3.2 mm thick for steel and magnesium and 4 mm thick for aluminium), namely steels, stainless steels, aluminium, magnesium, nickel, nickel alloys, bronze and brass. Some dissimilar metals can be spot welded, but with difficulty. However, the method is quite difficult to use for highly conducting materials like Al and Mg.

(*a*) The diameter of spot welds,

$$d = 1.2\,t + 4 \text{ mm, for } t < 3 \text{ mm}$$
$$= 1.5\,t + 5 \text{ mm, for } t > 3 \text{ mm}$$

where t = Thickness of parts being welded.

(*b*) Spacing of spot welds = $3d$.

Spot Welding Machines. There are three types of spot welding machines which are in use:

I. Standard machines.

II. Special multiple-electrode machines.

III. Portable welders.

I. Standard Machines. These machines have an upper and lower horn (arm) carrying the electrodes and extending from an upright frame. The upper horn is movable (up and down) while the lower horn is fixed. There are two designs of standard machines :

(*i*) Rocker arm type (*ii*) Press type spot or projection welders.

(*i*) **Rocker Arm Type Machine.** In this machine, the movable horn is mounted on a rocker arm which is pivoted in the frame. It is tilted upward to open the gap and downward to apply the pressure, by the rocking action of the upper arm (the lower electrode arm being stationary). This design is adaptable to a wide range of work, usually for light production work with lighter gauges. The throat depth (distance from the face of the upright frame to the end of the horn) is about 1.2 m for these machines. These machines are of two types :

(*a*) **Foot Type.** When the foot treadle is pressed down, the two electrodes are brought together. The welding pressure is then transmitted to the work through a compression spring on the rear rod (see Fig. 3.29, where an air cylinder is shown in place of a rod, the other construction of the machine is same). After a suitable amount of compression is obtained, the welding circuit is closed by means of a mechanical switch. As the compression movement of the rear rod follows through, the switch trip opens and welding current ceases to flow. Releasing the pressure on the foot treadle opens the electrodes and resets the trip switch.

(*b*) **Air Operated Machines.** In this design, a double acting air cylinder replaces the rear rod and compression spring, (Fig. 3.29).

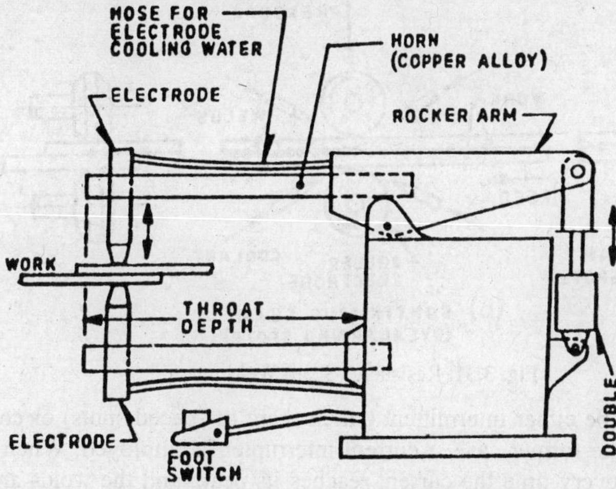

Fig. 3.29. Rocker-Arm Spot Welding Machine.

(*ii*) **Press Type Machine.** The name comes from straight line vertical stroke of the upper welding head. The straight line motion is provided by an air or hydraulic cylinder, (Fig. 3.30). These machines are used for mass production. The throat depth for these machines is about 1.5 m.

II. Special Multiple Electrode Machines. These machines are employed for high-production jobs.

(*i*) In one system, there is one common welding transformer and a hydraulic cylinder for each electrode pair. The electrodes are brought in contact with the work one pair at a time.

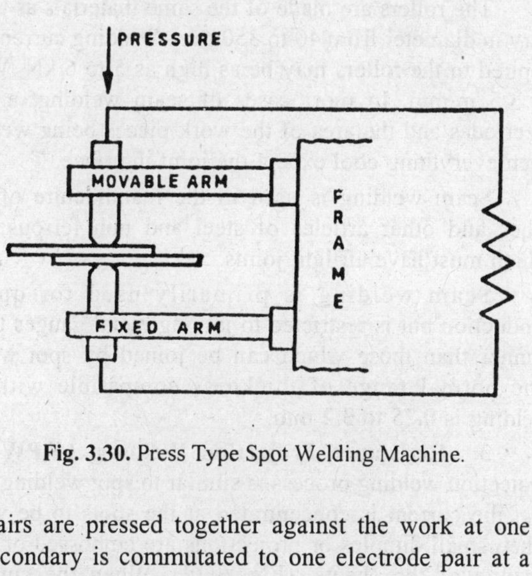

Fig. 3.30. Press Type Spot Welding Machine.

(*ii*) In another system, all electrode pairs are pressed together against the work at one time, and the current from the transformer secondary is commutated to one electrode pair at a time. Some machines have a separate transfer and controls for each electrode pair.

III. Portable Spot Welding Guns. These welders are used when the work is too large to be moved to a welding machine, for example, trucks and railroad cars. The gun consists of a transformer and controls in a case, usually hung from the ceiling, with leads and an air line to a pair of jaws. This permits bringing the welding unit and the operator to work: The jaws carry the electrodes and can be moved about in the work area. The jaws are closed by an air cylinder and the electrodes bite like teeth on the work and pass current to make the weld. These guns ca make 200 spots/minute.

2. Resistance Seam Welding, (RSEW). To obtain a series of spot welds along a line by the RSW method, an interrupted work movement will be necessary. The same result can be achieved much more conveniently and rapidly in the resistance seam welding where the electrodes are in the form of rotating disc electrodes, with the working being welded moving continuously by the electrodes, (Fig. 3.31).

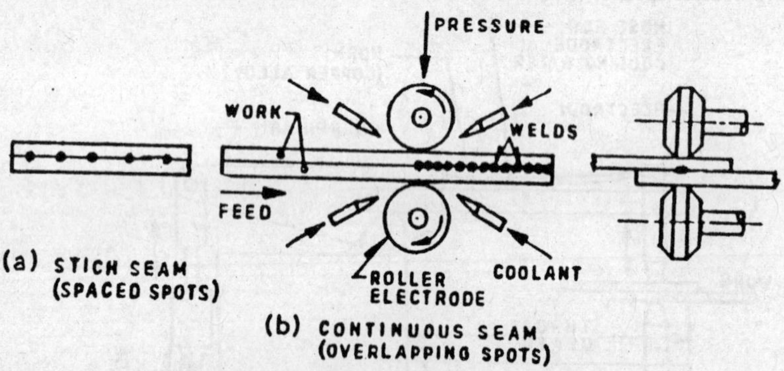

Fig. 3.31. Resistance Seam Welding.

Seam welding may be either intermittent (stitch seam or spaced spots) or continuous seam (over lapping seam). In the former case, a current interrupter is employed. When the a.c. is left on, a spot weld is made every time the current reaches its peak, and the welds are spaced close enough to give a gas and liquid- tight joint.

The rollers are made of the same materials as the electrodes for sport welding. Rollers may vary in diameter from 40 to 350 mm. Welding currents range from 2000 to 5000A while the force applied to the rollers may be as high as 5 to 6 kN. Welding speeds range is commonly from 0.5 to 3.5 m/min. In most cases of seam welding, a stream of water is directed over the disc electrodes and the area of the work pieces being welded to keep everything cool except the joint interface.

Seam welding is used in the manufacture of tanks, tubes and other articles of steel and non ferrous metals which must have airtight joints.

Seam welding is primarily used for quantity production but is restricted to joining metal gauges that are thinner than those which can be joined by spot welding. The normal range of thickness compatible with seam welding is 0.25 to 3.2 mm.

3. Resistance Projection Welding, (RPW). The projection welding process is similar to spot welding except that the current is concentrated at the spots to be welded, when small dimples or projections are embossed or coined on one of the sheets, (Fig. 3.32). When the current is applied, the projections soften and are pushed back in place

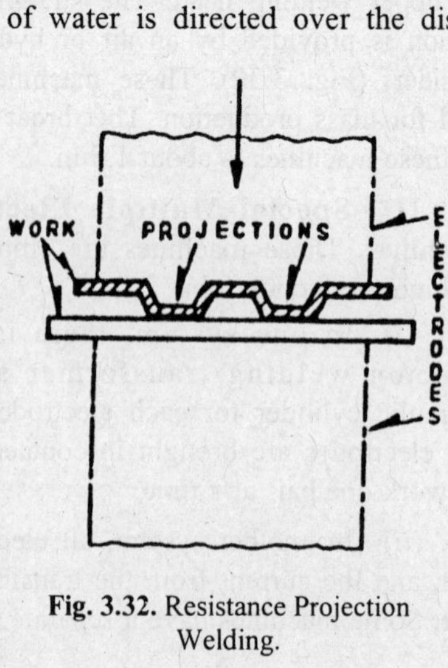

Fig. 3.32. Resistance Projection Welding.

by the electrode pressure as the weld nuggets form. Electrodes are relatively large and several welds can be made simultaneously with a single electrode, resulting in a lower cost of operation. Also, because larger diameter elec- trodes can be used with a greater heat capacity, the need for water cooling is often eliminated. The projections tend to localize the heat, permitting thicker materials to be welded resulting in a stronger weld structure than that obtained with spot welding. Unlike spot welding, the method leaves no depression on the free surface.

Projection welding is used extensively by Automobile industry for joining nuts, bolts and studs to steel plates in car bodies. Because of their higher thermal conductivity, the method is especially suitable for copper, aluminium, brass etc.

Projections at the point of contact should be approximately = 10 to 80% of the sheet thickness.

If pieces of unequal thickness are to be welded, jobs of bigger thickness should have the projections to develop uniform heating of the two pieces.

4. Electric Resistance Butt Welding. There are two types of electric resistance butt welding processes : the 'upset' and the 'flash'.

(*i*) **Upset Welding, (UW).** In this process, the ends of the two parts to be joined together are clamped in position in the electrodes. The movable head is moved towards the fixed head until the abutting surfaces of the work pieces are in light contact. Then the proper current is made to flow across the interface for a preset time, while the light pressure between the two parts is maintained. When the interface has been heated to the welding temperature (plastic state), the current is switch off and the welding pressure is increased to form an "upset", (Fig. 3.33). This results in the lateral flow of the surface oxide layers, bringing clean metal surfaces in contact. Metal is not melted when upset butt welded and there is no spatter.

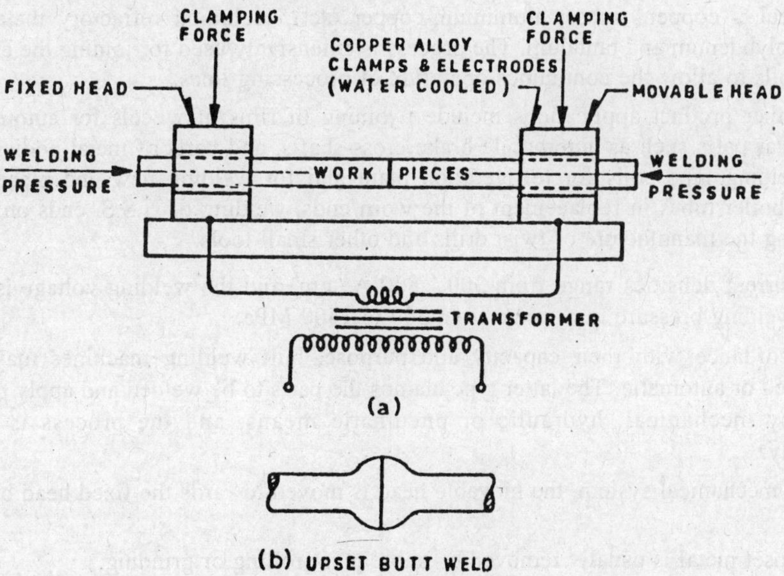

Fig. 3.33. Upset Butt Welding.

The welding voltage may vary from 5 to 15V while the current densities range from 300 to 800 A/cm^2. Because the contact resistance is inversely proportional to the pressure, therefore, the pressure is less at the start and is then increased to whatever is necessary to effect the weld. Final pressures range from 17 to 55 MPa. Upset butt welding is extensively used in the fabrication of tubular sections, pipe and heavy steel rings. It is also used for joining small ferrous and non ferrous strips. Areas upto 0.05 m^2 have been successfully welded, but generally the process is limited to smaller areas (upto 1000 mm^2) because of current limitations.

(*ii*) **Flash Butt Welding, (FW).** In this process, the parts to be welded approach towards each other and come into contact with the current switched on. The procedure of butt welding is as follows : After the parts are properly positioned and correct current, head speed and time are selected, a cycle start button is actuated. This makes the movable head to approach the fixed head. As the abutting surfaces come very near to each other, extremely rapid heating takes place when surface asperities first make contact. Molten metal is violently expelled and burns in air with considerable force and "Sparking" or "arcing", thus giving the process its name "flash butt". As this process continues for a few seconds, a very thin layer at the interface is melted. The current is then shut off and the two parts are rapidly pressed together causing a small upset, (Fig. 3.34). This squeezes out liquid metal and oxides formed on the joint surfaces and the solid clean metal faces of the abutted parts are welded together.

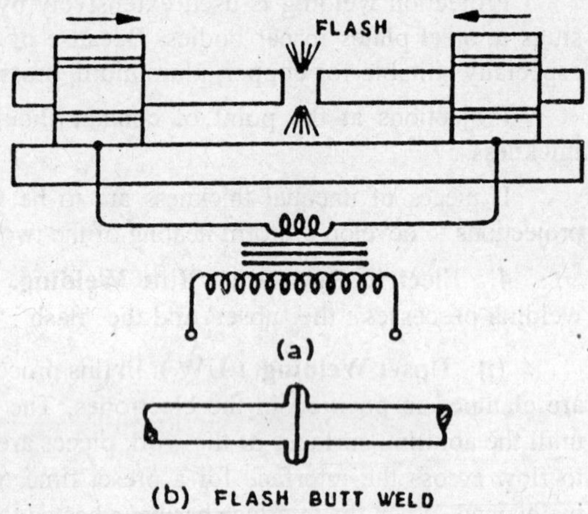

Fig. 3.34. Resistance Flash Welding.

This process is used for joining thin- walled tubes, rails, chain links, tools (low carbon steel shank to H.S.S. tool bit), press manufactured parts. Also, dissimilar metals can readily be welded (such as steel, - copper, -brass; aluminium-copper etc.), including refractory metals such as tungsten, molybdenum and tantalum. The process is extensively used for joining the ends of wire and sheet coils to allow the continuous operation of processing lines.

The other product applications include : joining of rims of wheels for automobiles and cycles, tubular parts such as automobile brake cross-shafts, and parts of metal bodies for motor cars, for welding tube coils for refrigeration plant and for welding new end piece on to the locomotive boiler tubes in replacement of the worn ends, welding of H.S.S. ends on to C- steel shanks during the manufacture of twist drills and other small tools.

The current densities range from 300 - 800 A / cm^2 and the welding voltage is from 5 to 15 V. The welding pressure may range from 350 to 1400 MPa.

In accordance with their capacity and purpose, butt welding machines may be either hand-operated or automatic. The latter type clamps the parts to be welded and apply pressure for upsetting by mechanical, hydraulic or pneumatic means, and the process is controlled automatically.

In the mechanical system, the movable head is moved towards the fixed head by means of a cam drive.

The upset metal is usually removed by subsequent rolling or grinding.

(*iii*) **Percussion Butt Welding, (PEW).** This process uses the heating effect of a high-intensity electric arc produced by a rapid discharge of electric energy, in combination with a percussion blow to forge weld metal pieces together. When the movable clamp is released, it moves rapidly alongwith one of the workpiece (Fig. 3.35). When the two parts are about 1.6 mm apart, there is a sudden discharge of electric energy. This causes intense arcing over the surfaces and brings them to high temperature. The arc is extinguished by the percussion blow of the two parts coming together with sufficient force to effect the weld.

Percussion welding makes it possible to weld parts having considerable difference in mass and in thermal and electrical conductivity, for example, welding of copper to steels, steel to magnesium alloys and so on.

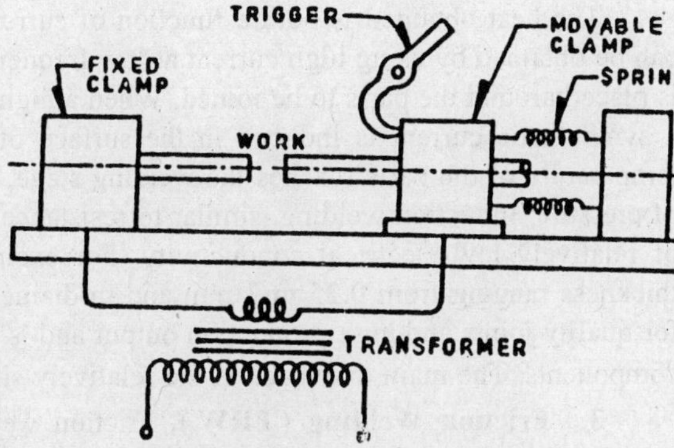

Fig. 3.35. Percussion Butt Welding.

Percussion welding is used for special joining situations, for example, joining dissimilar metals that can not be welded economically by flash welding. The method is also used to weld pins, studs, bolts, and so on, to other components as well as to join sections of pipe, rod, or tube to each other or to flat sections. Since control of the path of an arc is difficult and also because of the impact required, the method is limited to small pieces (about 150 to 300 mm^2).

Commercial product applications include : Butt welding steel or other metals without flash or upset, joining aluminium rods, tubes or bars to copper, joining corrosion-resistant alloys to steel or non-ferrous pieces, joining of threaded steel studs to aluminium or magnesium alloys or forgings, joining of silver contact tips to copper studs and joining of stellite tips to steel or non-ferrous metals.

3.5. MISCELLANEOUS WELDING PROCESSES

The welding processes discussed below are listed under miscellaneous processes because they are not used as extensively as those previously described.

1. Forge Welding (FOW). Forge welding is defined as "A solid state welding process wherein coalescence is produced by heating and by applying pressure or blows sufficient to cause permanent deformation at the interface," (Fig. 3.36 (*a*)). This is the oldest industrial welding

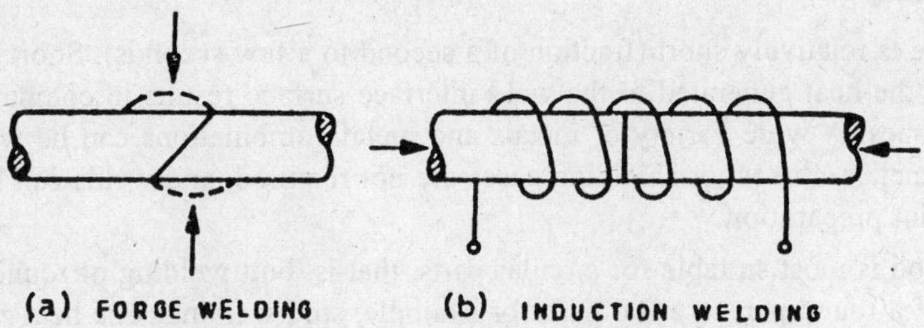

Fig. 3.36. Forge Welding, Induction Welding.

process. The parts to be welded are heated in a forge or some other furnace to within the hot-working temperature range and then forged together by hand or power hammering or pressing. During forging, oxides, slag and other contaminants are squeezed out, ensuring inter-atomic bonding. The workpieces commonly forge welded are of wrought iron and steel. The commonly used forge welding processes are : (*i*) Hammer welding (*ii*) Die welding, and (*iii*) Roll welding.

Roll welding is employed in the cladding of dissimilar materials. Die welding is used in the manufacture of seamless piping.

2. Induction Welding (IW). Induction welding is defined as "A welding process wherein coalescence is produced by the heat obtained from resistance of the work to the induced electric current with or without the application of pressure", (Fig. 3.36 (*b*)).

The heat obtained is both a function of current and frequency. A given degree of heating can be obtained by using high current at low frequency or a low current at high frequency. A coil is placed around the parts to be joined. When a high frequency current is passed through the coil, a sympathetic current is induced in the surface of the parts, which generates heat. When the temperature of the parts reaches the welding stage, the parts are welded with or without the use of pressure. Induction welding, similar to resistance welding, is primarily applicable to materials of relatively low electrical conductivity. For example, for steel the method is successful for thickness ranging from 0.25 to 3 mm and in diameter upto 75 mm. Induction welding provides for quality joints and high production output and is used for PVC plastic tubes and polyethylene components. The main drawback is the relatively small thickness of the parts joined ($\not>$ 5mm).

3. Friction Welding (FRW). Friction welding is defined as "A solid state welding process wherein coalescence is produced by heat obtained from mechanically induced sliding motion between rubbing surfaces. The parts are held together under pressure". One part is rotated at relatively high speeds and under pressure against the second part which is held stationary. The frictional work at the contacting surfaces is transformed into heat. The contacting surfaces are thus heated to a high temperature below the melting temperature. At this point, the relative motion between the two is stopped. The weld is then completed (by upsetting) due to the pressure which still continues or additional pressure may be necessary, (Fig. 3.37), to produce a solid state weld. Axial pressure, when rotating, may be upto 70 MPa and at the end it may reach 140 MPa. Relative motion is stopped by :

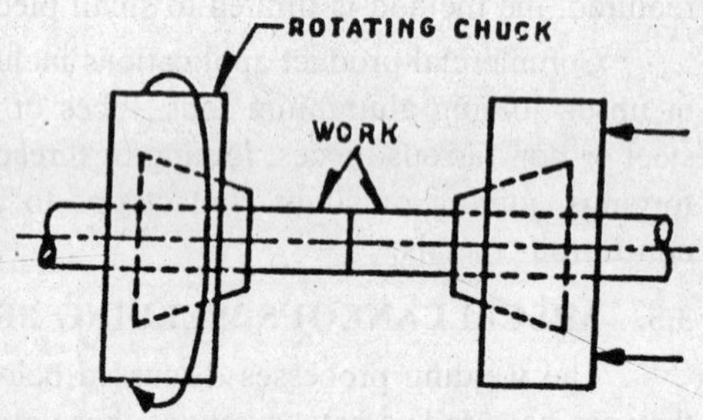

Fig. 3.37. Friction Welding.

(*i*) a braking system to the machine spindle.

(*ii*) an arrangement which releases the stationary part to permit the two parts to rotate together.

Weld time is relatively short (fraction of a second to a few seconds). Short weld times and confinement of the heat generated to the weld interface surface results in comparatively narrow heat-affected zones. A wide variety of metals and metal combinations can be welded with this process. Filler metals, fluxes, or shielding gases are not required, and welds can be made with a minimum of joint preparation.

The method is most suitable for circular parts, that is, butt welding of round bars or tubes. Also welding of a round part to a flat part, for example, stud welding. The heat zone being very thin, therefore, dissimilar metals are easily joined, for example, aluminium to steel, copper to steel, copper to aluminium, aluminium to brass, and mild steel shanks to high speed steel tools. The rotational speed is proportional to thermal conductivity of the metal. The method is limited to smaller components.

Besides the advantages of the process given above, it also has the following additional advantages :

1. Low initial capital cost.
2. Low - cost power requirements.
3. The process is clean.
4. High quality welds.
5. Annealing of weld zone is not necessary.
6. Very little loss of material through exclusions.

7. The process welds the whole surface of contact, unlike OAW and arc welding which weld only round the periphery.

For welding a round bar to a flat plate, the thickness of the plate should not exceed one quarter of the bar diameter. This combination of dimensions provides an equal balance of heat.

The high quality of welds obtained by FRW is due to the following factors :

(*i*) As the surfaces rub against each other, all oxide and adsorbed films and foreign particles are removed into flash, because the metal is deformed in radial directions. Also, no oxides are formed during welding due to the intimate contact between the rubbing surfaces.

(*ii*) The metal at the joint and in the heat-affected area acquires a structure having equi-axial and fine grains.

Product applications. The components successfully produced include : H.S.S. Twist drills, flanges upto 15 cm diameter pipes, gas turbine shafts, Aero - engine drive shafts and valves, refrigerator tubes of dissimilar metals, hydraulic tubes, steering columns, welding of sintered products, and sulphur - bearing steels.

Other applications are given below :

1. Manufacture of measuring tools and running centres for lathes. In the past, plug and screw-thread gauges were forged in several passes from Chromium steel. By FRW, a gauge blank is made up from two parts, (Fig. 3.38). An automatic machine joins a common-steel shank to the gauging members from chromium steel. This results in saving a sizeable amount of the expensive chromium steel and greater output.

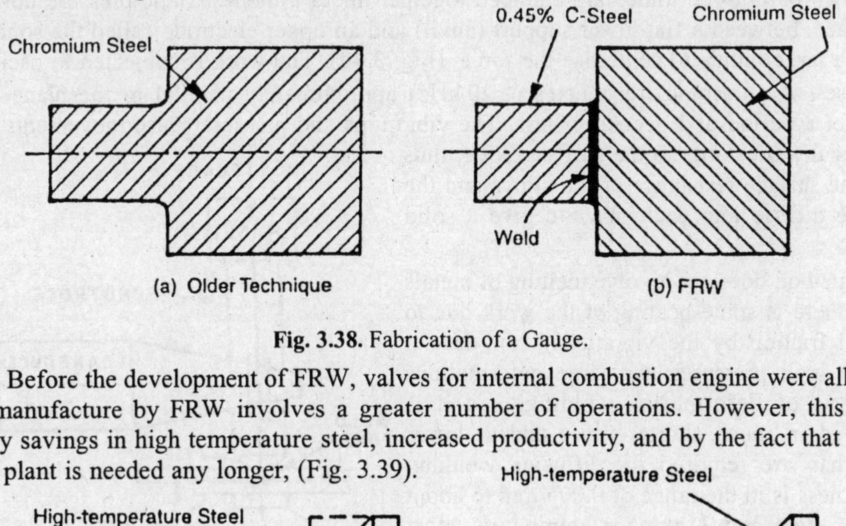

Fig. 3.38. Fabrication of a Gauge.

2. Before the development of FRW, valves for internal combustion engine were allstamped: Valve manufacture by FRW involves a greater number of operations. However, this factor is offset by savings in high temperature steel, increased productivity, and by the fact that no heavy forging plant is needed any longer, (Fig. 3.39).

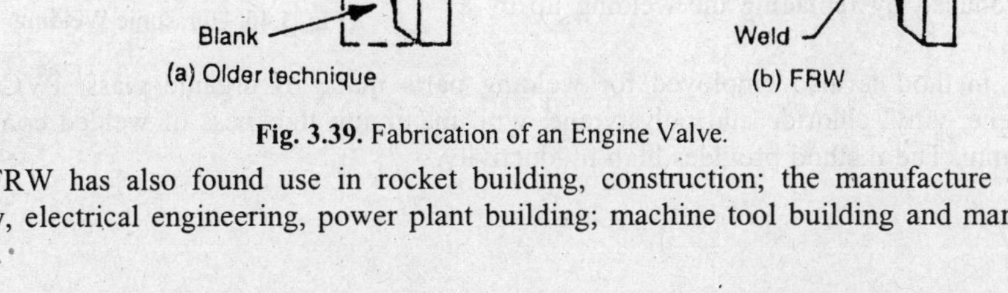

Fig. 3.39. Fabrication of an Engine Valve.

3. FRW has also found use in rocket building, construction; the manufacture of farm machinery, electrical engineering, power plant building; machine tool building and many other industries.

In FRW, which is also known as ''Continuous-drive FRW'', friction is produced by rotating one of the pieces, relative to the other (Fig. 5.37), or both, while they are clamped together by a force. Due to very short cycle time, the process is very fast and only electric resistance welding can vie with it in this respect. Power consumption in FRW is one-fifth to one-tenth of that used for electric resistance welding.

Inertia Welding. Inertia welding or ''Inertia-Drive FRW'', differs from the conventional FRW (continuous-drive) that the spinning part is now attached to a rotating flywheel. The flywheel is brought to a specified rotational speed and is then detached from the driving motor. After that, the rotating assembly is pressed against the stationary part. The kinetic energy of the flywheel is converted into heat energy, as the two parts rub against each other. The welding operation is completed as the flywheel comes to rest, with the two parts remaining pressed together. The kinetic energy and hence the rotational speed of the flywheel is calculated, so that the weld is completed when rotation stops.

The method of FW is very convenient, where a small bar (75 mm ϕ × 300 mm) is to be welded to a large blank (200 mm ϕ × 50 mm). The larger blank is held in the lathe chuck and rotated at high speed. The smaller bar is held in the tool post and kept pressed against the rotating blank. Heat generated due to friction welds the two surfaces. During the process of welding, the molten metal flies out and proper guards should be provided. Each cycle takes about 5 to 10 minutes and gives a very strong joint.

4. Ultrasonic Welding (USW). Ultrasonic welding is defined as ''A solid state welding process wherein coalescence is produced by the local application of high frequency vibratory energy as the work parts are held together under pressure''.

The two parts to be joined are clamped together under a modest static pressure normal to their interface, between a flat lower support (anvil) and an upper electrode called the sonotrode, which has a large radius to centralise the force, (Fig. 3.40). The work is subjected to oscillating shear stresses of ultrasonic frequencies (≥ 20 kHz) approximately parallel to the plane of the interface, for a period of 1 second or less. The vibrations and pressure cause movements of the metal molecules breaking up the surface oxide, thus lowering the surface resistance of the alloys and the two surface diffuse into each other to give a solid state weld.

The method does not involve melting of metals (of course there is some heating of the work due to the internal friction by the vibrations), nor does it involve the high pressures and large deformations characteristic of deformation welding, and it is accomplished in much shorter times and at lower pressures than are required for diffusion welding. Metal thickness is in the range of 0.005 mm to about 1.0 mm for steel and 3 mm for aluminium. Most metals can be welded and also dissimilar metals. The process is suitable for lap welding foils and delicate instrument and electronic components. Seam welds can be produced by replacing the welding tip by a roller.

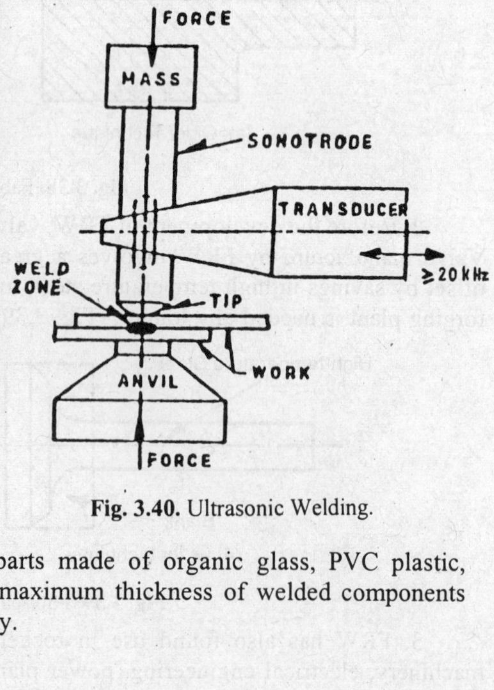

Fig. 3.40. Ultrasonic Welding.

The method is also employed for welding parts made of organic glass, PVC plastic, polyethylene, vinyl chloride and polystyrene, with maximum thickness of welded components being 10 mm. The method provides high productivity.

Other advantages of the process include :

1. Welds are strong and free from foreign inclusions.
2. Very little preparation is needed for the weld ; usually, it invol-ves degreasing.
3. Post cleaning of welds is usually not necessary.
4. Thin pieces can be welded to thicker pieces.

Product applications. The range of applications of the process can meet many of the requirements of the atomic and electronic sections of industry. The process is used for :

Air - craft construction, for the building of transistors, for fine - wire assemblies, bimetal electrical contacts, light - gauge metal for the production of containers and for the welding of aluminium foil for packaging requirements.

5. Explosive Welding (EXW). Explosive welding or Explosion welding is defined as "A solid state welding process wherein coalescence is effected by high velocity movements produced by a controlled detonation". The technique eliminates the problems generally associated with fusion welding-melting, heat effects and microstructural changes. The process can join many combinations of metals not possible or extremely difficult or expensive to combine using standard methods.

A schematic description of a typical explosive-bonding set up for joining two flat plates is shown in Fig. 3.41. The two plates can be placed at angle to each other also. A single electric detonation cap can be used to ignite the explosive. When the explosive mat placed on top of the cladding sheet (flyplate) is detonated, the sheet joins to the base metal by forming tight whirls or vortexes at the interface.

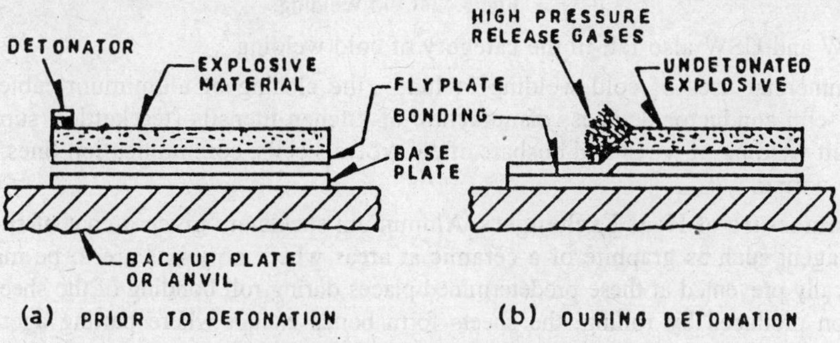

Fig. 3.41. Explosive Welding.

The most extensive use of flat clad products has been in coinage and in heat exchangers. Clad tube sheets for heat exchangers are produced by explosive cladding as an adjunct to the more expensive weld overlay process. Composite armour and metal-matrix reinforced composites are also being produced as well as clad plates and sheets for use in the fabrication of chemical processing equipment.

Explosive cladding is finding use in the die-casting industry for nozzles, die-casting biscuits, and other components. Work is underway for the cladding of components for the glass industry, and many potential applications exist for the automotive industry, such as bimetallic disk brakes, clad piston faces and motor-block repairs. The method augments other processes for the conduct of jobs difficult or impossible to accomplish by other means.

6. Flow Welding. In flow welding, coalescence is produced by heating with molten filler metal, poured over the surfaces to be welded until the welding temperature is attained and until the required filler metal has been added. This process is not used to any extent.

7. Cold Welding (CW). Cold welding is defined as "A solid state welding process wherein coalescence is produced by the external application of mechanical force alone". Since it is carried out at room temperature, the bonding of the molecules at the interface may not be of very high quality.

The process is quite similar to electric resistance welding with the exception that no heat is used to remove the oxide and gas layers but rather severe plastic deformation is relied upon to get metal to metal contact. The equipment for this process is very simple and inexpensive : light presses with simple "dimpling dies" being required for cold welding thick sections and special pliers similar to bolt cutters being sufficient for thin sections.

Fig. 3.42 (*a*) shows the lap welding of sheets. It relies on a 50 to 90% expansion of surfaces when indentors penetrate the sheets to be welded. Shoulders on indentors promote welding and limit distortion. Butt welding of wires is shown in Fig. 3.42 (*b*). Roll bonding, (Fig. 3.42 (*c*)) is highly effective because large extensions can be obtained.

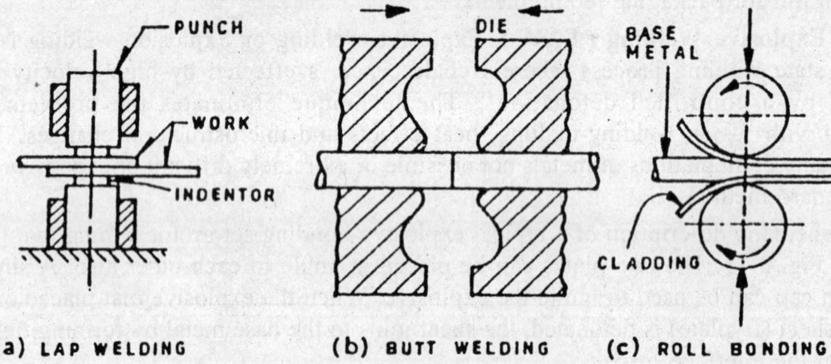

Fig. 3.42. Cold Welding.

EXW and USW also fall in the category of cold welding.

Commercial uses of cold welding include : the closing of aluminium cable sheaths and cases for semi-conductor devices, manufacture of kitchen utensils (tea kettles, supports, racks), lap and butt welding of wires and busbars of electrolysis cells, communication lines, trolley wires and home wiring.

Manufacture of Heat Exchangers. Aluminium sheets are given a coat of special paint or a parting agent such as graphite or a ceramic at areas where no joints are to be made. Bonding will be locally prevented at these predetermined places during roll bonding of the sheets. Due to the deformation produced by rolling, the sheets form bonds except where parting agents have been applied. The bonded sheets are then annealed and the parting agent is evaporated. Pressurised air or a fluid is passed through the areas formerly coated with parting agent. These get inflated or expanded to form parts such as evaporator plates, and heat exchanger tubes for refrigerators.

8. Diffusion Welding (DFW). Generally, better bonding is obtained when the temperature is high enough to ensure diffusion, typically above 0.5 times the melting point. Diffusion welding is defined as "A solid state welding process wherein coalescence is produced at the faying surfaces by the application of pressure and elevated temperatures. The process does not involve macroscopic deformation or relative motion of parts. A solid filler metal may or may not be inserted".

The method has been used for centuries by goldsmiths. Filled gold is obtained by placing gold face sheets over a silver or copper core. When a weight is placed on top of the sandwich and it is held in a furnace for a sufficient long time, a permanent bond is established. The required pressure can also be generated by : a press or by restraining the assembly with a fixture made of a lower expansion material, for example, Molyb- denum.

Since 1970s, the method has been extended to airframe construction. Parts of complex shapes are made of titanium.

9. Electron Beam Welding (EBW). EBW can be defined as "A welding process wherein coalescence is produced by the heat obtained from a concentrated beam composed primarily of high velocity electrons impinging upon the joint to be welded. The kinetic energy of the electrons is changed into heat on impact with work, giving intense local heating".

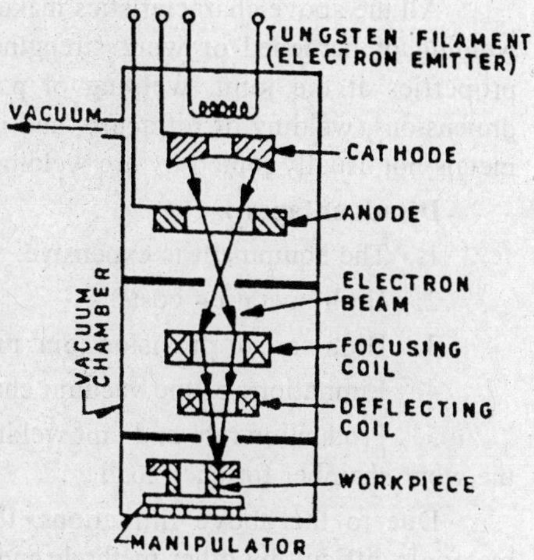

Fig. 3.43. Electron Beam Welding.

The electron beam is produced in a high vacuum environment by an electron gun, usually consisting of a tungsten or tantalum cathode, a grid or forming electrode and an anode. A stream of electrons is given off from a tungsten filament heated to about 2200°C, (Fig. 3.43). The electrons are gathered, accelerated to high velocity and shaped into a beam by the potential difference between cathode and anode. The beam is collimated and focused by passing through the field of an electromagnetic focusing coil or "magnetic lens". Beams typically are focused to about 0.25 to 1 mm diameter and have a power density of about 10 kW / mm^2, which is sufficient to melt and vaporize any metal. The operation is carried out in a vacuum, which enables the beam source to be at a distance of upto about 1 m from the work. Deep penetration with a very narrow heat affected zone is achieved by this process. Aluminium can be fused upto a depth of about 40 mm and stainless steel upto about 30 mm, with a width of fusion area of about one-tenth of the penetration.

The equipment comprises :

(*i*) an electron beam gun

(*ii*) a vacuum chamber

(*iii*) a means of moving the work under the beam while welding

(*iv*) A microscope for viewing the work.

Electron beam guns are rated at about 30 to 175 kV and a beam current of about 50 to 1000 mA.

The use of vacuum chamber is :

(*a*) to prevent the tungsten filament of gun from oxidising.

(*b*) it prevents the electrons from colliding with molecules of air and becoming scattered, with a loss of power.

(*c*) enables the welding to be done in a very pure atmosphere eliminating the need for a shielding gas.

Other advantages of EBW are :

(*i*) Ability to make welds that are deeper, narrower and less tapered than arc welds with a total heat input much lower than in arc welding.

(*ii*) Superior control over penetration and other weld dimensions and properties.

(*iii*) Higher welding speeds.

(*iv*) Clean and sound welds.

(*v*) Energy conversion efficiency is high, about 65%.

All the above characteristics make possible : minimize distortion and shrinkage in welding, welding of hardened or work strengthened metals, no significant deterioration in mechanical properties at the joint, welding of parts that have already been finished to final assembly dimensions, welding of refractory and reactive metals and welding of combination of dissimilar metals not usually joined by arc welding.

Disadvantages :

1. The equipment is expensive.
2. High operating costs.
3. High cost of precision joint preparation and precision tooling.
4. Limitations of the vacuum chamber. Work size is limited by the size of the chamber.
5. Production rate and unit welding cost are adversely affected by the need to pump down the work chamber for each load.

Due to the above limitations, the application of EBW is confined to the welding of materials difficult by other methods and to work where exceptionally good results are important.

10. Laser Beam Welding (LBW). The word 'LASER' stands for : Light amplification by stimulated emission of radiation. Laser welding is performed by focusing the coherent monochromatic light beam emitted by the laser source onto metal parts which are welded by surface heating and thermal conduction through the metal. There are two types of lasers in use : solid state lasers, for example, ruby (an Al_2O_3 crystal with Cr ions) and Gas lagers. Here we shall deal with solid laser.

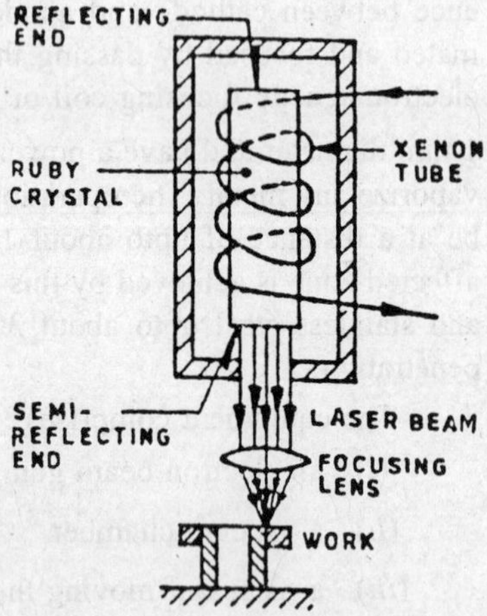

Fig. 3.44. Laser Beam Welding.

The laser crystal (Ruby) is in the form of a cylinder, (Fig. 3.44), the ends being flat and parallel to a high degree of accuracy and silvered to give mirror reflecting surfaces. There is a small aperture on the axis of the crystal, through the mirror at the output end. When the crystal is pumped with highintensity white light from a xenon or krypton lamp, the Cr ions in the crystal get excited. The excited ions possess more energy and some of it is given as a red fluorescent light. This light is reflected backward and forward in the crystal between the two ends (mirrors), striking more Cr ions on the path. These ions affected by the collisions are each caused to emit their quota of red light. There is a cumulative effect of the increasing red light exciting more and more Cr ions, until the number of collisions is high enough to cause a burst of red light through the small aperture in the mirror at the output end of the crystal. The beam produced is extremely narrow and can be focused to a pin point area by an optical lens.

The method is finding growing application, particularly for thin gauge metals, upto about 0.5 to 1.5 mm thick. LBW provides as high quality weld as EBW ; it can be remote controlled and requires no vacuum to operate in.

11. Thermit Welding (TW). Thermit welding is "A group of welding processes wherein coalescence is produced by heating with superheated liquid metal and slag resulting from a chemical reaction between a metal oxide and aluminium with or without the application of pressure. Filler metal, when used, is obtained from liquid metal".

Thermit is a mixture of powdered aluminium (1 part) and iron oxide in the form of roll scale (3 parts). When the thermit mixture is heated or ignited with a torch or ignition powder, to the temperature of 1100°C, it burns according to the following reaction :

$$8Al + 3Fe_3O_4 \rightarrow 9Fe + 4Al_2O_3 + heat$$

The reaction is nonexplosive and requires about a minute to go to completion. A temperature of about 2760°C and higher is reached in the combustion of thermit mixture. The reaction results in the formation of liquid reduced iron and slag (Al_2O_3) which are used for welding the pieces.

There are two types of thermit welding processes :

(*i*) plastic or pressure welding process.

(*ii*) fusion or non-pressure welding process.

Pressure Welding Process. The parts to be welded are butted tightly together and enclosed in a removable mould which provides a space between the inner surfaces of the mould and the parts. The products of the reaction are then poured into the mould in such a manner that slag enters the mould first. On contact with the metal parts, the slag cools rapidly and provides a layer of brittle, glasslike material so that the thermit iron which flows does not fuse with or adhere to the surface of the parts. The heavier thermit iron sinks to the lower half of the mould. This and the aluminium slag on top will both give up their heat to the pieces being welded. When the ends of the pieces to be welded reach the welding heat, they are forced together by means of clamps to make a pressure butt weld. The mould is then removed and the thermit iron and slag are knocked off from around the weld.

Non - pressure Welding Process. Here, the sections to be joined are lined up and a parallel gap or vee joint is cut at the fractured interface. A wax pattern is then formed between and around the ends which are to be joined. Then a sand mould complete with sprue, gate and riser is built around the joint area. The mould is then heated to melt the wax, causing it to flow into the sand mould and leaving a void space at the joint. The thermit powder is ignited in a crucible placed on top of the mould and the resulting iron is tapped-through a bottom hole of the crucible, directly into the mould. This highly superheated iron results in a deep penetration weld between the sections to be joined, (Fig. 3.45). After solidification, the mould is destroyed and the still hot excess iron is chiselled off.

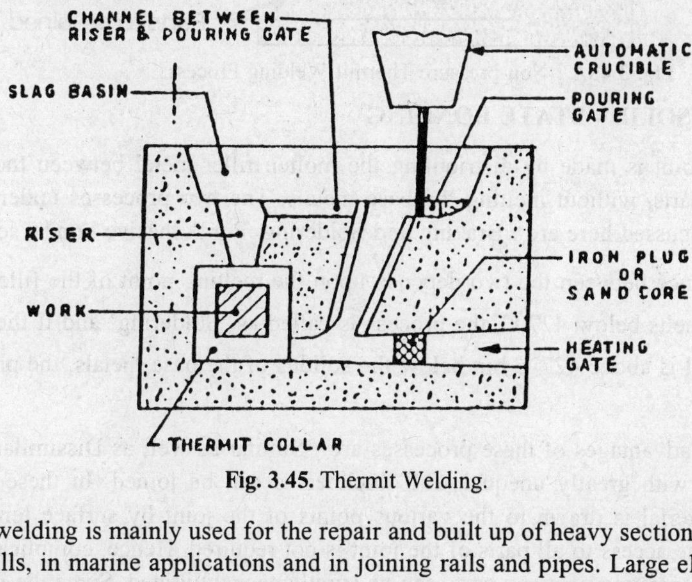

Fig. 3.45. Thermit Welding.

Thermit welding is mainly used for the repair and built up of heavy sections on crankshafts, steel rolling mills, in marine applications and in joining rails and pipes. Large electrical bus bars are similarly welded, using aluminium and Cu_2O. The chemical reaction being :

$$3CH_2O + 2Al \longrightarrow 6Cu + Al_2O_3 + Heat$$

It is clear that the thermal reaction (discussed above) results in temperature such higher than necessary. Again, the end product is iron and not steel, which must be the case when steel parts are to be joined by the non-pressure thermit welding process. For this, the reaction mix is added with Carbon, ferro-manganese and ferro-silicon. Carbon is added to convert iron to steel at the end of the reaction. Ferro-manganese and ferro-silicon are added as de-oxidising agents. They will help in cooling-off the reaction and getting an adequate final temperature. The reaction is started by burning a magnesium ribbon dipped in the reaction mix.

Fig. 3.45A shows a non-pressure thermit welding process for joining thick sections like rails etc. As already noted, the sand mould is in one piece. For this, wax is placed in the joint space and around. The gating system, also in wax, is attached. The complete assembly is embedded is moulding sand. The mould is inverted and heated. The wax melts and pours out leaving the cavity around the metal parts to be joined and also the pouring basin, runner and gating system.

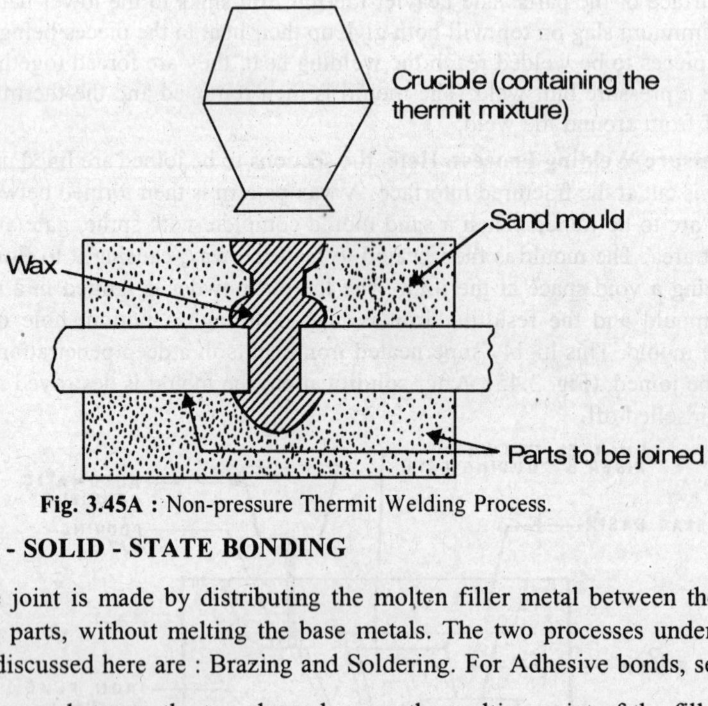

Fig. 3.45A : Non-pressure Thermit Welding Process.

3.6. LIQUID - SOLID - STATE BONDING

Here, the joint is made by distributing the molten filler metal between the closely fitted surfaces of the parts, without melting the base metals. The two processes under this category, which will be discussed here are : Brazing and Soldering. For Adhesive bonds, see Chapter 12.

The difference between the two depends upon the melting point of the filler metal. When the filler metal melts below 427°C, the process is called as 'Soldering' and if the melting point of the filler metal is above 427°C but below the solidus of the base metals, the process is called as 'Brazing'.

The major advantages of these processes are : Similar as well as Dissimilar materials can be joined. Parts with greatly unequal wall thicknesses can be joined. In these processes, the molten filler material is drawn to the various points of the joint by surface tension (capillary attraction). So, the access to all parts of the joint is not required. Hence, complicated assemblies and assemblies consisting of many parts can be simultaneously joined. Since the molten metal is distributed in the closely spaced joint by capillary action, the strength of the brazed or soldered joint is markedly affected by the width of the gap between the parts being joined together.

Closing the gap will improve the strength of the joint. However, too narrow a gap may spoil the capillary flow of the molten filler metal. The optimum gap width depends on the type of filler and base metals. In brazing steel parts with copper or Silver fillers, the optimum gap width ranges between 0.03 and 0.15 mm. In soldering with a binary tin-lead solder, it is from 0.05 to 0.2 mm. It should be understood that the strength of the brazed joint increases upon a certain gap between the two joining surfaces beyond which it decreases.

In these processes, no post joining heat treatment is normally necessary and joints are virtually free of internal stresses. The processes lend themselves to automation. However, these processes have the following limitations :

1. Relatively lower strength, particularly in the case of the lower melting point alloys.
2. Temperature resistance is limited by the melting point of the filler material.
3. Inspection of joint could be difficult.
4. Cost of joint could be high.
5. The process calls for narrow and uniformly spaced joints, due to which large parts/assemblies, allthough apparently brazable or solderable, are seldom made in this way.
6. Compared to welding, these processes require more accurate surface preparation (for better spreading and wetting action of filler metal) and assembly before the job can be done.

The two processes find extensive use in : the manufacture of automobile radiators, plate and tube heat exchangers, impellers, fans, fuel and oil pumps, appliance parts and for the joining of wires.

Brazing makes stronger joints as compared to soldering. The strength of a brazed joint ranges from 3 to 4.5 N/mm^2 while that of the soldered joint ranges from 0.3 to 1.0 N/mm^2. The term 'brazing' implies the use of brass as the filler material, but nowadays a number of other alloys are also in use. The filler metal is called 'spelter'.

3.6.1. Brazing. The brazing process can be defined as the process to join two metal pieces heated to suitable temperatures by using a filler metal having a liquidus above 427°C and below the solidus of the base metals. The filler metal is distributed between the closely fitted surfaces of the joint by "capillary attraction".

During brazing, the base metal of the two pieces to be joined is not melted. Some diffusion or alloying of the filler metal with the base metal takes place even though the base metal does not reach its solidus temperature. The greater the degree of adhesion and inter-diffusion between the molten filler metal and the base metals, the higher the mechanical strength of the joint will be. To achieve this and to obtain a strong joint, the basic requirement is that the filler metal must thoroughly wet the base metal surfaces. Therefore, the surfaces must be cleaned and free of contaminants that would prevent adhesion. Thus the scale is removed by mechanical (with a steel wire brush or emery cloth) or chemical (pickling in acids) means and heavy oily residues (oil, grease, paint etc.) are removed by degreasing with hot alkaline solutions or organic solvents. Again, when the assembly is

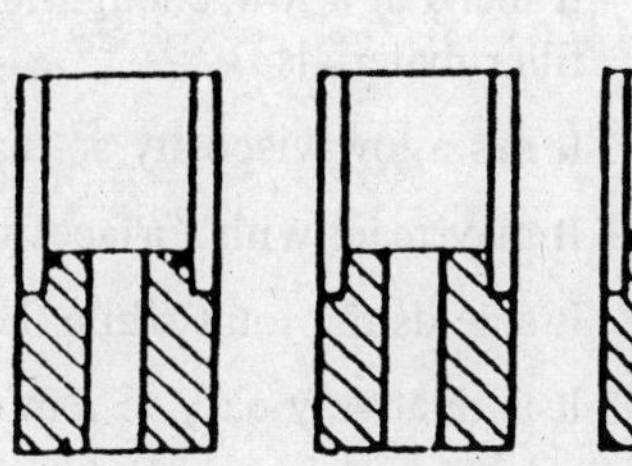

Fig. 3.46. Brazing Joints.

heated to melt the filler metal, oxides may form which will prevent wetting of the surfaces by the molten filler metal. This can be overcome by :

1. performing brazing is vacuum or an appropriate (neutral or reducing) atmosphere.
2. Wetting of the surfaces is also ensured by the application of fluxes.

Steps in Brazing.

(*i*) The surfaces to be joined are cleaned (and subsequently rinsed and dried) and fitted closely together.

(*ii*) A flux is applied to all surfaces where the filler metal is to flow.

(*iii*) After that, the joint is heated to the proper brazing temperature. Solid filler metal may be preplaced on the metal pieces and thus melted as the metal pieces are heated, or it may be applied to the metal pieces after the brazing temperature is reached. Only a small amount of filler metal is needed to fill the joint completely, (Fig. 3.46), which shows the various ways of placing the filler metal at the joint.

Some other commonly used joints in brazing are shown in Fig. 3.47.

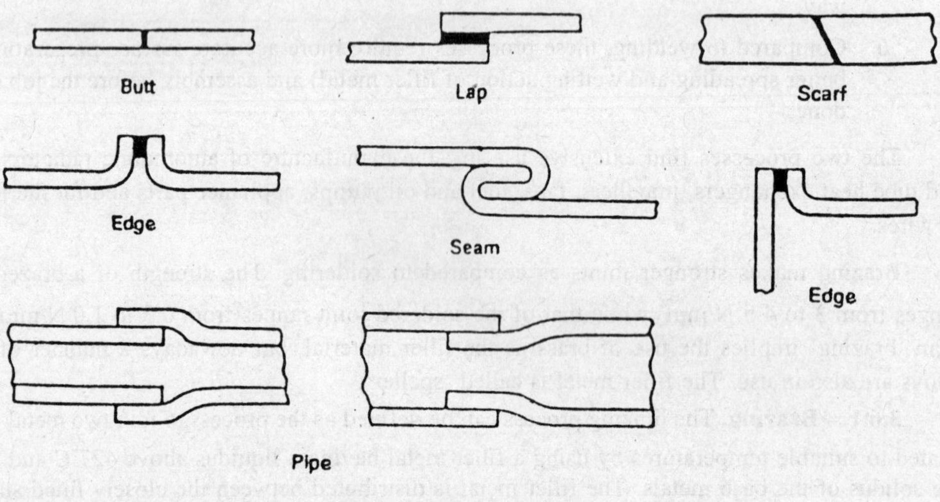

Fig. 3.47. Brazing Joints.

Fluxes. Fluxes are used to prevent oxidation of the base metal and the filler metal during brazing, form a fusible slag of any oxides which may be present or formed, and promote the free flowing of the filler metal by capillary attraction. The requirements of a good flux are :

1. It melts at a low enough temperature to prevent oxidation of the base metals and the filler materials.
2. It has a low viscosity so that it is replaced by the molten filler metal.
3. It may react with surfaces to facilitate wetting.
4. It shields the joint while the filler is still liquid.
5. It is relatively easy to remove after the solidification of the filler.
6. It lowers the surface tension of the molten filler metal and thus promotes its flow into the joint.

Common fluxes are : compositions of borates, fluorides, chlorides, borax and boric acid in various proportions according to specific requirements. Fluxes are used in form of powder, paste, or slurry. Borax is used as fused borax, because water in it will cause bubbling during heating. Alcohol can be mixed with fused borax to form a paste. Most fluxes are corrosive and the residue should be removed from the work after the brazing is completed.

Filler Materials. The filler metal or the braze metal must :

1. wet the surfaces of the base metals at the joint.
2. have high fluidity to penetrate crevices. For capillary attraction to exist, the clearance between the parts being joined must be quite small (0.03 to 0.05 mm), otherwise the filler metal would run out of the joint. A wider clearance in a joint leads to its reduced strength.
3. preferably have a narrow melting range.
4. not lead to galvanic corrosion during service.

The filler metal is applied in the form of wire, strip, preforms, powder or paste to the joint area as noted above in step (*iii*) under "steps in brazing". Alternatively, the filler metal is pre-applied to the surface of one of the contacting parts as a coating or cladding, often by rolling, electrolytic deposition or hot dipping.

Typical filler metals are listed in Table 3.3, for various classes of base metals. Because the base metals are not melted, the brazing metal is always different in composition from the base metals. In general, brazing metals of higher melting points will give higher strength, but this may affect the strength of the base metals. These brazing metals are based on Ag, Cu and Ni and are also called as 'Hard solders'.

Table 3.3. Brazing Materials

Braze metal	*Composition, %*	*Brazing process*	*Base metal*	*Brazing Temp.* °C
Brazing Brass	Cu : 60, Zn : 40	Torch, Dip Furnace	Steel, Cu, High Cu alloys, Ni alloys, Stainless steel	900
Manganese Bronze	Cu : 58.5, Sn : 1, Fe : 1, Mn : 0.25, Zn : 39.5	Torch	-do-	-do-
Nickel Silver	Ni : 18, Cu : 55 to 65, Zn : 27 to 17	Torch Induction	Steel, Ni, Ni alloys	-do-
Copper Silicon	Si : 1.5, Mn : 0.25, Cu : 98.25, or Si : 1.5, Zn : 1, Cu : 97.5	Torch	Steel	-do-
Silver alloys	Ag : 5 to 80, Cu = 15 to 52 balance : Zn + Sn + Cd	Torch, Furnace, Induction, Resistance, Dip	Steel, Cu, Cu alloys, Ni, Ni alloys, Stainless Steel	700
Silver alloys with P	Ag : 15, P : 5, Cu = 80	-do-	Cu, Cu alloys	-do-
Copper phosphorous	Cu : 93, P : 7	-do-	-do-	-do-

Brazing methods. Depending upon the heat source used to melt the brazing metal, the various brazing methods are given below :

1. **Torch brazing.** A joining process that may employ acetylene, natural gas, butane or propane in combination with air or oxygen to supply the heat required to melt the filler rod and diffuse it into the surface of the base metal. This technique is not extensively used for continuous mass production. Flux used is in the form of paste or powder.

2. **Furnace brazing.** This is a high production process where the heat is supplied by gas or electric heating coils. The furnaces are of box type (batch type) or continuous type. The latter employs a wire mesh belt to transport the parts to be brazed. In this process, preformed shapes of filler metal be placed on the parts to be joined prior to entering the furnace. The process is well suited to high production. The use of flux can be avoided if an inert atmosphere is maintained in the furnace. The flux is applied in paste form.

3. **Induction brazing.** Here also, the filler metal is used in the preformed shape. The parts are heated by placing within the field of high frequency induction coil. The parts are heated by eddy currents because the parts to be joined offer electric- magnetic resistance to the changing induction field. Heating is very rapid, and by properly shaping the induction coils, the heat can be applied in the local area of the joint to be brazed. The process can be mechanised. Flux is applied in the form of paste.

4. **Dip brazing.** dip brazing is of two kinds :

(*i*) In chemical dip brazing, the parts with preformed filler metal are placed into a molten bath of flux.

(*ii*) In the molten metal bath process, the assembled parts are first prefluxed and then immersed into a molten bath of filler metal.

The first method is more adaptable for joining large parts, whereas the latter method is restricted to small parts.

5. **Resistance brazing.** Here, the same equipment is used as for resistance welding, excepting that a filler metal is placed in the joint. Rapid heating minimizes oxidation and the heat affected zone (HAZ) is small.

6. **Laser brazing and electron beam brazing.** These high cost techniques are justified only for precision work of high value and relatively high-temperature materials.

Only the first four brazing methods are of industrial importance.

In the conventional Brazing and soldering methods, the use of flux can be avoided by doing the process in inert gas (say Ar) atmosphere or in vacuum (As in EB brazing).

Braze welding. It differs from the conventional brazing in that much wider gap is filled with brazing brass with the help of torch. Here, capillary action plays no part in making the joint. The technique is also used for the repair of iron and steel castings. The grooves and fillets can also be filled by using bronze rods and then the process can be called as "Bronze welding". The method is particularly useful for repairing cast iron castings. Straight welding of the casting requires its complete preheating, which is not always desirable or feasible. The brasses and bronzes have greater strength and ductility than cast iron and form a strong and satisfactory bond with the base metal. Since the base metal is not fused or highly heated, internal stresses are kept to a minimum and cracking is not likely upon cooling.

Silver brazing. When silver alloys are used for brazing, the method is called as "Silver brazing" which is often referred to as "Silver soldering" or "Hard soldering". The method is more expensive than plain brazing because the filler metal is more expensive. The method is preferred where a superior joint is needed and where the quantity of filler metal required is small. It is very popular in small area repair work and is used widely in the electrical and refrigeration industries.

Advantages of Brazing :

1. Lower temperatures are involved in brazing than in fusion welding. This minimizes distortion and allows joining at lower preheat temperature than required for fusion welding. Thus, medium carbon, and high-carbon steels and C.I. which harden in the heat affected zone and may be susceptible to cracking may be brazed or braze welded with little or no serious hardening.
2. Almost all the common engineering materials may be satisfactorily brazed or braze welded.
3. Inaccessible joint areas which could not be made by MIG or TIG processes, and spot or seam welding can be formed by brazing.
4. Thin-walled tubes and light-gauge sheet metal assemblies not joinable by welding can be joined by brazing.
5. Brazing can join dissimilar metals such as brass to stainless steels and carbon steels to alloy steels.
6. Multiple joints can be made at one time, as in furnace brazing.
7. Can be employed for joints which may have to be later disassembled.
8. Neat appearing joint requiring a minimum of cleaning.
9. The process can be done more quickly and more economically.
10. Many brazing operations can be mechanised.
11. Because of lower temperatures involved as compared to fusion welding, metallurgical damage to the base metal may be avoided.

Limitations :

1. Limited size of parts. Very large assemblies, although brazable, can be made more economically by welding because of the cost and availability of large equipment such as brazing furnaces.
2. Machining of the joint edges for getting the desired fit is costly.
3. Degree of skill required to perform the brazing operations is high.

Uses of Brazing. Assembly of pipes to fittings, carbide tips to tools, radiators, heat exchangers, electrical parts and repair of castings. Leak-tight joints for pressurized and vacuum systems are readily joined by brazing. The copper-brazed joints in steel and the nickel-alloy-brazed joints in stainless steels are made for high-temperature service. Corrosion resistance can be provided for infood-service equipment which employs silver-brazing alloys for joining stainless steels. The chemical industry uses nickle alloy brazing for brazing stainless steels subject to corrosive service.

3.6.2. Soldering. In soldering, two parts are joined by the use of a molten filler metal whose melting point is below the solidus (melting point of the base metals) and in all cases below 427°C.

Soldered joints are weaker than brazed joints. Because of lower working temperatures, good wetting is more critical than in brazing. The soldering process comprises the same steps as the brazing process, that is,

(*i*) Surface preparation which involves fitting the surfaces to each other, cleaning them mechanically and chemically and covering the cleaned surfaces with a flux. The clearance in a joint is about 0.05 to 0.20 mm (for steel).

(*ii*) After this, soldering proper is done.

Commonly used soldering joints are : Lap, butt, seam and pipe joints.

The fluxes used in soldering are : corrosive type and non corrosive type. The common corrosive fluxes are : Zinc chloride, mixtures of zinc chloride and ammonium chloride. The flux must be washed off after soldering to prevent corrosion. Common non corrosive fluxes are rosin and rosin plus alcohol. These are essential for electrical connections where corrosion can create local high resistance and even loss of conduction.

The most widely used solders are alloys of tin and lead in various proportions. Small quantities of some other metals may be added. These filler metals used in soldering are called "Soft solders".

(*i*) 63/37 Tin lead solder has the lowest melting point and solidifies at a constant temperature. It is most suitable for electrical connections.

(*ii*) 70/30 Tin lead : Good alloy pretinning.

(*iii*) 60/40 Tin lead : Good electric grade solder.

(*iv*) 50/50 Tin lead : General purpose solder.

(*v*) 32/68 Tin lead : Plumber's solder.

(*vi*) Tin-lead-silver solder : High temperature electrical solder for instruments.

(*vii*) 96.5/3.5 Tin-silver : High temperature electrical instrument solder.

(*viii*) A low (< 5%) tin content gives higher strength and is suitable for automotive radiators and lock seam cans, and tubes made of tin plate.

(*ix*) 35/65 Tin Lead : Its wide freezing range makes it ideal as a wiping solder for the joining of copper tubes.

(*x*) Lead silver solders are used for higher temperature service. Ag : 1.5 to 3.5%. Certain other metals such as cadmium, bismuth and indium may be added for some specific purposes.

Lead being toxic has adverse effects on the environment. Due to this, lead free solders are being developed and are now in wider use. They are being used in connection with supplying drinking water and other applications. Typical examples are :

1. Tin-silver (given above) : for electronics.
2. Tin-Bismuth (42%, 58%) : for electronics.
3. Tin-zinc alloys with 9 to 100%. Zinc for soldering of aluminium in conjunction with special fluxes.
4. Zinc-aluminium : for corrosion resistance, soldering of aluminium.
5. Cadmium-silver : for strength at high temperatures.

Soldering techniques. The various techniques used for brazing can also be employed for soldering (flame, hot dip, resistance, induction). Ultrasonic soldering (similar to ultrasonic welding) is used for fluxless soldering of aluminium and some nonferrous metals such as silicon and germanium.

In 'Radiant heating (infrared heating)', special quartziodine incandescent tubular lamps are used which can be easily focused to concentrate the heat at the area to be soldered. Radiant heating can also be used for brazing. In "wave soldering" fresh, hot solder is lifted to the connections by one or more standing waves of molten solder. The wave is formed by continuously pumping the solder through a narrow slot, to the surface. All joints are formed as the assembly passes over the waves. The operation is continuous and limited in speed only by time and heat requirements. This method is applicable to such items as the terminals of transformers and the lips of cans. The method is of great value in assembling printed-circuit boards.

Because of the lower temperatures involved, a hand-type heated copper or iron-plated copper bit (soldering iron) is extensively used. The tip of the iron is tinned before applying it to the work. The most common iron is of electric type and is used in many fields like radio and electronic industries. To do a soldering job, the iron is heated to 250 – 300°C, the tip is dipped into the flux, tinned with the solder and then applied to the parts prepared for soldering to heat them at the joint. At the same time, solder is applied to the joint where it is melted by the iron moving along the joint and enters the clearance. Here the solder cools and forms the seam.

3.7. WELDABILITY

Weldability of metals refers to their capacity of being welded into reliable permanent joints (sound welded joints) having specified properties, for example, definite weld strength, proper structure etc. In other words, weldability relates to the ease with which a metal can be welded and to the properties of the joint compared with those of the parent metal. Ease of welding depends upon the melting temperature, thermal conductivity, thermal expansion and the surface condition of the work. The weld quality is related to the possibility of the occurrence of weld defects, the partial loss of constituent elements from the metal due to vaporization and to changes in mechanical properties due to grain growth.

Effects of various elements on Weldability

1. **Carbon.** Carbon has a strong influence on the quality of the weld. An increase in the carbon content affects the strength, hardness and toughness of the weld. When steels having carbon content more that about 0.3% are welded and then cooled suddenly, a brittle zone results along the weld. The effect produced by carbon is much less in arcwelding than in gaswelding process.

2. **Manganese.** Manganese increases the tendency for hardening. When its content exceeds 0.6%, it lowers the ease of welding especially if the percentage of carbon is also high and such steels are subjected to cracking. On the other hand, a very low percentage of manganese tends to promote internal porosity with cracking. Optimum percentage is 0.4 to 0.6%.

3. **Silicon.** A high percentage of silicon especially alongwith a high carbon content may result in excessive cracking.

4. **Phosphorus.** Phosphorus in steel in excess of 0.04% has an unfavourable effect on the mechanical properties of weld since brittleness is increased.

5. **Sulphur.** Sulphur is added to some steels to give easy machining properties but too high a content has an extremely detrimental effect on the weldability of steels since it causes redshortness and tends to cause cracking at a sulphur content over 0.04%.

Elements such as nickel, chromium and vanadium in steels will lead to the formation of hard brittle zones. To prevent this, heat treatment processes will be required.

Weldability of Various Materials

1. **Low Carbon Steels or Mild Steel (C content upto 0.30%).** weld easily, no tendency to harden except those having the upper limit of carbon content.

2. **Medium Carbon Steels (C content 0.3 to 0.55%).** Some care is necessary. Hardening effect of the higher C-content can be prevented by preheating to about 250°C and then slow cooling of the weld.

3. **High Carbon Steels (C content 0.55 to 1.20%).** These steels can be welded but preheating and post weld treatments are necessary. With the higher carbon contents, a brazing operation is preferred because brazing temperature has little effect on the metallurgical condition of steel.

4. Low Alloy Steels. These steels have good weldability, especially if the carbon content does not exceed about 0.18%. With carbon content greater than this amount, special electrodes are necessary and preheating with slow after cooling will generally be required.

5. Cast Iron. The presence of free graphite and the high carbon content in cast iron lead to considerable difficulties in welding. Also, since its ductility is low, cracks may form in the weldment when cast iron is heated and due to uneven expansion and contraction caused by local heating. Cast iron may be welded either hot or cold.

Hot Welding. The work is preheated to 600 – 650°C and this temperature is maintained throughout the whole welding process. A considerable volume of liquid metal is produced at the weld. This retards the cooling of the molten metal and thereby prevents the chilling. The hot welding of cast iron can be done with either an electric arc or a gas flame and with either complete or restricted preheating of work. For welding small parts of intricate shape, complete preheating is done and is accomplished in metal boxes heated by charcoal. After completing the weld, the part is covered with asbestos or coal dust or embedded in hot sand to retard cooling.

The welding flame or arc tends to burn the silicon out of the weld, allowing the carbon to assume the combined form and render the metal hard when cold. This is overcome by using cast iron welding rods rich in silicon (upto 3.6% Si) so that the loss is replaced by the rod.

Cold Welding. The more widespread method is cold welding of cast iron with an electric arc. No preheating is necessary, because the heat is much more highly concentrated and localised. The finish weld is covered with sheet asbestos or embedded in hot sand to retard cooling.

6. Stainless Steels. Stainless steels always contain chromium, which forms an extremely dense Cr_2O_3 film, which tends to reduce the corrosion resistance of these steels. So, these steels need shielding during welding.

7. Aluminium Alloys. Aluminium oxidises easily and the film of aluminium oxide that forms on its surface has a melting point higher than the parent metal and impedes the fusion of the edges. So, aluminium and its alloys need to be protected against oxidation when welding. Before welding with OAW, the material must be covered with a flux. Preferred welding methods are : TIG and MIG. Also, aluminium and its alloys possess high heat conductivity and a high latent heat of fusion, therefore, they require more heat for welding.

8. Magnesium Alloys. Magnesium alloys are highly inflammable, have low melting point and form very refractory magnesium oxide on the surface of the welding pool. They get easily oxidised and also combine with nitrogen to form magnesium nitride which weakens the weld. Magnesium also dissolves hydrogen which leads to porosity of the weld. Due to all these factors, these alloys need perfect shielding during welding. The welding techniques are similar to those used for aluminium alloys.

9. Copper and Copper Alloys. Copper and its alloys can be welded by taking certain measures. Electrolytic copper can be welded with a gas flame or a carbon arc. The filler rod used in gas welding is of pure electrolytic copper or of copper containing upto 0.2% P and upto 0.3% Si. These two elements are intensive deoxidisers of welding pool. Because of the high heat conductivity of copper, the torch used should be from 1.5 to 2 times larger than for welding steel.

Brasses are welded by oxidising flame to prevent the fuming away of zinc. Similarly Bronzes are also welded by oxidising flame or arc and argon-arc methods.

10. Nickel Alloys. Most of these alloys are easily oxidised at the welding temperature and so shielding is necessary for good welds. Coated electrodes and shielded arcs can be used.

11. Titanium and Zirconium alloys. These alloys readily combine with N_2, O_2, C and H_2, especially when in the molten state. Therefore effective shielding by an inert gas is essential.

12. Refractory metal alloys. Tungsten, Molybdenum and Niobium can be welded, but the volatility of the oxides makes special techniques (*e.g.* EBM) mandatory.

Table 3.4 illustrates the suitability of various welding processes for various materials.

Table 3.4. Suitability of Various Welding Processes

Materials	*Welding / Brazing Process*							
	SMAW	*SAW*	*TIG*	*MIG*	*Spot*	*OAW*	*Furnace Braze*	*Torch Braze*
L-C Steels	R	R	S	S	R	R	R	S
M-C Steels	R	R	S	S	R	R	R	S
L-alloy steels	R	R	S	S	R	S	S	NR
Autenitic Stainless Steels	R	R	R	R	R	S	S	S
Ferritic and Martensitic Stainless Steels	R	S	S	S	S	S	S	S
High temperature, high strength alloys	R	S	S	S	S	S	NR	NR
C.I.	S	NR	S	NR	US	R	NR	R
Al and alloys	S	NR	R	R	R	S	R	R
Mg and alloys	US	US	R	S	S	NR	NR	NR
Ni, Zr and alloys	R	S	R	R	R	S	R	R
Ti and alloys	US	US	R	S	S	US	S	S
Cu and alloys	NR	NR	R	R	S	S	S	R

R : recommended, S: satisfactory

NR : not recommended, US : unsuitable

3.8. TESTING AND INSPECTION OF WELDED JOINTS

Testing and inspection of welded joints is done on the same lines as for castings. The tests fall under two categories : Destructive testing and Non-destructive testing.

Destructive Testing. These tests are done on a sample to improve the design of the weld, welding technique etc. and to know the mechanical properties of the weldment. These mechanical tests included : Tensile test, Bend test, impact strength test, hardness test and relative elongation test. The shape and size of the test specimen are selected to comply with state standards.

Non - Destructive Testing (NDT). The tests under this category include :

(1) **Visual Inspection.** Visual inspection of the weld and checking of its dimensions can reveal : shape of profile, uniformity of surface, undercuts, surface cavities and slag inclusions, cracks, porosity, unfilled craters etc.

(2) Hydraulic tests (pressure tests) are applied to weldments that are to operate under pressure.

(3) Air (pressure) tests are done to check the air tightness of the work.

The other NDT include :

— Radiographic inspection of the weld (x-ray and γ-ray testing) will reveal such defects as porosity, blowholes, cracks, poor fusion and slag inclusion.

— Fluorescent penetrant inspection to reveal fine surface cracks.

— Magnetic inspection reveals fine cracks and pores in the weld.

— Ammonia penetrant test : This is a leak test. The welded vessel is filled with compressed air to which 1% ammonia has been added. The welded joint is then covered with paper impregnated with a 5% solution of mercuric nitrate. Black spots appear on the paper in case of leakage.

3.8.1. Welding Defects.

When a weld is completed, the parent (base) metal is hottest at the point of fusion. The temperature of the base metal progressively decreases further away from the centre of fusion zone. Due to this temperature gradient, a fusion weld and the metal adjacent to it contain various zones having distinctly different grain structures and hence different properties.

Various zones of a typical weld are shown is Fig 3.48. These are :-

(*i*) *Fusion Zone* :- It is the portion of weld in which the base (parent) metal has been fused (melted).

(*ii*) *Heat Affectel Zone (HAZ)* : - This is the zone wherein the base metal is metallurgically affected by the heat of welding, but is not melted.

(*iii*) *Weld Metal Zone (WMZ)* : - This is the portion of the weld which consists of weld metal. This zone includes any base metal which is melted and resolidified, as well as any filler metal which did likewise.

(*iv*) Weld Zone : - Weld Zone = WMZ + HAZ

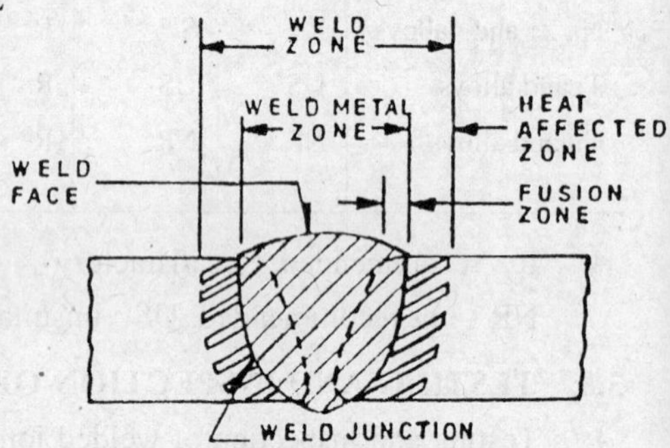

Fig. 3.48. Welding Zones.

The Welded joint should be free from any defect that would make it unsuitable for its imtended purpose. The principal defects to be found in welded joints are as follows :

1. *Cracks* :- Cracks that occur in or near a weld act as stress raisers which make the service life of the weldment very unreliable. Welding cracks are of the following types :

(*a*) *Micro-cracks* :- They are very small and are revealed only under a microscope.

(*b*) *Macro-cracks* :- These cracks can be seen by the unaided eye or by the use of a low power (5 to 10 X) magnifier.

(*c*) *Fissures* :- These are wide cracks which emerge to the surface of the metal.

Cracks may occur at the following locations :-

(*i*) In the weld (Fuse) Metal Zone

(*ii*) In the Base metal zone

(*iii*) Sometimes, the cracks originate in one Zoneand then spreal to the other Zone.

Cracks in the base metal usually occur at low temperatures (around 200°C) and are known as Cold cracks, On the other hand, the cracks in the weld metal zone occur while the metal is still very hot. Such cracks are called Hot cracks.

Cracks in the weld may be due to structural stresses in the metal (for example, the formation of martensite), heavy shrinkage, extra high amount of sulphur, phosphorous or carbon in the metal, excessively rigid clamping of the parts being weldel or the presence of gases in the weld metal.

Cold cracking can occur due any to the following several factors :- improper welding conditions, the presence of gas and other impurities in the weld, wrong choice of filler rod and metallurgical factors such as excessive cooling rate resulting in the formation of martensite and formation of brittle phases in the weld when cold or the formation of phases which are brittle at high temperatures [Allotropic transformations].

Cracks is the base metal can occur due to the following reasons :- Corrosion, Base metal defects, Base metal composition variations, hydrogen embrittlement and internal stresses set up due to restrained shrinking after welding.

The phonomenon of cracks can be explained as below :-

With thick parts, the weld zone is normally a small spot within a large mass of cooler metal. Heat flows off rapidly into the surrounding metal. When the welding is completed, what is called a "mass quench" results. This causes hard martensite to form in hardenable steel in part or all of the zone heated above the critical temperature. The martensite is brittle and does not yield but breaks when the stresses in the weld becomes high enough. Formation of martensite must be avoided. The more hardenable the steel, the slower must be the rate of cooling. The parts should also be preheated to retard the cooling rate.

Heating and cooling cause expansion and contraction in the weld. The metal in the joint is hotter and tends to shrink more on cooling than the bulk of the metal on either side. If the members being welded are restrained in a fixture or by a structure to which they are attached, high stresses are induced across the weld and thus cause cracking. Even if cracking does not occur, residual stresses may impair fatigue strength and cause warpage or distortion when they are unbalanced by later machining. These can be prevented by : using as thin material and as little filler metal as possible, preheating to minimize temperature difference between the weld and the base metal, and welding from the inside or confined portion of a structure to the outside or points of most freedom, a minimum number of welds and the maximum use of intermittent welds.

Gases such as O_2, N_2 and H_2 are readily absorbed by the molten weld metal. As the molten metal cools, its solubility for these gases diminishes and the gases try to escape. If they are not able to escape and remain within the weld, they will cause porosity and gas pockets in the weld. If they are able to reach just under the outer surface and are not able to escape (if the metal has cooled), they exert pressure and crack the metal. H_2 is a major cause of cracking in steel. Cracking due to gases can be avoided by preventing the evolution of these gases during welding (by thoroughly cleaning the joint from moisture, oil, grease, paint etc.) and absorbing of these gases from the atmosphere (by proper shielding). Slow cooling also eliminates cracking as the gases are able to escape.

2. Udercut. An undercut is a groove melted into the base metal adjacent to the toe of the weld, Fig. 3.49. The reasons for this can be : non-uniform feed of the filler rod, improper position of the electrode or torch tip or excessive heating. Undercutting can sometimes be corrected by adding additional metal to fill the groove.

3. Porosity or voids. Blowholes and gas pockets weaken welds appreciably and act as stress raisers from which cracks spread. The causes of these defects and the remedies have already been discussed above under cracking defect.

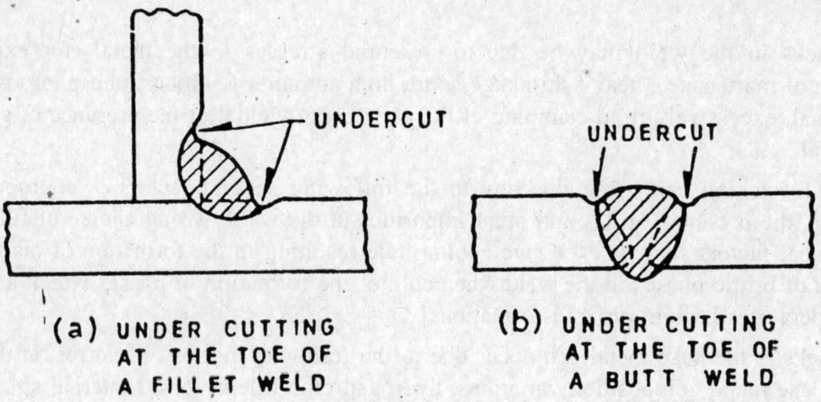

Fig. 3.49. Unercut.

4. Inclusions. Inclusions usually refer to slag but may be scale or dirt entrapped in the weld deposit during welding. The inclusions are due to the contamination of the base metal and the deposited metal by oxides, non uniform melting of the electrode coating, high melting point and high viscosity of the slag or insufficient deoxidising of the metal in the weld.

5. Lack of fusion. It is the lack of coalescence between the deposited and the base metal or incomplete penetration of the weld metal into the base metal. The usual cause is in ability to raise the temperature of the base metal to its melting point, and faulty welding conditions and/or techniques.

6. Burning. It is oxidation of the metal in the weld and adjacent base metal. It is caused by a strong oxidising medium, an arc of excessive length or excessively high welding current.

7. Internal residual stresses and distortion. Internal residual stresses and distortion or warping have already been discussed under ''cracking''.

Welding Metallurgy : It may be defined as the science dealing with the factors which affect the properties of the finished weld.

Since most welding is fusion welding, both the base metal welded and any filler rod which may be used, are melted and then resolidified during the process. Consequently, there is a close relationship between casting metallurgy and welding metallurgy. Both are concerned with the relationship of properties to identifiable micro-structural features, such as dendrites, segregation, columnar grains, centre-line weakness, porosity and inclusions. The problems associated with the solidification of castings, such as, hot tearing, shrinkage stresses, cracking, warping and dimensional changes, also occur in weldments.

The topic of metal working, both hot and cold, is another concern of welding metallurgy. Welds may be worked hot or cold by peening to obtain the sometimes benefitial effects of these processes. The microstructural characteristics of grain size, grain shape, phenomenon of slip, strain hardening, recrystallization and growth are the business of both metal working metallurgy and welding metallurgy.

Heat treatment accompanies every thermal welding process. Both the weld metal itself and the adjacent base metal (HAZ) undergo such heat treating processes as annealing, austenizing, preheating, postheating, normalizing, overheating, quench hardening, tempering, and recrystallization, either by deliberate attempt (intent), or as unavoidable *accompanying effects* of the welding process.

Thermal Effects on Weldment, Heat Affected Zone, Grain Size and its Control : As has been mentioned above when a weld is completed, the parent (Base) metal is hottest at the point of fusion. The temperature of the base metal progressively decreases further away from the centre of fusion zone, Fig. 3.49A. Due to this temperature gradient, a fusion weld and the metal adjacent to it contain various zones having distinctly different grain structures and hence different properties.

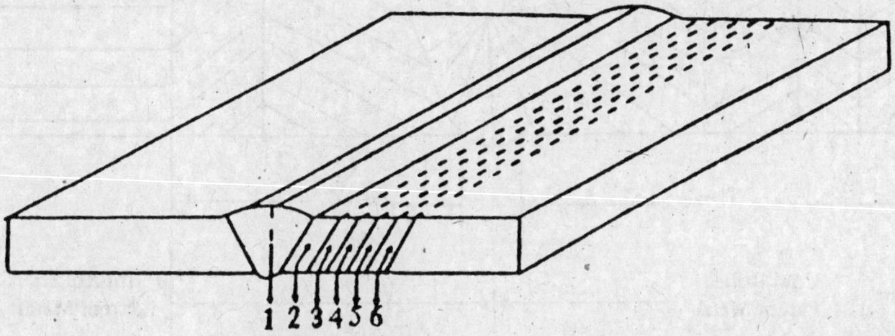

ZONE	Approximate Temperature °C
1	Above 1500
2	1400 1500
3	900 1400
4	600 900
5	300 600
6	150 300

Fig. 3.49A. Temperature Gradient in the Welding Zone (for steel).

As already noted the weld zone (WZ) comprises of weld metal zone (WMZ) and heat affected zone (HAZ), that is,

$$WZ = WMZ + HAZ$$

HAZ extends from the fusion line (which is the junction between weld dilution and the unmelted base metal) to the base metal which has not been sufficiently heated to alter the original structure.

After the completion of the welding process, the molten weld joint is allowed to cool to ambinet temperature. The solidification process is similar to that in a casting. Thus, in this zone (WMZ or fusion zone), the grain structure has many of the characteristics of a casting. The solidification process beings with the formation of columnar grains. These grains are relatively long (their lateral growth being restricted due to the neighbouring grains growing longitudinally) and form parallel to the direction of heat flow. Because the molten metal has cooled slowly, it generally generates coarse grains. Thus, the grain structure in the fusion zone has low strength, toughness and ductility.

The HAZ can be divided into sub-zones as shown in Fig. 3.49B. The first sub-zone of HAZ, next outward from the WMZ is the grossly HAZ or overheated zone. Here, the temperature of the metal is somewhat less than its melting point. Overheating causes severe grain growth which is very pronounced throughout this zone. It results in coarse grain structure and reduced mechanical properties in comparison with base metal. However granular refinement can be carried out by post-weld heat treatment.

The next sub-zone of HAZ, attains a temperature just above the critical temperature, AC_3 and long enough for the complete allotropic transformation to austenite. It is followed by a cooling rate fast enough to prevent grain growth from taking place. This results in a fine grain structure. This zone is called "refining zone" or zone of allotropic refinement".

The nature of the grain structure beyond the refining zone will depend on whether the parent metal was originally in the hot rolled or cold-rolled condition. In the case of hot-rolled condition, the next zone will consist of metal with the unchanged structure of the original base metal and is called as "Unaffected Zone".

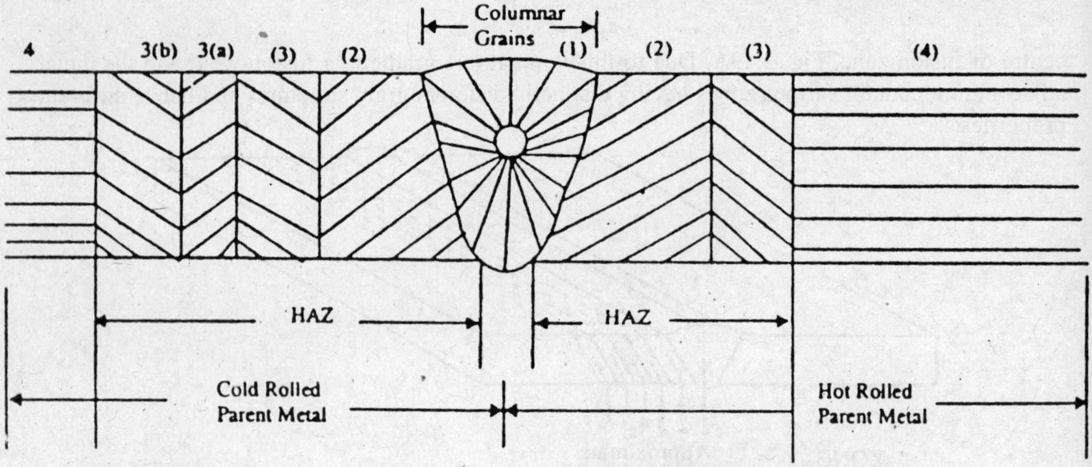

(1) WMZ
(2) Zone of extreme grain growth
(3) Zone of allotropic refinement
(3a) Zone of grain growth after recrystallization
(3b) Zone of recrystallized grains
(4) Unaffected zone

Fig. 3.49B. Grain Structures in Various Zones of a Welded Joint.

If in the cold worked condition, another metallurgical phenomenon will take place, of "recrystallization". The coarse and elongated grains resulting from cold working, will start breaking into fine, equiaxed grains. This sub-zone (of HAZ) of recrystallized fine grains will be separated from the "refining zone" by a small sub-zone containing grains of larger size than those which remained fine after recrystallization. This is caused by the growth of recrystallized grains nearer the high temperature zone (refining zone). Beyond the recrystallized zone, will be the unaffected zone.

3.9. WELDING COST

Estimates of welding cost may be done for one or more of the following reasons :

(*i*) To compare the economics of welding process with other processes of fabrication or manufacturing.

(*ii*) To compare the economics of various methods of welding process.

(*iii*) To determine the selling price of a product for a quotation, so as to ensure a reasonable profit to the company.

(*iv*) To check the quotations supplied by vendors.

(*v*) To decide whether a part or assembly is economical to be fabricated in the plant or is to be purchased from outside.

Elements of Welding Cost. The elements of cost in welding operations are : Labour cost, material cost and overheads.

The Labour Cost will depend upon the total time spent to make a weldment. The total time will consist of two parts : preparation time and actual welding time. The preparation time includes the following items :

(*i*) Time for preparing edges of the parts to be joined, by gas cutting, shearing, planing, shaping and grinding etc.

(*ii*) Time for assembly.

(*iii*) Time for tackup and positioning in welding fixture (if used).

(*iv*) Down time or lost time for changing electrodes or welding rods, moving to location and changing settings etc.

(*v*) Time for personal needs of the welder (going to lavatory, taking a cup of tea, smoking etc.)

(*vi*) Time for cleaning the welded joint.

(*vii*) Time for stress relieving (if required).

Operation factor, OF, is defined as :

$$\text{OF, \%} = \frac{\text{Actual welding time}}{\text{Total time}} \times 100$$

$$= \frac{\text{Actual welding time}}{\text{Actual welding time + Preparation time}} \times 100$$

$$\text{Actual welding time} = \frac{\text{Total length of job to be welded, m}}{\text{Welding speed, m/hr.}}$$

$$\therefore \text{ Labour cost} = (\text{Total welding time, hr.}) \times (\text{labour rate/hr}) \quad ...(a)$$

Material Cost will consist of : In case of OAW, the cost of welding rods, flux and gases (O_2 and C_2H_2). In case of arc welding, the items to be considered will be electrodes, flux or shielding gas, filler metal (if used) and power.

(*i*) **OAW.**

Cost of welding rods = welding rod consumption per m of weld × total length of weld, m × welding rod price

Cost of flux = Amount of flux used per kg of welding rod × Total kg of welding rod consumed × flux price ...(*b*)

Note. If flux is coated on the welding rods, then the cost of flux will be included in the welding rod price.

Cost of fuel gases = Amount of gases used, litres × gas price ...(*c*)

(*ii*) **Arc Welding**

Cost of electrodes = Electrode consumption per m of weld × total length of weld, m × electrode price

Cost of flux or shielding gas = flux or shielding gas used per kg of electrode × Total kg of electrode consumed × flux or shielding gas price ... (*d*)

Electrode consumption per metre of weld can be calculated as,

$$m_e = \frac{m}{1 - f}$$

where m_e = mass of electrode required, kg

m = mass of the metal deposited, kg

= area of weld, m^2 × length of weld, m × density of weld material, kg/m^3.

f = percentage of electrode mass lost as stub ends, spatter etc.

For SMAW, the value of f can be more than 30 %.

Note. In case of SMAW and FCAW, the cost of flux is not calculated separately, as it is included in the electrode price.

$$\text{Power cost} = \frac{V \times I}{1000} \times \frac{T}{60} \times \frac{1}{\eta} \times \text{rate per unit (kWh)} \qquad \text{...}(e)$$

where V = volts

I = current, amps.

T = welding time in minutes

η = efficiency of the welding machine

= 0.6 for a welding transformer (a.c.)

= 0.25 to 0.30 for a welding generator (d.c.)

Overheads. There is no direct method of calculating overheads. It is calculated indirectly as proportional to labour cost, as

Overhead cost = labour cost × overhead rate, where ...(f)

$$\text{Overhead rate} = \frac{\text{Total overheads of the company, for a period}}{\text{Total labour charges of the company, for a period}}$$

Overheads include : equipment depreciation and maintenance, taxes, insurance, technical supervision etc.

The sum of the items (a), (b), (c) and (d) will give the total welding cost.

For various welding processes, the cost items appear in different proportions. For example, for SMAW (manual),

Labour and overheads = 80 to 86%

Electrodes = 8 to 15%

Power and Equipment = 2%

On the other hand, for RW processes, the cost may be mostly for power and equipment and little for labour and non for consumable electrodes.

3.10. DESIGN ASPECTS

The following points should be kept in mind when designing a weldment :

1. Weldments should be designed to require a minimum of weld metal.
2. Thermal contraction of metal, which has been heated by welding, may cause internal residual stresses and distortion. These can be controlled or reduced by :

   (a) Preheating (b) Minimum number of welds (c) Smallest size of weld that fulfills requirements (d) Maximum use of intermittent welds (e) Slow after cooling.
3. Sharp discontinuities in metal should be kept at a minimum since these cause stress concentration.
4. An important strength weld should not be located where much of it may be removed later by machining.
5. Welds should be located so that adequate strength will be provided at the proper places on a structure or part.
6. As far as possible, a straight line force pattern should be provided.
7. Laps, straps and stiffening angles should be avoided except as required for strength.
8. Lap welds and lap strap welds are not recommended for elements over 10 mm thick.

9. Where ever possible, use butt welds.
10. The ends to be welded should be of equal thickness.
11. Welds at the vulnerable cross-sections should be avoided.
12. The use of welding fixtures should be avoided as far as possible.
13. Welds should not be subjected to bending.
14. A weld should not be located at the point of maximum deformation.
15. Ribs should be designed correctly and these should be used with care.
16. Provide for easy access to welds so that they are accessible for inspection.
17. Distribute heavy loading over long welds in the longitudinal direction.
18. Avoid large flat walls, which tend to bulge and flex.
19. The joint should have properly prepared grooves.
20. If alternating stresses are involved, avoid running a weld at right angles to the direction of maximum principal stress owing to the low fatigue resistance offered by welds.
21. Whenever possible, the design should provide for welding in the flat or horizontal position, not overhead.

General characteristics of common welding processes are given in Table 3.5, below :

Table 3.5. Characteristics of welding processes

Welding Process	*Materials to be welded*		*Min. Thickness mm*	*Costs*			*operator skill*	*Operation*
	General	*Preferred*		*Equipment*	*Labour*	*Finishing*		
OAW	All but refractory	C.I., Steels	0.6	D-E*	A	A	A	Manual
SMAW	All but Zn	Steels	1.5 - 2.0	D	A	A - B	A	-do-
FCAW	All Steels	L - C Steels	1.5	B - D	A - D	A - C	A - D	All
SAW	All Steels	-do-	5	B - C	B - D	A - C	C - D	Automatic
MIG	All but Zn	Steels, Al, Cu	0.5	B - C	A - C	B - D	A - D	All
TIG	-do-	All but Zn	0.2	B - C	A - C	B - E	A - D	All
EBM	-do-	-do-	0.05	A	A - D	C - E	A - D	All

* A indicates the highest value and E indicates the lowest value.

3.11. HARD FACING OF METALS

Hard facing or surfacing is the technique of depositing a layer of hard material on a component to increase the hardness, strength and wear resistance of the base metal. The technique is widely used in bearings, camshafts, valves and valve seats, hot extrusion dies, closed dies especially for abrasive powders, earth handling and mining equipment of many types such as rock drills, stone crushers etc., hammer mills, shear blades, and many types of trimming and cutting dies.

The composition of the surfacing metal differs from that of the base metal. Hard facing materials include stellite and other cutting and wear resistant alloys. Tips or rods from 5 to 10 mm thick, cast from stellite alloy, are used in the hard facing of tools by welding techniques. The cutting tool materials with very hard phases (mostly carbides) have such a high alloying element concentration that they cannot be manufactured into welding rods. The ingredients are incorporated in the flux coating or packed inside tubular rods, and the alloy is formed in the welding process itself.

Hard facing is done by means of gas, arc, or shielded arc welding techniques. Gas and shielded arc welding ensures a more uniform composition of the deposited layer. Surfacing by ordinary arc welding is faster and cheaper, but there is greater danger of dilution of the surfacing metal with the base metal. Deposition of tungsten carbide by an electric arc is called "spark hardening", useful for cutting tools. When thick layers are deposited, one speaks of 'weld overlays'. However, the thickness of the deposit should not, as a rule, exceed 2 mm, because the susceptibility to cracking increases with thicker layers.

The hard facing techniques and conditions should ensure a strong bond of the deposit with the base metal, restrict their mixing and avoid the formation of cracks and other defects in the deposited layer. Parts to be hard faced are first preheated to 350 to 500°C, the hard faced parts are to be cooled slowly.

Hard facing increases the service life of certain parts by 3 or 4 times and enables worn parts to be repeatedly restored.

3.12. CHARACTERISTICS OF WELDING PROCESSES

The characteristics of the main welding processes are given next page in Table 3.6.

Table 3.6. Main Welding Processes and their Characteristics

S. No.	*Welding process*	*Welded Materials*	*Recommended thickness or cross-section of components welded*	*Type of welded joint*	*Methods of cleaning components before welding*
1.	Gas welding	Steel, Al alloys, Cu alloys, hard alloys	< 2 mm – 10 mm	Butt, Edge	By steel-wire brush
2.	PGW	Steel	< 25,000 mm^2	Butt	Machining ends with cutting
3.	SMAW	Steel, Al Alloys	> 1.5 – 2 mm	Butt, Lap, T-, and Edge	By steel-wire brush
4.	Carbon-arc Welding	Low C-Steel, Al, Cu	> 4 – 12.0 mm	Butt and Edge	-do-

5.	SAW	Steel	> 2 – 2.5 mm	Butt, Lap, T-, and Edge	-do- or gas flame
6.	Argon-shielded arc welding	Stainless steel, Al and Mg alloys	< 4 mm	Butt, T and Edge	By steel wire brush
7.	Resistance spot welding	Low-carbon, alloyed and stainless steels; Al and Cu alloys	< 12 mm, < 10 mm, < 6 mm, < 2.5 mm	Lap	Cold-drawn steel without cleaning; hot rolled steel cleaning by etching, sand blasting or machining
8.	Flash butt welding	Steel and alloys		Butt	By steel-brush
9.	Upset Butt Welding	Steel, Al Alloys and Cu alloys	upto 1000 mm^2	-do-	Ends cleaning by machining
10.	Atomic hydrogen welding	Alloyed steels	Bars < 10 mm	Butt, T- and Edge	By steel-wire brush
11.	Cold welding (Roll welding)	Low C-, alloyed, and stainless steels; Al and Cu alloys	< 8 mm, < 2 mm	Lap	Sand blasting or machining
12.	LBW	Chemically aggressive and refractory metals and alloys	< 1 mm	Lap	Careful cleaning and degreasing not essential
13.	EBW	-do-		Butt	-do-
14.	USW	Similar and dissimilar metals and alloys	< 10 mm, < 0.05 – 0.5 mm	Lap	-do-
15.	Friction Welding	Carbon and alloys steels, non-ferrous alloys	Round section components < 40 mm dia.	Butt	By steel-wire brush or sand blasting.
16.	ESW	Carbon and alloyed steels	20 – 600 mm	Butt	-do-

3.13. ADHESIVE BONDS

Permanent joints can be made by means of adhesives applied in a thin layer between the connected parts. They are employed to fasten metal elements, metal and non-metallic (textile laminate, foam plastic) meterials, and two non-metallic elements.

Advantages

1. Reliable connection of parts made of very thin sheet materials.
2. Dissimilar materials can be joined.
3. The process is used to reduce production costs.
4. It is used to lessen the mass of parts.
5. Highly skilled labour is not required for bonding.
6. The process provides for tight and corrosion - free joints.
7. Smooth bonded surfaces.
8. Exterior surfaces remain smooth.
9. Only low temperatures are involved, therefore, absence of stresses or their lower concentration.
10. Heat sensitive materials can be joined without any damage.
11. Complex assemblies can be made at low cost.
12. Adhesives contribute towards absorption of shocks and vibrations.
13. Adhesive bonds can tolerate the thermal stresses of differential expansion and contraction.
14. Because the adhesives bond the entire joint area, good load distribution and fatigue resistance are obtained.
15. The joints are sufficiently strong in shear and withstand dynamic and variable loads.
16. Compared to welded, soldered and riveted joints, adhesive bonded parts have uniformly spread stresses and do not tend to warp.
17. The process is very fast.
18. Stress Concentrations are reduced or entirely absent.
19. Adhesives are indispensable where Welding, Soldering or Brazing fails or where bolts and rivets cannot be used for fear of stress concentration e.g., honeycomb structures.

Disadvantages

1. Comparatively low operational temperatures (maximum upto about 100°C for most adhesives).
2. Low resistance to tear - off.
3. Reduced strength of some adhesives in the course of timing (ageing).
4. Tendency to creep, if subjected to long standing and heavy loads.
5. The need for extended polymerization time.

Applications

Adhesive bonds have been particularly developed in the air-craft industry : The appearance of honeycomb elements is entirely due to this method. The method is used in critical structures such as control surfaces in aircraft and entire airraft body. In machine tools, adhesives are employed to bond carriage - guide - ways to beds, and in automobile industry to fasten friction linings to clutch - disks and brake - bands. Adhesive bonds are also used in appliance and consumer goods fields and also for sealing, vibration damping and insulating etc.

3.13.1. Adhesive-bonded Joints. Adhesive - bonding of parts is effected on the following types of surfaces :

1. On cylindrical surfaces, for example, placing bushings into holes in housing - type parts and discs onto shafts, coupling pipes together, fitting plugs, and fastening linings to brake blocks etc.

2. On flat surfaces, for example, lap - type joining of sheet parts with one or two straps and so on.

Typical adhesive-bonded joints are shown in Fig. 3.50.

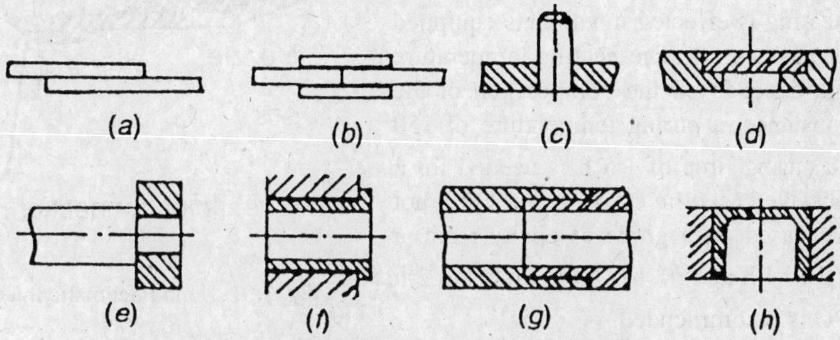

Fig. 3.50. Main types of Adhesive-bonded Joints.

The strength of an adhesive joint is dependent on the amount of clearance in it, which is normally kept at 0.05 to 1.5 mm. With increased clearance, the strength of the joint decreases. As the length of over lapping of the joint increases, the force needed to break down the joint increases asymptotically approaching a certain limit. Surface roughness of the parts bonded should be held to within 6.3 to 1 μm Ra. Increase incuring time has a favourable effect on the strength of the adhesive-bonded joint. With cold - curing, the strength grows continuously over a long period of time. The strength of joints bonded with cold curing adhesives increases if the polymerization process is accompanied by heating. Heating also greatly reduces the curing time.

Making an adhesive - bonded joint

An adhesive - bonding process comprises the following steps :

(*i*) Preparation of part surfaces.

(*ii*) Preparation and application of the adhesive.

(*iii*) Assembly of parts under a pressure determined by the grade of the adhesive.

(*iv*) Heating of the assembled product.

The surfaces to be bonded must be cleaned and degreased. Cleaning is done with wiping wastes, brushes or in a sand blaster. The substances used for degreasing are : acetone, trichloroethylene, Carbon tetra chloride and other organic solvents. Aluminium- alloy parts are prepared by pickling. Where necessary, the surfaces to be joined are machined to obtain a surface finish that provides for better holding of the adhesive.

Adhesives are prepared in special poly chloro ethylene or metallic containers ; the latter need to be chrome-plated or coated with silicone varnish. Hot curing adhesives can be stored in closed containers for a long time. Cold - curing adhesives are prepared just before use as their pot life is only 30 to 40 min.

The method of application of an adhesive depends on its viscosity. Liquid adhesives, that can be applied with brushes or sprayers, are used most commonly. Some grades of adhesives are convenient to apply with spatulas, rollers or injectors.

The adhesive is spread in an even thin layer (0.1 – 0.2 mm) with a bristle brush or a spatula. To prevent frothing, the adhesive must be applied moving the brush in one direction.

In hand pneumatic injectors, (Fig. 3.51), compressed air is supplied through an inlet connection. The air extrudes the adhesive by means of a piston through a nozzle having a diameter of 1 mm.

After the application of adhesive, the parts are assembled in special fixtures and clamped by means of lever mechanisms, springs, or pneumatic clamping devices. Clamping force must ensure a unit pressure of 0.05 to 1.0 MN/m^2.

Lastly, heating is effected in cabinets equipped with electric or gas heaters. The heating temperature and curing time depends on the composition of the adhesive. For instance, a curing temperature of 150 to 160°C and a curing time of 1.5 h is needed for a cold-curing adhesive based on epoxy resin. For a hot curing adhesive based on a grade of epoxy resin, a curing time of 3 to 4 h at 150° to 160°C or 1.5 to 2 h at 180° to 190°C is recommended.

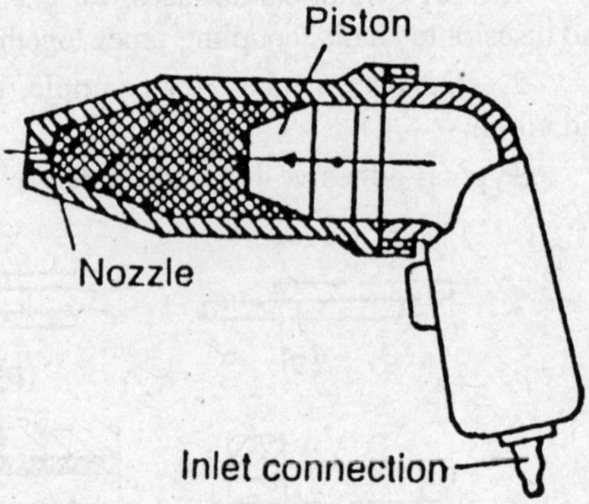

Fig. 3.51. Hand Pneumatic Injector.

Adhesives should be handled very carefully as their constituents are toxic. The work, therefore, should be done with gloves on, under proper exhaust ventilation.

3.13.2. Adhesives. There is a large variety of adhesives available for bonding metals with metals and metals with structural non - metallic materials. They can be classified into the following main groups :

1. Adhesives based on epoxy resins. The available epoxy resin- based adhesives are both cold and hot curing. These are used for cold and hot joining of metals, ceramics, plastics, wood and other materials.

In cold curing adhesives, a curing agent, such as polyethylene polymide (8 to 10 parts by mass) or hexamethylene diamine (20 parts by mass) is added to 100 parts by mass of the resin. Maleic anhydride (40 parts by mass) is added as a curing agent to the resin in making hot curing adhesives. The various epoxy resins used as adhesives are given below with the curing temperatures given within brackets :

Epoxy (room temperature cure, 16 to 32°C)

Epoxy (elevated temperature cure, 93 to 177°C)

Epoxy nylon (121 to 177°C)

Epoxy phenolic (121 to 177°C).

2. Phenol - resin based resins. These are modified by various compounds. Curing takes place at a temperature of about 150°C with the jointed components being held against each other. Phenol polyvinyl acetate adhesives are available readymade without subsequent introduction of curing agent. These adhesives can sustain temperatures upto 70°C. Phenolic rubber and phenolic resin-based adhesives modified by organic polymers and silicone compounds feature high temperature resistance. Phenol formaldehyde is used to bond foam plastics, textile laminate etc. The other common adhesives in this group are :

Neoprene - phenolic (135 – 177°C)

Nitrile - phenolic (135 – 177°C)

Butyral - phenolic (135 – 177°C)

3. Polyurethane adhesives. These adhesives have resistance to temperature of 100 to 120°C and the same strength as Phenol polyvinyl acetate adhesives.

4. Special grade adhesives. These are used for higher temperature resistance and possess high shearing strength.

Bonding Plastic parts. The above mentioned adhesives and special- purpose adhesives are used to bond plastic parts. For many thermoplasts, their solvents serve as adhesives, for example, dichloroethane for organic glass, benzol for polystrene, acetone for viniplast etc. The scope of automation of adhesive-bonding processes is the application of the adhesive to mating surfaces, assembly and accurate location of the parts bonded, and subsequent curing. Adhesives can be applied with rollers, or fed with an injector into the clearance between the mating parts ; dipping the mating parts into it is also practicable.

PROBLEMS

1. Define the "welding process".
2. Give the applications of the "welding process".
3. Write the advantages and drawbacks of the "welding process".
4. How the "welding process" may be classified ?
5. Differentiate between "autogeneous", "homogeneous", and "hetro-geneous" welding processes.
6. Sketch the various weld joints.
7. Sketch the various types of welds used in making a joint.
8. Sketch and write on the various edge preparations used for welded joints.
9. Sketch and write on the various "welding positions".
10. Why the cleaning of a joint is important before welding ?
11. What is meant by "fluxing" ? Why it is done ? What are the properties which a good flux should possess ?
12. Define : OAW and PGW processes.
13. Sketch and compare the two systems of OAW process.
14. Write on the "Gas welding equipment".
15. Sketch the three types of flames used in OAW process. Give the uses of each.
16. Write the steps in lighting up the OAW flame.
17. Write the closing down procedure of OAW flame.
18. Define power of blow pipe and nozzles.
19. Sketch and compare the two welding techniques used in OAW process.
20. Sketch and explain PGW process.
21. Write about the other gases used in OW process, in place of acetylene.
22. Write on "oxy-acetylene flame cutting". How the cutting tip differs from a welding tip ?
23. Define "electric arc welding".
24. List the principal advantages of :
    (*a*) Arc welding over gas welding. (*b*) Gas welding over arc welding.
    (*c*) D.C. arc welding over a.c. arc welding.
25. Sketch the two polarities of d.c. supply and compare these for welding process.
26. Write on the different types of electrodes used in arc welding.
27. What is the purpose of coating on an arc welding electrode ?
28. Write the constituents of a "coating" and write the function of each.
29. Write on "coding" of electric arc electrodes.
30. Name the ten methods of arc welding. Of these methods, which is the most widely used ?
31. Explain the following electric arc welding processes with the help of neat sketches :

a. SMAW *f.* Atomic hydrogen welding
b. FCAW *g.* Electro-slag welding
c. GTAW *h.* PAW *d.* GMAW *i.* Stud welding *e.* SAW

32. Name the shielding gases used in "Inert gas arc welding" methods and compare them.
33. What metals can be joined by "shielded arc welding" ? What are some of the most important advantages gained by this type of welding ?
34. Write a note on "electric arc cutting".
35. Define "Resistance welding process".
36. Name six types of resistance welding methods. For what kind of production is resistance welding mainly employed ?
37. With the help of a neat sketch explain the "spot - welding method".
38. What metals may be spot - welded ? Can dissimilar metals be spot - welded ?
39. Sketch and write on the various spot welding machines.
40. How does "seam welding" differ from "spot welding" ?
41. From what materials are spot welding electrodes usually made ?
42. What are the special features of "resistance projection welding" ?
43. With the help of neat sketches explain the following welding methods :
    a. Upset butt welding *b.* Flash butt welding
    c. Percussion butt welding.
44. What is the difference between "flash" and "upset welding" ?
45. With the help of neat sketches explain the following welding methods :
    a. Forge welding *f.* Cold welding
    b. Induction welding *g.* Electron-beam welding
    c. Friction welding *h.* Laser-beam welding
    d. Ultra-sonic welding *i.* Thermit welding
    e. Explosive welding.
46. For what commercial applications can the EBW process be economical ?
47. What are "flow welding" and "diffusion welding" processes ?
48. Distinguish between "welding", "brazing" and "soldering" processes.
49. Which method of resistance welding is used to join dissimilar metals ?
50. Distinguish between "brazing" and "braze welding".
51. Write the steps to be taken in brazing process.
52. Write about the various fluxes used in brazing process.
53. Write about the filler materials used in brazing process.
54. Write a note on the various brazing methods.
55. Write the advantages and limitations of brazing process.
56. Write the common uses of brazing process.
57. Distinguish between "soft solder" and "hard solder".
58. Write about the various soldering techniques used.
59. Define "weldability".
60. What effect does carbon content of steel have on weldability ?

61. What are the effects of the following elements on weldability ? Mn, Si, P, and S.
62. What is the purpose of preheating a part to be welded ? How would you select a preheat temperature ?
63. Write briefly on "Testing and Inspection of welded joints".
64. What causes weldments to crack ? Explain the reasons and suggest the remedies.
65. Write the various welding defects. Give their reasons and suggest the remedies.
66. Write a note on welding costs.
67. Give the design aspects of a weldment.
68. Write on "hard facing of metals".
69. List the safety measures to be taken in welding shop.
70. Explain the principle of Arc welding.
71. List the work preparation for welding operation.
72. Identify various welded joints by BIS symbols.
73. List the major advantages and limitations of OAW.
74. Write about selection of filler rod and flux for OAW.
75. List the functions performed by the electrode coatings.
76. How electrodes are selected for a job ?
77. List the common solders used in the soldering method.
78. Write the various fluxes used in the soldering method.
79. State applications of soldering process.
80. State the advantages and limitations of soldering process.
81. Explain fusion as it relates to welding processes.
82. Explain the chemical reaction that takes place in an OAW torch.
83. What is the level of temperatures obtained in the three types of flames of an OAW tore ?
84. Why, for welding Brasses and Bronzes, an oxidizing flame is desirable?
85. Why is SMAW a commonly used method of electric arc welding ? Why is it also called stick welding and MMAW ?
86. Why is the quality of SAW very good ?
87. Why a flux is not needed in GTAW and GMAW ?
88. Why is tungsten the preferred material for non-consumable electrodes ?
89. What is the advantage of EBW and LBW as compared to arc welding ?
90. What are : Weld Zone, Fusion zone, Weld metal Zone and Heat affected Zone ?
91. Discuss the metalurgical effects due to thermal gradients in HAZ.
92. Why is OAW limited to rather thin sections ?
93. What is Arc stability ? How is it achieved ?
94. What is Arc blow ? How is it avoided ?
95. What are the sources of weld spatter ? How can it be controlled ?
96. List a few products that can be fabricated by resistance welding processes.
97. Explain the principle of Electric Resistance Welding.
98. Describe the features of a fusion weld. Identify the different zones.
99. Differentiate between Hot welding and Cold welding.
100. How can cracking in weldments be avoided ?

101. The voltage – arc length characteristic of a D.C. arc is given by V = (20 + 40L) volts, where L is the arc length in cm. The static – volt ampere characteristic of the power source is approximated by a straight line with no load voltage of 80V and a short circuit current of 1000 A. Determine the optimum arc length and the corresponding Arc power. (**Ans.** 0.5 cm, 20 kVA)

102. The voltage – arc length of ad.c. source is given by

V = (20 + 4L) volts, where L = length of the arc in mm. The arc length is expected to vary between 4 mm and 6 mm and it is desired to limit the current in the range 450 to 550 A. Assuming a linear power source characteristic determine the open circuit voltage and the short circuit current of the power source. (**Ans.** 80 V, 1000 A)

103. Write a short note on Adhesive bonds.

4

Mechanical Working of Metals

4.1. GENERAL

Mechanical working processes are based on permanent changes in the shape of a body, that is, on the plastic deformation under the action of external forces. Mechanical working processes include : rolling, forging, extrusion, drawing and press working (sheet metal working). The stresses induced in the part (due to the application of external forces) are greater than the yield strength and less than the fracture strength of the material, except in shearing, piercing and blanking sheet metal working processes, where the stresses induced in the part are equal to or greater than the fracture strength of the material.

Mechanical working processes exclude the machining (metal cutting) processes, where, even though external forces are used and the stresses induced in the part are equal to or greater than the fracture strength of the material, the part is obtained by removing the material in the form of chips. In mechanical working processes, no chips are produced. In shearing, piercing and blanking sheet metal working processes, even though the part is obtained by cutting it from the sheet metal, no chips are produced. That is why, these processes are also called 'chipless manufacturing processes'. In all other mechanical working processes, the size or shape of a part is obtained by deforming and displacing the material under the action of large external forces. Hence these processes are also called as metal working or 'metal forming processes'.

Starting materials for the mechanical working processes are: cast ingots and billets of various cross-sections and weights, cast preforms, powdered-metal bars, bar stock, powdered-metal preforms or blanks (for press working processes).

Advantages of Mechanical Working Processes over other manu facturing processes.

A high-rate of the output of metals and alloys is one of the mainstay of the present stage of the scientific and technical revolution, and despite the appearance of a great number of new non-metallic materials, mechanical working of metals is certainly to develop at an ever increasing rate in the decades to come. 90% of all metal produced are worked today by rolling, forging, extrusion, drawing etc.

Metal working processes possess certain special advantages as compared to other manufacturing processes. A high effectiveness of mechanical working of metals is due to the following advantages :

1. Higher productivity, as compared to other manufacturing processes. Modern rolling and forging units fabricate hundreds of tonnes of end products per working shift. Hence these processes are high production rate processes and are most suitable for mass production.

2. The possibility of fabrication items with preset specific physical and mechanical properties and required structure, which, in the final analysis, ensures high-standard quality of fabricated products, one which cannot be equalled by any other alternative technique.

This is due to the fact that the fibre lines in the material are not interrupted, but are made to follow the contour of the formed part (The strength of a component is maximum parallel to the direction of the fibre lines, and minimum perpendicular to them). In case of casting, the fibre lines get destroyed on melting of the material. A cast component will have the typical cast dendritic structure resulting in very low mechanical properties. Also, the 'yield' of casting processes varies between 60 to 70%. A machine component has the drawbacks that during manufacture, the fibre lines are also cut and get interrupted resulting in poor mechanical properties of the component. Also, lot of metal is removed to get the final shape of the component, which goes waste in the form of chips. Again, cut fibres are exposed to atmosphere resulting in corrosion of the machined component.

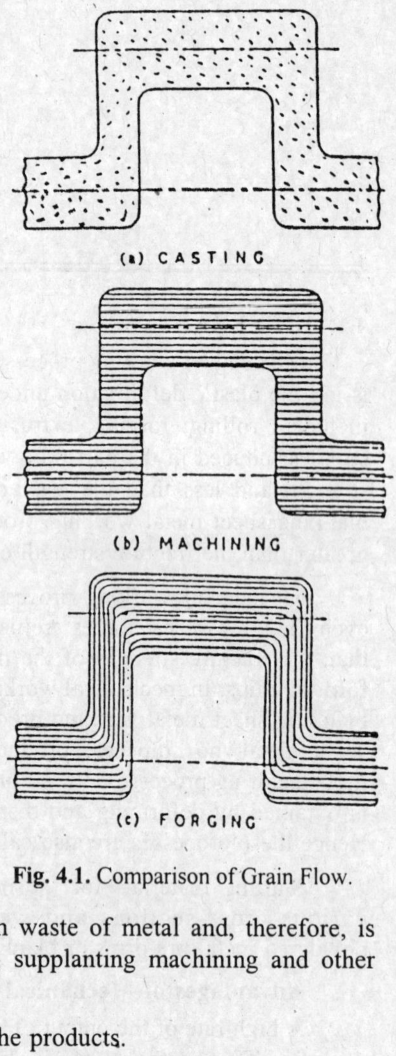

Fig. 4.1. Comparison of Grain Flow.

This is best explained by the example of manufacture of a crank shaft. The casting, (Fig. 4.1 (*a*)) has no grain flow and so has the poorest mechanical properties. In (Fig. 4.1 (*b*)), the crank shaft has been produced by machining from a bar stock and the fibre of the metal gets cut and interrupted and for this reason, the mechanical properties of this shaft will be poorer than those of the crank shaft made by forging, (Fig. 4.1 (*c*)), where the fibre of the metal has not been interrupted and continues along the entire length of shaft.

3. It is clear from the above discussion that the formed compo nents will have higher strength, and better corrosion and wear resistance, as compared to casting and machining processes. Also, formed components have high strength-to-weight ratio.

4. Mechanical working of metals involves minimum waste of metal and, therefore, is used on an ever wider scale in various engineering fields, supplanting machining and other methods that involve loss of metal to chips. etc.

5. High dimensional accuracy and surface finish of the products.

6. By controlling the end forming temperature and the degree of deformation, it is possible to impart any strength to the component within the permissible range. This is a unique advantage of metal working processes as compared to other processes.

7. Many an item cannot simply be fabricated by any other alternative means. Extra-thin foil, wire, sheet steel and other products, which are indispensable in modern civilization, originated with the advent and development of mechanical working.

8. During mechanical working of metal, the grains of the material get elongated in the direction of metal flow. So, they would be able to offer more resistance to stresses across them. As a result, the mechanically, worked metals called wrought products would be able to achieve better mechanical strength in specific orientation, that of the flow direction.

Classification of Metal - Working Processes. There are many ways of classifying metal working processes :

1. Working Temperature. Depending upon the temperature at which a material is mechanically worked or formed, the metal working process can be classified as : Cold forming, Hot forming and Warm or Semi - hot forming.

Cold forming or cold working can be defined as the plastic deforming of metals and alloys under conditions of temperature and strain rate, such that the work - hardening or strain - hardening is not relieved. Theoretically, the working temperature for cold working is below the recrystallization temperature of the metal / alloy (which is about one - half the absolute melting temperature). However, in practice cold working is carried out at room temperature or at temperatures less than 0.3 × melting point of the metal.

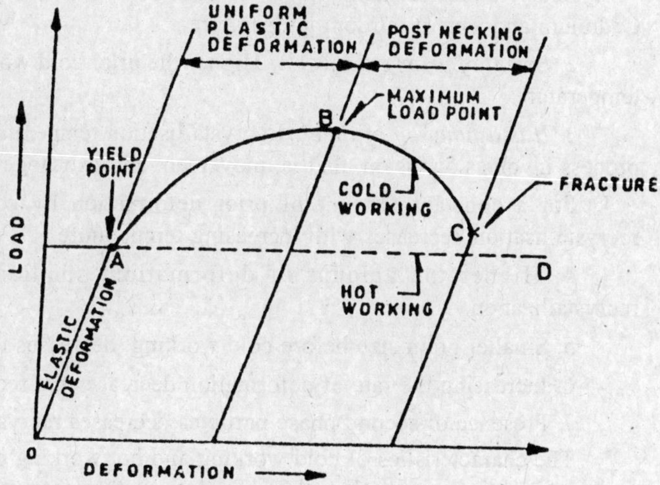

Fig. 4.2. Load Deformation Curve.

Hot working or hot forming can be defined as the plastic deformation of metals and alloys under conditions of temperature and strain rate, such that recovery and recrystallization takes place simultaneously with the deformation. The hot working is carried out above the recrystallization temperature of the material (typically 0.7 to 0.9 times the melting point temperature), and after hot working, a fine refined grained recrystallized structure is obtained.

The reaystallisation temperature is discussed in some detail as below :

Recrystallisation Temperature:

Recrystallisation temperature is very important from the point of view of mechanical working of metals and also regarding process annealing operation.

When a metal is heated and deformed under mechanical force, an energy level will be reached when the old gain structure (which is coarse due to previous cold working) starts disintegrating. Simultaneously, an entirely new grain structure (equi-axed, stress free) with reduced grain size starts forming. This phenomenon is known as "recrystallisation" and the temperature at which this phenomenon starts is called "Recrystallisation temperature". It takes some time for this phenomenon to get completed. According to ASM (American Society of Metals), the recrystallisation temperature is defined as "the approximate minimum temperature at which the complete recrystallisation of a cold worked metal occurs within a specified period of approximately one hour". Recrystallisation decreases the strength and raises the ductility of the metal.

Factors affecting Recrystallisation Temperature:

There are many factors that influence the recrystallisation temperature. These factors are discussed below:

1. Type of Metal : It is lower for pure metals as compared to alloys.

For pure metals, $T_{cr} = 0.3\ T_m$

and for alloys, $T_{cr} \cong 0.5\ T_m$

where T_m = Melting point of the metal

In practice, Tcr is taken as equal to $\left(\frac{1}{3} \text{ to } \frac{1}{2}\right) \times T_m$.

Recrystallisation temperature of lead and Tin is below room temperature and that for Cadmium and Zinc it is room temperature.

2. *Extent of prior cold work :* Higher the prior cold work, lower will be the recrystallisation temperature.

3. *It is a function of time :* Recrystallisation temperature is a function of time, because the process involves diffusion, that is, movement and exchange of atoms across grain boundaries.

For a constant amount of prior deformation by cold working, the time required for recrystallisation decreases with increasing temperature.

4. Higher the amount of deformation, smaller the grain size becomes during recrystallization.

5. Smaller grain size before cold working, decreases recrystallisation temperature.

6. Increasing the rate of deformation decreases the recrystallization temperature.

7. Presence of second phase particles decreases recrystallisation temperature.

The characteristics of cold working and hot working of metals can be understood from the load deformation curve, (Fig. 4.2). To deform the metal plastically for a permanent change in shape of the work-material, the applied load / stress has to be equal to (atleast) or greater than the yield point of the work material (the elastic deformation being completely recoverable, once the load is taken off). The curve OABC represents the cold working, whereas the curve OAD represents the hot working process. The portion AB represents the uniform plastic deformation region, that is the work hardening or strain hardening region (which means that increasing loads will have to the applied to get increasing deformations). At point 'B' (maximum load point) local necking of the material starts and the portion BC represents the post necking deformation, ultimately resulting in fracture at point C.

It is clear that in cold working, the force increases with increased deformation, whereas, in hot working, once the forces equals the yield point load, the deformation proceeds at almost constant value of force.

(*a*) **Characteristics of Cold - working.** It is clear from the Fig. 4.2, that due to work hardening, the strength of the metal increases but its ability to deform further (formability or ductility) decreases with cold working (See Fig. 4.3 (*a*)). So if the material is excessively

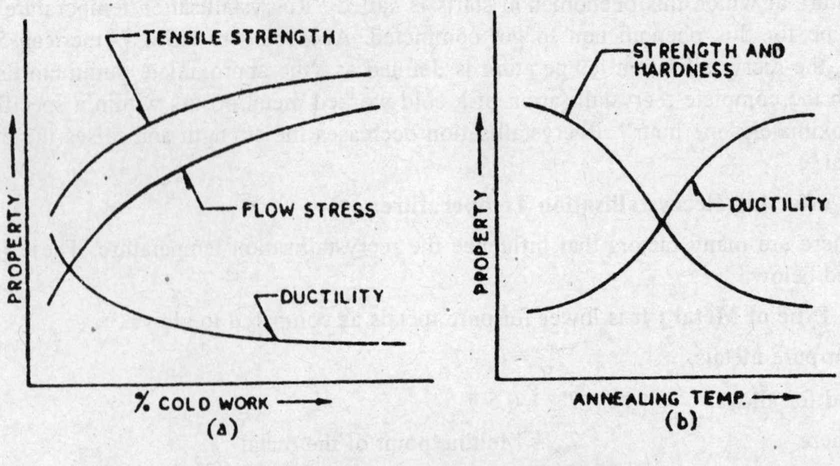

Fig. 4.3. Effects of Cold Working and Annealing.

deformed, it may fracture before it is formed. To avoid this, large deformations in cold working are obtained in several stages, with intermediate annealing. This will soften the cold worked material and restore its ductility and hence the formability. (Fig. 4.3 (*b*)).

Advantages of Cold - working :

1. Since cold working is done at room temperature or low temperatures, no oxidation and scaling of the work material occurs. This results in reduced material loss.

2. Surface defects are removed.

3. Excellent surface finish which reduces or completely eliminates subsequent machining resulting in enormous saving in material.

4. High dimensional accuracy.

5. Highly suitable for mass production and automation, because of low working temperatures.

6. Thin gauge sheets can be produced by cold working.

7. Heavy work hardening occurs and so the inherent strength of the material is permanently increased. This makes it possible to use inferior materials whose properties are enhanced by work hardening and preferential flow.

8. The physical properties of metals that do not respond to heat treatment can be improved by cold working.

Draw backs of Cold - working :

1. At low temperatures, the strength of a metal is very high. So, large forces are needed for deformation. For this, high capacity equipment is required which is costly.

2. The ductility / formability of metals is low at low temperatures. Hence for large deformation cold working requires several stages with interstage annealing, which increases the cost of production.

3. Also, due to limited ductility at room temperature, the complexity of shapes that can be readily produced is limited.

4. Since low reductions are required to attain the required parameters, more sensitive controls are needed in cold working than in hot working.

5. Due to very high forces, tool pressures and power requirements are high too. So, the tooling must be specially designed, which increases the tool cost.

6. Severe stresses are set up in the metal during cold working. This requires stress relieving, which again increases the cost.

7. Due to the above factors, normally, only the ductile metals are cold worked.

Materials for Cold - working. In principle, any material can be cold- worked. In practice, however, the choice is limited by the following two factors :-

1. The ability of the tool material to withstand the required pressures for cold -working of a material. Obviously, the tool material must have a mechanical strength greater than that of the material to be cold - worked. Also, from the point of view of economy, the tool (die etc.) must have a reasonable working life, that is, it must be able to withstand the developed working stresses for a reasonable length of time.

2. The economic requirement that the maximum possible deformation of the material should be obtained in a single working operation (single). This will depend upon the cold ductility and cold flowability of the material.

Thus, the two principal limitations to cold - working of a material are : the permissible stress placed on the tool material and the ductility of the material to be cold - worked.

Both cold ductility and cold flowability of a material depend closely on its chemical composition. As for steel, with an increase in the percentage of carbon or alloying constituents, its deformability decreases and the resistance to deformation increases. The maximum limit is usually 0.45% carbon for steels used in cold extrusion and 1.6% carbon for other cold forging operations. Impurities such as S, P, O_2 and N_2, also impair the cold workability of the steel. For cold working, the micro-structure of the material, also plays an important role. Soft annealing known as spheroidize annealing of steel before cold working, improves its cold workability. The grain size is also an important factor. Large grain is easier to cold work, while the parts made from fine grained material are stronger. A good guiding rule for forging steels used to produce forgings is that the stress on the die should not exceed 2500 N/mm^2 and the material must allow at least a 25% deformation in a single step.

As a general rule, the requirements for a material to be cold worked are :

1. Yield stress curve of gentle slope.
2. Early yield point.
3. Then great elongation with pronounced necking before fracture.

The materials commonly used for cold working include : low and medium carbon steel (0.25 to 0.45% C), low alloy steels, copper and light alloys such as Aluminium, Magnesium, Titanium, and Berrylium.

Cold forming is most suitable for axisymmetric components such as shaft components, flanged components, finished gears and bearing races etc.

Characteristics of Hot - working. It is clear from Fig. 4.2 that hot working of the metal occurs at an essentially constant load/flow stress, that is, there is no work-hardening of the metal.

Advantages of Hot - working :

1. The strength of the metals is low at high temperatures. Hence low tonnage equipments are adequate for hot working.
2. Very large workpieces can be deformed with equipment of reasonable size.
3. Because of high ductiliy at high temperatures and absence of work hardening, large deformations can be undertaken in a single stage, and, complex parts can be fabricated.
4. Interstage annealing and stress relieving are not required.
5. Blow holes and porosities are eliminated by welding action at high temperatures and pressure.
6. Grain size can be controlled to be minimum (Fig. 4.4). This makes the metal tougher.

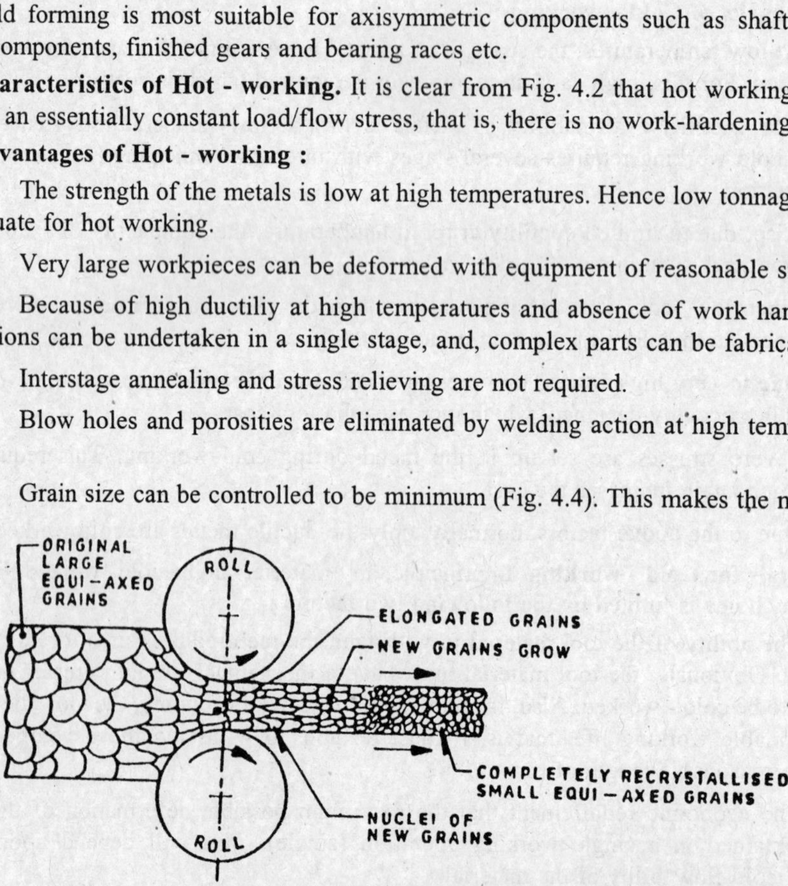

Fig. 4.4. Micro-Structural Changes in a Hot working Process (Rolling.)

7. Inclusions within the metal are broken up and elongated into fibres or threads with definite orientation. This again makes the metal tougher.

8. Segregation may be reduced or eliminated, since hot working promotes diffusion of constituents.

Typical hot working temperatures are :-

Steels	: 1100 to 1260°C
Cu and its alloys	: 760 to 925°C
Magnesium	: 315°C
Aluminium and its alloys	: 370 to 455°C

Drawbacks of Hot - working :

1. Due to oxidation and scaling, there is heavy material loss. This also results in poor dimensional accuracy and surface finish.
2. Automation is difficult due to high working temperatures.
3. Thin parts (sheets, wires etc.) can not be produced due to loss of ductility because of high rate of loss of heat.
4. High energy costs to heat the metals to high temperature.
5. Surface decarbonisation in steels, reduces strength and hardness on the surface.
6. Due to high working temperatures, these is a serious problem of surface reactions between the metal and the furnace atmosphere, more so in the case of reactive metals like Ti, calling for inert atmosphere.

Note 1. It is clear from the above discussion that hot working is mainly employed to produce large deformations in the material. The final dimensional tolerances, accuracy and surface finish are obtained by cold working involving only a small deformation in the material.

2. It should be noted that the difference between cold working and hot working depends only upon the temperature of recrystallization and not on any arbitrary temperature of deformation. Lead, tin and zinc recrystallize rapidly at room temperature (Below 27°C) after large deformations. Hence, the working of these metals at room temperature will constitute their hot working. Similarly working of tungsten at about 1090°C will be termed its cold working, because it has recrystallization temperature above this value.

However, in every day use, cold working means working at room temperature and hot working means working of a preheated material above its recrystallization temperature.

Warm working or Semi-hot working. It can be defined as plastic deforming of a metal or alloy under conditions of temperature and strain rate, such that the drawbacks of both cold working and hot working are eliminated and their advantages are combined together. For this, the selection of proper temperature for warm working is very important. This depends upon the following factors :

1. Yield or flow strength of the metal or alloy.
2. Ductility of the material.
3. Dimensional tolerance on the components.
4. Oxidation and scaling losses.

The variations of the above properties relative to the working temperature can be studied to arrive at the proper working temperature for warm working. For example, for 0.13% C steel ; the following observations are made :

Yield-Strength. In general, yield strength decreases with increase in temperature. However, in the temperature range of 150°C to 350°C and 800°C to 900°C, it increases with, increases in temperature. The first temperature range is called blue brittleness range of steel and in the second range, structural changes occur in steel. Both these ranges are brittle ranges and if steel is worked in these temperature ranges, it will fracture. So, the best temperature ranges from the yield strength point of view are 400°C to 750°C and above 900°C.

Ductility. In general, the ductility or formability of a material increases with increase in temperature. However, in temperature ranges of 250°C to 350°C and 800°C to 900°C, it decreases with increase in temp-erature. Thus, from the ductility point of view, the best temperature ranges for the above mentioned steel are : 400°C to 750°C and above 900°.

Dimensional tolerance and Scaling and Oxidation losses. The dimensional tolerances increase rapidly above 700°C. Similarly, scaling and oxidation losses, which are negligible upto 700°C, increases very rapidly above this temperature.

From the above discussion, it is clear that the best temperature range for working the above-mentioned steel is 400°C to 700°C. Working of steel within this temperature range is called warm working or Semi-hot working.

In general, warm working is done at a temperature of 0.3 T_m to 0.5 T_m, where T_m is the melting point of the material.

2. Type of Stress applied to the workpiece. On the basis of the type of stress applied to the workpiece as it is formed into shape, the processes can be classified as :

(*i*) Direct compression-type processes.

(*ii*) Tension type processes.

(*iii*) Combined stress (tension and compression) type processes.

(*iv*) Bending type processes.

(*v*) Shear Stress type processes.

Processes in which the material is subjected principally to compres-sive stress are : forging, rolling, extrusion, coining, spin forging, swaging and kneading etc. The metal flows at right angles to the direction of the applied stress. These processes are also called as "Squeezing processes".

Processes involving the use of pure tensile stresses are : Stretch forming, vacuum forming and creep forming.

In combined stress type processes, the primary applied forces are frequently tensile with the indirect compressive stresses coming into play due to the reaction of the workpiece with the die. The processes include : Drawing processes (wire, rod and tube), deep drawing and embossing etc.

Bending involves the application of bending moments to the sheet. The processes include : straight bending or flanging, V-bending, stretch flanging (concave flanging), shrink flanging (convex flanging) and seaming etc.

Processes involving shear stresses refer to the application of shear forces of sufficient magnitude to rupture the metal in the plane of shear (chipless forming). The processes include : Piercing, blanking, press shearing and cutting off, notching and nibbling etc. Fig. 4.5 shows typical metal working operations.

3. Steady or Non-steady state forming. In steady state processes, the zone in which plastic deformation is enforced, remains fixed in shape and size and does not alter with time. Examples are : drawing processes (wire, rod, tube) and rolling. In non-steady processes, the geometry and zone of plastic deformation continuously change, for example, forging process. Extrusion process is transitory, that is, the deformation is nonsteady at the beginning and end but acquires steady state conditions while the greater part of the billet is extruded.

4. Primary and Secondary metal forming processes. Primary metal working processes are involved in the initial breakdown of the original material (ingot, bloom, billet) to shapes that are then processed for the final product, for example, rolling, forging, extrusion.

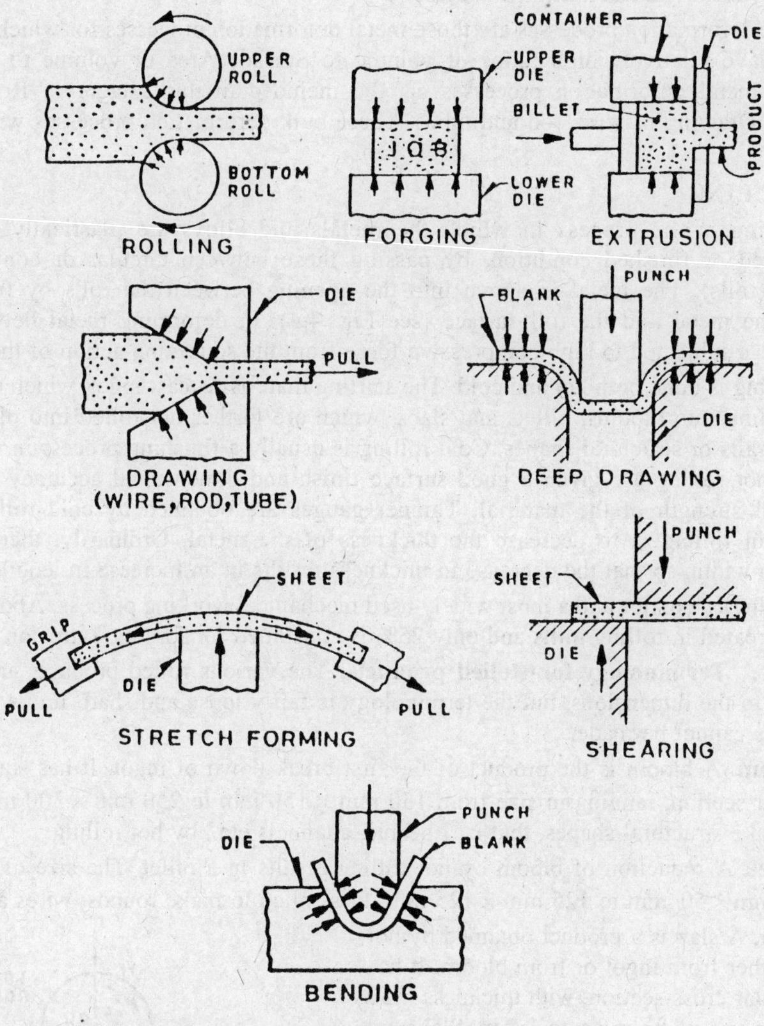

Fig. 4.5. Typical Metal Working Processes.

Secondary processes are those which take the products of some primary processes and change their geometry and properties to the semi-finished or finished stage, for example, drawing processes, spinning, swaging, coining, embossing, stretching, bending, deep drawing, rubber forming and sheet metal forming etc.

5. Extent of plastic deformation zone. As per this criterion, the metal working processes can be classified as : Bulk deformation processes and Sheet metal working processes.

In bulk deformation processes, the thickness (diameter or other major dimension) of the workpiece is substantially reduced (changed). These may be steady state or non-steady state processes, for example, forging, extrusion, rolling, drawing etc.

In sheet metal working processes, any change in sheet thickness is fairly limited. But the starting material, that is the sheet, is the product of a bulk deformation process, that is, rolling.

6. According to low or high rate of strain. Conventional metal working processes and non-conventional (HERF, HVF) processes.

7. Shape of workpiece or finished part (sheet metal, bar stock etc.)

4.1.1. Bulk – Deformation Processes :-

Bulk deformation processes are those metal deformation processes for which workpieces or products have relatively high ratios of volume to Surface Area or volume to thickness. The following metal deformation processes can be included in this catagory : Rolling, Forging, Extrusion, Drawing of wire, rod and tubes. These bulk deformation processes will be discussed ahead.

4.2. ROLLING

Rolling is the process in which the metals and alloys are plastically deformed into semifinished or finished condition, by passing these between circular or contoured rotating cylinders (rolls). The metal is drawn into the opening between the rolls by frictional forces between the metal and the roll surface (see Fig. 4.4). In deforming metal between rolls, the workpiece is subjected to high compressive force from the squeezing action of the rolls.

Rolling is done both hot and cold. The starting material is cast ingot, which is broken down by hot rolling into blooms, billets and slabs, which are further hot rolled into plate, sheet, rod, bar, pipe, rails or structural shapes. Cold rolling is usually a finishing process in which products made by hot rolling are given a good surface finish and dimensional accuracy with increased mechanical strength of the material. Thinner gauges are obtained by cold rolling. The main objective in rolling is to decrease the thickness of the metal. Ordinarily, there is negligible increase in width, so that the decrease in thickness results in an increase in length.

Rolling is a major and a most widely used mechanical working process. About 75% of steel output is treated in rolling mills and only 25% is consumed for forging extrusion and founding.

4.2.1. Terminology for Rolled products. The various rolled products are given names according to the dimensions, but the terminology is fairly loose and sharp limits with respect to dimensions cannot be made.

Bloom. A bloom is the product of the first break down of ingot. It has square or slightly rectangular section, ranging in size from 150 mm × 150 mm to 250 mm × 300 mm. A bloom is used to make structural shapes, that is, I beams, channels etc., by hot rolling.

Billet. A reduction of bloom by hot rolling results in a billet. The size of a billet ranges from 50 mm × 50 mm to 125 mm × 125 mm. It is rolled to make rounds, wires and bars.

Slab. A slab is a product obtained by hot rolling, either from ingot or from bloom. It has a rectangular cross-section, with thickness = 50 to 150 mm and width = 0.6 to 1.5 m. Slabs are further rolled to get plates, sheets, strips, coil and skelp.

Plate. A plate is a finished or semi-finished product with a minimum thickness of 6.35 mm. Its width will be equal to the width of the roll and the length equal to the maximum which can be handled or shipped.

Sheet. A sheet is a thin partner of plate with a maximum thickness of 6.35 mm.

Strip. A strip is a narrow sheet and has a maximum width of 600 mm with a maximum thickness of 6.35 mm. Since it is normally handled in coil form, its length can be considerable and is limited only by the manufacturing and handling facilities.

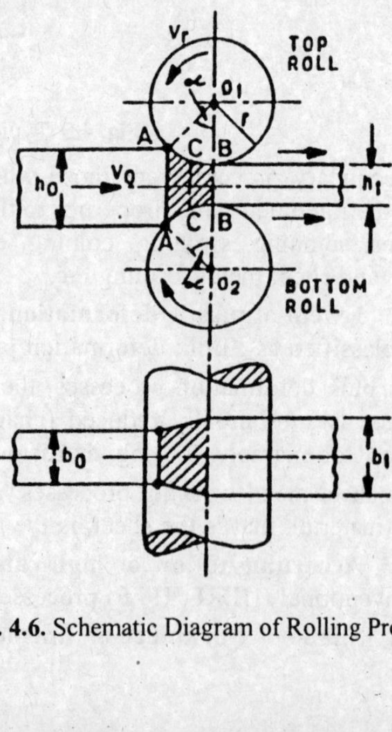

Fig. 4.6. Schematic Diagram of Rolling Process.

Foil. It is a thin strip with a maximum width of 300 mm and a maximum thickness of 1.5 mm. It is available in coil form.

Bar. It is a long, straight, symmetrical piece of uniform cross-section. It may be round, square, or of another configuration. A circular bar is called a rod.

Wire. A wire is a thin variety of bar, available in coil form and not normally so identified over 9.5 mm cross-section.

Billet are normally further rolled to make rounds, bars and wires. They can also be rolled into flats and sections. These intermediate products are further processed to get final products as shown below. Billets are also hot forged to get final products as shown below :-

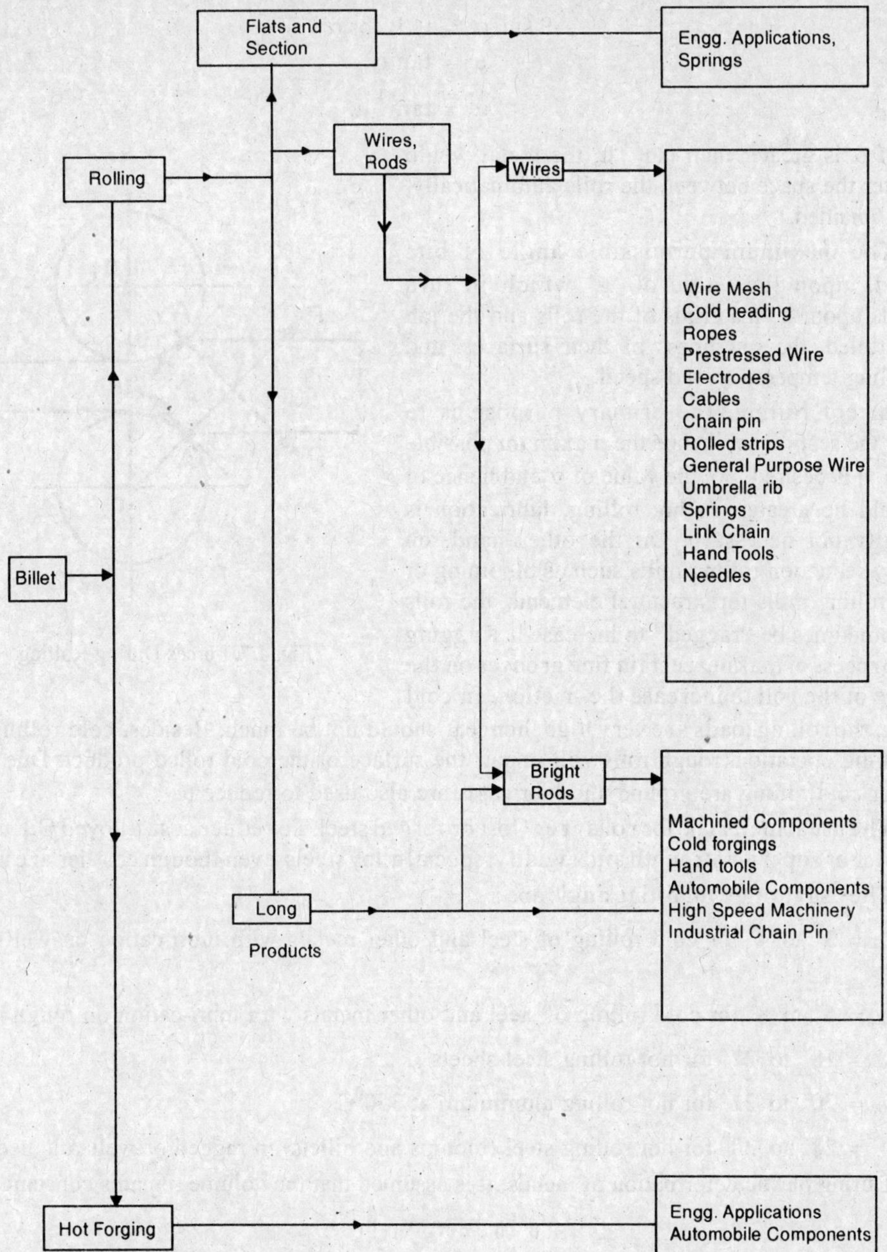

4.2.2. Mechanism of Rolling. Fig. 4.6 shows a schematic diagram of the rolling process. The metal contacts each of the two rolls along the arc AB, which is known as the 'flat strip' arc of contact. This arc corresponds to the central angle α, called the "angle of contact or bite". The process of metal rolling is made possible by the friction that occurs between the contact surfaces of the rolls and the part being rolled. At the moment of bite, two forces act on the metal from the side of each roll, normal force P and the tangential force μP (Fig. 4.7) where μ is the co-efficient of friction between the metal and the roll surfaces. The part would be dragged in if the resultant of horizontal component of the normal force P and tangential force (frictional force) μP is directed in that direction. In the limiting case,

$$P \sin \alpha = \mu \; P \cos \alpha$$

$$\mu = \tan \alpha$$

or $$\alpha = \tan^{-1} \mu.$$

If α is greater than $\tan^{-1} \mu$, the metal would not enter the space between the rolls automatically, that is, unaided.

The maximum permissible angle of bite depends upon the valve of 'μ' which in turn depends upon the materials of the rolls and the job being rolled, the roughness of their surfaces, and the rolling temperature and speed.

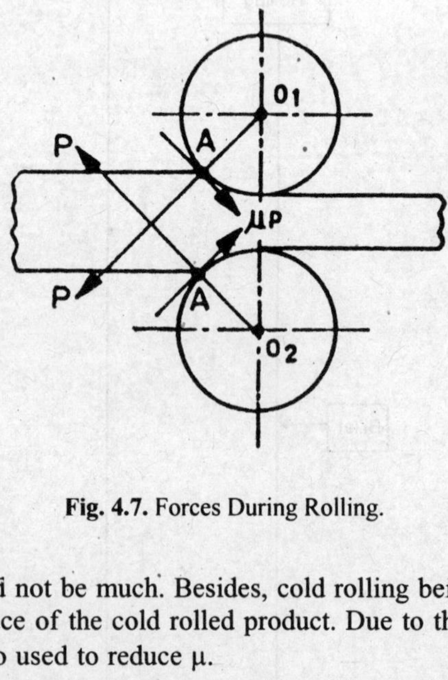

Fig. 4.7. Forces During Rolling.

In hot rolling, the primary purpose is to reduce the section and hence the maximum possible reduction is desired. So, the value of α and hence of μ should be greater. In hot rolling, lubrication is generally not necessary. On the other hand, on primary reduction rolling mills such as blooming or rough rolling mills for structural elements, the rolls may sometimes be "ragged" to increase μ. Ragging is the process of making certain fine grooves on the surface of the roll to increase the friction. In cold rolling, the rolling loads are very high, hence μ should not be much. Besides, cold rolling being a finishing operation, rough rolls will impair the surface of the cold rolled product. Due to this, rolls for cold rolling are ground and lubricants are also used to reduce μ.

The usual materials for rolls are : Cost or forged steel. To reduce cost, alloyed C.I. can also be used. For superior strength and rigidity, special alloy steels, even though costlier are used.

The usual values of biting angle are :

$\alpha = 3^o$ to 4° for cold rolling of steel and other metals with lubri-cation on well-ground rolls.

$= 6^o$ to 8° for cold rolling of steel and other metals with lubri-cation on rough rolls.

$= 18^o$ to 22° for hot rolling steel sheets

$= 20^o$ to 22° for hot rolling aluminium at 350^oC

$= 28^o$ to 30° for hot rolling steel (blooms and billets) in ragged or well-roughed rolls.

During plastic deformation of metals, it is assumed that the volume remains constant, that is,

$$V_0 \; h_0 \; b_0 = V_1 \; h_1 \; b_1$$

Since $$b_0 \approx b_1$$

$$V_1 = V_0 \cdot \frac{h_0}{h_1}$$

Now $h_1 < h_0$, $\therefore$ $V_1 > V_0$, and so we have

$V_1 > V_r > V_0$, V_r is the velocity of the roll.

Hence, the speed at which the metal is delivered by the rolls, V_1, is higher and the metal entrance speed V_0 is lower than the peripheral speed of the rolls V_r. At a section C - C in the deformation zone (shown hatched), the velocity of metal will equal the velocity of rolls. This section is called "neutral or no slip section". To the left of neutral section, the deformation Zone is called "Lagging Zone" ($V_0 < V_r$) and to the right of the section, it is called "Leading Zone", ($V_1 > V_r$). We have two definitions,

$$\text{Forward slip} = \frac{V_1 - V_r}{V_r} \times 100\,; \quad \text{Backward slip} = \frac{V_r - V_0}{V_r} \times 100$$

(*i*) Absolute draught, $\Delta h = (h_0 - h_1)$ mm ; relative draught $= \frac{\Delta h}{h_0} \times 100$

(*ii*) Absolute elongation, $\Delta l = (l_1 - l_0)$ mm ; Co-efficient of elongation $= \frac{l_1}{l_0}$

(*iii*) Absolute spread $= (b_1 - b_0)$ mm

Spread increases with an increase in roll diameter and co- efficient of friction, as well as with a fall in temperature of the metal in the course of hot rolling. Spread is proportional to the draught and depends upon the thickness and width of the job.

During cold working, the change in the properties of metals depend upon the percentage of cold work used, which is defined as

$$\% \text{ of cold work} = \frac{\text{Original area of cross--section-- Final area of cross section}}{\text{original area of cross section}} \times 100$$

$$= \frac{A_0 - A_f}{A_o} \times 100$$

$\therefore$ For cold rolling of sheets, % of cold working $= \frac{h_0 b_0 - h_1 b_1}{h_0 b_0} \times 100$

$$= \frac{h_0 - h_1}{h_0} \times 100 = \left(1 - \frac{h_1}{h_0}\right) \times 100 \qquad (b_0 \cong b_1)$$

4.2.3. Rolling mills and their Classification. A rolling mill is a complex of machines for deforming metal in rotary rolls and performing auxiliary operations such as transportation of stock to rolls, disposal after rolling, cutting, cooling, melting, piling or coiling etc. A set of rolls in their massive housing is called a "stand". Rolling mills are classified according to the number of rolls in the working stand, the product rolled, arrangement of stands etc.

1. Classification based on number of rolls in the stand. According to this criteria, rolling mills are of the following types :

(*i*) two high rolling mill ; (*ii*) three high rolling mill

(*iii*) four high rolling mill; (*iv*) multi-roll rolling mill; (*v*) universal rolling mill

(*i*) **Two high rolling mill.** As the name implies, it has two rolls with a constant direction of rotation about the horizontal axes, (Fig. 4.8 (*a*)). For successive reductions, the stock is returned to the entrance of the rolls by hand carrying or by means of a platform which can be raised to pass the stock over the rolls. The upper roll may be raised or lowered to change the

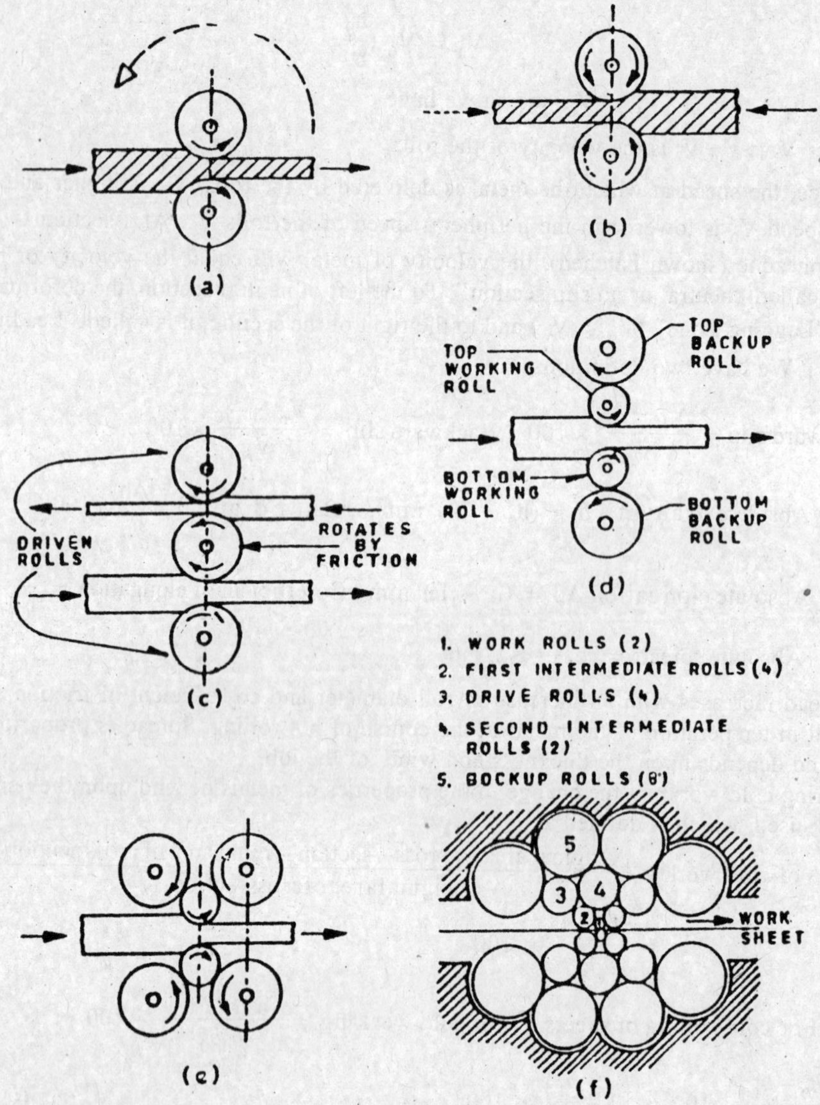

Fig. 4.8. Rolling Mills.

distance between the rolls. This method of successive reductions slows down the process. The faster method is to pass the stock through a series of rolls for successive reduction, but this method requires more investment in equipment. The alternative procedure is to use ''two high reversing'' rolling mill, (Fig. 4.8 (*b*)), where the direction of rotation of the rolls is reversed after each pass. The rolls are brought closer together after each pass and the bar or plate reciprocates many times between the rolls before the final thickness is obtained. This design is limited by the length that can be rolled and by the inertial forces that must be over come each time a reversal is made. Also, much more power is consumed as compared to simple two high stand. Some time is also lost in reversing the mill.

A two high non-reversing mills are the most common for both hot and cold rolling since they are least expensive. Such stands are widely used in mills through which the bar passes only once (continuous mills). Normally, the job movement is facilitated by providing support rolls at the entry and exit of the rolling mill.

Two high reversing mills are often used for the first rolling of an ingot into blooms and slabs. These stands are also employed as roughing stands of plate, universal, rail and structural and other mills.

(*ii*) **Three high rolling mill.** The disadvantages of a two high reversing mill are over come by using a three-high rolling mill in which three rolls with a constant direction of rotation are arranged in a single vertical plane, (Fig. 4.8 (*c*)). Lifting tables are provided on one or both sides of the stand to raise and lower the bar after each pass. In these mills, the top and bottom rolls are drive rolls and the middle roll rotates by friction. These mills are employed as blooming mills, for billet rolling and finish rolling. These stands also find extensive use in open-train section mills.

(*iii*) **Four-high rolling mill.** For a given width of plate/sheet and reduction per pass, the bending of the rolls will be less as their diameter is increased. However, increasing the roll diameter will increase the arc of contact between the roll and the plate/sheet (see Fig. 4.6). This will result in an increase in roll separating force (due to the reaction of the metal on to the rolls), which is given as,

P = mean specific pressure between metal and roll × contact area.

∴ In plate/sheet rolling (especially cold rolling), the diameter of work rolls should be as small as possible to reduce P and power requirements. To avoid bending of work rolls (due to their low strength and rigidity), large diameter back up rolls are installed, Fig. 4.8 (*d*) resulting in a four-high rolling mill. Four-high rolling mills are used in reversing mills for the hot rolling of armour and other plate, as well as for the hot and cold rolling of sheet steel in continuous mills.

Note. The rolling load can be reduced by applying tensile force to the workpiece in the horizontal direction. This will lower the compressive yield strength of the material. Both "back tension" and "front tension" can be applied.

(*iv*) **Multiple - roll mills.** In four-high rolling mill, the diameter of back up rolls can not be greater than 2 to 3 times that of the work rolls. As the diameter of work rolls is decreased more and more to accommodate processes with exceedingly high rolling loads, the size of the back up rolls must also decrease. A point is reached when the back up rolls themselves begin to bend and must be supported, hence the ultimate-design- cluster mill.

Stands with a cluster of six (Fig. 4.8 (*e*)), 12 and 20 rolls are intended to manufacture strips down to 0.001 mm thick and upto 2000 mm wide. In 12 or 20 rolls stands, the very small diameter of working rolls (10 to 30 mm) allows no coupling of driving spindles, and the work rolls are idle running. They are supported by a row of driving rolls which, in turn, are supported by a row of back up rolls. The work rolls are thus rotated by friction against the driving intermediate back up rolls.

Sendzimir mill. This mill is a modification of the cluster mill with 20 rolls which is very well adapted to rolling thin sheet or foil from High-Strength alloys (*e.g.* Stainless Steel). Only the outer rolls are driven. The remaining rolls get the motion due to friction, from the back up rolls, (Fig. 4.8 (*f*)). High C-steel sheets of 5 to 50 μm can be made to an accuracy of 1 to 5 μm. This type of mill has been set up in Salem Stainless Steel Complex in Tamil Nadu.

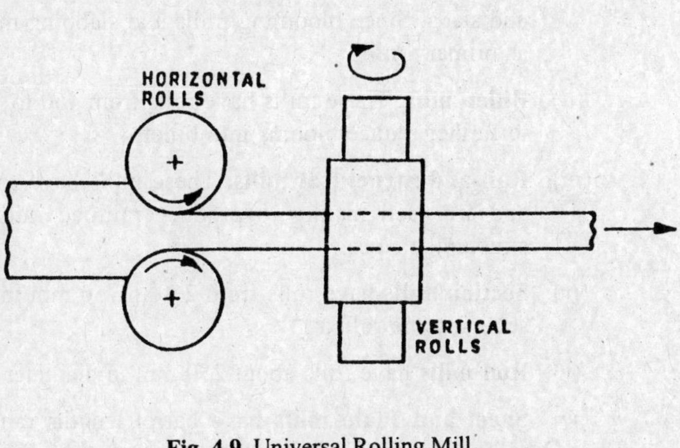

Fig. 4.9. Universal Rolling Mill.

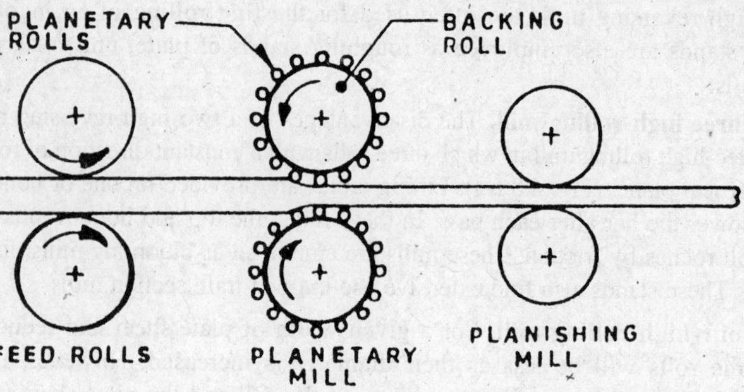

Fig. 4.10. Planetary Rolling Mill.

(*v*) **Universal Rolling mill.** In this rolling mill, the metal is reduced by both horizontal and vertical rolls, (Fig. 4.9). The vertical rolls roll the edges of the bar even and smooth. Vertical rolls are mounted either on one side (front or back) or on both sides of the horizontal roll stand. The horizontal rolls may be either two-, three-, or four-high arrangement. These mills are used for rolling wide strip, sheets, plates and slabs that require rolled edges and also for rolling of beams and H- sections.

Planetary Rolling mill. This mill consists of a pair of heavy backing rolls surrounded by a large number of small planetary rolls, Fig. 4.10. The main feature of this mill is that it hot reduces a slab to coiled strip in a single pass. Each pair of planetary rolls gives an almost constant reduction to the slab as it sweeps out a circular path between the slab and the backing roll. The total reduction is the sum of a series of such small reductions following each other in rapid succession. The feed rolls push the slab through a guide into planetary rolls. On its exit side, a 2- or 4-high planishing mill is installed to improve the surface finish and is followed by a coiler.

Steckel rolling. In all the above methods, the rolls are powered and the job is pulled in due to the frictional force between the metal and the roll surface. In steckel rolling, the strip is drawn through idler rollers by front tension only, the torque on the rolls being zero.

2. Classification based on the products rolled. According to the products rolled, the rolling mills may be classified as follows :

(*i*) **Blooming and slabbing mills.** These are heavy mills with rolls from 800 to 1400 mm in diameter. They are designed to roll ingots (2 to 25 tonnes in weight) into blooms and slabs. Since blooming mills and slabbing mills are quite similar, these are called as primary mills.

(*ii*) **Billet mills.** These mills have rolls from 450 to 850 mm in diameter and are designed to further reduce blooms into billets.

(*iii*) **Rail and Structural mills.** These mills have rolls from 750 to 800 mm in diameter and are used mainly to produce railroad rails, beams, channels and other heavy structural shapes.

(*iv*) **Section mills** have rolls from 250 to 750 mm in diameter, depending upon shape and section to be rolled.

(*v*) **Rod mills** have rolls about 250 mm in diameter and are used to produce wire rod.

(*vi*) **Sheet and Plate mills** have barrel lengths ranging from 800 to 5000 mm for hot rolling and from 300 to 2800 mm for cold rolling.

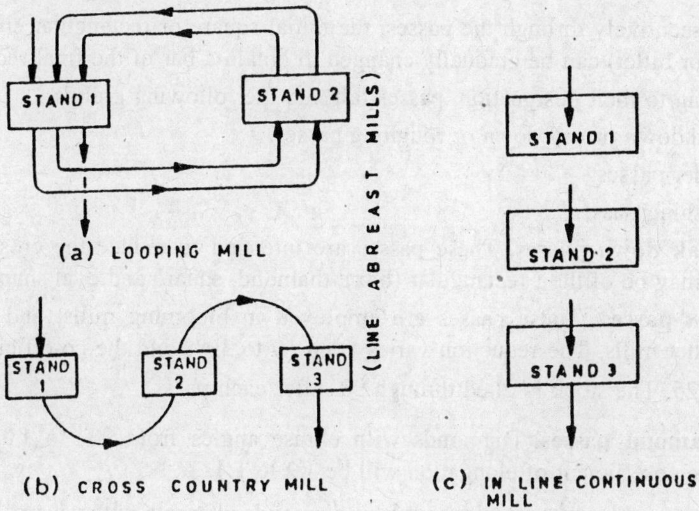

Fig. 4.11. Arrangement of Rolling Stands.

(*vii*) **Seamless Tube mills** produce seamless tubes as discussed later.

(*viii*) **Tyre and Wheel mills** are used to manufacture rail-road wheels and tyres.

3. Arrangement of Rolling stands. Single stand mills are used singly or in various floor-layout combinations for rolling metals.

(*i*) **Looping mill.** (Fig. 4.11 *a*). The products are looped from one mill stand to another by arranging the stands in a "train" in which the rolls of one mill stand are driven from the ends of the rolls in the adjacent stand.

(*ii*) **Cross-country mill.** , (Fig. 4.11 *b*). It consists of individual roll stands placed some distance apart so that the workpiece must leave one set of rolls before it enters the next set. To save spare and avoid complicating the mill drive mechanism, the roll stands are generally arranged in one or more parallel lines. Transfer and skid tables are used to reverse the direction of travel of the workpiece and to convey it from set of roll stands to another.

(*iii*) **Continuous mill.** , (Fig. 4.11 *c*). It employs a number of individual roll stands arranged one behind the other in a straight line (in tandem). The work pieces continuously pass through the various roll stands and emerge from the last roll stand as a finished shape. The rolls of each successive stand must turn faster than those of the preceding one by a precise amount so as to accommodate the increased length which accompanies each successive reduction in thickness.

All the above three mills require a large capital outlay and are only justified for mass production. Looping and cross country mills require the work piece to be bent or turned between the stands and are, therefore, used for rolling rods, rails or sections. Continuous mills are used for plates, strips or sheets. Looping mills and cross-country mills are also called as "Line abreast continuous mills" and continuous mills are called as "In- line continuous mills".

4.2.4. Roll pass design. As already discussed, because of the limitations in the equipment and workability of the metal, rolling is accomplished progressively in many steps. Plate, sheet and strip are rolled between rolls having a smooth, cylindrical, slightly cambered (convex) or concave working surface. Bars, rods and special purpose shapes (I-beams, channels and rails etc.) are rolled between grooved rolls. The shape cut into one roll is called the "groove". The shape formed when the grooves of the mating rolls are matched together is called the "pass". By rolling the metal consecutively through the passes, the initial square or rectangular cross-section of th ingot (bloom or billet) can be gradually changed to obtain a bar of the final required shape.

According to their designation, passes fall into the following groups :

(1) Breakdown or roll down or roughing passes.

(2) Leader passes.

(3) Finishing passes.

1. **Break down passes.** These passes are intended to reduce the cross-sectional areas These passes may be of the : rectangular (box), diamond, square and oval shapes, (Fig. 4.12).

(*i*) **Box passes.** These passes are employed in blooming mills, and in roughing an continuous billet mills. The reduction varies from 10 to 30% and the co-efficient of elongatio from 1.1 to 1.25. The stock is tilted through 90° after each pass.

(*ii*) **Diamond passes.** Diamonds with obtuse angles from 90° to 100° are commonl employed. The co-efficient of elongation will be 1.2 to 1.4.

(*iii*) **Diamond - square series** in which diamonds alternate with squares is used chiefly i rolling square bars and flats. Diamonds with obtuse angles from 100° to 125° are employed i this series. The co-effi-cient of elongation ranges from 1.25 to 1.50. The shortcoming of the serie is deep passes which weaken the rolls.

(*iv*) **Oval - square series.** This is the most effective and wide-spread series. Its chie advantage is larger co-efficient of elongation (1.5 to 2.5 or even higher). Large elongatior

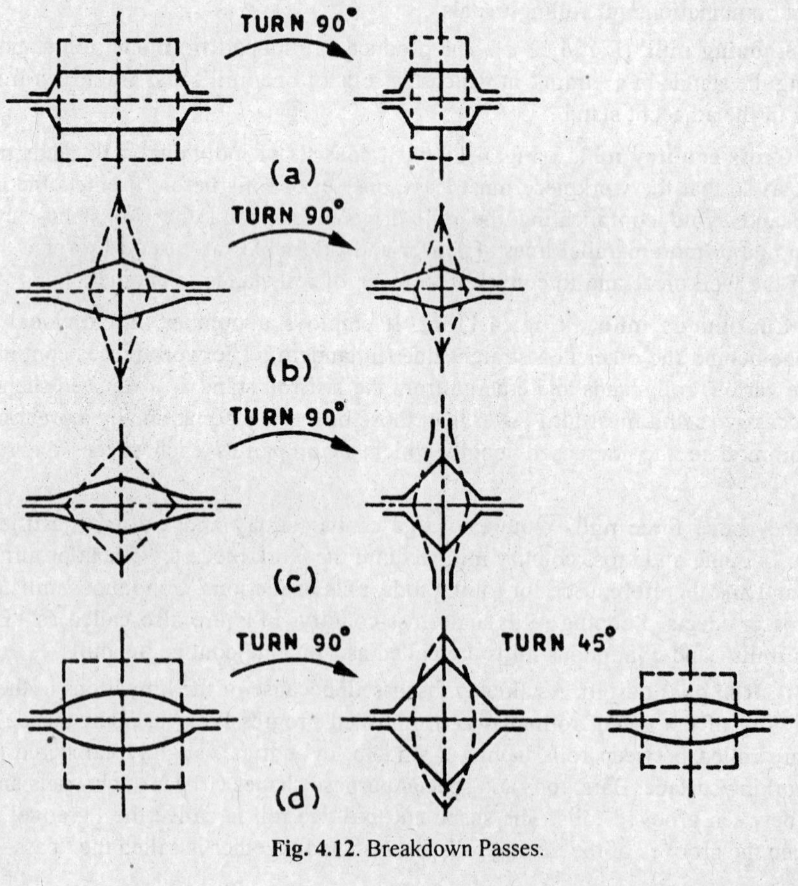

Fig. 4.12. Breakdown Passes.

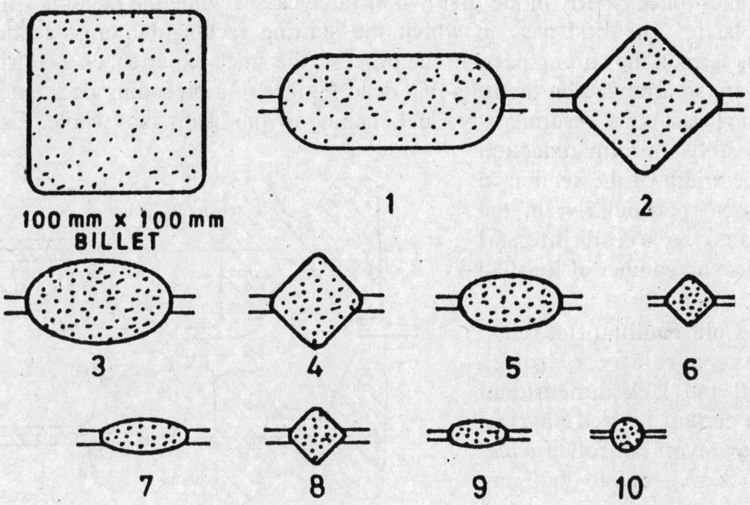

Fig. 4.13. Reduction of a Billet to a Rod.

nable very small sections to be rolled and, therefore, this series is the main sequence used in roducing wire-rod and other small sections.

2. Leader passes. The function of these passes is to gradually bring the cross section of ıe bar to the final shape. The finishing pass for rolling rounds is preceded by an oval leader pass, ıe finishing pass for rolling squares is preceded by a diamond leader pass etc.

3. Finishing pass. The shape of the last or finishing pass should be identical to that of the nished rolled stock with due consideration for the linear co-efficient of thermal expansion and ɔr the dimensional tolerance of the final product.

Fig. 4.13 illustrates the 10 steps used to reduce a 100 mm × 100 mm billet to a 12.7 mm iameter rod. From 8 to 10 steps are needed to complete most commercial shapes, such as beams, channels and rails from blooms.

4.2.5. Shaped Section Roll Passes. Despite the great variety of shaped section stock ɔnfigurations the feature is common as to their roll pass design is a non-uniform deformation ıroughout the width of sections. Intricate sections are rolled generally in closed passes of a shape etermined by section's configuration.

Consider roll pass design and deformation conditions in beam passes intended to roll beams.

Sections are shaped inside closed flange-type ısses (Fig. 4.14), and the deformation conditions of ection's elements vary considerably both in character ıd in duration.

Section web (Fig. 4.14) is reduced directly with ıe effect that the overall width of the section increases. ı the open part of the pass or in the live flange, the etal as if enters a narrowing slot, i.e., experiences a de reduction with the effect that the flange gradually ins out and in some instances increases in height orced spread due to side reduction). In a dead flange, e metal undergoes a small side reduction and is duced in height as well.

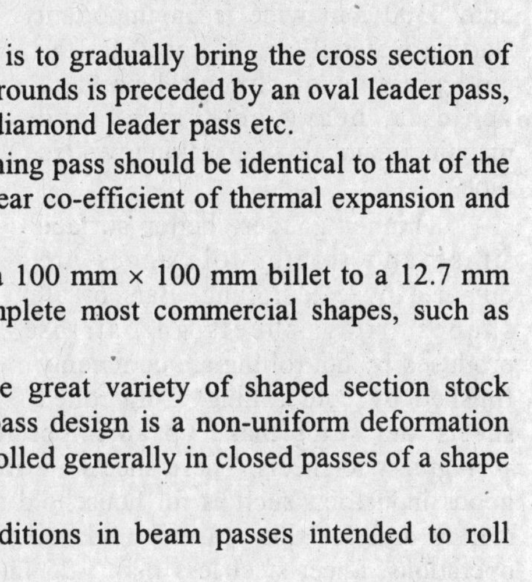

Fig. 4.14. Division of beam pass into component elements

Figure 4.15 shows a general I-beam rolling flowsheet. Commonly, most of the non-uniform deformation takes place chiefly in the first two or three passes while the metal is still hot and thus sufficiently plastic. The first pass in which the starting rectangular cross-section undergoes deformation is termed the slitting pass. In the process, the stock experiences a considerable draft. Beam passes are so arranged on the rolls that dead flanges of a preceding pass enter live flanges of the following pass, this ensuring a gradual shaping of the beam as a whole. Care is taken to achieve a possibly uniform reduction throughout the width of the section in the final passes (particularly, in the finishing pass) to avoid internal stresses, cracks and camber of finished sections.

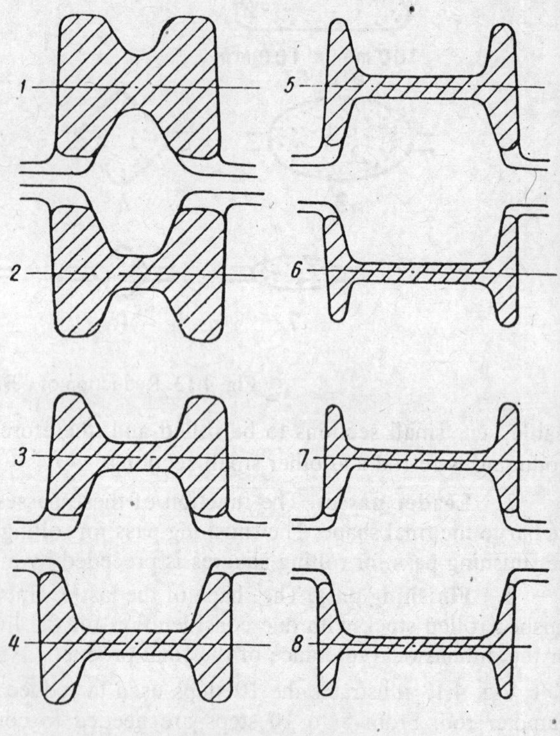

Fig. 4.15. Rolling I-beam

4.2.6. Cold Rolling. Hot rolled products have a relatively rough surface finish and lack dimensional accuracy and certain desired physical properties. How ever, hot-rolled plate, over 6 mm thick and 1800 to 5000 mm wide, is an important starting material in ship building, boiler making, and high-rise welded machine structures. Hot rolled sheet, 0.8 to 6 mm thick and upto 2300 mm wide is an important starting material (in coil form) for the cold pressing of structural parts of vehicles, heavy equipment, and machinery and also for making welded tube.

Thinner gauges, better surface finish and tighter tolerances are obtained by cold rolling. Bars of all shapes, rods, sheets and strips produced by hot rolling are commonly finished by cold rolling. Cold rolled sheets and strips make up an important part of total steel production. Cold rolling gives *springiness* to sheets. These sheets are major raw material for some high- production consumer goods industries, such as for household appliances. Machinability of steel is improved by cold rolling and for this reason, cold rolled stock is widely used for fast automatic machining operations. Sheet steel less than 1.25 mm thick is cold rolled because it cools too rapidly for practical hot rolling. Actually, shapes with wall thickness less than 3 mm are very difficult to be hot rolled. Prior to cold rolling, steel is pickled to clean the surface and remove scale.

Large quantities of steel are rolled to about 0.75 mm for automotive and appliance bodies and down to 0.15 mm for food and beverage cans. Copper is rolled to various gauges for roofing, containers, cooking vessels and down to 0.075 mm for radiator fin stock. Aluminium alloy sheet of about 1.0 mm thickness is extensively used in aircraft fuselages, automotive components, and trailer construction. Aluminium foil of down to 0.8 μm gauge is used in large quantities for packaging. Foils of down to 0.3 μm thickness are produced on special.mills in all materials.

Ralling speed may range upto 25 m/s for coldrolling. It can be higher with highly automated and computer controlled facilities.

4.2.7. Rolling Defects:

There are two types of defects which can be observed in rolled products:

1. Surface defects
2. Internal structural defects

1. *Surface defects :* These defects include defects such as: Scale, rust, scratches, cracks, pits and gouges etc. These defects occur due to the impurities and inclusions in the original cast material or various other conditions related to material preparation and rolling operation.

2. *Internal structural depicts :* These defects include defects like:

(*a*) Wavy edges
(*b*) Zipper cracks
(*c*) Edge cracks
(*d*) Alligatoring
(*e*) Folds
(*f*) Lamitations

Defects due to bending of Rolls : The defects of wavy edges and zipper cracks occur due to bending of rolls as explained below:

Rolls act as straight beams loaded transversely (due to rolling loads) and so undergo deflection, Fig. 4.16 (*a*). Due to this, the edges of the job get compressed more than the central portion, that is, the edges are thinner than the central portion, giving the sheet a "Crown". Since the reduction in thickness is converted into increase in length of the strip, the strip elongates more at the edges relative to the centre. However, the material is continuous and to maintain this continuity, strains within the material adjust. There are compressive strains on the edges and tensile strains at the centre, Fig. 4.16 (*b*). Because the edges are restrained from expanding freely in the longitudinal direction wavy edges on the sheet result, Fig. 4.16 (*c*). Under other conditions of the ratio of the mean thickness to the length of the deformation zone, the strain distribution might lead to cracks in the centre of the sheet (known as zipper cracks), Fig. 4.16 (*d*), Zipper cracks can also occur due to poor material ductility at the rolling temperature.

The remedy is to provide a "Camber" to the rolls, that is, their diameter is made slightly larger at the centre than at the edges. fig. 4.16. (*e*). When under load, the rolls will get flattened along the containing surface and will provide a straight uniform gap to the strip, Fig. 4.16 (*f*).

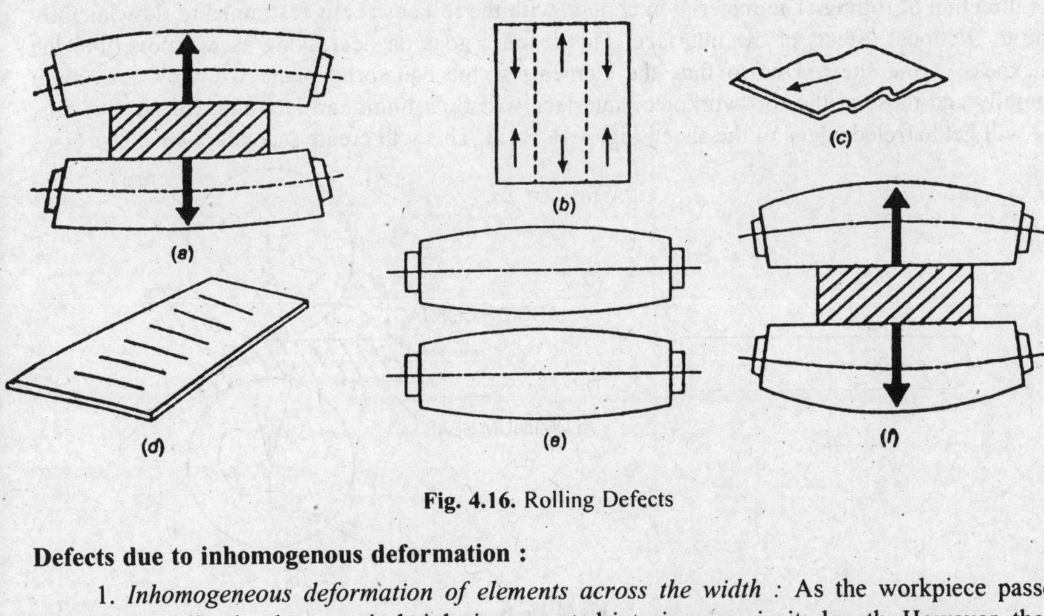

Fig. 4.16. Rolling Defects

Defects due to inhomogenous deformation :

1. *Inhomogeneous deformation of elements across the width :* As the workpiece passes through the rolls, the decrease in height is converted into increase in its length. However, there

is a tendency for lateral spread also (in the transverse direction to the rolling direction). This outword lateral spread is countered by the inward shear/frictional forces. Due to friction hill, these forces are maximum towards the centre of the sheet. So, for the elements near the centre of sheet, the lateral spread will be less than for the elements near the edges. Hence, the decrease in thickness for the elements near the centre will be mainly converted into increase in length. But for the elements near the edges, part of the decrease in thickness is converted into lateral spread, so the increase in length will be less than for the elements near the centre. This will result in rounding of the ends of the sheet in the direction of rolling, Fig. 4.16 A (*a*). However, due to continuity of material elements, the elements near the edges will be under tension and the elements near the centre will be under compression. Such a situation can lead to edge cracks, Fig. 4.16 A (*b*). If the conditions are very severe, the sheet might split along the centre (Centre Split), Fig. 4.16 A (*c*).

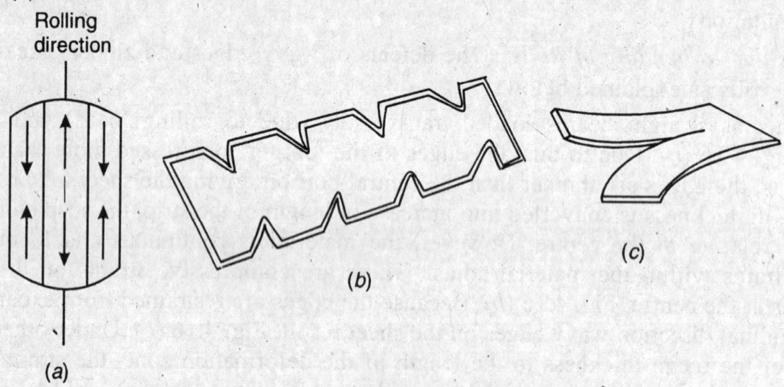

Fig. 4.16. A. Rolling Defect

2. *Inhomogeneous deformation in the thickness direction :*

During rolling, the reduction in height is converted into increase in length of the sheet. But all the elements along the thickness of the sheet do not undergo the same lateral deformation in the direction of rolling. The material in contact with the roll surface is restrained to flow laterally due to frictional effects at the interface. This restrain goes on decreasing as we move into the thickness of the sheet. Due to this, the elements at the equatorial plane will flow maximum laterally and those at the roll-workpiece interface will flow minimum laterally. Because of this, we will get barreled edges of the sheet, Fig. 4.16 B (a). This will create secondary tensile stresses

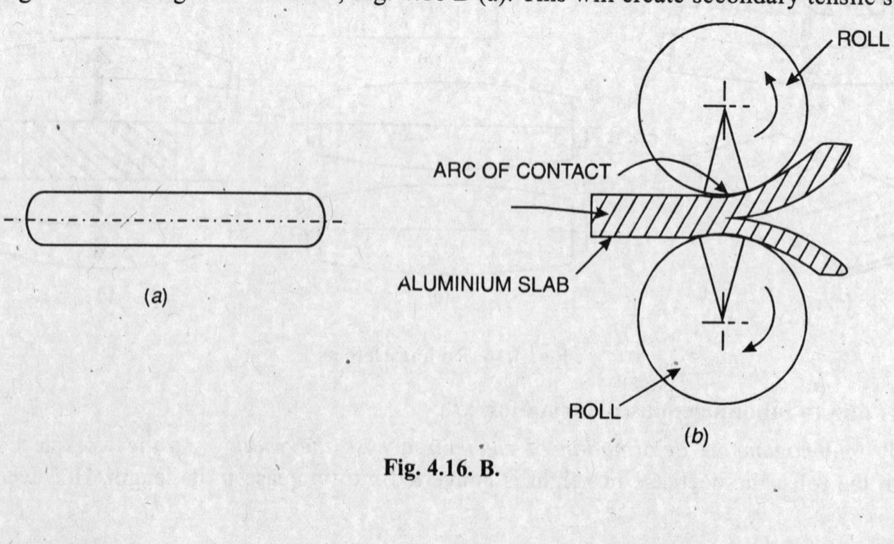

Fig. 4.16. B.

at the equatorial plane. As the extent of barreling increases, cracks will appear on the edges at the equitorial plane. With more severe conditions, the sheet may rupture at the equatiorial plane and follow the path the their respective rolls, resulting in a defect known as "Alligatoring", Fig. 4.16 B (*b*)

Folds: These defects are encountered during plate rolling if the reduction per pass is very small.

Laminations : Internal defects such as fissures are encountered due to incomplete welding of pipe and blow-holes during the rolling process. Some defects like longitudinal stringes of non-metallic inclusions are introduced at the time of ingot productions. Under severe reduction, these defects can result into small cracks or laminations along the thickness direction. Due to this, the strength along the thickness direction can get drastically reduced.

4.2.8. Main Process Variables in Rolling Process:

The rolling load can be calculated as,

$$P = l.b.\ p_m$$

where l = Roll-strip contact length

b = width of sheet

p_m = mean specific pressure

As is clear from Fig. 4.6, l depends on roll diameter and angle of bite. It is approximately given as

$$l \cong \sqrt{R\ .\ \Delta h}$$

Where R = Roll diameter,

and Δh = draft = h0 – h1 (See Fig. 4.6)

p_m depends upon R, h0, h1, and yield strength of the work material.

∴ Main process variables in Rolling process are:–

1. Roll diameter
2. Angle of bite
3. Temperature
4. Strength of work material
5. Speed of rolling
6. Roll gap or draft (h0 – h1)
7. Co. efficient of friction (that is lubrication).
8. Dimensions of sheet, i.e., h0, h1 and b.

4.3. FORGING

Forging may be defined as a metal working process by which metals and alloys are plastically deformed (reduced) to the desired shapes by the application of a compressive force. Forging may be done either hot or cold. However, forging is always understood to be hot working, unless stated otherwise. It differs from rolling process which also employs compressive force to deform the metal, in that the pressure is applied intermittently instead of continuous pressure as in rolling. The products generally are discreet pieces rather than a flowing mass as in rolling. Again, the dies used to apply pressure to the material is not a cylindrical roll (as in rolling process) but usually a rectangular block of certain thickness. Also, the dies have a linear working motion instead of rotary motion of rolls in rolling.

Depending upon the mode of application of compressive force, the forging equipment used may be grouped under two main categories :

1. Hammers. Hammers impart stress on the material by impact and they operate in a vertical position.

2. Forging Presses and Forging Machines (Upsetters). In these equipments, the compressive force is applied continuously and the material is gradually pressed or squeezed into shape. The forging press operates in a vertical position like a hammer. The forging machines or the upsetters operate in horizontal position.

Forgings may be produced in either open or closed dies. Dies are made in pairs. In open die forging, the material is struck or pressed between two flat surfaces and the compressive force is progressively applied locally on different parts of the metal stock. In closed-die forging process the force is applied to the entire surface, causing the metal to flow into a die cavity that has been cut to a specific shape. Open-dies are less costly than closed dies and are more economical for a few parts but can be used for simple shapes parts only. On the other hand, for more complex and accurate parts and with increased production rates, closed-dies are employed. Also, less skill is required with closed-dies than with open dies. Most hammers and presses may be used for open or closed-die forgings.

Both types of forging equipment mentioned above have their advantages and limitations with respect to the shapes to be produced and the materials to be used. Open-die hammers are used for production of simple shaped components and for low outputs where the cost of complex impression or closed dies will not be justified. Closed-die hammers are used for production of close- tolerance parts where the quantity of production is comparatively small (upto 2000 parts per die run). For close limit forgings and for large quantity production (5000 to 6000 forgings per die run), presses are employed. Presses are used for large jobs of symmetrical shape, whereas hammers and forging machines are used for smaller parts. When there is a decrease in section, hammer is employed and when there is an increase in section, forging machine or upsetter is used.

In general, the forging process has got the following advantages and drawbacks as compared to other manufacturing processes.

Advantages :

1. Impurities in the metal in the form of inclusions are broken up and distributed throughout the metal, resulting in structural homogeneity.
2. Porosity, voids and blowholes in the metal are largely eliminated, resulting in internal soundness.
3. Coarse or columnar grains are refined, resulting in increase in yield and ultimate strengths.
4. Better mechanical properties, especially if the fibre flow lines are properly directed.
5. Greater strength per unit of cross-sectional area under static loads, that is, minimum weight per unit strength; and better resistance to shock.
6. Ability of the forging to withstand unpredictable loads.
7. Superior machining qualities. The uniform structure permits higher machining speeds; the freedom from imbedded impurities allows longer tool life and fewer grindings; and homogeneity reduces machining scrap.
8. Minimum metal removal in machining.
9. Relatively smooth surface of the forging.
10. Forgings can be held to within fairly close tolerances.
11. Rapid duplication of components.
12. Wide range of forgeable metals is available.
13. Forgings are easily welded.

Drawbacks :

1. In hot forging, due to high temperature of metal, there is rapid oxidation or scaling of the surface resulting in poor surface finish.
2. Tooling and handling costs are high.
3. Many intricate and cored shapes possible by casting process cannot be forged.
4. Usually, forgings cost more than castings.

The forging temperatures for hot forging of various metals are as given below :-

Ferrous metals = 930 to 1375°C

Copper, Brasses, Bronzes = 595 to 930°C

Aluminium and Magnesium alloys = 345 to 485°C.

The conventional forging processes described in this chapter are intended to apply only to hot forging methods.

4.3.1. Basic Forging Operations

1. Upsetting. When a piece of stock is worked in such a way that its length is shortened and either or both its thickness and width (or diameter of a circular stock) increased, the piece is said to be upset and the operation is known as upsetting.

2. Heading. When the upsetting is done in such a way that the section of the stock is increased only on one end of the stock, the operation is called heading.

3. Fullering. It is the operation of reducing the stock between the two ends of the stock at a central place, so as to increase its length. The inclined surfaces of the die prevents material movement in the width direction (spread) because there is a pressure component acting in the direction of material flow. Repeated strokes, with the work piece rotated around its axis between strokes, allow substantial material redistribution.

4. Edging. The function of edging or rolling operation is to distribute the metal longitudinally by moving metal from the portion of the stock where it is in excess to the portion which is deficient in metal. For example, for forging a connecting rod from a bar stock, the metal will have to be moved longitudinally since the section of the rod varies along its length. This is an important operation in closed die forging.

5. Drawing down. Drawing down or "Cogging" is an operation similar to fullering with the difference that the stock is reduced at only one end (and its length increased) instead of at a central place as in fullering.

6. Bending. Bending operation makes the longitudinal axis of the stock in two or more places. This operation is done after the stock has been edged or fullered and edged so that the stock is brought into a proper relation with the shape of the finishing impression (in closed die forging) where such a section is of non-symmetrical section.

7. Flattening. This operation is used to flatten the stock so that it fits properly into the finishing impression of a closed die.

8. Blocking. It is a forging operation which imparts to the forging its general but not exact or final shape. This operation is done just prior to finishing operation.

9. Cut-off. A pair of blades, either milled in the corner of a pair of forging dies, or inserted in the dies, used to cut away a forging from the bar after the finishing blow.

10. Piercing. Piercing is the operation done with the help of a punch to obtain blind or through holes in the metal. The pierced billet is further processed.

11. Punching. This is the operation of shearing out a slug in a forging to produce a hole.

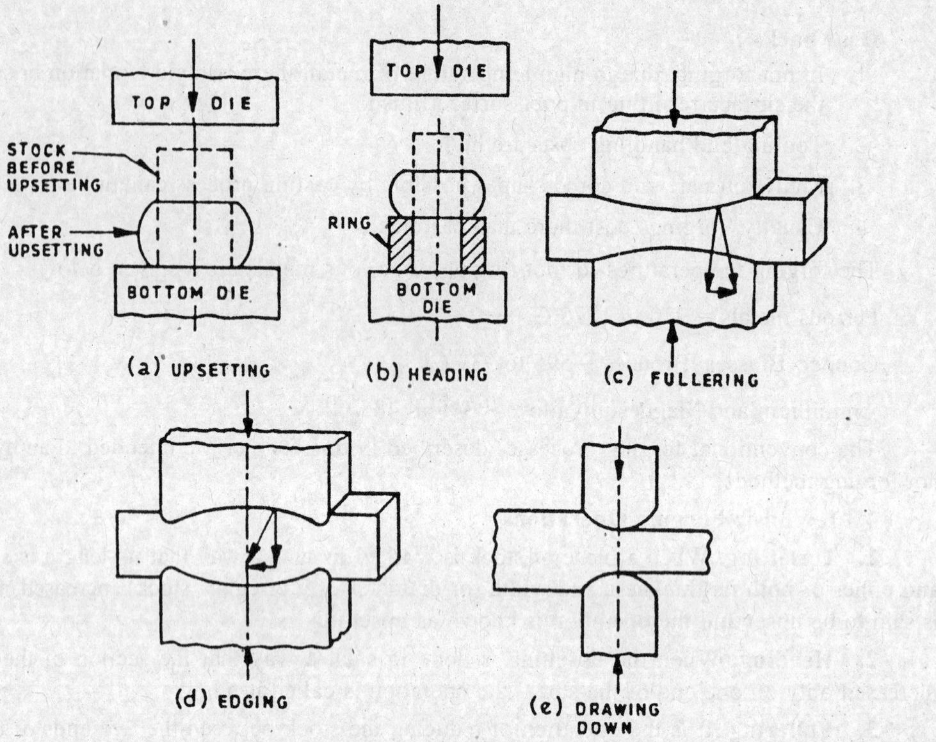

Fig. 4.17. Basic Forging Operations.

12. Swaging. It is the operation of reducing or changing the cross-section area of diameters by revolving the stock under fast impact blows.

13. Coining. It is a cold closed die forging operation (no flash) to obtain closer tolerances and smoother surfaces.

4.3.2. Forging Processes. Forging processes may be grouped under two headings : open die forging and closed or impression die forging :

4.3.2.1. Open-Die Forging. In open-die forging which is also called 'Hammer forging' or 'Flat-die forging', the work piece is struck or pressed between two flat surfaces. Open-die forging is used where number of components to be forged is too small to justify the cost of impression-dies or where the sizes are too large and too irregular to be contained in usual impression dies. The shapes most commonly used by open-die forging are : Bars, slabs or billets with rectangular, circular, hexagonal or octagonal cross-sections, weldless rings and many other components of simple shapes. The forging equipment used for open-die forging may be : Hammers or presses.

(*a*) **Hammers.** A hammer consists of four main parts :

(1) Falling weight *i.e.* ram or tup

(2) Frame or guide for the falling weight.

(3) Base or anvil.

(4) Lifting mechanism for the ram.

The working surfaces of the pair of dies are flat and horizontal. One die, called the upper die, is fastened to the bottom of the ram and the other die, called the lower die, is fitted on the top of the anvil. The work-material is positioned on the bottom die and is shaped by the falling weight. As the weight falls by gravity, the heavier the falling weight, and greater the height and velocity of the fall, the greater will be the force of the blow of the hammer. There is little, if any, lateral confinement of the work-piece and the desired shape is obtained by manipulating the forging in between blows. If the workpiece is too heavy to be manipulated by hand, mechanical manipulators are used to hold and handle work. The capacity or the tonnage of a hammer is specified by the weight of its falling parts (piston rod, piston and ram including ram die or upper die). For instance, if its falling parts weigh 50 kN, the hammer will be called as a 50 kN hammer. Forgings made by hammer forging may weigh from a few newtons to over 2000 kN.

The hammer is an energy-restricted equipment. The total K.E. attained by the falling parts when the upper die touches the workpiece positioned on the lower die, during a working stroke, is dissipated by plastic deformation of the material and by elastic deformation of the ram and anvil. This K.E. depends both on the weight of the falling parts and the velocity they attain at the end of the downward stroke $\left(\text{K.E.} = \frac{1}{2} mv^2\right)$. Again, v, depends upon the stroke of the hammer ($v = \sqrt{2gh}$, h = stroke). Therefore, it is necessary to rate the capacities of the hammers in terms of kinetic energy, *e.g.,* metre-kilograms or metre- tonnes. The practice of specifying a hammer by its falling weight is not useful. That can be regarded only as a specification and not the capacity of the hammer.

We know that in hammer forging, the hammer is lifted up to a certain height and then it is allowed to fall by gravity. Depending upon the lifting mechanism, the hammers may be classified as : Spring hammers, Pneumatic hammers and steam-and- air hammers. Steam and air hammers are most widely used in forging practice and they operate either on steam or on compressed air.

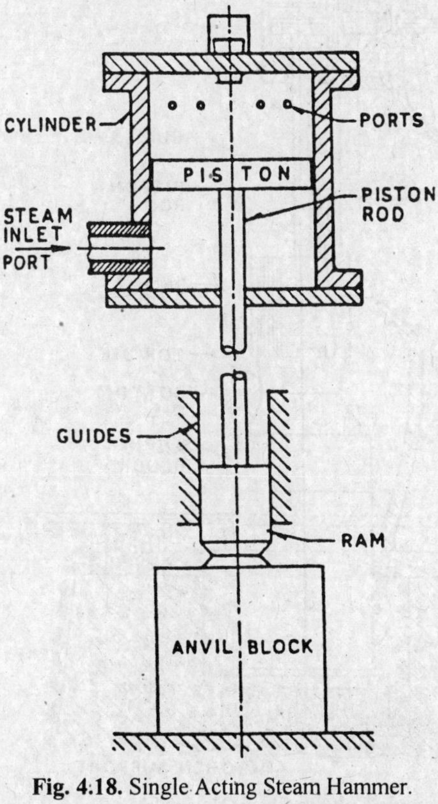

Fig. 4.18. Single Acting Steam Hammer.

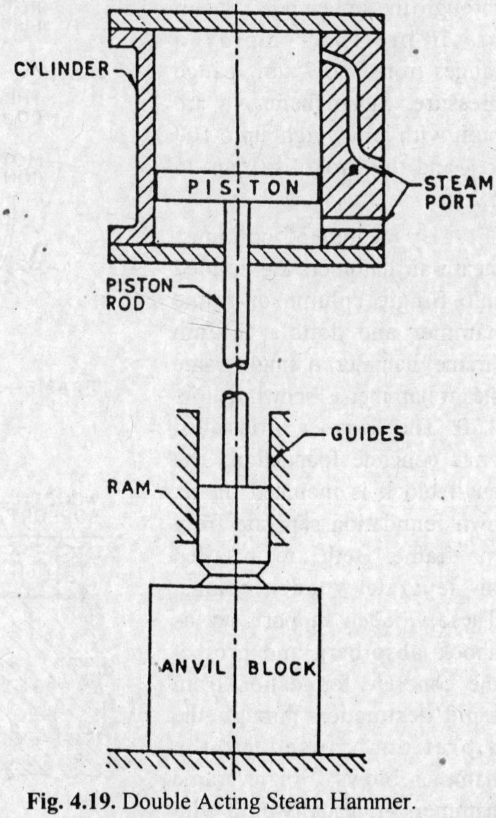

Fig. 4.19. Double Acting Steam Hammer.

Steam and air hammers are of two types : single acting and double acting. In single acting hammer, the steam or air pressure is used only to lift the ram which then falls by gravity. In double acting hammer, the steam or air is employed both for raising the ram and for increasing the force of its impact on its downward stroke. In 'single acting steam-air hammer', steam or compressed air is delivered to the cylinder through inlet port (Fig. 4.18), thus raising the piston and the ram connected to it. The force necessary to ensure quick lift up of the ram can be 3 to 5 times the ram weight. When the ram has been lifted to a sufficient height, the steam or air is cut off which escapes through the inlet port. The ram drops down under its own weight and strikes the workpiece positioned on the anvil block. In order to slow down the speed of the piston as it reaches its upper position, several ports are provided in the upper section of the cylinder. As these ports are covered by the piston on its upward travel, the air in the cylinder above the piston cannot escape and will start getting compressed. This compres-sed air will act as a buffer and slow down the upward motion of the piston. Single acting steam-air hammers are seldom used now.

In double acting steam hammer (Fig. 4.19), ports are provided both at the upper section and lower section of the cylinder. On the downward stroke of the ram, steam enters the cylinder from the upper port and acts on the piston to increase the impact force of the ram. For lifting the ram, steam enters the cylinder from the bottom port. As the ram moves upward, the steam above the piston escapes through the upper port. Steam or air pressure employed ranges from 7 to 9 atm., gauge pressure. Steam hammers are built with ram weight upto 100 kN, and the stroke is 0.9 m to 1.2 m.

By design of columns, steam-air hammers are divided into Single column or frame hammer and double column (frame) hammer. A single frame steam hammer is shown in Fig. 4.20. The hammer is installed on a concrete foundation. The anvil block is mounted on its own foundation separate from the frame. Both are installed on separate wooden beams. These wooden supports act as shock absorbers and protect the concrete foundation from rapid destruction through the vibrations caused by the hammer blows. Single frame hammers are usually built with

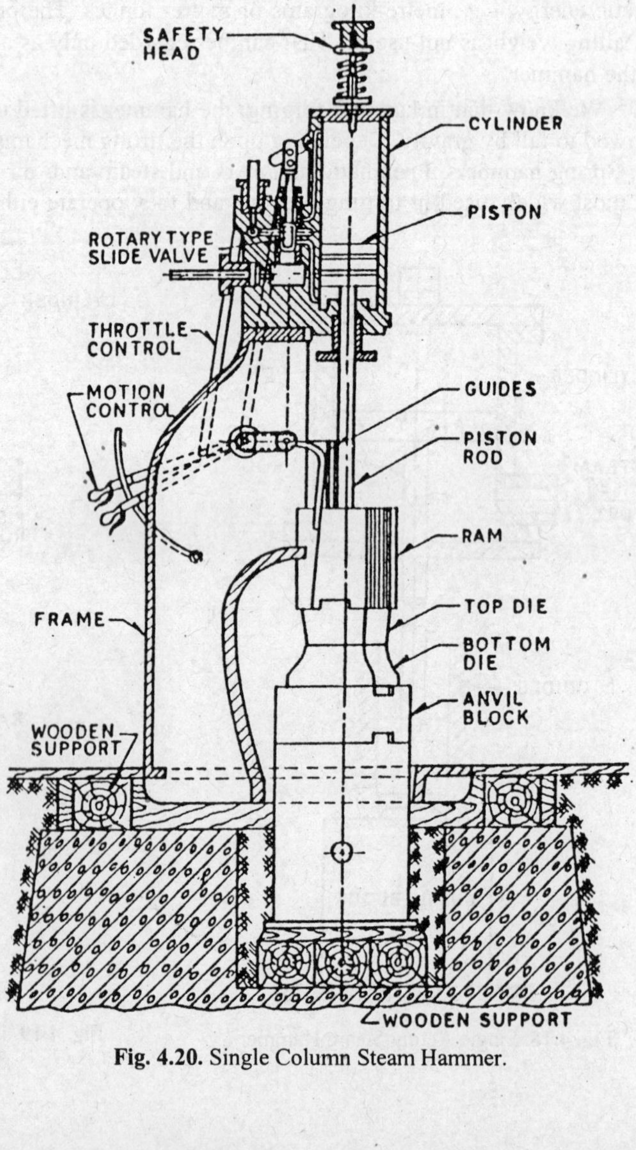

Fig. 4.20. Single Column Steam Hammer.

falling parts weighing 2.5 to 10 kN. A double frame steam hammer is shown in Fig. 4.21. Steam is supplied through the slide valve into the cylinder under or above piston. The piston is connected by piston rod to ram and top die. Bottom die is secured to anvil which is mounted on separate foundation. Such hammers are built with falling parts weight from 10 to 50 and even 100 kN, but more frequently, from 30 to 50 kN. A double frame hammer is more powerful than a single frame hammer, but it does not permit the operator more freedom of motion around the dies. In steam hammers, the anvil block normally weighs at least 15 to 20 times the total falling weight of the hammer.

The air-or-steam hammer can operate with either single strokes or continuous blows of the ram die on the forging. If required, the forging can be held down (squeezed) to the anvil or the ram can be held up. The single acting hammers are also called as "Steam/air lift, gravity drop hammers" and double acting hammers are called as "Steam/Air lift, Steam/Air drop hammers".

Pneumatic Hammer. Pneumatic power forging hammers are usually used to smith forge small parts. In these hammers, the weight of Anvil is equal to (15 to 20) times the falling weight. A pneumatic forging hammer, (Fig. 4.22), has two cylinders : a compressor cylinder and a ram cylinder. Air is compressed on both upward and downward strokes of the piston in the compressor cylinder and is delivered to the ram cylinder where it actuates the ram of the ram cylinder, delivering the forging blows to the work. The reciprocation of the compressor piston is obtained from a crank drive which is powered from an electric motor through a reducing gear.

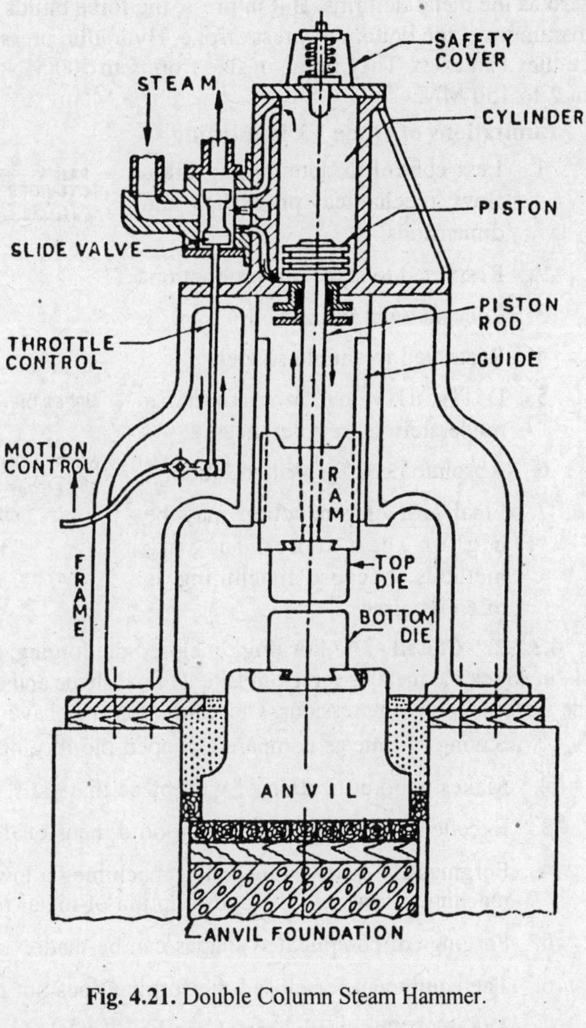

Fig. 4.21. Double Column Steam Hammer.

The distribution of air between compressor cylinder and ram cylinder is controlled by means of two rotary valves with ports through which air passes into the ram cylinder, below and above the ram, alternately. The valves are actuated by depressing foot treadle or operating a hand lever. By controlling air distribution the required ram movement can be attained : either continuous blows are delivered to the forging, or the forging is held down on the anvil, or the ram is held in its upper position. The upper die is secured to the bottom of the ram and the bottom die is secured to a steel cap by means of taper keys in a dovetail joint. The steel cap is mounted on a heavy anvil which is separate from the hammer frame.

The size of the pneumatic hammer, *i.e.*, the weight of its falling parts (weight of the ram piston, ram and upper die) usually varies from 0.5 to 10 kN. Such hammers operate at 80 to 200 blows per minute. In these hammers, the ram falls not only under its own weight but the downward force is also increased by the action of compressed air above the ram.

(*b*) **Hydraulic Presses.** For the fabrication of heavy forgings, hydraulic presses are employed. In presses, the material is worked upto a greater depth as compared to hammers. The action of a press is opposite to that of the hammer. In hammer forging, the maximum amount of energy is available at the instant the upper die touches the work-piece and is gradually reduced to zero as the metal deforms. But in press, the force builds up from minimum at the start of stroke to maximum at the bottom of press stroke. Hydraulic presses are rated according to the maximum force they can exert. They range in size from 5 to 500 MN. For open-die forging, their size ranges from 2 to 150 MN.

Limitations of Open - Die Forging

1. Less control in determining grain flow, mechanical properties and dimensions.
2. Restricted to short run production.
3. Poor material utilization.
4. Restricted to simple shapes.
5. Difficulty of maintaining moderately close tolerances.
6. Absolute need for skilled labour.
7. Final cost of production may be higher than other forging methods, because machining is often required.

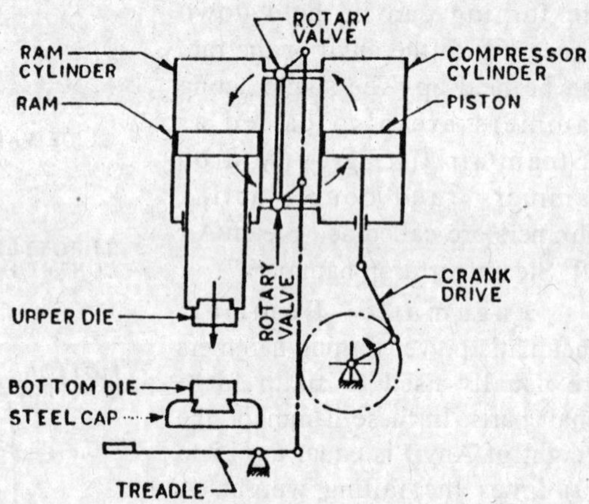

Fig. 4.22. Pneumatic hammer.

4.3.2.2. Closed - Die Forging. In closed-die forging, cavities or impressions are cut in the die block, in which the metal is forced to take its final shape and dimensions. The flow of metal is limited by the surfaces of the impressions. Closed-die forgings have the following characteristics :-

1. Saving of time as compared to open-die forging.
2. Makes good utilization of workpiece materials.
3. Excellent reproductivity with good dimensional accuracy.
4. Forgings are made with smaller machining allowances, thus reducing considerably the machining time and the consumption of metal required for the forging.
5. Forgings of complicated shapes can be made.
6. The equipment for closed-die forging does not require highly skilled workers.
7. The grain flow of the metal can be controlled ensuring high mechanical properties.
8. Method is suited for rapid production rate.
9. Cost of tooling is high, therefore, suitable for large production runs.

(*a*) **Classification of Dies.** The dies used in closed-die forging or impression-die forging may be classified into two groups :

(*i*) **Single Impression Die.** This die contains only one cavity or impression which is the finishing impression. The preliminary forging operations are done by hand or on forge hammers, forging rolls etc., and only final finishing operation is done in the die-cavity.

(*ii*) **Multi - Impression Die.** This die contains finishing operation and one or more auxiliary impressions for preliminary forging operations. The final shape of a part is progressively developed over a series of steps from one die impression to the next. Generally, multi-impression dies are very expensive to make and are employed only when the quantity to be made is sufficiently large.

Advantages of Multi - Impression Die

(1) Complete sequence of forging operations can be carried out on one equipment only, avoiding the use of auxiliary forging equipment.

(2) Use of multi-impression dies is particularly suited for production of small and medium sized forgings in large quantities as this method gives 2 to 3 times the production compared with the method of production using a single die. This is because the time of preparation of the 'use' on auxiliary equipment is reduced or eliminated.

(3) All the preliminary operations can be performed on these dies with good ease. The 'use' can be prepared to fairly accurate dimensions. Besides this more accurate forgings are prepared.

(4) Wastage of forging metal is reduced.

(5) 'Use' may not be reheated for the finishing impression.

(6) Initial die cost becomes insignificant in case of high output.

(7) Finishing impression lasts long, because, much of the load is taken by blocking impression.

(*b*) **Classification of Closed - Die Forging.** Closed-die forging is of three types :

(*i*) **Drop Forging.** In this process, the forging is made by hammering a heated bar or billet into aligned die cavities. The forging equipment used for this is 'Drop Hammer'. Drop hammers are very similar in design and arrangement to the steam hammers already described. They have the added provision of larger and more accurate guides for the vertical movement of the ram, so that the top die meets the bottom die in close alignment. Also the frame of a drop forging hammer is secured to the anvil block and the total hammer structure is mounted on single foundation. This is to keep upper and lower dies aligned. The two most commonly used drop forging hammers are : Steam-Air-drop hammer and Board drop hammer.

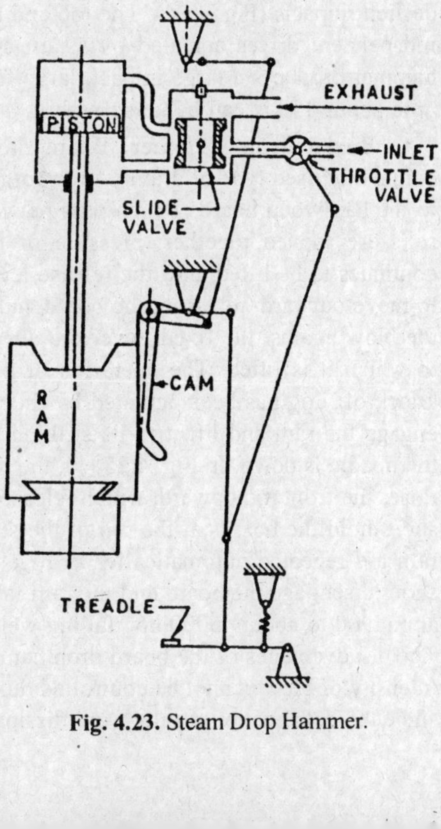

Fig. 4.23. Steam Drop Hammer.

Steam-Air Drop Hammer. Steam-air drop hammer has two column frame which is secured to the anvil block by spring loaded bolts, which dampen impacts. A steam drop hammer is not truly a drop hammer, since steam pressure is used to raise the ram as well as to control the strike force. A steam hammer ranges in capacities from 2.25 to 225 kN. It works rapidly and can deliver upto 300 blows per minute. The steam or air pressure used ranges from 7 to 9 gauge atmospheres. The impact of the striking blow can be varied from a light tap to full blow. The operation is controlled by a foot treadle. The control diagram of a steam drop hammer is shown in Fig. 4.23. The treadle is used to deliver the striking blow by controlling the throttle valves. The treadle must be depressed for each blow that is delivered by the steam drop hammer. The magnitude of the blow is dependent on the speed and distance the treadle is depressed. When the treadle is pressed down, the throttle valve becomes wide open and steam enters through the inlet to the centre of the slide valve and flows to the top of the piston to exert pressure. When the treadle is released, steam enters at the bottom portion of the cylinder and lifts the piston and ram. As the ram moves upward, there is a slight movement of the cam which raises the slide valve slightly. Due to this, a small amount of steam is admitted to the upper side of the piston, which exhausts from the lower side. As the ram travels further up, the slide valve moves slightly downward. Due to this action of cam when the treadle is up, ram oscillates in the idling position, thus ensuring a constant supply of steam through the cylinder and avoiding the possibility of condensation.

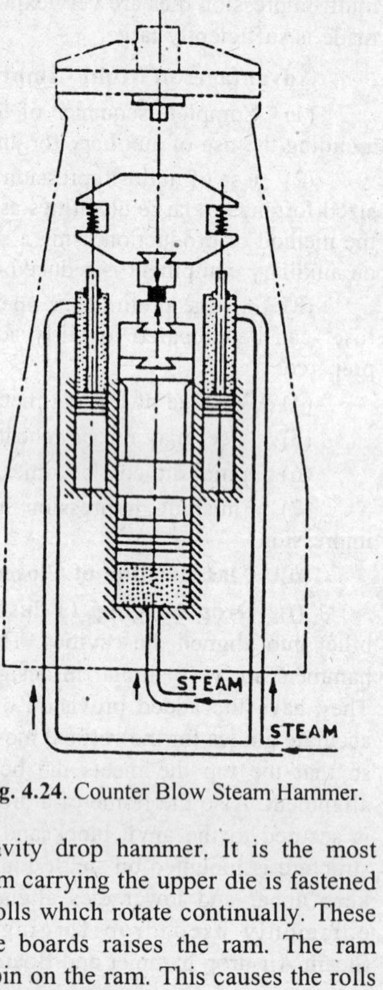

Fig. 4.24. Counter Blow Steam Hammer.

A variation of the hammer discussed above is the 'gravity drop hammer'. In this design, the steam or air that is used to lift the ram is suddenly released to allow the ram to fall freely under gravity. The chief drawback of steam-air drop hammer is a need for large foundations, and despite the precaution, appreciable vibration of soil and of production shop building. This trouble is eliminated by using a 'Counterblow Steam Hammer'. This hammer has two opposed vertically moving rams, which move in opposite directions and meet together for the desired impacts (Fig. 4.24). The top and bottom dies have independent drives actuated by steam cylinders. As this hammer produces a large impact, large forgings weighing upto about 45 kN can be made with it.

Board drop Hammer. Board drop hammer is a gravity drop hammer. It is the most commonly used type of gravity-drop forging hammer. The ram carrying the upper die is fastened to the hardwood boards which pass between a pair of steel rolls which rotate continually. These rolls are moved together, pressure of the rolls against the boards raises the ram. The ram continues to be lifted until the release lever is actuated by a pin on the ram. This causes the rolls to move outward, releasing the board and allowing the ram to fall freely by gravity. The force of the blow against the fixed lower die depends upon the weight of the falling ram and the height to which it is lifted. The operation of a board drop hammer is explained in Fig. 4.25 (*a*), the knock off dog has been actuated by the ram at the bottom of its stroke. This causes the rolls to engage the ram and lift it to its striking position. The brake shoes do not contact the boards as the treadle is down. In Fig. 4.25 (*b*), the release lever has been actuated by a pin on the ram. This raises the front rod upward, which releases the rolls from the boards. If the treadle is up, the break shoes hold the boards at the top of the stroke. However, if the treadle is down, the blows of the ram are repeated automatically. In Fig. 4.25 (*c*), when the treadle is pressed down, the brake shoes disengage the board and the ram falls by gravity. The board drop hammer sizes range from about 900 to about 45000 kN. falling weight. The height of fall ranges from about 0.75 to 1.5 m. The disadvantages of the board drop hammer are : the boards are liable to frequent breakage, the intensity of blow cannot be controlled during the stroke. Hammers are energy restricted machines since the deformation results from dissipating the kinetic energy of the ram.

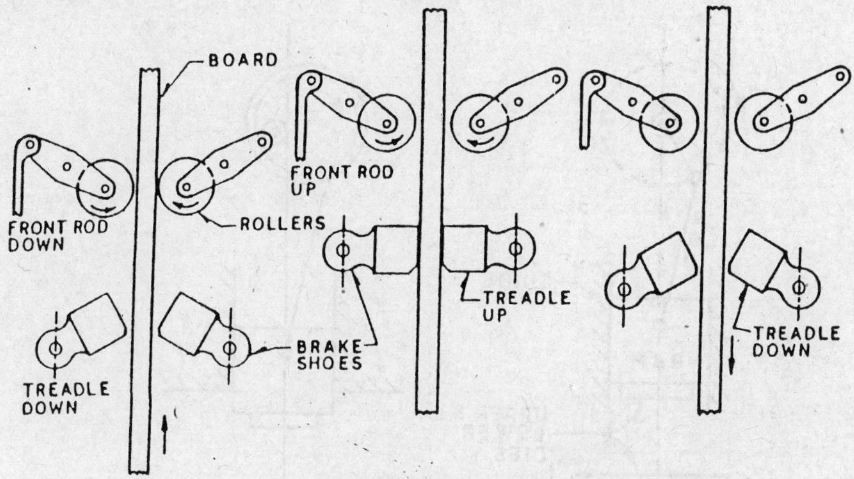

Fig. 4.25. Board Drop Hammer.

The other types of drop hammers employed in modern forge shops are :

1. Belt drop hammers. In these hammers, the ram is suspended from a belt. The mechanism of belt drop hammers is operated by 5 to 15 hp electric motors. Generally, these hammers are made with ram weights ranging from 1 kN to 1.5 kN. The main disadvantage of these hammers is the rapid wear of the belts which are in constant touch with rotating pulley.

2. Rope drop hammers. Here the ram is suspended from a rope.

3. Chain drop hammers. Here, the ram is suspended from a chain.

(*ii*) **Press Forging.** Like drop forging, press forging also uses impression dies but the parts are made by plastically deforming a metal blank into die-cavities by a slow squeezing action. The forging pressure builds up from the start to the end of stroke, resulting in maximum penetration and in improved grain flow throughout the entire forging. Completed forgings are ejected manually or mechanically from the die cavities.

The presses used for press forging are : Mechanical Presses (crank, eccentric, knuckle or screw type) and upright hydraulic presses (see Fig. 4.26). The power flow diagram of a crank type mechanical press is shown in Fig. 4.27. An electric motor drives the flywheel mounted on the counter shaft by means of a belt drive. Torque from the counter shaft is transmitted to the crankshaft through gearing. From the crankshaft, reciprocating motion is given to the ram with the help of a connecting rod. The bottom die is locked in position by means of a wedge mechanism. Disk clutch is used to start and stop motion of ram which is brought to a gradual stop by means of a brake. Mechanical presses are faster than hydraulic presses and operate at about 25 to 100 strokes per minute. Their capacities range from 900 kN 110 MN and they are generally employed for non-ferrous alloys weighting up to about 135 N. Hydraulic presses are larger having capacities upto 450 MN.

The initial cost of a hydraulic press is higher than that of a mechanical press of equivalent capacity. However, in a hydraulic press, pressure can be changed as desired at any point in the stroke by adjusting the pressure control valve. This will help in controlling the rates of deformation according to the metals being forged. But in hydraulic presses, the contact time between the workpiece and the dies is more as compared to a mechanical press. Due to this the die life in hydraulic presses is sometimes shortened because of heat transfer from hot workpiece to the dies.

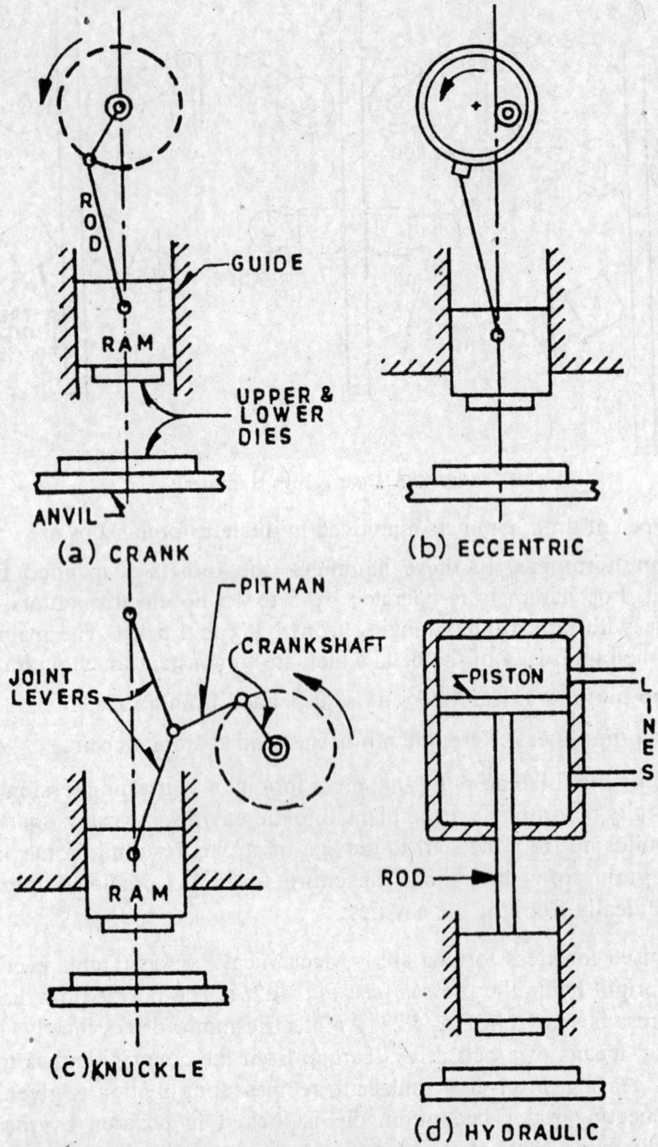

Fig. 4.26. Presses.

Hydraulic presses are of two types :

1. Direct-drive hydraulic presses. These presses usually employ hydraulic oil or water as the working medium which is pressurised directly by high pressure pumps.

2. Accumulator-driven hydraulic presses. These presses usually employ a water oil emulsion as the working fluid supplied from accumulators which are in turn pressurized by high pressure pumps.

The larger presses (10,000 to 50,000 t) are of this type. The accumulator drive is more economical if one accumulator system can be used by several presses, that is, multiple ram forging, thereby, leading to enhanced production rates.

Mechanical presses are stroke-restricted machines since the length of the press stroke and the available load at various positions of the strokes represent their capabilities. Hydraulic presses are load-restricted machines because their capacity to do a forging process depends upon their maximum rating. The ram stroke in mechanical presses (the most commonly used is eccentric press) is shorter than in a forging hammer or hydraulic press, therefore, these presses are best suited for low profile forgings.

It may be of interest that every 10 MN of crank mechanical press approximately to 1 tonne of falling parts of a hammer. The cost of a crank-type press is higher than that of an equivalent hammer. However, crank type presses are economically advantageous even in low volume production runs, provided equipment is utilized to no less than 35 to 40%. Hydraulic presses are used to produce slender forgings and forgings from poorly plastic alloys and to effect operations requiring large-stroke tools. This technique is applicable to the manufacture of large levers, flanges, toothed wheels, hollow bodies, railway wheel disks, crankshafts, tank bottoms, propellers, ribbed panels and other parts of aircraft and rocket bodies.

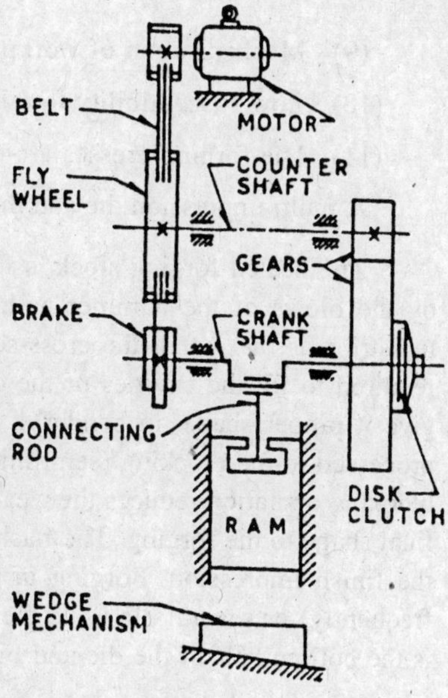

Fig. 4.27. Crank Type Mechanical Press.

In friction screw presses, the blow is midway between that of a hammer and mechanical hydraulic press. These machines are used extensively for hot brass work, but can also be used for steel drop-forging of small sizes.

Advantages of Press Forging over Drop Forging :

Die forging presses were introduced in the early thirties, for large scale production of die forgings. These forging presses have been replacing the drop forging hammers and today are widely used through out the die forging industry. The special advantages of die forging presses are as given below :

(1) Press forging is considerably quieter operation than drop forging.

(2) Press forging is normally faster than drop forging since only one squeeze is needed at each die impression.

(3) Alignment of the two die halves can be more easily maintained than with hammering.

(4) Structural quality of the product is superior to drop forging.

(5) With ejectors in the top and bottom dies, it is possible to handle reduced die drafts. Forgings obtained will thus be more accurate. Also, reduced die drafts reduce the weight of the charge and subsequent machining of the workpiece.

(6) High longitudinal and transverse stability of the press contributes to fine accuracies and this results in uniform forgings with exacting tolerances and low machining allowances.

(7) The number of strokes and ram speed are high (for mechanical press) and the ejectors release the forgings from the dies immediately after the deformation has taken place. The shorter contact periods, thus obtainable in mechanical die forging presses, increase die life.

(8) Simple handling enables high output even with unskilled operators.

(9) Mechanisation of workpiece transfer further increases productivity.

(10) Low susceptibility to failure and simple maintenance.

(11) Die forging presses are ecologically safer than forging hammers and screw presses.

A multi impression die and the forging sequence for a connecting rod is shown in Fig. 4.28.

The heated forging stock is first placed into the fullering impression 1 where it is reduced by the blows of the hammer with an increase in its length. The next operation "edging", 2, redistributes the metal, the cross-section is increased at certain places and reduced at others as required to fill the cavities of the die. After this, the bending operation is done on the stock to give it proper shape, in bender 4. Before being forged in the finish impression 3, the stock is processed in the blócking (semifinish) impression 5. By giving the forging its general shape, the blocking operation reduces the wear of the finishing operation. The finish impression, imparts the final shape to the forging. The flash gutter 6 (to collect the excess metal) is provided only around the finish impression. Forging in the blocking and finishing impressions is done in one (less frequently) or several blows of the hammer ram. Fig. 4.29 (*a*) is the part drawing, Fig. 4.29 (*b*) is the bottom half of the die and Fig. 4.29 (*c*) illustrates the forging sequence.

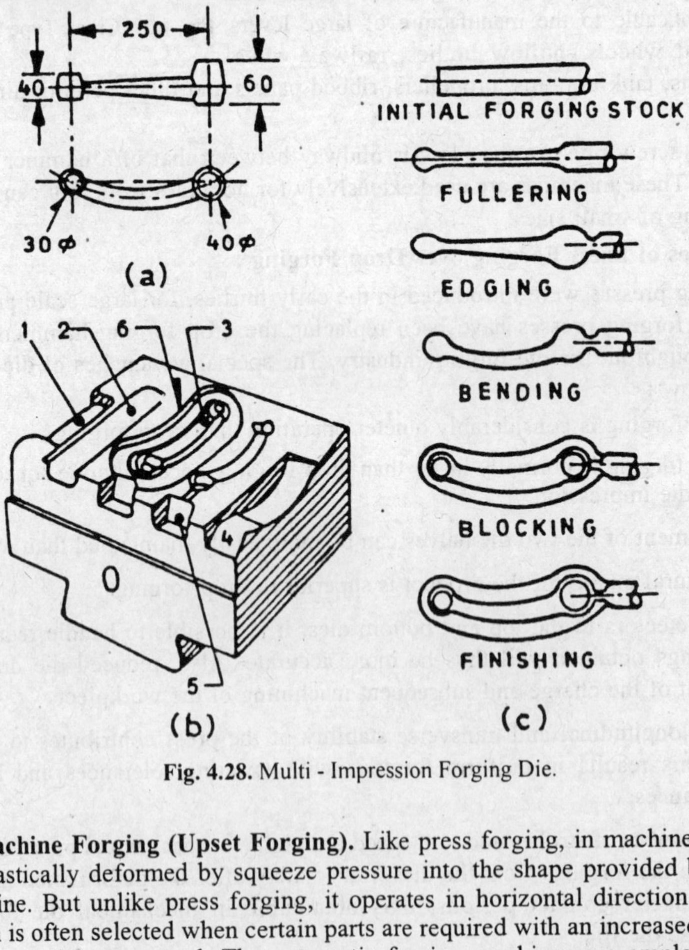

Fig. 4.28. Multi - Impression Forging Die.

(*iii*) **Machine Forging (Upset Forging).** Like press forging, in machine forging also, the material is plastically deformed by squeeze pressure into the shape provided by the dies in the forging machine. But unlike press forging, it operates in horizontal direction. It is the forging method which is often selected when certain parts are required with an increased volume of metal at the centre or only at one end. The automatic forging machines are employed for the mass production of nuts, bolts, rivets, wood screws, roller and ball bearing balls, gear blanks, valve stems, axles, couplings and other axially symmetrical items. The initial blanks are standard rolled stock, such as sections or rods cut to required lengths. The forging machine consists of a heavy cast steel body in which its three main components, stationary die, moving die and heading tool are properly secured. The operation of the machine can be explained with the help of Fig. 4.29. Fig. 4.29 (*a*) shows the parts to be forged. For this purpose, barstock of the correct size is selected, one of its ends heated and stock placed in position in the stationary die and pushed forward until it contacts the stock gauge. As the lever is pressed down, the moving or gripper die closes against the stationary die so that the stock is tightly gripped. At the same time, automatic stock-gauge moves aside, as shown in Fig. 4.29 (*c*). Then header slide moves forward and forces the heading tool against the projecting plastic stock, moving the stock into the cavities of the dies (Fig. 4.29 (*d*)). After the metal completely fills the die impression, both the moving die and the heading tool start moving back to their original positions, as shown in Fig. 4.29 (*e*). Horizontal forging machines are specified or rated for size by the diameter of the largest bar size they could handle. These machines can handle bar stock ranging in diameter from 6.35 mm. to 254 mm. with a maximum metal deforming pressure ranging from 500 kN to 30 MN.

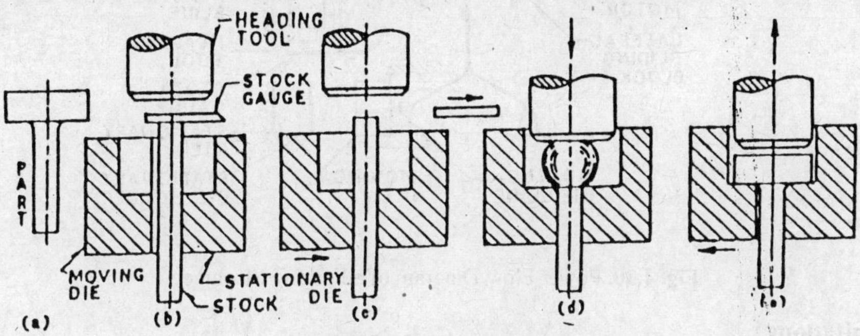

Fig. 4.29. Upset Forging.

The power flow diagram of a forging machine is shown in Fig. 4.30. An electric motor rotates flywheel which is mounted on a drive shaft. This shaft, through a system of gears. rotates crank shaft, connected to sliding block through connecting rod. Sliding block carries heading tool slide in which heading tool is inserted. Connecting rod connects the sliding block only when the foot treadle is pressed down. The free end of the connecting rod then enters the recess of the sliding block, thereby actuating the latter. Through levers and with the aid of eccentrics, lateral sliding block actuates the moving die slide which carries the moving die. The stationary die is secured in stationary die holder. The movements of sliding block and moving die slide are adjusted so that the stock to be upset, when inserted in the machine upto stock gauge, is gripped between the moving die and the stationary die and only after this does the heading tool, secured in he heading tool slide, commences to move. Stock gauge is now withdrawn from the range of travel of the slide by the roll. During its return stroke, the heading tool releases the work.

Advantages of Machine Forging

(1) The quality of the forging is better than obtained by drop forging or press forging.

(2) Forging in forging machine is accompanied by little or no flash, whereas in drop forging, the flash may amount from 10 to 75% or even upto 200% of the weight of the finished forging.

(3) Forging machines have a higher productivity and their maintenance is much cheaper as compared to drop forging hammers.

(4) There is saving in material and also machining expenses as no or little draft is needed on forging made by upsetters.

(5) The upsetting process can be automated.

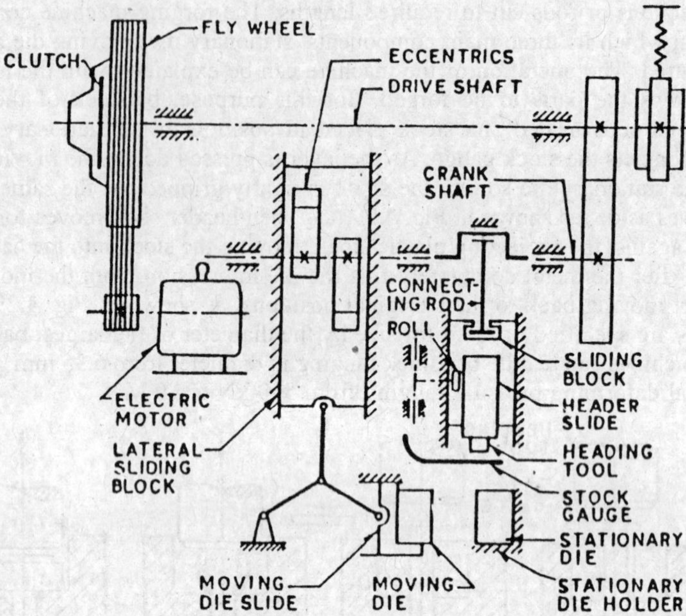

Fig. 4.30. Power Flow Diagram of a Forging Machine.

Limitations

(1) It is not convenient to forge heavier jobs due to the material handling difficulties.

(2) The maximum diameter of the stock which can be upset is limited (Max about 25 cm).

(3) Intricate, nonsymmetric and heavy jobs are difficult to be forged on a forging machine.

(4) The tooling cost may be high.

Rules for Machine Forging

In completing a forging in the forging machine, one or more steps comprise the sequence so that metal may not fold upon itself producing a cold shut or cracks in the forgings. These steps are commonly termed as 'passes', 'blows'; or 'shots'. To use good forging design, it is necessary to understand the practical laws that govern the action of metal in the forging machine and cover the majority of the forging machine die design problems.

For practical purposes the laws have been translated into series of rules for use in forging machine practice so as to be certain that the forgings do not develop cold shuts and other injurious defects. The rules apply to all diameter stock to be upset.

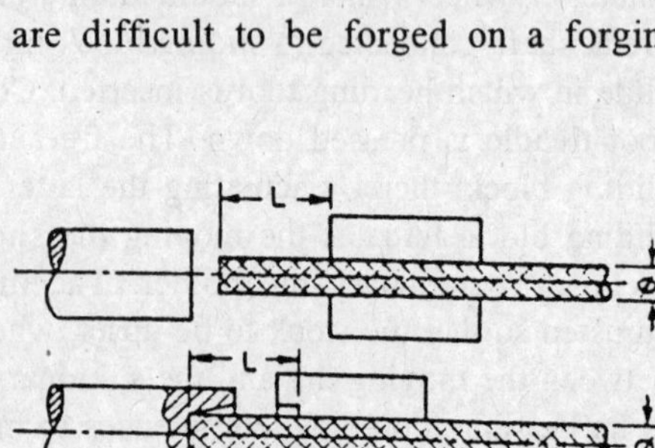

Fig. 4.31. Rule 1.

Rule No. 1. The limit of length of unsupported stock that can be gathered or upset in one blow without injurious buckling is not more than 3 times the diameter of the bar, (Fig. 4.31). If an attempt be made to upset a length of stock longer than 3 times the bar diameter, instead of upsetting uniformly, the stock will buckle at a point near the middle and result in more of the upset forming on one side of the centre of the stock. In practice, it is better that the length of the unsupported stock is within $2\frac{1}{2}$ times the bar diameter.

Rule No. 2. Length of stock more than 3 times bar diameter can be successfully upset in one blow, provided the diameter of the upset made is not more than 1.5 times bar diameter. If this is kept more than 1.5*d*, the buckling will be excessive and the stock will fold in. In practice, it is advisable not to exceed 1.3 times bar diameter (Fig. 4.32).

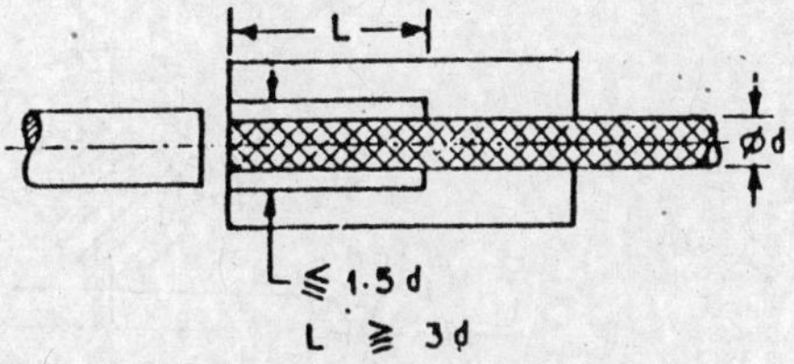

Fig. 4.32. Rule No. 2.

Rule No. 3. In an upset requiring more than 3*d* in length then the diameter of the upset is 1.5*d*, the amount of unsupported stock beyond the face of the die must not exceed one diameter of bar, (Fig. 4.33). However if the diameter of the hole in the die is reduced below $1\frac{1}{2}d$, then the length of unsupported stock beyond face of the die can be correspondingly increased.

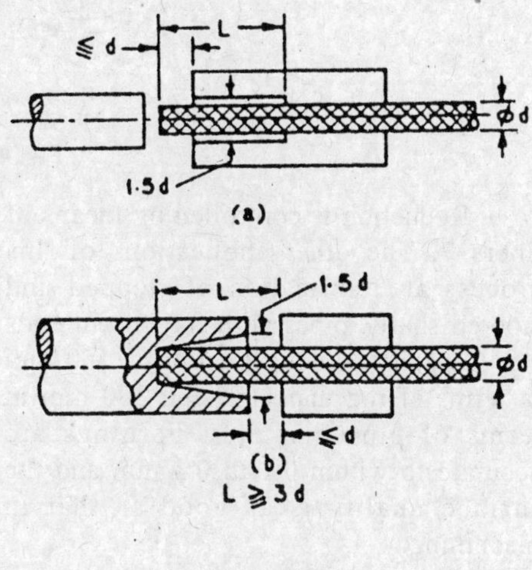

Fig. 4.33. Rule No. 3.

Speed range of forging equpment :–

Equipment	Speed range, m/s
Hydraulic Press	0.06 — 0.30
Mechanical Press	0.06 — 1.50
Screw Press	0.6 — 1.20
Gravity drop hammer	3.6 — 4.8
Power drop hammer	3.0 — 9.0
Counter blow hammer	4.5 — 9.0

4.3.3. Roll Forging. A variety of shapes having straight or tapered reduced sections may be forged with the aid of rolls. Here, the rolls are not completely cylindrical but are sector shaped to permit the stock to enter between the rolls. These roll segments have one or more sets of grooves to impart a desired shape to the piece being forged. When the rolls are in the open position, the heated stock is advanced upto a stop, (Fig. 4.34). As the rolls rotate, they grip and roll down the stock. The stock is transferred to a second set of grooves. The rolls turn again and so on until the piece is finished. Like this, the bar stock can be increased in length, reduced in diameter and changed in section as desired.

Roll forging being a rapid process, it can be efficiently used for the preliminary reduction of forging stock as a preparation for subsequent die forging, as well as for the production of certain straight and tapered forgings, for example, levers, leaf springs, axles and arrow and spear type parts (cutlery and scissors and knife blades). Pressure on rolls may be as high as 1 MN.

4.3.4. Rotary Forging or Swaging. Round bars or tubings can be reduced in their diameters by a process called "rotary forging or swaging". The process can be done both in the hot and cold state. However, cold working is preferable because of the greater ease of handling and better surface finish obtained. Fig. 4.35 schematically illustrates the rotary forging process. Small rams 1 with dies 2 inside the spindle head 3 slot are brought together to reduce the work piece and separate again under the action of centrifugal force or springs. The spindle head is housed inside cage 4 with rollers 5 which rest from the outside on annular ring 6. As the spindle rotates, the small rams run against the rollers and so strike blows from various directions upon the job fed axially which is reduced in diameter by drawing out process.

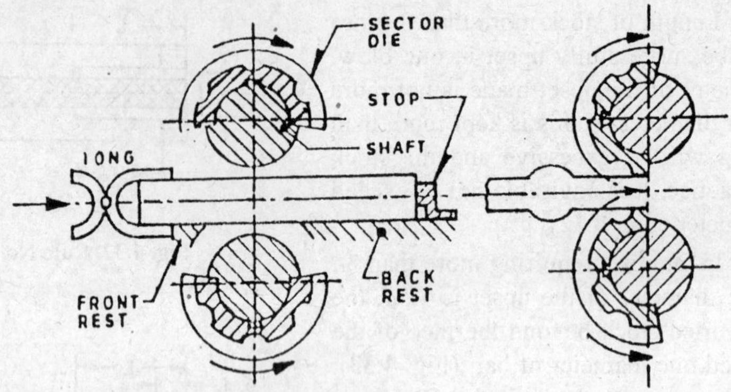

Fig. 4.34. Roll Forging.

Reduction is controlled by means of liners 7. The other applications of this process are fabrication of stepped and tapered shafts, pipes with forged out ends etc. Capacity of hot reduction is 150 mm in terms of rod diameter, and 300 mm in terms of pipe diameter. Forgings are accurate to within 0.1 to 0.3 mm and the surface quality is as good as that in machining.

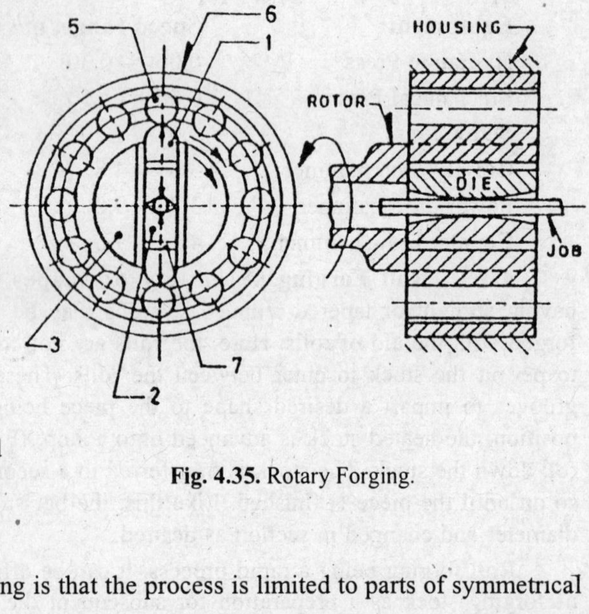

Fig. 4.35. Rotary Forging.

The other advantages of this process are :

1. Low initial investment.
2. Easy maintenance.
3. Low tool cost
4. Low labour cost (semi-skilled labour).
5. Rapid production.
6. Consistency of the product.

The major limitation of rotary swaging is that the process is limited to parts of symmetrical cross-section only.

4.3.5. Trimming of flash and straightening of the forging. After a forging is produced in dies, the flash around the edges of the forging, must always be removed. The operation of flash removal, *i.e.*, trimming is done on special trimming dies of crank or eccentric type. Trimming can be done with or without previous heating. Forging of aluminium alloys, copper, brass, bronze and of mild steel (C-content 0.20 and 0.25 percent) are usually trimmed in unheated state. However, the forgings of harder steels as well as those of larger dimensions should always be heated before trimming.

A trimming die, (Fig. 4.36), consists of a trimmer blade and a punch. The trimmer blade is fastened to a shoe which is bolted to the table of the press. The punch is secured to the press slide by means of its dovetail. The blade is made of tool steel and has a cavity both in contour and dimensions to the finish impression of the forging die. The clearance between the trimmer blade and punch is 0.5 to 1 mm. For trimming the flash, the forging is placed over the cavity in the trimmer blade, the flash serving as the support for the forging. When the press is started, punch travels downwards and cut off the flash. The forging falls down into the shoe and the flash remains on the top of trimmer blade.

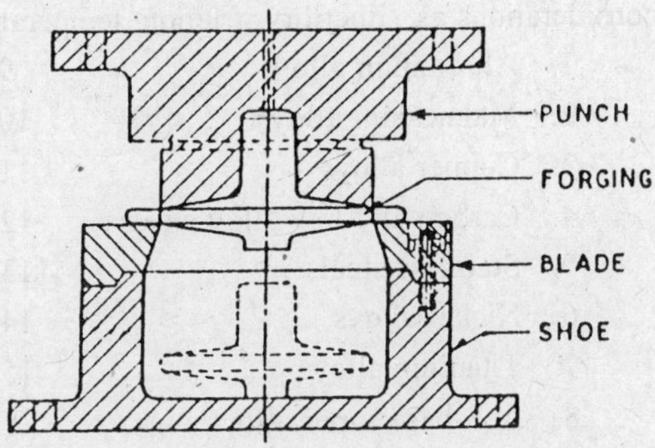

Fig. 4.36. Trimming Die.

Straightening. After the trimming operation, the forging may get slightly distorted or deformed, and must be straightened. This operation is done on a hammer or in a special press immediately after the trimming operation. Straightening can be done cold or hot as the case may be. Straightening under the hammer is done with one light blow in the finish forging impression of the die or in a special straighting impression.

4.3.6. Forgeability : It is a metallurgical fact that not all metals can be successfully forged. The term "forgeability" is the word used to express this forging ability (or lack of it) both qualitatively and quantitatively. These is no simple way, of defining this term. However, the forgeability of a metal can be defined as the metal's tolerance or relative ability to deform before cracks appear in the metal. A metal's forgeability may be excellent, zero or some what in-between. Many metals can be forged a little, or in certain ways such as in the closed-die forging process, or at room temperature. These same metals may crack under continued forging or when unconfined or when hot forged or cold forged. The degree of forgeability of a metal depends upon the crystallographic structure, metallurgical factors (purity phases present and grain size) and such factors as : the melting point, the yield strength, strain rate and die friction. These factors are discussed below :

1. Crystallographic Structure :-

(*a*) F.C.C. metals are the most forgeable, followed by B.C.C. and hexagonal close packed.

(*b*) Forgerbility increases with billet temperature upto the point at which a second phase appears or where there is incipient melting of grain boundaries. or grain growth becomes excessive.

2. Purity :- The forgeability of a metal is lowered by the impurities present, specially when they are segregated in groups or in grain boundaries.

3. Phases present :- The forgeability is affected by the number, amount and composition of phases and their location in the structure. Alloys which are present in the solid solution form at the forging temperature, generally possess good forgeability.

4. Grain Size : - Generally, smaller the grain size the better the forgeability.

5. Melting point :- Higher the melting point of a metal, the lower its forgeability.

6. Yield strength :- Higher the yield strength of a metal, the lower its forgeability.

7. Strain rate :-

(*i*) Metals that have low ductility have reduced forgeability at higher strain rate.

(*ii*) Highly ductile materials are not so strongly affected by increasing strain rate.

8. Die friction :- The forgeability of a metal decreases as the die-friction increases.

The following is a list of materials in order of decreasing forgeability in terms of such considerations as : ductility, friction, temperature, strength and quality of forging :

1. Aluminium alloys
2. Magnesium alloys
3. Copper alloys
4. Carbon and low alloy steels
5. Stainless steels
6. Nickel alloys
7. Titanium alloys
8. Iron-base super alloys
9. Cobalt-base super alloys
10. Columbium alloys
11. Tantalum alloys
12. Molybdenum alloys
13. Nickel-base super alloys
14. Tungsten alloys
15. Beryllium

Two classes of metal in particular, tend to have poor forgeability. These are : (1) metals with small but significant amounts of second phases which melt at the usual forging temperature of the basic alloy, and (2) metals which are brittle and tend to segregate in the grain boundaries thus giving brittle effects to the alloy as a whole , by establishing a semi-continuous brittle phase.

An example of the first of these two categories is leaded free cutting copper alloys in which the lead remains in the elemental form as a separate phase even when heated to the usual forging temperature range for copper alloys. During the forging of such an alloy, tremendously high local hydraulic pressures are developed in the melted lead liquid phase. these pressures tend to cause the breaking open of channels between grains as the high pressure liquid seeks to force its way outward to relieve the high pressure gradient. These channels are really cracks. Oddly, somewhat lesser, but still significant amounts of lead may improve forgeability to the point of providing an alloy with the benefits from forging, and yet retain the excellent machining qualities given by the lead.

Some metals which exhibit poor forgeability when processed without lateral restriction to confine the spreading metal, such as a hammer forging, become relatively forgeable when used in the closed-die processes of forging. This is due to the fact that ductility of a metal increases by isostatic compression (metal being confined in the die-cavity).

It is important to distinguish between forgeability and ease of forging. As previously stated, forgeability refers to the property of being deformable by forging without cracking. Ease of forging, on the other hand, refers to the power or force requirement for accomplishing the deformation. Usually when commonly forged metal is heated into its forging temperature range, it exhibits both higher forgeability and greater ease of forging than at lower temperatures.

4.3.7. Defects in Forgings. Forging defects may result from the following causes : poor quality of the stock, incorrect die design, improper heating, incorrect forging conditions, wrong forging methods, uneven cooling of stock after forging etc. The common defects in forgings are discussed below :

1. Cold shuts or laps. Cold shuts or laps are short cracks which usually occur at corners and at right angles to the surface. They are caused by metal surface folding against itself during forging. Sharp corners in dies can result in hindered metal flow which can produce laps.

2. Pitting. Pitting of the forging surface is caused by scale, which if not removed thoroughly from the die cavities is worked into the surface of forging. When this scale is cleaned from the forging, depressions remain which are known as 'scale pits'. Pitting should be avoided as much as possible during manufacture of forging by proper control of furnace and frequent cleaning of dies.

3. Die shift. Die shift is caused by misalignment between the top and bottom forging dies. This may be caused due to loose wedges.

4. Incomplete filling of dies. This defect may be caused by : wrong amount of metal, insufficient number of blows during forging, forging the stock at too low a temperature when it has practically lost its plasticity, poor forging design or incorrect die design.

5. Dimension of forging do not correspond to those specified in the drawing. The causes can be : worn out dies, incorrect dies or as a result of a mismatched forging.

6. Dents. Dents are the result of careless work.

7. Burnt and overheated metal. This defect is caused by improper heating conditions and soaking the metal too long.

8. Fins and rags. Fins and rags are small projections or loose metal driven into the surface of the forging.

9. Ruptured Fibre structure. This is a discontinuity in the flow lines of the forging which is revealed only when observing the macrostructure. This defect is caused by : working some of the alloys too rapidly during the forging operation, inadequate stock size or improper die design.

10. Cracks. Cracks which occur on the forging surface may be longitudinal or transverse. Their occurrence may be due to : bad quality of ingot, improper heating, forging at low temperature or incorrect cooling of alloy steel forgings.

11. Hair Cracks. These are very fine surface cracks not exceeding a fraction of a millimetre in width. These can occur due to : defects in metal ingot or too rapid cooling of the forging.

12. Slags, Sand and Porosity. These defects may occur either on the surface of, or inside the forging. These defects may be due to defects in the ingots or incomplete discard of the ingot head.

13. Internal cracks. Internal cracks in forgings can result from too drastic a change in shape of the raw stock at too fast a rate.

14. Flakes. Flakes are internal breaks or ruptures occurring in some grades of alloy steel. These are caused by too rapid a cooling from forging temperature.

15. Decarburization. If the raw stock is subjected to too high a temperature for too long a period, it can produce decarburized surface on the forgings, particularly in high carbon steels.

The chief defects mentioned above may be divided into two groups :

(*a*) **Irremovable defects.** Such as deep cracks, cavities, burnt metal and decarburization.

(*b*) **Removable defects.** Shallow cracks, overheating, coarse grained structure, internal stresses and distortion.

Defects in forgings can be remedied as follows :

(*i*) Surface cracks and decarburized areas are removed from important forging by grinding on special machines.

(*ii*) Shallow cracks and cavities can be removed by chipping out of the cold forgings with pneumatic chisels or with hot sets during the forging process.

(*iii*) Distorted forgings are straightened in presses.

(*iv*) The mechanical properties of the forging can be improved and internal stresses removed by annealing or normalising.

4.3.8. Cleaning and Finishing of Forgings

1. Removal of Oxide scale. A thin layer of scale (iron oxide) which is caused by contact of heated steel with air, is formed on the surface of steel forgings. Amount of scale depends upon the forging temperature and length of time of the operation. The simple way to remove the scale is by employing steam or compressed air to blow away the scale, from the surface of the forging as well as from the die cavities.

2. Cleaning by pickling. Pickling process is used to remove hard scale from the surface of the forgings. It consists of immersing the forgings in a tank filled with an acid solution, which is 12 to 15% concentrate of sulphuric acid in water. The solution acts to loosen the hard scale from the forging surface and remove it. The acid solution should not react with the clean metal while removing the scale. For this, an inhibitor agent is added to the acid solution.

3. Tumbling process. Tumbling process is used to remove scale and for the general cleaning of the forgings. The forgings along with some abrasive materials such as coarse sand or small metallic particles are placed in a barrel. The tilted barrel is rotated at low speeds. The forgings and the abrasives roll over themselves. This action loosens the scale from the surface of the forgings and removes it alongwith affecting a general cleaning of the forgings.

4. Blast cleaning. Blast cleaning process consists of directing a jet of sand, grit or metallic shots against the forgings. The blasting force is obtained from compressed air or centrifugal force through suitably designed apparatus. By this process the scale is removed and a smooth surface finish is imparted to the forging.

4.3.9. Heat Treatment of Forgings. After their manufacture, forgings are heat treated before and after machining. The initial treatment is done to improve the grain structure and also to make the machining of the forging easier. The final treatment is done to impart certain properties to the forgings for the service intended. For example to make the forging hard and tough, it is hardened and tempered. So the heat treatment of the forgings is done for the following purposes :

(1) To relieve the forgings of stresses set up during forging and subsequent cooling.

(2) To improve the mechanical properties of the product.

(3) To equalise the structure of the metal of the forging.

(4) To impart that degree of hardness to the forging which makes it most easy to machine.

Annealing. Annealing is a form of treatment which is used to accomplish the following results : remove internal stresses, improve mechanical properties, make the metal softer for machining or grinding. The annealing process consists in heating the forging in an annealing furnace upto or beyond the critical temperature of metal. For steel forgings, this temperature ranges from 750 to 900°C depending upon the carbon content of steel. The forging is then allowed to cool slowly in the furnace.

Normalising. Normalising consists in heating the forging as in annealing and then allowing the forging to cool in air at atmospheric temperature. The following results are achieved by normalising : a fine grade structure better than obtained in annealing, improved mechanical properties-increased tensile strength and ductility and removal of internal stresses.

Hardening. Hardening heat treatment process consists in heating the forgings slightly above the critical temperature of the metal and cooling it rapidly. The cooling medium is usually water, brine or oil.

Tempering. During hardening process. severe stresses are set up in the forging due to rapid cooling. These stresses must be removed, otherwise the forging may crack. This is done by tempering. Tempering consists in heating the hardened forging to a temperature between room temperature and the critical temperature of the metal. The forging is kept at this temperature for a definite period and then the forging is allowed to cool in air.

4.3.10. Metallurgy of Forging. The reason of high strength-to- weight ratio of forged parts is that these have fibrous structure and the grain structure or the flow lines of the metal are not interrupted but are made to follow the contour of the forged part. The main objective of good forging design is to control the lines of metal grain flow so that a part with greatest strength and resistance of fracture is produced. This is clear if we compare the three methods of manufacturing crankshaft : casting, machining and forging (Fig. 4.37). The crankshaft produced by casting (Fig. 4.37 *a*) has no grain flow and so has the poorest mechanical properties. In Fig. 4.37 (*b*), the crankshaft has been made by machining from a bar stock and the fibre of the metal gets interrupted and for this reason the mechanical properties of this shaft will be poorer than those of the crankshaft made by forging [Fig. 4.37 (*c*)], where the fibre of the metal has not been interrupted and continues along the entire length of the shaft.

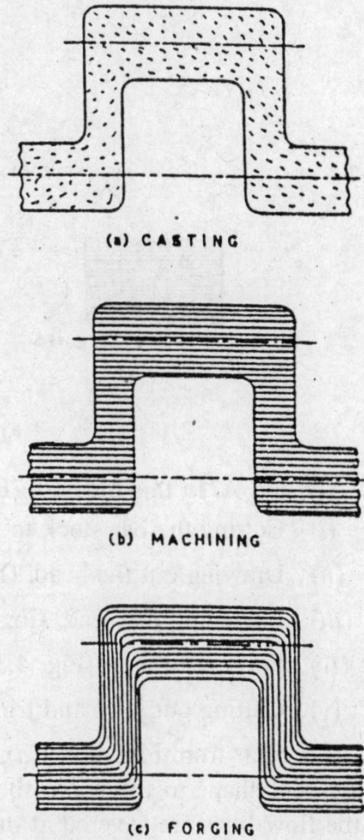

Fig. 4.37. Comparison of Grain Flow.

It is clear from Fig. 4.37 (*c*), that the effect of flow lines is to produce marked directional properties in the material and here (in the case of crankshaft), these qualities add toughness to the webs and where they join the round parts of the shaft. Controlling the directions of fibrous structure within forgings will develop the maximum mechanical properties for applications where shock and fatigue are encountered. For static loads, the direction of these fibre lines is not so important. Below, we will explain how flow lines are used to advantage in forging typical components.

(*i*) **Gear Blank.** Flat forgings of gear blanks are usually made by upsetting the bar stock. The pattern of the flow lines in the bar stock (which has been produced through forging or rolling) is shown in Fig. 4.38 (*a*). When this bar stock is upset on ends (that is, the bar stock is placed in dies such that the flow lines are vertical), the grain pattern produced will be radial and the gear blank will have flow lines concentrated where they will give greatest strength to the teeth, Fig. 4.38 (*b*), and all the teeth will be equally strong. This is desired because each tooth can be considered as a cantilever beam and the fibre flow lines will be parallel with the expected tensile and compressive stresses along the face of the gear teeth. However, if the bar stock is placed in the dies for upsetting in such a manner that its flow lines are horizontal, then the grain pattern produced with be as shown in Fig. 4.38 (*c*). The teeth cut on such a blank will not be of the same strength. Teeth 1 and 2 in Fig. 4.38 (*c*) will withstand less stress than teeth 3 and 4, because they have fibres running perpendicular to the expected stresses.

(*ii*) **Coupling Hooks.** Coupling hooks are subjected to severe service conditions and the safety of the railway traffic depends upon the strength of the hook. Therefore, the coupling hooks should be forged in such a manner that : (*i*) due consideration is given to the direction of the fibre of the material (*ii*) the fibre of the material is not severed during the forging process. Also, it must be ensured that the highest possible tensile strength is developed in the hook. Below, we compare the two methods of forging a coupling hook :

temperature and the critical temperature of the metal. The forging is kept at this temperature for a definite period and then the forging is allowed to cool in air.

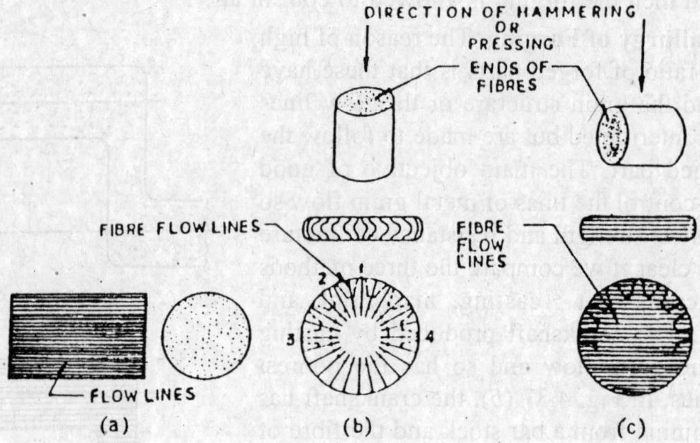

Fig. 4.38. Forging of Gear Blank.

Method A. In this method, (Fig. 4.39-I), the sequence of operations is as follows :

(*i*) Cutting the bar stock to proper length, (Fig. 4.39 I(*a*)).

(*ii*) Drawing out the head, (Fig. 4.39 I(*b*)).

(*iii*) Flattening the head, (Fig. 4.39 I(*c*)).

(*iv*) Forging in dies, (Fig. 4.39 I(*d*)).

(*v*) Cutting out jaws and trimming the flash.

It is clear from Fig. 4.39 I(*d*), that in order to give bent shape to the hook, the metal is cut and the flow lines are severed at the nose of the hook which is the most dangerous section. Since the tractive force is directed along and not across the flow lines, there is danger that failure will occur at this section.

Method B. In this method, (Fig. 4.39. - II), after drawing out the head of the hook, the material is bent and then forged in dies. In this way the flow lines will follow the contour of the hook which will result in maximum tensile strength of the hook. The mechanical properties of the hook produced by this method will be much higher than of the hook produced by the first method.

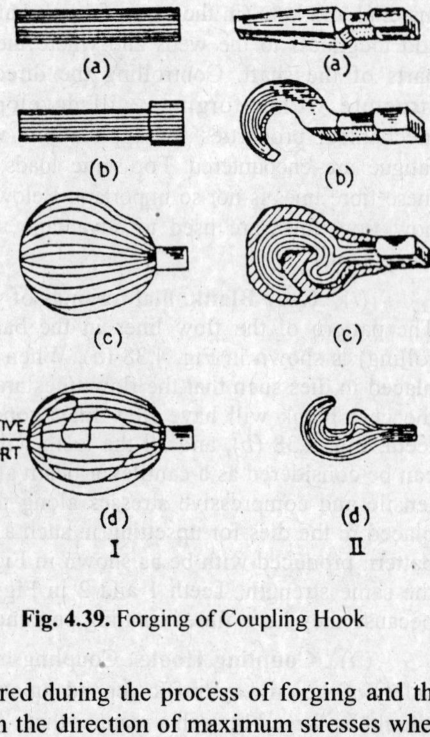

Fig. 4.39. Forging of Coupling Hook

It is clear from above that while designing a forging and forging tools, it must be ensured that : the fibre flow lines of the material are not severed during the process of forging and the fibres in the forging should follow the contour or lie in the direction of maximum stresses when the part is in service, since the metal forgings with marked fibre have much more ductility and toughness when tested in the direction of fibre than when tested across the fibre.

4.3.11. Cold Forging. Limited formability limits the use of cold forging to : low and medium carbon steels (0.25 to 0.45% C), low alloy steels, copper and light metals such as aluminium, magnesium, titanium and beryllium. Also, due to limited formability and heavy work hardening, several stages with intermediate annealing are needed to get a large deformation, which increases cost. The process is most suitable for axisymmetric components such as shaft components, flanged components, finished gears and bearing races etc.

Various cold forging techniques employed in forging practice are : sizing, coining and heading etc. The aim of sizing operation is to obtain closer dimensional tolerances on portions of a forging. In plane sizing, the forging is pressed between two flat dies. The operation straightens the forging, enables accurate dimensions to be obtained and improves the surface finish. Three dimensional sizing imparts the final size to the whole forging in impression dies with the formation of flash which is subsequently trimmed. If the sizing operation is such that the metal flow is restricted in a space confined by the die impression (that is no flash is formed), the sizing operation is called " coining".

Sizing and coining operations are commonly done on knuckle joint presses, (Fig. 4.26 (*c*)). The crankshaft driven by motor reciprocates the ram (slide) by means of a linkage consisting of the pitman and two knuckle joint levers. The upper die is fastened to the slide. This arrangement enables a high tonnage to be exerted at the bottom of the stroke with a comparatively low torque on the drive shaft. Coining presses range in ratings from 0.1 MN to 8 MN and even more.

The "cold heading" process was first introduced for the cold forming of heads on rivets, bolts and screws. At present, the process has found widespread application in the mass production of small parts and fastenings : bolts, screws, rivets, nuts, nails, links and pins etc. The process is also well suited to the production of small steel balls for use as ball bearings and rollers for use as roller bearings. The stock for these parts is wire or cold drawn bars from 0.6 to 40 mm in diameter of carbon steel, non ferrous metals and their alloys.

A cold heading operation sequence is shown in Fig. 4.40. The stock is fed through a die to a stop with the help of rolls. Next, the die is shifted to cut off the blank and to carry it over to the heading line where the required head is formed by a blow of a punch mounted on the heading gate. Cold heading in one or several strokes is done chiefly in automatic cold headers in which all operations including wire (or bar) feed, cut off, carry over and ejection of the finished parts are completely automatic. These machines have an output from 20 to 250 pieces per minute.

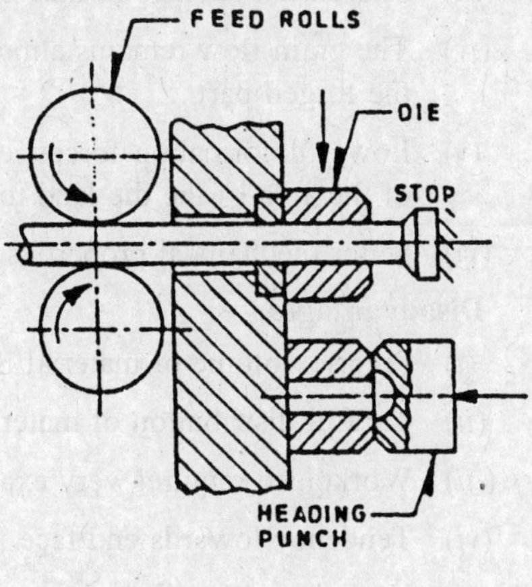

Fig. 4.40. Cold Heading.

4.3.12. New trends in Forging Technology. The development of advanced forging techniques have come from the aerospace industry and its rapid development over the past 40 years. The need to manufacture highly stressed airframe and engine parts led very rapidly to the combination of specific primary forming and forging processes. In particular, spectacular advances have been made through the combination of Powder Metallurgy (P/M) and forging processes. This trend found applications also in automobile and machine tool manufacture due to the following reason:

The cost of the finished components has been increasing due to increase in cost of raw materials, power and labour. This has focussed attention of cost reduction by better utilization of raw materials and production of close tolerances components necessitating minimum finish machining.

The above mentioned factors have resulted in the development of new forging technologies which will be discussed below :-

1. Precision and High Precision forging (No draft forging).
2. Closed-die forging without flash.

'No draft forging' also called 'precision' or 'close-tolerance forging' is a special variation of conventional press forging. As the name implies, the die walls have little or no draft. The part produced by this method can be used "as forged" or after a very little amount of machining. For this, the following factors are of prime requirement :

(*i*) The accurately sized blank is prepared after an extensive sequence of preforming steps.

(*ii*) The die cavity is made to exacting standards.

(*iii*) There is increased control over preheating both the dies and blank, descaling of die surfaces, lubrication, die material and use of special die inserts, punches, pins etc.

No draft forging is currently used for aluminium, copper and magnesium. Its use is being extended to some steels and titanium. Examples of parts made by this process are : complex shaped turbine blade and spiral teeth bevel gears. Because of high tooling cost, this method is justified only for mass production of complex and close tolerance parts. The advantages of this method are :

(*i*) Savings of 10 to 30% or more in weight of material used.

(*ii*) No trimming required.

(*iii*) Substantial savings in machining costs.

(*iv*) The grain flow remains almost undisturbed as no or very little machining is needed of the forged part.

(*v*) Lower deformation forces, especially in the case of smaller forgings where the ratio of the area of the die land to that of the die itself is large.

(*vi*) better mechanical properties through the elimination of residual flash.

Disadvantages

(*i*) Precise volume of material is required.

(*ii*) Careful distribution of material required.

(*iii*) Workpiece requires very exact positioning in die cavity.

(*vi*) Tendency towards end face flash formation.

(*v*) Variations in wall thickness.

3. **Warm Forging.** As discussed earlier, the aim of warm forging is to combine the advantages of both hot and cold forging. The forging temperature affects both the behaviour of the material during deformation as the properties of the finished part. In comparison to hot forging, warm forging requires higher forces because of greater flow stress, thus making increased demands on the forging tools and equipment. On the other hand, greater part accuracy in terms of both dimensions and surface finish can be achieved. The optional temperature for a particular warm forging operation always represents a compromise. Its lower limit is set by the force which can be produced by the forging press and by the forgeability of the material, the upper limit is determined by the amount of oxidation and thus the amount of scale which can be tolerated.

4. **Hot Die Forging.** In conventional forging, the workpiece tends to cool during the process because the dies are at a lower temperature as compared to that of the workpiece (the

difference in temperature between workpiece and die may be upto 1000°C and more). This seriously inhibits the complete filling of the dies. In drop forging, the die forging is also adversely affected by higher resistance to deformation since higher strain rates are involved. These difficulties can be overcome by reducing the temperature difference between the workpiece and the die (hot die forging) or by keeping the workpiece and the die at the same temperature (isothermal forging). Dies are preheated to the desired temperature by flame, induction or resistance heating. However, induction heating is more popular as it gives more flexibility for using different coil sizes with the same power source.

In hot die process, the difference in temperature is only about 200°C. This method was developed in order to be able to use Ni-based material for die manufacture. At operating temperatures of 750 to 850°C, these alloys require no vacuum or inert gas atmosphere. Ti alloys are forged at temperatures of about 950°C. The process may be seen as a compromise between isothermal and conventional forging, although in fact it is more closely related to isothermal forging.

5. Isothermal Forging. This process is a further development of hot die forging. Here, both the dies and the forging stock are kept at the same temperature (forging temperature) during the forging process. The forging stock for this process can be either as used in conventional forging or a preform obtained by P/M. The materials most used are Ti and high temperature resistant Ni base alloys which because of their high flow stress are difficult to forge. Parts made of these materials have been forged by this process, for the automotive and aircraft industry. The process can be used economically for batch sizes between 100 and 1000 parts. This process requires control over the deformation and, therefore, hydraulic presses are employed for isothermal forging. The heated billet is placed in the bottom die and the load exerted by the press is maintained at a preselected low value. The downward speed of the press slide is automatically controlled by the resistance of the work piece to deformation. The die materials depend upon the level of temperature required. The material should possess the following properties : toughness, high hot hardness, high temperature strength and resistance to wear and thermal fatigue. The die material most commonly used for forging Ti alloys (generally without a surrounding inert gas atmosphere) is the Ni-base alloy. Ti alloys are generally forged at 950°C. When higher temperatures are required such as for components of Ni-base alloys, Molybdenum base die materials are needed. Since Mo reacts strongly with O_2 at high temperature, it is necessary to protect the dies from the surrounding air by either an inert atmosphere or a vacuum, otherwise they would be oxidised to pieces in a few minutes.

Isothermal forging offers the following advantages:

(*i*) Complex and close tolerance parts can be produced from comparatively simple preforms, generally in a single stroke of the press and a single die set.

(*ii*) Greater material utilization and less machining.

(*iii*) Uniform wrought properties of the product.

(*vi*) The speed of the press slide especially towards the end of the forging stroke is quite lower as compared to conventional forging and consequently lower forging loads are needed in isothermal forging, about $\frac{1}{4}$ to $\frac{2}{5}$ of that required for conven tional forging of the same material.

(*v*) Lower production cost due to reduced number of forging operations.

(*vi*) Reductions in cross-section of upto 80% can be achieved in single forging operation, thus facilitating the forging of parts with fins and ribs.

6. Powder Forging. Powder forging also called powder metal forging (P/M forging) or hot forming combines the conventional forging and powder metallurgy processes. In conventional forging, a major limitation is to obtain a uniform flow of the raw solid material within a forging die. This limits the complexity of the forging or requires more steps in the forging process, with resulting multiple and costly dies or substantial post forging machining. Therefore, the more nearly the raw material shape conforms to the shape of the forging, the easier it is to obtain a desired product. In P/M forging, the preforms closely resembling the forging, made by P/M method are hot forged in close impression dies, often in a single forging step. The steps in the process can be written as :-

1. Powder
2. Compacting the preform
3. Sintering the preform
4. Finish forge

The powder forging method thus combines dimensional precision, weight control and high material utilization of powder metallurgy with the mechanical properties of conventional forging, at reasonable production cost. This technology is being widely employed for aluminium, steels, titanium, nickel base and many other structural alloys. The automotive industry is the principal user of this process. Examples of powder-forged production parts include : gears and pinions, bearings, pistons, connecting rods, crankshaft, camshaft, cylinder liners, piston rings, clutch discs and plates and clutch knobs etc. This technique can be used either with conventional forging or isothermal forging or hot die forging.

The material structure of sinter forged parts is very fine grained and homogeneous. The mechanical properties are thus largely isotropic (non-directional grain flow), which in some cases has a positive effect on the fatigue strength of sinter- forged parts making them superior to conventionally forged parts. While the fatigue strength of conventionally forged parts (steel) differs by about 100% in the longitudinal and transversal directions, the corresponding values for powder-forged steel are the same in whatever direction they are measured. This isotropic property is a great advantage in many components for example, increased fatigue strength of powder forged connecting rods and the reduction in the tendency of teeth to fail in fatigue tests on powder forged gear wheels. These are just two examples of part families for which powder forging offers advantages over traditional methods, particularly in the case of components which have to withstand high dynamic stresses. When in addition to stresses, the design of the parts poses problems of material distribution, powder forging is almost unrivalled. The only alternative methods of manufacture are either expensive mechanical machining or casting techniques, whereby in the later case the parts are seldom able to withstand very high dynamic stresses.

The precision powder forged parts have the following advantages over the wrought counterparts :

(*i*) Almost 100 per cent material utilization.

(*ii*) Better surface finish, of the order of 5 to 10 μ m.

(*iii*) Material homogeneity and fine grain size.

(*vi*) Almost complete elimination of machining.

(*v*) Close dimensional tolerances in the range of qualities IT8 to IT10.

(*vi*) Density increases upto about 99%.

The main drawback of this process is that the preform contains about 15 to 30 per cent porosity. This reduces the ductility of the preform during the forging operation. Also, due to internal gaps the preform is susceptible to internal oxidation and decarburization at the forming temperature.

7. Liquid Metal Forging (Squeeze Casting). The liquid metal forging process is hybrid between conventional casting and forging methods and produces complex shaped components from molten metal in a single step. The various stages in liquid metal forging process are :

(*a*) Pouring the molten metal into the bottom forging die.

(*b*) Allowing the metal to solidify partially.

(*c*) Closing the upper and bottom dies, applying pressure and maintaining it for a fixed time until solidification is completed.

(*d*) Opening the dies and ejecting the forged component from the bottom die.

The metal may be poured manually into the bottom die for small forgings (upto 10 kgs) and for heavier parts, it is poured mechanically. The dies are kept at a temperature of 100 to 250°C. The successful production of a forging by this process depends upon the proper selection of the following variables : Pouring temperature of metal, Temperature of dies, Delay time to apply pressure, Press speed, Pressure level, Period of pressure application, metal volume and die coatings for facilitating the ejection of the forging. The liquid metal forging process has the following advantages over conventional casting and forging processes :

(*i*) Ability to produce thinner and more complex components as compared to those produced by either casting or forging.

(*ii*) There are no gas and shrinkage porosities. These get closed up during the process.

(*iii*) Pressure needed is quite less as compared to conventional forging. Therefore, low capacity presses are needed for this process.

(*iv*) Production rates are higher as compared with any conventional casting process. However, the rate would be lower than that for conventional forging. But this difference can be made up by the lower machining requirements of squeeze castings.

(*v*) Both cast and wrought alloys such as cast iron, aluminium and supper alloys can be used.

(*vi*) The process is economical from the material point of view. There is better material utilization and also scrap can be used. Gates and risers are not needed. So, yield is higher in comparison to sand castings.

(*vii*) Mechanical properties of the product are comparable to those by conventional forging. And they are considerably better than sand castings.

(*viii*) The tooling and equipment needed are basically simple, low cost and readily available.

Like any other process, the liquid forging process has got the following limitations :-

(*i*) All the component shapes cannot be obtained by this process.

(*ii*) If there is a large difference in the dimensions of thinnest and thickest sections of the component, the pressure distribution will not be uniform and also there will be chances of shrinkage porosity and cracking of the component as the thinner sections will solidify first.

(*iii*) cooling rates being faster, the effect on microstructure of the component will be different as compared to casting or forging. This may necessitate different heat treatments for liquid forged parts.

8. Gatorizing. This new forging technique involves the following steps :

(*i*) The forging stock is preconditioned in inert atmosphere to obtain a temporary condition of fine grained microstructure resulting in low strength and high ductility. For this, the work-piece is mechanically worked at temperature slightly below the recrystallization temperature of the material.

(*ii*) After preconditioning, the isothermal forging operation is carried out at slightly below the recrystallization temperature of the material.

(*iii*) Lastly, certain heat treatment operations are done on the forgings to get normal high strength and hardness.

At present, this method is being used for making aircraft components of nickel and titanium based alloys.

9. Orbital Forging. Orbital forging or rotaforming is a cold forming process. In this method, the pressure of the upper die of the work-piece is concentrated on a small area at any time, and not on the total workpiece area as in conventional forging. The upper die is slightly inclined to the vertical axis of the machine and it imparts a high frequency circular rocking motion across the top surface of the workpiece (Fig. 4.41). At the same time, the workpiece is slowly moved hydraulically upward and pressed against the orbiting upper die. The forging operation get completed when the hydraulic ram touches a preset stop. The ram is then lowered and the hydraulically operated ejector ejects the forging from the lower die. All the metals that can be cold worked can be forged by orbital forging. Examples of parts produced by this process include : parts flanged with indented or crown shapes and disks etc. To avoid tilting of the workpiece due to orbital forging, its height to diameter ratio should be less than 2. The advantages of orbital forging are :

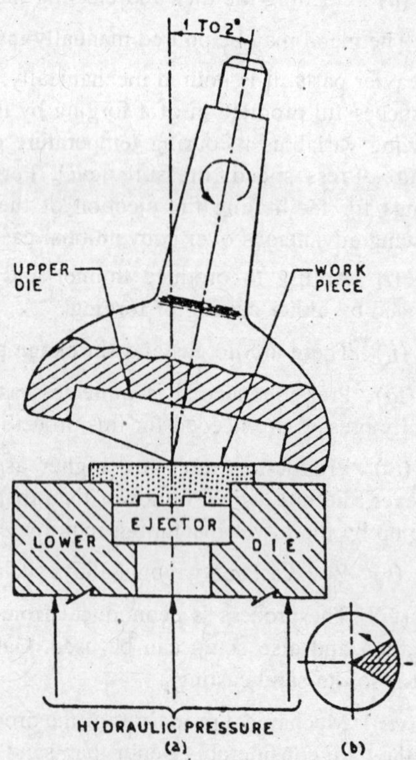

Fig. 4.41. Orbital Forging.

(*i*) Press capacity requirements are only 5 to 10% of that needed by conventional forging, because only a small portion of die actually contacts the workpiece

(*ii*) Better surface finish.

(*iii*) The method requires smaller and lower cost equipment.

The method has the limitation in that the forging is obtained by filling the lower die and so the bottom surface of the upper die should be flat and smooth.

10. Hot Isostatic Pressing (HIP). In this method the metal powder is sealed in a metal or ceramic container that has the shape of the desired part. The container is placed in a special pressure vessel that has the capability of simultaneously heating the container and subjecting it to a hydrostatic pressure (gas). The powder is compacted, densified and sintered in one step. Upon removal of the container, a finished part close to the final shape is obtained. The H.I.P.

method is particularly suited to producing parts from high temperature alloys that are difficult to forge and machine. For nickel based alloys, the pressure and temperature employed are 100 MPa and 1260°C respectively. For pressurisation, both helium and argon have been used. This method can also produce preforms for subsequent forging. At present, this method is being tried to forge exotic materials for aircraft industry. A 35% saving in the material needed to make a gas turbine disk has been reported for the HIP process.

11. Incremental Forging. In this method, very big forgings are made by working different areas of the forging into shape, one at a time. Since only a limited area is worked at any time, the forging equipment can be much smaller in capacity as compared to conventional forging. This makes it possible to forge huge parts on presses of modest capacity. This method is particularly suited to aircraft industry. A major drawback of this method is that the workpiece tends to cool below the forging temperature as it moves from incremental step to step. Reheating the part to its original forging temperature may destroy the thermomechanical work already done on it in the case of many alloys. This can be prevented by two methods : The workpiece is progressively reheated to a lower temperature after its temperature falls below the forging temperature. Secondly, insulating blankets and other coatings are used to cover the billet so as to minimize the heat loss.

12. Super - Plastic Forging. This process is carried out at even lower strain rates than isothermal forging. This is accompanied by an increase in the pressure contact time and a further reduction in the deformation force. The preconditions for super plastic behaviour of a material are :-

(*i*) Very fine structure with grain size of just a few μ m. Changes in structure due to grain growth during forging are not permissible.

(*ii*) The forging temperature must lie above Tm/2.

(*iii*) Strain rates must be very small 10^{-1} to $10^{-5}s^{-1}$

Super plasticity occurs in principle with all pure metals. However, preference is generally given to eutectic or eutectoid alloys since they also usually fulfill the first precondition, namely, that of a fine grain structure. The temperatures required for super plastic forging of high-temperature resistant Ni-and Ti-based alloys between 800 and 1150°C. In this range, due to high thermal and long term stresses, it is necessary to make dies from Mo based alloys. Super-plastic forging is mainly used to manufacture high temperature resistant engine components (impeller wheel etc.) of Ni-base alloys.

4.4. Extrusion

Extrusion may be defined as the manufacturing process in which a block of metal enclosed in a container is forced to flow through the opening of a die. The metal is subjected to plastic deformation and it undergoes reduction and elongation during extrusion. The section of the product will depend upon the shape of the die opening. This process could well be considered an adaptation of closed die forging, the difference being that in a forging, the main body of the metal is the product and flash is cut away and discarded ; in extrusion, the flash (metal flowing out of the die) is the product and the slug remaining in the die is not used.

Extrusion is more widely used in the manufacture of solid and hollow sections from non-ferrous metals and their alloys (aluminium alloys, copper, brass and bronze etc.), but steel and other ferrous alloys can also be successfully processed with the development of molten-glass lubricants. The initial material in extrusion is cast or rolled billets. The range of extruded items is very wide : rods from 3 to 250 mm in diameter, pipes of 20 to 400 mm in diameter and wall thickness of 1 mm and above, and more complicated shapes which cannot be obtained by other mechanical working methods, (Fig. 4.42).

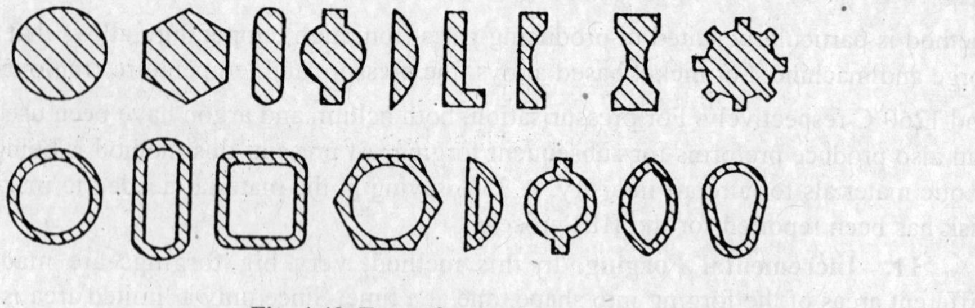

Fig. 4.42. Type of extrusions.

Extrusion may be done hot as well as cold. The two methods employed for extrusion are : Direct or forward extrusion and Indirect or backward extrusion. In direct extrusion, the product emerges from the die in the same direction as the direction of application of pressure, whereas in indirect extrusion, the product travels against the direction of application of pressure.

4.4.1. Hot Extrusion. Fig. 4.43 illustrates the hot extrusion methods for producing solid and hollow products by forward extrusion (Fig. 4.43 *a*) and by backward extrusion (Fig. 4.43 *b*)).

Direct Extrusion. In Fig. 4.43 *a* (*i*), the heated billet is placed in the container. It is pushed by a ram towards the die. The metal is subjected to plastic deformation, slides along the walls of the container and is forced to flow through the die opening. At the end of the extruding operation a small piece of metal, called butt-end scrap, remains in the container and cannot be extruded. To obtain a hollow section, a hollow billet is often used around a mandrel. In some instances, the billet is solid and is pierced by the mandrel or an axial hole is drilled in it by means of a drilling machine before the extrusion cycle begins. The mandrel may be fastened to the ram. As the ram moves in the direction towards the die, the metal is forced out through the annular clearance and form a tube, (Fig. 4.43 *a* (*ii*)). With a modern extrusion press, the piercing mandrel is actuated by a separate hydraulic cylinder from the one which operates the ram, (Fig. 4.43 *a* (*iii*)). The billet is first upset with the ram with the piercing mandrel withdrawn. Next the billet is pierced with the pointed mandrel ejecting a plug through the die. Then the ram advances and extrudes the billet over the mandrel to produce a tube. The outside diameter of the extruded tube equals the diameter of the die hole and the inside diameter equals that of the steel mandrel. This method is called as "Tubular extrusion".

Indirect Extrusion. For the production of a solid part, the die is mounted on the end of a hollow ram and enters the container, (Fig. 4.43 *b* (*i*)) the other end of container being closed by a closure plate. As the ram travels, the die applies pressure on the billet and the deformed metal flows through the die opening in the direction opposite to the ram motions, and the product is extruded through the hollow ram. Hollow section is extruded inside an annular gap between the container and the ram, (Fig. 4.43 *b* (*ii*)). In indirect extrusion, there is practically no slip of billet with respect to the container walls.

Comparison of Direct and Indirect Extrusion. The direct extrusion, is the simplest, but it is limited by the fact that as the ram moves, the billet must slide or shear at the interface between billet and container. These large friction forces must be overcome by very high ram forces, which produce very high residual stresses on the container. In the indirect method (also called reverse, back or inverted), the billet proper does not move relative to the container, instead the die moves. The friction involved is only between the die and container and this is independent of the billet length. The friction forces are lower and the power required for extrusion is less than for direct extrusion. Extruding force is 25 to 30% less than in direct extrusion. However, a long hollow ram is required and this limits the loads which can be applied. Due to this, and complex design of tools, the indirect extrusion finds only limited application. The scrap or the process waste which is equal to the poorly shaped leading end of the section plus the unextruded butt is 18 to 20% of

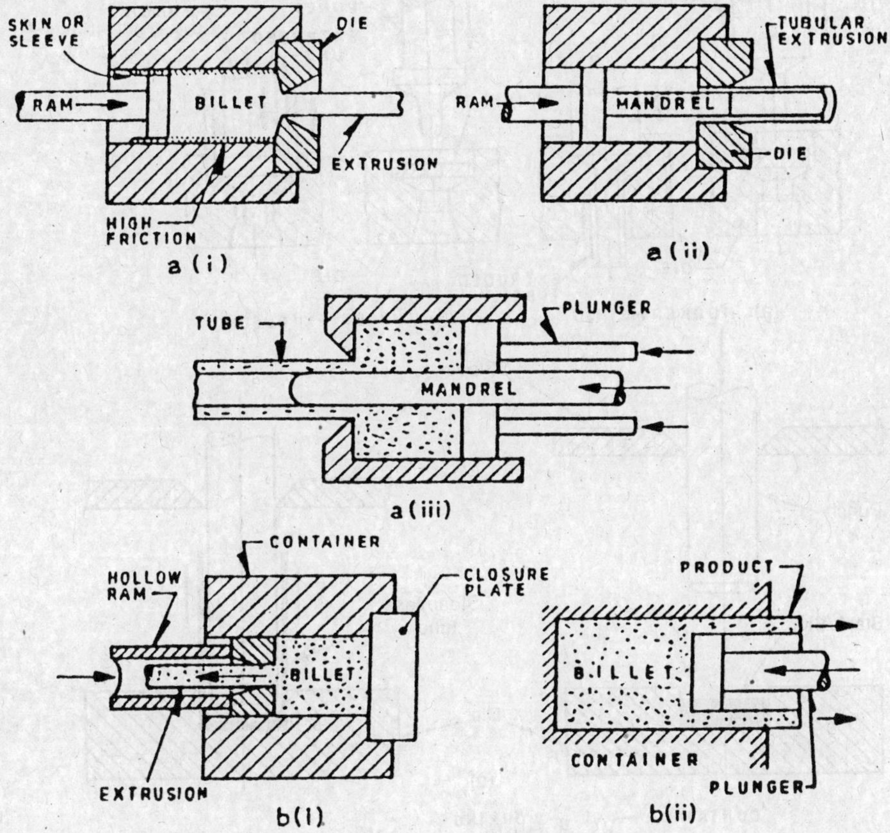

Fig. 4.43. Hot extrusion.

billet weight in direct extrusion, whereas it is only 5 to 6% of billet weight in the cast of indirect extrusion.

Extrusion ratio. The extrusion ratio is defined as the ratio of the cross-sectional area of the billet to the cross-sectional area of the product. It reaches about 40 : 1 for hot extrusion of steel and may be as high as 400 : 1 for aluminium.

Machines for extrusion. Both mechanical and hydraulic presses can be used for extrusion provided they possess high kinetic energy and long ram guides. Hydraulic presses with either vertical or horizontal plungers are most commonly used. Mechanical presses are used to a lesser extent. The pressure of the working fluid (water or emulsion) may reach 400 atm in hydraulic presses and produce a force of 300 MN. The plunger transmits the pressure to the ram and the pressure plate which force the metal through the die. To reduce the friction between the metal and the container walls and to achieve a more uniform flow of the metal, lubricants are used. Lubricants include a mixture of machine oil and graphite, molten glass (for steel) and other lubricating materials.

Extrusion speed is of the order of 0.5 m/s for light alloys and of the order of 4.5 m/s for Cu alloys.

4.4.2. Cold Extrusion. This process is similar to hot extrusion except that the metals worked possess the plasticity necessary for successful forming without heating them. Usually, these metals have a high degree of ductility. Cold extrusion is also done to improve the physical properties of a metal and to produce a finished part (sizing operation). The widely employed cold extrusion methods are : The Hooker or extrusion down method, the impact or extrusion up method and hydrostatic extrusion. Cold extrusion is done mostly on vertical mechanical presses

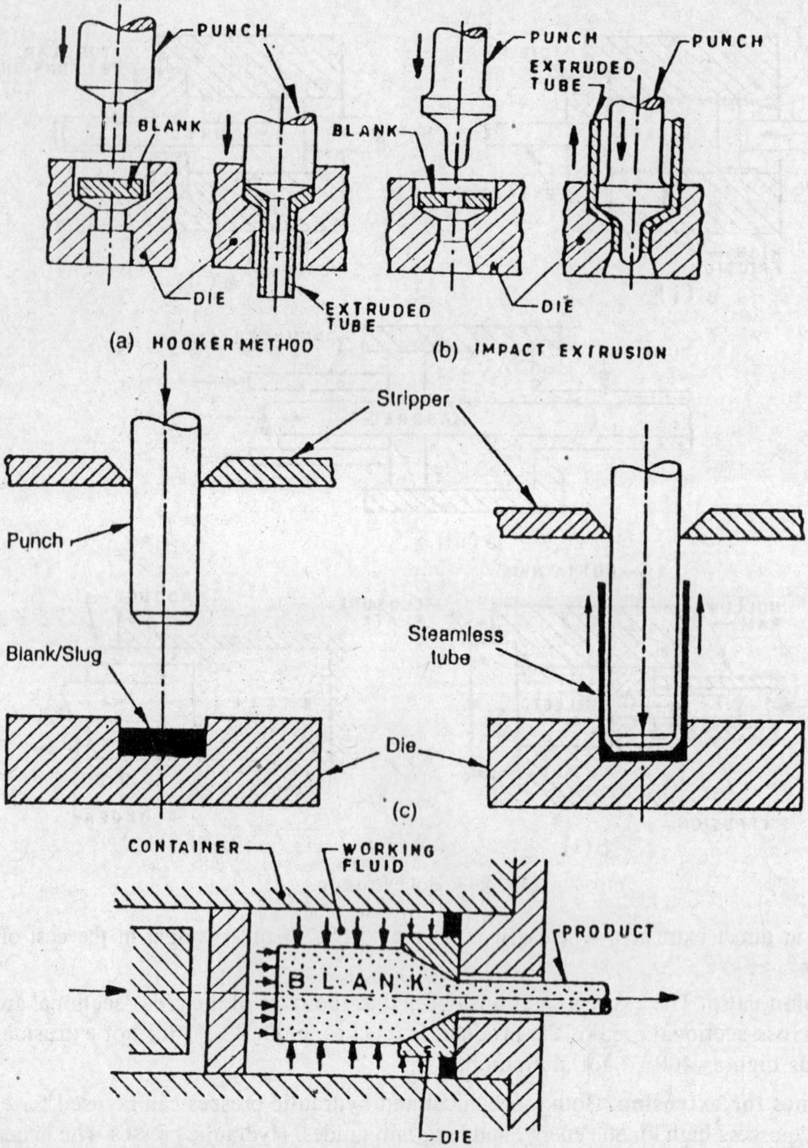

Fig. 4.44. Cold Extrusion.

because they are fast and simple. The method is fast, wastes no or little material and gives higher accuracy and tolerances.

1. The Hooker Method. In the Hooker or extrusion down method, the ram/punch has a shoulder and acts as a mandrel, [Fig. 4.44 (*a*)]. A flat blank of specified diameter and thickness is placed in a suitable die and is forced through the opening of the die with the punch, when the punch starts downward movement.

Pressure is exerted by the shoulder of the punch, the metal being forced to flow through the restricted annular space between the punch and the opening in the bottom of the die. If the tube sticks to the punch on its upward stroke, a stripper [see Fig. 4.44 (*c*)] will strip it from the punch. Small copper tubes and cartridge cases are extruded by this method. In place of a flat solid blank, a hollow slug can also be used.

2. Cold Impact Extrusion. The cold impact extrusion method consists in placing a flat blank of specified diameter in the die cavity and striking it by a punch with a powerful blow. The material gets heated up and becomes plastic and is forced to squirt up around the punch, [Fig. 4.44 (*b*)]. Thin walled tubes of low flow strength materials (tin, lead, aluminium etc.) are rapidly formed by this method. The end of the tube will correspond to the shape of the die cavity and also of the punch. The outside diameter of the tube takes the shape of the die and the wall thickness is equal to the clearance between the punch and the die. The operation is fully automatic and the production rate is as high as 50 tubes or more per minute. When the punch is on its upward stroke, the tube sticks to it. To effect release, either a stripper [See Fig. 4.44 (*c*)] or compressed air is directed against the tube, thus stripping it from the punch. Threads may be formed at the end of these collapsible tubes by retractable die portions or by other methods. These collapsible tubes are used for cosmetics (cream, shaving cream), tooth paste, grease etc.

Other product applications include : Cans, fire extinguisher cases, radio shields, food containers, boxes for condensers and cigarette lighter cases (if symmetrical).

A variation of the method where a flat solid blank or slug (that is, without a hole) is used in the die cavity (the bottom of the die cavity is closed) is shown in Fig. 4.44 (*c*). The method is also being employed to make thick walled products such as paint-spray-gun containers and grease-gun containers, by using large presses.

Impact extruxion speed is 100 to 350 mm/s

3. Hydrostatic Extrusion. In hydrostatic extrusion, [Fig. 4.44 (*d*)], a high pressure liquid medium is used for the transmission of the force to the billet/blank. The liquid completely envelops the blank and enters die canals. Due to the hydrostatic pressure, the ductility of the material is increased. Even brittle materials like tungsten, cast iron and stainless steel etc. can be extruded. This also permits the extrusion of very long billets or even wire, accompanied by large reductions. There is no container friction and the pressurised fluid also acts as a lubricant and because of this, the extruded product has a good surface finish and dimensional accuracy. However, the absence of container friction combined with reduced die friction can increase the tendency to internal crack formation. The pressure transmitting fluids commonly used for hydrostatic extrusion are : Glycerin, Ethylglycol, SAE 30 mineral lubricating oil, castor oil with 10 per cent alcohol and isopentane. The hydrostatic pressure ranges from 1100 to 3150 N/mm^2. The main commercial applications of this process are : Cladding of metals, making wires of less ductile materials and extrusion of nuclear reactor fuel rods.

4.4.3. Advantages of Extrusion

1. The range of extruded items is very wide. Cross-sectional shapes not possible by rolling can be extruded, such as those with re-entrant sections.
2. No time is lost when changing shapes since the dies may be readily removed and replaced.
3. Dimensional accuracy of extruded parts is generally superior to that of rolled ones.
4. In extrusion, the ductility of the metals is higher as the metal in the container is in composite compression, this advantage being of particular importance in working poorly plastic metals and alloys.
5. Very large reductions are possible as compared to rolling, for which the reduction per pass is generally ≤ 2.
6. Automation in extrusion is simpler as items are produced in a single passing.
7. Small parts in large quantities can be made. For example, to produce a simple pump gear, a long gear is extruded and then sliced into a number of individual gears.
8. Extrusions are lighter, more sound and stronger than castings.

9. They do not need draft or flash to trim and needless machining as they are more accurate than forgings.

4.4.4. Drawbacks of Extrusions

1. Process waste in extrusion is higher than in rolling, where it is only 1 to 3%.
2. Inhomogeneity in structure and properties of an extruded product is greater due to different flows of the axial and the outer layers of blanks.
3. Service life of extrusion tooling is shorter because of high contact stresses and slip rates,
4. Relatively high tooling costs, being made from costly alloy steels.
5. In productivity, extrusion is much inferior to rolling, particularly to its continuous varieties.
6. Costs of extrusion are generally greater as compared to other techniques.
7. Only the shapes with constant cross – section (die-outlet cross-section) can be produced.

Due to the above factors, the main fields of application of extrusion process are :-

(*a*) working of poorly plastic and non ferrous metals and alloys.
(*b*) manufacture of sections and pipes of complex configuration.
(*c*) medium and small batch production.
(*d*) manufacture of parts of high dimensional accuracy.

4.4.5 Main Process variable in extrusion process:

Total extrusion pressure is given as,

$$pt = \sigma_{x_0} + p_f$$

where σ_{x_0} = extrusion pressure to deform the material

$$= \frac{\sigma_0 (1 + B)}{B} [1 - R^B]$$

and p_f = ram pressure required by container friction.

$$= \frac{4 \tau_i L}{D}$$

Now σ_0 = yield strength of work material

R = extrusion ratio

$B = \mu \cot \alpha$

μ = Co-efficient of friction

α = half Die angle

τ_i = Interface shear between billet and container wall.

L = Length of billetin container

D = Internal diameter of container

∴ Main process variables in Extrusion process are:

1. Strength of billet material
2. Extrusion ratio
3. Co-efficient of friction/lubrication
4. Die-angle
5. Size of billet (Length and diameter)
6. Temperature
7. Speed of extrusion

4.4.6. Some Other Extrusion Processes

1. Side Extrusion. In this process, which is also known as "Lateral extrusion", the material flows at right angle to the direction of ram motion. The method can be : solid side extrusion and hollow side extrusion. In solid side extrusion, a solid body, with a solid protrusion of any profile is extruded. In hollow side extrusion, a workpiece with a hollow protrusion of any profile, is extruded. The tool opening is determined by the split die and the mandrel.

The extrusion force required in this process is very high. Due to this, the method is mainly used for non-ferrous metals and highly plastic materials, like lead. Fig. 4.45 shows the principle of side extrusion process for cable sheathing.

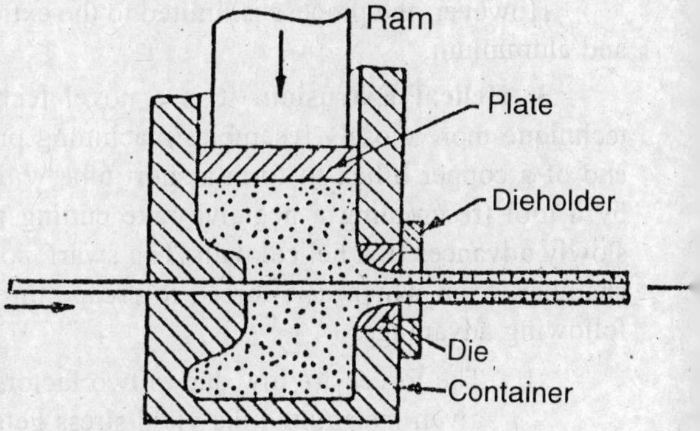

Fig. 4.45. Principle of Side Extrusion.

2. Continuous Extrusion. In conventional extrusion process, we start with a finite length of billet and get a finite-length product. However, in continuous extrusion process, continuous feed stock is converted to continuous product. This is achieved by applying the extrusion pressure on the periphery of the feedstock, instead of on the back, as in conventional extrusion. Again, whereas in conventional extrusion, there is a need to reduce friction between the billet and the container, in continuous extrusion, it is the friction force which provides the driving force for the extrusion. The impetus for the development of continuous extrusion process has been the need for the production of large tonnage of products continuously.

The most successful technique of continuous extrusion is the "Conform process" illustrated in Fig. 4.46. Continuous feedstock in the form of rod is inserted between the grooved extrusion wheel and the mating die shoe. As the wheel rotates, the rod is carried round the wheel

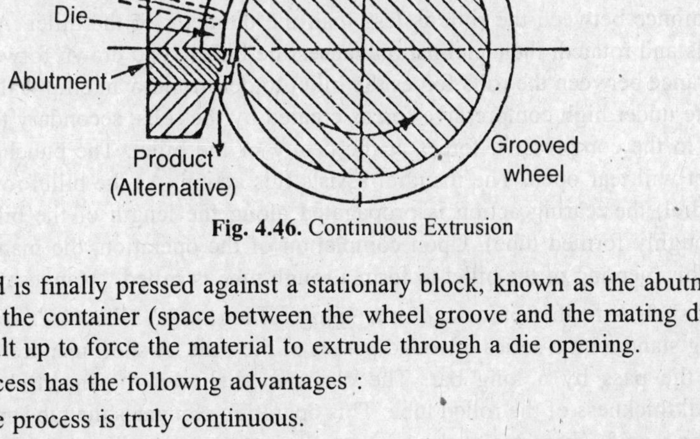

Fig. 4.46. Continuous Extrusion

by friction and is finally pressed against a stationary block, known as the abutmen. It gets upset to conform to the container (space between the wheel groove and the mating die shoe). Enough pressure is built up to force the material to extrude through a die opening.

The process has the followng advantages :

1. The process is truly continuous.
2. Raw stock, in the form of rod, powder or machined swarf, can be used.
3. Metallic and non-metallic powders can be intimately mixed and extruded.

4. Polymeric materials and even fibre reinforced plastics can be extruded.
5. Low capital and operating costs.
6. If the process is run fast, enough heat is generated due to friction and the product emerges in an annealed condition. No further intermediate heat treatment is needed and the product can be put to use directly.

However, the process is limited to the extrusion of non-ferrous metals, mainly copper alloys and aluminium.

3. Helical Extrusion. It is a novel technique for the production of wire or tube. The technique more closely resembles machining process. A conical punch is driven slowly into the end of a copper billet, to form a short tube with an annular face. This is then steadily deformed by a tool (resembling a negative rake cutting tool) that rotates about the axes of the built and slowly advances in a helical path. The swarf, so produced, is not wasted but is trapped in a small chamber, from which it is forced to escape through an orifice, as a wire. The technique has the following advantages :

1. The forces are low due to two factors : only a small volume of metal is deformed at a given instant and the yield stress gets reduced, because of severe local heating.
2. Very large reductions in cross-section are possible (upto about 150 times in diameter).

4.5. PIPE AND TUBE MANUFACTURING

The pipes and tubes can be seamless or with seam. Tubes with seam are manufactured by the various welding techniques. Seamless tubes are made by : extrusion and piercing methods. The extrusion method, "Tubular extrusion" has already been discussed under Art 4.4. Seamless tubes are used in high temperature and pressure conditions/applications as well as for transporting gas and chemical liquids. Extruded tubes are used for gun barrels. Upto 400 mm diameter steel pipes have been made.

4.5.1. Piercing Method of making Seamless tubes. Seamless tubes are made in two stages :

1. Manufacture from round blank or billet of a thick walled shell in a piercing mill.
2. Rolling of shell into a pipe of given diameter and wall thickness.

A heated round billet with its leading end centre punched, pierced or drilled, is pushed longitudinally in between two large convex shaped (tapered) rolls that revolve in the same direction, their axes being inclined at opposite angles of about 6° from the axis of the billet, Fig. 4.47 (*a*). The clearance between the rolls is less than the diameter of the billet. As the billet is caught by the rolls and rotated, their inclination causes the billet to be drawn forward into them. The reduced clearance between the rolls forces the billet to deform into an elliptical shape. As the billet tries to rotate under high compressive forces created by the rolls, secondary tensile stresses are set up at 90° to the compressive forces at the centre of the billet. The punched hole at the centre of the billet will tear open. The mandrel assists this action. As the billet rotates and feed towards the mandrel, the tearing action is propagated along the length of the billet forming a seamless shell (roughly formed tube). Upon completion of the operation, the mandrel is forced out of the shell. This piercing of the billet to form a rough tube is called "Mannesmann process".

The pierced shell is further processed in a "plug rolling mill", (Fig. 4.47 (*b*)), which is a two-high reversing stand with a series of round passes in the rolls. A short mandrel (plug) is held in the centre of the pass by a long bar. The clearance between the mandrel and the pass determines the wall thickness of the rolled tube. This operation elongates the tube and reduces the wall thickness. The outside diameter of the tube is only approximate, however. The tube is still in a rough state. Therefore, another operation, called "reeling" is performed between reelers and

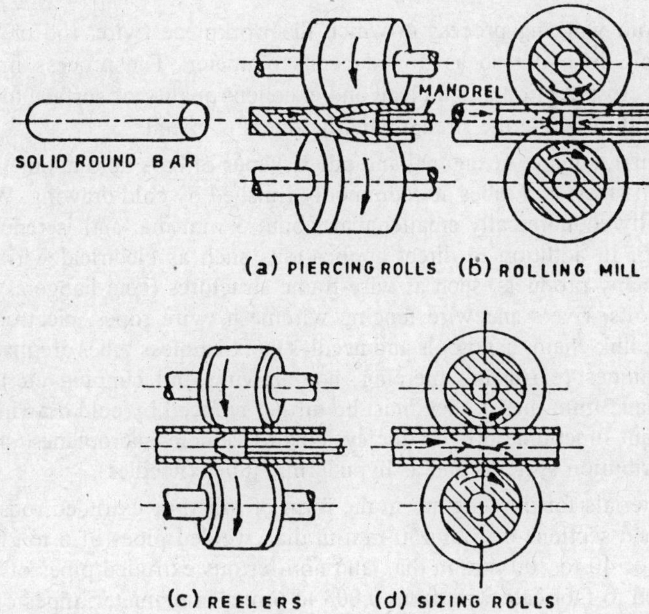

Fig. 4.47. Piercing Methods.

over a mandrel as shown in Fig. 4.47 (*c*). The reeling operation improves the finish of the inside and outside surfaces, eliminates irregularities, scratches and out of roundness, and decreases the differences in wall thickness. Finally, a sizing operation is performed between sizing rolls without use of a mandrel, (Fig. 4.47 (*d*)) The final operations of reeling and sizing are often conducted on cooled tubes in order to improve their finish and size.

Continuous Seamless Tube Mill. A most effective and advanced method for pipe rolling is the manufacture of thick walled shells in a piercing mill and then subsequent rolling in several continuous mills. These mills consist of several stands located one after another, the pipe being reduced in grooves and the direction of reduction changing by 90° in the adjacent stands. Rolling speeds can be adjusted in stands by means of electric motors, this allowing tension rolling and so increasing the reduction per passing and so thinning out pipe wall.

A typical continuous mill for manufacturing pipes from carbon and alloy steels in diameters from 30 to 102 mm and wall thickness of 1.75 to 8 mm from blanks 140 mm in diameter, consists of : furnace for rapid heating of blanks/billets, a piercing mill, multi-stand (7 to 9) mill for rolling rough pipes on a long mandrel, a device for extracting the mandrel, an induction furnace for heating rough pipes prior to subsequent rolling, an 11-stand continuous reducing mill in which the pipe is rolled without the use of a mandrel but with tension (when wall thickness is to be reduced) or without tension (when no change in wall thickness is desired) and a 19-stand continuous reducing mill for rolling with high tension in order to sharply reduce the diameter and the wall thickness of the pipe. After rolling, pipes are cut by saws and finished in continuous automated production lines.

Extensively used are automatic varieties of mills. An automatic mill is a standard non-reversible two-high stand with grooved rolls. Pipes from 57 to 425 mm in dia. with wall thickness of 3 to 30 mm can be rolled on these mills.

4.6. DRAWING OF WIRE, ROD AND TUBE

Drawing is a cold working process in which the workpiece (wire, rod or tube) is pulled through a tapered hole in a die so as to reduce its diameter. The process imparts accurate dimensions, specified cross-section and a clean and excellent quality of surface to the work. The process may appreciably increase the strength and hardness of metal.

Round, rectangular, square, hexagonal and other shapes of bars upto about 10 cm across or in diameter, wire of all sizes, and tubes are commonly finished by cold drawing. Wire can not be hot rolled commercially/economically smaller than about 5 mm dia. and is reduced to smaller sizes by cold drawing. In addition to direct applications such as electrical wiring, wire is the starting material for many products, such as wire-frame structures (coat hangers, shopping carts etc.), nails, screws, bolts, rivets and wire fencing wiremesh, wire ropes, electrodes, chain pin, umbrella ribs, springs, link chain, handtools and needles etc. Seamless tubes are made by a variety of hot working techniques (extrusion, piercing, hot drawing and cupping etc.), but below a minimum size (less than 3 mm thick), they must be further reduced by cold drawing. Cold drawn tubes perform important functions in hydraulic systems of vehicles, aeroplanes, ships, industrial machinery, water distribution systems and as hypodermic tubes (needles).

The starting materials for drawing are in the form of rolled or extruded rods of 5 to 9 mm in dia., rolled or formed sections of 5 to 150 mm in dia., welded pipes of 6 to 200 mm in dia., seamless rolled pipes of 40 to 200 mm in dia., and non-ferrous extruded pipes of 20 to 400 mm in dia. Drawing is used to fabricate wire from 0.008 to 6 mm in diameter, pipes of 1 to 360 mm in outside dia. and wall thickness of 0.1 to 10 mm, etc.

4.6.1. Wire and Rod drawing. Fig. 4.48 (*a*) shows a section of a typical circular draw die. The die geometry is typified by a bell mouthed entrance Zone (angle β is usually about 40°), a conical working zone with half die angle α (usually 6° to 12°) and a short sizing land. The entrance zone allows the introduction of lubricant into the working zone and also it protects the

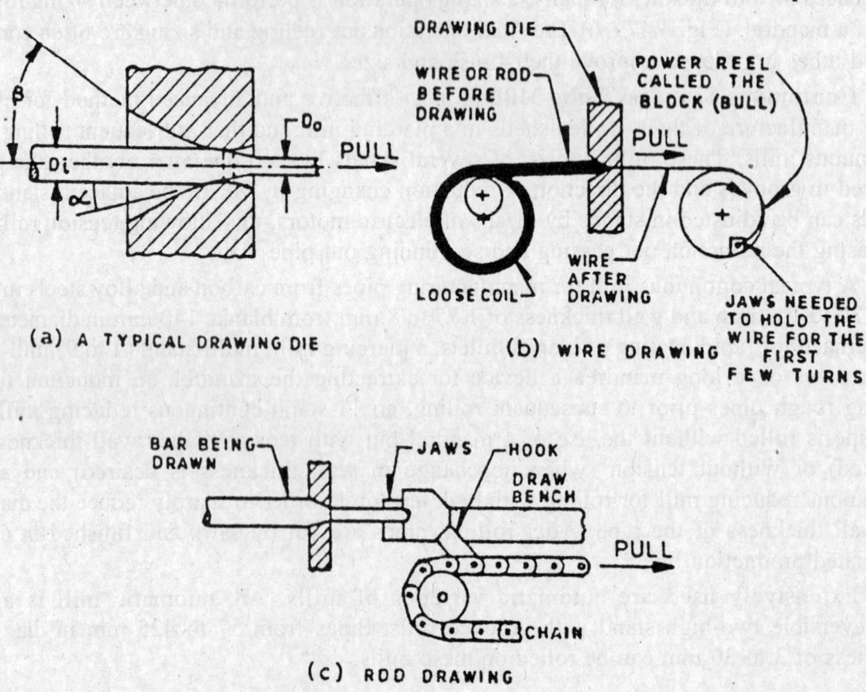

Fig. 4.48. Wire and Rod Drawing.

work material against scoring by the die edges. The actual reduction takes place in the conical working zone. The land (a few mm long) serves to guide the wire or rod as it comes out of the working zone of the die. It ensures accuracy of dimensions and of sections shape. The exit zone prevents damage to die bearing and scoring of finished product. Very large dies (for drawing rods and tubes of large cross-section) are made of high carbon or high-speed steel, the moderate-size dies of tungsten carbide, and the dies for the finest wires (down to 0.05 mm dia.) are made of diamond. The deformation force is supplied by the pulling action but the material is deformed in compression due to the reaction of the die on the material. Therefore, drawing is often called indirect compression deformation process.

1. Wire drawing. Materials below about 16 mm diameter are handled in coil form. At the start of wire drawing, the end of the rod or wire to be drawn is pointed (by swaging etc.) so that it freely enters the die orifice and sticks out behind the die. This end is gripped by the jaws of the plier, or carriage which pulls the rod through all zones of the die hole where it undergoes deformation, that is, reduction and elongation. The wire is pulled sufficiently so that its end can be attached to the power reel. After that, the power reel rotates at the proper speed and pulls the wire through the die with the wire getting continuously wound on the reel, (Fig. 4.48 (*b*)). For fine wire, the material may be passed through a number of dies, receiving successive reductions in diameter, before being coiled. Such successive drawing in general requires driving rolls between the dies. The drawing arrangement used in wire drawing is called "Bull-block drawbench". The diameter of the drawing block in machines of various sizes may vary from 150 to 1000 mm and the available power ranges from 5 kW to 110 kW.

2. Rod drawing. In rod or bar drawing, the product must remain straight. The maximum length of the rod which can be drawn depends upon the maximum travel of the carriage which pulls it through the die (it may be 15 m to 30 m). The front pointed end of the rod protruding from the die is gripped by the jaws of the plier (drawing Carriage) and the hook is lowered to engage the moving chain. This drawing arrangement used in rod drawing is called "chain drawbench". The pull capacity of the drawbench, (Fig. 4.48 (*c*)), ranges from 10 kN to 1500 kN.

Drawing speed may be 0.15 m/s for the largest size rods and 1.5 m/s for small rods and coils and upto 25 m/s for very fine wires.

Preparing the metal for drawing. Before drawing, the rod is cleaned of scale. Descaling is commonly done by pickling in acid solutions. Steel rods are pickled in sulphuric or hydrochloric acid, or a mixture of acids ; copper and brass are treated in sulphuric acid ; nickel and its alloys, in a mixture of sulphuric acid and potassium bichromate.

After pickling, the metal is washed to remove all remains of acid and sludge from its surface. The metal surface is then conditioned by sulling, coppering, phosphating and liming.

In sulling or yellowing, the blank is given a thin coating of iron hydroxide, $Fe(OH)_3$, which combines with lime to serve as a filler for the lubricant. Phosphating consists of applying a film of phosphates of Mn, Fe or Zn. Lubricant sticks well to the phosphate coating, and the co-efficient of friction drops to 0.04 - 0.06. In copper coating, the metal is dipped into a weakly acidified solution of copper sulphate. Copper and tin provide a good bearing surface on steel. Liming is done by immersion for a short time in a boiling lime solution. Liming neutralizes the remaining acid and forms a vehicle film for the lubricant. The final operation before drawing is drying above 100°C. Drying drives off the moisture and a part of the hydrogen dissolved in the metal. A suitable lubricant is then applied to the dry surface of the rod. Lubrication reduces the required drawing force and energy consumption, enables a smooth surface to be obtained and increases die life. Various kinds of mineral and vegetable oils, animal fats, graphite, soap and certain emulsions are applied as drawing lubricants. Emulsion is supplied to the die by a hose or the rod is immersed in a bath of lubricant prior to or during drawing. Water may be circulated around the die to cool it as considerable heat is generated during wire drawing.

After pickling, the rod undergoes swaging or pointing of the leading end and is then drawn through the die.

Degree of drawing. The degree of drawing is measured in terms of "reduction of area (RA)" which is defined as the ratio of the difference in cross-sectional area before and after drawing to the initial cross-sectional area, expressed in per cent :

$$RA = \frac{D_i^2 - D_o^2}{D_i^2} = 1 - (D_o/D_i)^2$$

For fine wires, the reduction per pass of 15 to 25% is used, while for coarse wires this value per pass may be 30 to 35 per cent. For rods, the value can be upto 40 per cent.

4.6.2. Hydrostatic Wire Extrusion. Hydrostatic wire extrusion is now well established and is particularly suitable for brittle materials and composites. In hydrostatic extrusion, it is not necessary that the billet be straight or cylindrical and it need not have a constant cross- sectional area along its entire length. Thus, for example specially bent billets or even a coiled wire or multi-stepped workpiece can be extruded. Since there is no friction in the billet chamber, the billet length is dependent only on the dimensions of the pressure chamber. Thus, in hydrostatic wire extrusion, a wire coil placed in the pressure chamber, (Fig. 4.49) can be extruded upto a ratio of raw-wire length to raw-wire diameter of the order of 105. By applying a front tension, an efficient and well controlled extrusion-drawing process is available.

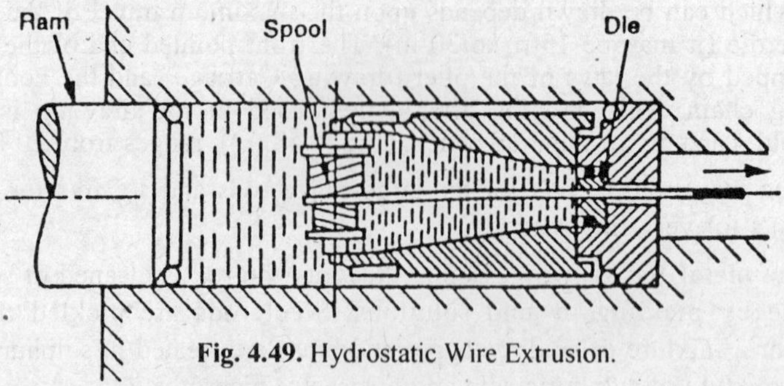

Fig. 4.49. Hydrostatic Wire Extrusion.

4.6.3. Tandem Wire Drawing. It is clear that for fine wire, the drawing involves passage through several dies in series, especially for copper wire, which is very ductile. This is called multi-stage or tandem wire drawing. The wire is passed through several dies in series (3 to 12), each held in water-cooled die- blocks. In this way, a higher deformation ratio can be achieved and the surface temperature of the wire is also reduced. The temperature gradients between the wire surface and the interior are reduced during the period the wire is between two dies. Additional cooling can also be used. If extensive deformation is required, intermediate anneals must be performed between the various stages of drawing.

To avoid excessive tension on the drawn wire, it is usual to wind one or two turns round a capstan between each pair of dies. The capstans are driven at carefully regulated speeds, so that there is a frictional drag at the capstan surface, which pulls the wire through the die, and the capstan also provides small back tension on the wire entering the next die. This back tension increases the drawing stress, but the die pressure is lowered, because from the conditions of yielding,

$\sigma_1 - \sigma_2 = \sigma_0$ (yield stress)

Now $\sigma_2 = -p$, the die pressure

$\sigma_1 + p = \sigma_0$

if the drawing stress σ_1 increases, p will decrease (σ_0 being constant for a material). This reduction in pressure improves die life, which is a very important factor in practical wire-drawing.

4.6.4. Tube drawing. Hollow cylinders and tubes or pipes which are made by hot-working processes such as extrusion, piercing and rolling, often are cold finished by drawing. Cold tube drawing is used :

1. To obtain closer dimensional tolerances.
2. To produce better surface finish.
3. To increase the mechanical properties of the tube by strain hardening.
4. To produce tubes with thinner walls or smaller diameters than can be obtained with hot forming methods.
5. To produce tubes of irregular shapes.

Like bar drawing, tube drawing is done in most cases with the use of a "drawbench".

The three common methods of tube drawing are : Tube sinking. Tube drawing with a plug and Tube drawing with a moving mandrel, the last two methods being more widely used because in tube sinking, (Fig. 4.50 (*a*)), the inside of the tube is not supported and so during drawing operation, the inner surface becomes uneven and there will be tendency for the wall thickness to increase slightly. In plug drawing and movable mandrel drawing, both the inner and outer surfaces of the tube are controlled and we get tubes of better dimensional accuracy as compared to Tube sinking. In plug drawing, the plug (which may be either cylindrical or conical) can either be fixed, (Fig. 4.50 *b* (*i*)), or floating, (Fig. 4.50 *b*(*ii*)). The friction with a fixed plug will be more than with a floating plug, so the reduction in area seldom exceeds 30% in this method. With a floating plug, this figure can be approximately 45% or with the same reduction, the drawing loads will be less with floating plug than with a fixed plug. The friction is minimized in Tube drawing with a movable mandrel [Eq. 4.50 (*c*)]. However, after tube drawing, the mandrel has to be removed by rolling which results in slightly increased tube diameter and reduced dimensional tolerances.

Tube sinking is used to only reduce the outside diameter of tube. If the wall thickness of the tube is also to be reduced, an internal die is also needed, which may be in the form of a plug or a mandrel. Mandrels are used for tubes from about 12.7 mm to 250 mm in diameter. Heavy walled tubes and those with less than 12.7 mm dia. are drawn without a mandrel. This is the procedure used for drawing hypodermic needles (starting from a 50 mm dia. tube) with outside diameter less than 0.02 mm and with inside diameter half of the outside diameter.

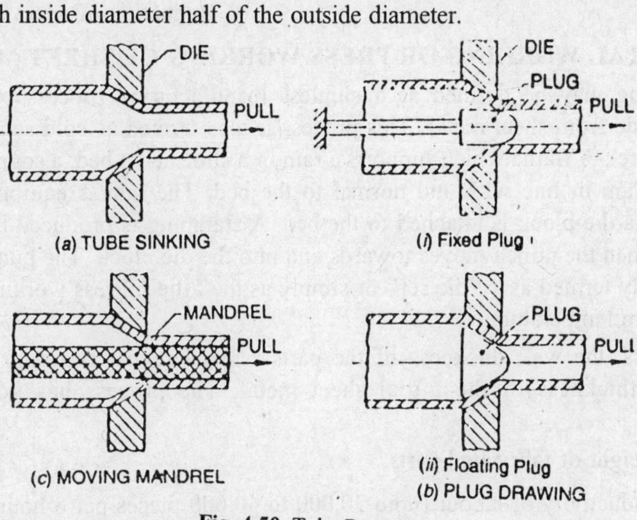

Fig. 4.50. Tube Drawing.

A number of high-tech industries such as electro-medical, bio- medical, orthodonitics, semi-conductors etc. use specially processed tubes such as hypodermic needles, hermetic seals, connectors, mercury weighted switches, silicon controlled rectifiers, electronic auto-ignition and speed control devices, photographic cartridges, numeric readout units etc. Dual tubes consisting of two concentric seamless tubes are used in the semi- conductor industry to handle hazardous gases such as Arsine, Phosphine, Silane, Diborane etc. Special shapes such as oval, rectangular and combinations thereoff are produced for instrumentation industry, bourdon tubing etc. The capillary tubes with OD of 4.5 mm and ID about 1.5 mm are used for transmission of pressure over long distances. They are used for liquid chromography. The seamless tubes are used extensively in aircraft hydraulic systems.

These special tubes are produced by drawing of seamless hollows. These tubes have to meet the following requirements :

(*i*) Very close dimensional tolerances.

(*ii*) Extremely good external and internal surface finishes.

(*iii*) Freedom from any chemical or other impurities.

(*iv*) Close tolerances on mechanical properties.

4.6.5. Main process variables in Drawing Process :

Drawing stress is given as:

$$\sigma_d = \frac{\sigma_0 (1 + B)}{B}\left[1 - \left(\frac{r_1}{r_0}\right)^{2B}\right]$$

where σ_0 = yield strength of metarial

$B = \mu \cot \alpha$

μ = co-efficient of friction

α = half die-angle

r_1 = radius of work piece at exist

r_0 = radius of work piece at entry.

∴ Main process variables of Drawing Process are :

1. Strength of work material
2. Tempearture
3. Speed of drawing
4. Co-efficient of friction/lubrication
5. Die-angle
6. Reduction of area

4.7. SHEET METAL WORKING OR PRESS WORKING OF SHEET METALS

Press working may be defined as a chipless manufacturing process by which various components are made from sheet metal. This process is also termed as cold stamping. The main features of a press are : A frame which supports a ram or a slide and a bed, a source of mechanism for operating the ram in line with and normal to the bed. The ram is equipped with suitable punch/punches and a die block is attached to the bed. A stamping is produced by the downward stroke of the ram when the punch moves towards and into the die block. The punch and die block assembly is generally termed as a "die set" or simply as the "die". Press working operations are usually done at room temperature.

In this process, the wall thickness of the parts remain almost constant and differs only slightly from the thickness of the initial sheet metal. The process has got the following advantages :

1. Small weight of fabricated parts.
2. High productivity of labour (upto 30,000 to 40,000 pieces per 8 hour working shift).

3. High efficiency of technique as regards the fabrication of items of diversified shapes, both simple and complex, such as washers, bushings, retainers (cages) of ball bearings, tanks and car bodies etc.

4. The parts obtained by cold sheet metal working are distinguished for their size accuracy (narrow tolerances with a high surface finish). In many cases, they require no subsequent machining and are delivered to the assembly shop.

All the above advantages have made sheet stamping very attractive to a host of industries, particularly, to automotive, aircraft, electrical engineering and others.

The initial material in cold press working is : low carbon steels, ductile alloy steels, copper and its alloys (brass, bronze), aluminium and its alloys, as well as, other ductile materials from tenth of a mm to about 6 or 8 mm thick. As discussed under cold rolling and wire drawing, the metal must be free from scale before sheet metal working. Scale is removed by pickling and other operations already discussed.

4.7.1. Elastic recovery or Springback. We know that in metal working processes, the total deformation imparted to a work piece will be the sum of elastic deformation and plastic deformation. We also know that elastic deformation is recoverable whereas plastic deformation is permanent. So, at the end of a metal working operation, when the pressure on the metal is released, there is an elastic recovery by the material and the total deformation will get reduced a little. This phenomenon is called as "Springback". This phenomenon is of more importance in cold working operations, especially in forming operations such as bending etc. Springback depends upon the yield point strength of a metal. The higher the yield point strength of a metal, the greater the springback. The amount of springback for a forming operation is difficult to predict and cut- and try methods are most satisfactory to account for it. To compensate for springback, the cold deformation must always be carried beyond the desired limit by an amount equal to the springback.

4.7.2. Press operations. The sheet metal operations done on a press may be grouped into two categories, cutting operations and forming operations. In cutting operations, the workpiece is stressed beyond its ultimate strength. The stresses caused in the metal by the applied forces will be shearing stresses. In forming operations, the stresses are below the ultimate strength of the metal. In this operation, there is no cutting of the metal but only the contour of the workpiece is changed to get the desired product. The cutting operations include : blanking, punching, notching, perforating, trimming, shaving, slitting and lancing etc. The forming operations include : bending, drawing, redrawing and squeezing. The stresses induced in the metal during bending and drawing operations are tensile and compressive and during the squeezing operation these are compressive.

Below we give the definitions of the various press operations :

(*a*) *Blanking.* Blanking is the operation of cutting a flat shape from sheet metal. The article punched out is called the 'blank' and is the required product of the operation. The hole and metal left behind is discarded as waste. It is usually the first step of series of operations, (Fig. 4.51 (*a*)).

Plain blanking is explained in Fig. 4.52.

(*b*) *Punching (Piercing).* It is a cutting operation by which various shaped holes are made in sheet metal. Punching is similar to blanking except that in punching, the hole is the desired product, the material punched out to form the hole being waste (Fig. 4.51 (*b*)).

(*c*) *Notching.* This is cutting operation by which metal pieces are cut from the edge of a sheet, strip or blank.

(*d*) *Perforating.* This is a process by which multiple holes which are very small and close together are cut in flat work material.

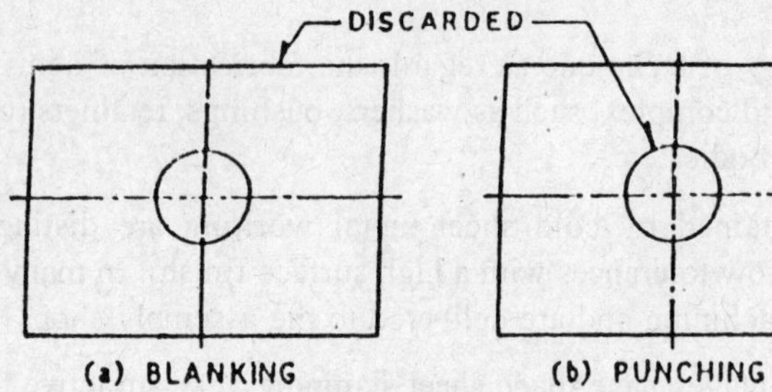

Fig. 4.51. Blanking and Punching.

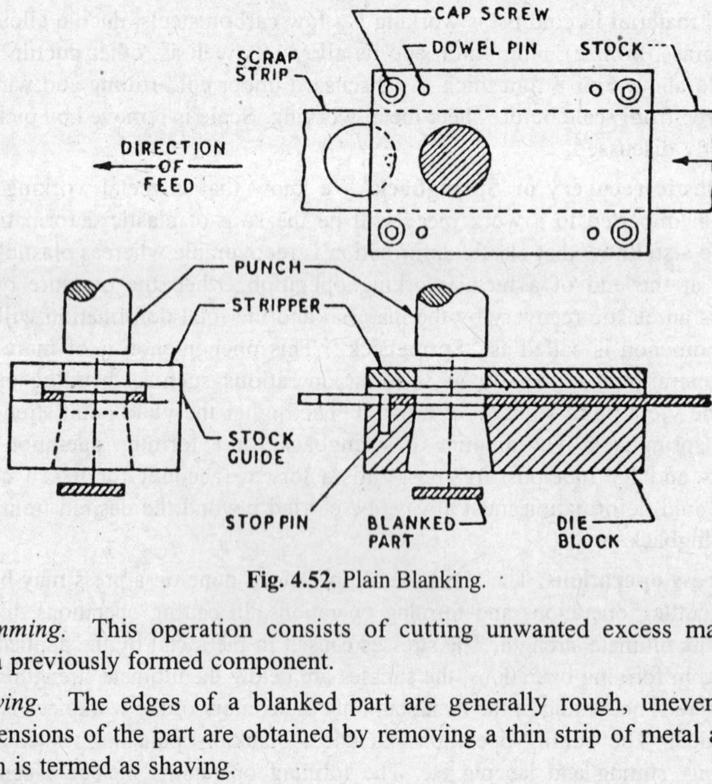

Fig. 4.52. Plain Blanking.

(*e*) *Trimming.* This operation consists of cutting unwanted excess material from the periphery of a previously formed component.

(*f*) *Shaving.* The edges of a blanked part are generally rough, uneven and unsquare. Accurate dimensions of the part are obtained by removing a thin strip of metal along the edges. This operation is termed as shaving.

(*g*) *Slitting.* It refers to the operation of making incomplete holes in a workpiece.

(*h*) *Lancing.* This is a cutting operation in which a hole is partially cut and then one side is bent down to form a sort of tab or louver. Since no metal is actually removed, there will be no scrap.

(*i*) *Nibbling.* The nibbling operation which is used for only small quantities of components, is designed for cutting out flat parts from sheet metal. The flat parts range from simple to complex contours. This operation is generally substituted for blanking. The part is usually moved and guided by hand as the continuously operating punch cuts away at the edge of the desired contour.

(*j*) *Bending.* In this operation, the material in the form of flat sheet or strip, is uniformly strained around a linear axis which lies in the neutral plane and perpendicular to the lengthwise direction of the sheet or metal.

(*k*) *Drawing.* This is a process of a forming a flat workpiece into a hollow shape by means of a punch which causes the blank to flow into a die cavity.

(*j*) *Squeezing.* Under this operation, the metal is caused to flow to all portions of a die cavity under the action of compressive forces.

Some of these operations are explained in Fig. 4.53.

4.7.3. Presses. Presses are classified in various ways. They may be classified according to :

(*i*) source of power

(*ii*) method of actuation of the rams (slides)

(*iii*) number of slides

(*iv*) type of frame

(*v*) the type of work for which the press has been designed.

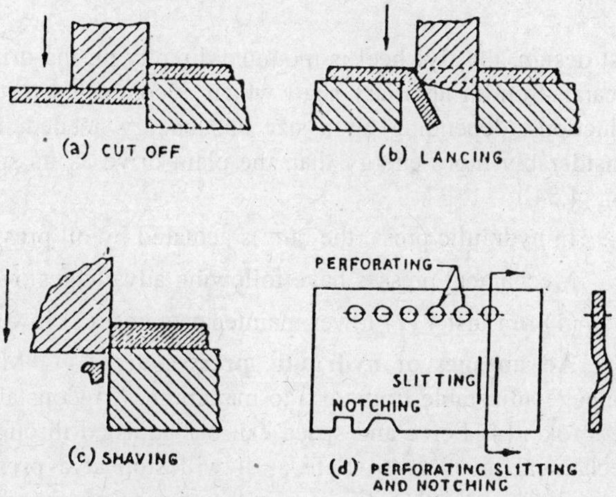

Fig. 4.53 Some Press Operations.

(*a*) **Source of power.** Two kinds of sources of power for applying force to the ram are : mechanical and hydraulic.

In mechanical presses, the energy of flywheel is utilized which is transmitted to the workpiece by gears, cranks, eccentrics or levers. The flywheel rotates freely on the crank shaft and is driven from an electric motor through gears or *V* belts. The motor runs continuously and stores energy in the flywheel. When the operator presses a foot treadle or actuates a button, the clutch gets engaged and the flywheel is connected to the crankshaft. The crankshaft (drive shaft) starts rotating and the stored up energy in the flywheel is transmitted to the ram on its downward stroke. The clutch to engage and disengage the flywheel to the drive shaft can be : a jaw clutch, an air-operated clutch or an electromagnetic clutch. In manually operated mechanical presses, the clutch is disengaged after each cycle. But in automatic presses in which the metal strip is fed to the die automatically, there is no need of single-stroke clutch disengaging mechanism and the ram moves up and down continuously. These presses can be classified as 'plain' and 'geared'. In the

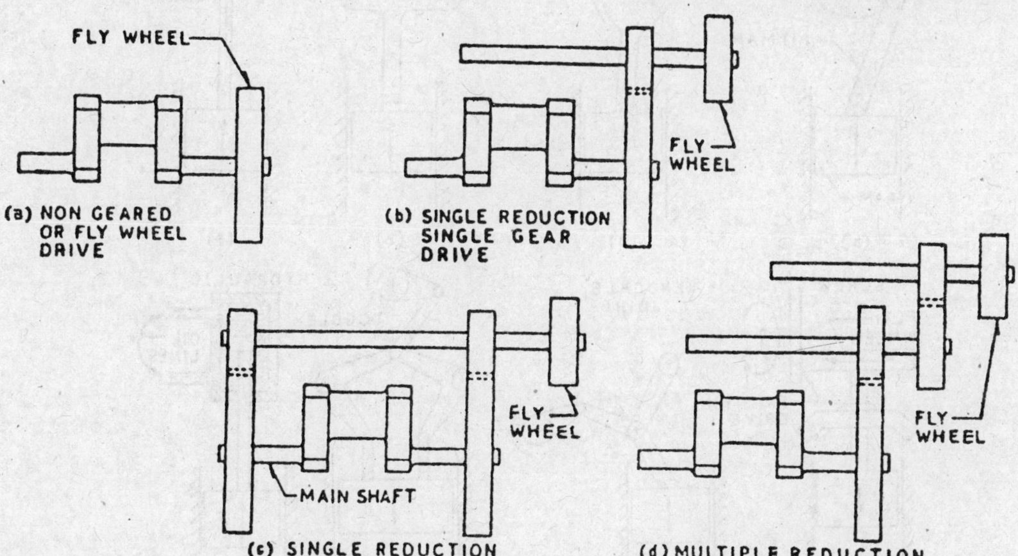

Fig. 4.54. Types of Mechanical Press Drives.

first design, the flywheel is mounted directly on the drive shaft. On a geared press, the flywheel is carried on an auxiliary shaft which is connected to the main shaft through one or more gear reductions, depending upon size and energy needed. In this arrangement, the flywheel stores considerably more energy than the plain drive as its speed is higher than the main drive shaft, (Fig. 4.54.)

In hydraulic press, the ram is actuated by oil pressure on a piston in a cylinder.

Mechanical presses have following advantages over the hydraulic presses :-

(1) run faster (2) lower maintenance cost (3) lower capital cost.

Advantages of hydraulic presses are : (1) More versatile and easier to operate. (2) Tonnage adjustable from zero to maximum. (3) Constant pressure can be maintained throughout the stroke. (4) Force and speed can be adjusted throughout the stroke. (5) More powerful than mechanical presses. (6) Safe as it will stop at a pressure setting. The main disadvantage of hydraulic press is that it is slower than a mechanical press.

A press is rated in tonnes of force, it is able to apply without undue strain. To keep the deflections small, it is a usual practice to choose a press rated 50 to 100% higher than the force required for an operation.

(*b*) **Method of actuation of slides.** The various methods used for transmitting power to the ram or slide are shown in Fig. 4.55. The drive (Fig. 4.55 *h*) is for hydraulic presses in which the ram is actuated from a piston by means of a connecting rod. The remaining arrangements are for mechanical presses. The most common drive is crankshaft. Crankshafts are used for longer strokes. For shorter strokes, eccentric drive is employed. Cam drive is similar to eccentric drive except that at the bottom of the stroke dwell is obtained. Rack and pinion drive is used where a very long stroke is required. In this drive, the movement of the slide is uniform and much slow as compared to crank drive.

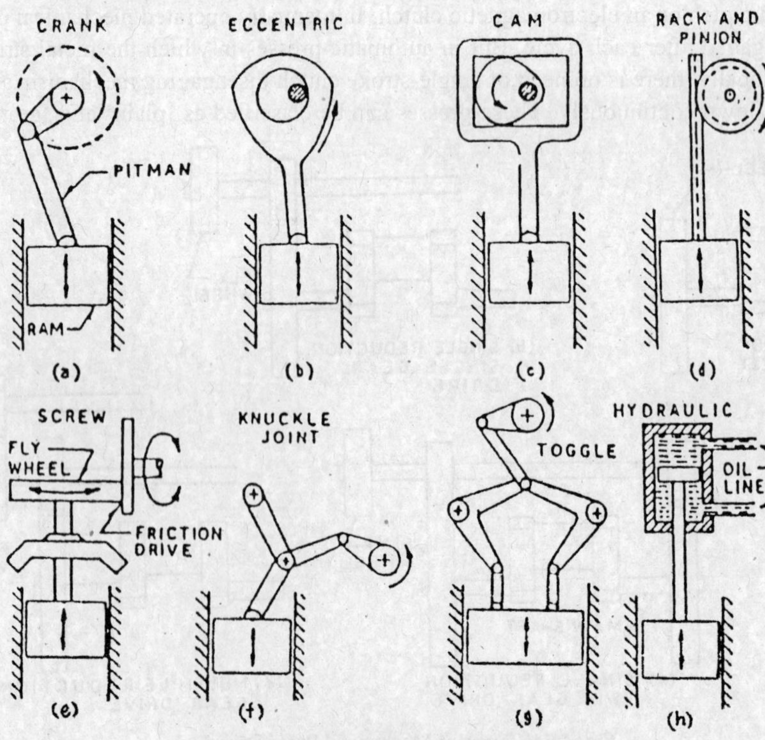

Fig. 4.55. Principal Trends of Press-Drivers.

In the screw drive, the slide is attached to the flywheel through a screw. As the flywheel is rotated by a friction drive, it moves down and the slide motion is continuously accelerated. At the bottom of the stroke, the entire energy stored in the flywheel is consumed by the work. The knuckle joint drive consists of an upper link and a lower link joined together with another horizontal link at the centre. The top end of the upper link is attached to a rigid portion of the frame. The horizontal link is driven by crankshaft. As the crankshaft rotates and completes its cycle, it pulls the horizontal link until the other two links assume a straight line vertical in position. This results in a short powerful movement of the slide with slow motion at the bottom of the stroke. Due to this reason, knuckle joint presses are used for heavy coining, embossing and sizing operations. The toggle drive is primarily used to drive the outer sides of a double action press. For example, in drawing operation, the outer ram should dwell to hold the blank during part of the stroke.

(*c*) **Number of Slides.** According to the number of slides, a press may be classified as : single-, double-, or triple-action press.

Single action press. This press has one plain ram or slide only which acts against a fixed bed. However, by using combination tools, more than one operation can be performed on this press at each stroke.

Double action press. This press has two slides moving in the same direction against a fixed bed. The outer slide is referred to as blankolder slide and the inner slide is called the punch or plunger slide, (Fig. 4.56). They are used chiefly for deep drawing operations.

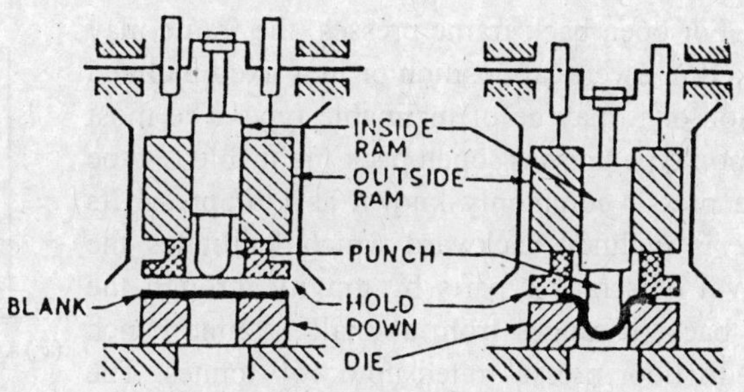

Fig. 4.56. Double Action Press.

Triple action press. This press has three slides, two slides moving in the same direction against the fixed bed as in double action press and the third or lower slide moving upward through the fixed bed, (Fig. 4.57). They are used only for large work such as motorcar body panels.

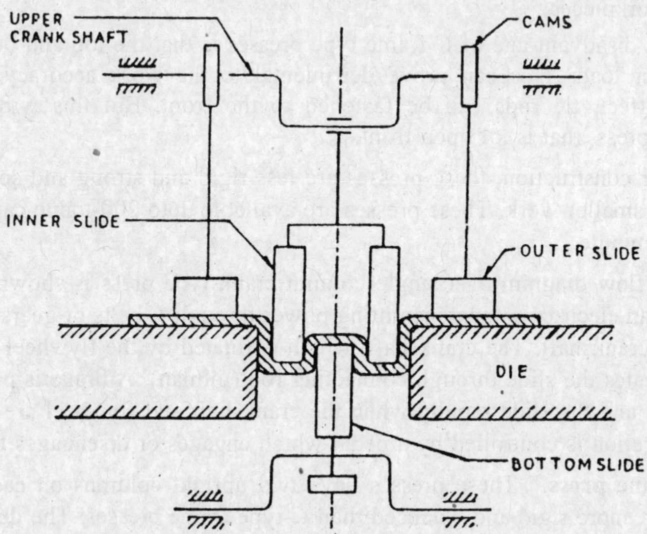

Fig. 4.57. Triple Action Press.

The double action press is known as double action cam press or double action toggle press, depending upon whether the outer slide is actuated by a cam located in the crankshaft or by a toggle drive. Triple action press is cam type for small work and toggle type for large work. The use of cams makes it easy to arrange that the holder descends ahead of the punch so that the blank is gripped before the drawing starts and also to keep the holder at rest during the drawing.

(*d*) **Type of frame.** According to the design of the frame, the presses may be classified as open frame presses and closed frame presses.

Open Frame Press. (Fig. 4.58) This press is also known as 'Gap frame' or '*C*' frame press. It has only one column to the rear of the die, whose shape resembles letter *C*. the front or die area is accessible from three sides, front, left side and right side. This speeds up production as the sheet metal can be fed in many ways. The back of such a press can be solid or open. Open back press permits feeding of the sheet metal from front to back or the ejection of the finished parts from the back. For open back frame presses, the frame may be fixed in a vertical position or in a fixed inclined position or it may be of inclinable type. The most common type press is 'open back inclinable *C*-type frame press', commonly known as *OBI* press. Its frame is inclined backward which facilitates the removal of scrap or parts by gravity through the open back. It ranges from a small 1 tonne bench press to floor presses rated upto 150 tonnes. The solid back type press in a fixed vertical position on integral base is more rigid than an inclinable press and has wider bed and arms. An 'adjustable bed press' or 'knee press' has bed that can be raised or lowered. This permits the accommodation of different sizes of dies and work, but loses some rigidity. In 'horn press', a large round post or mandrel projects out from the column in place of a bed. This mandrel is called 'horn' and supports round workpieces.

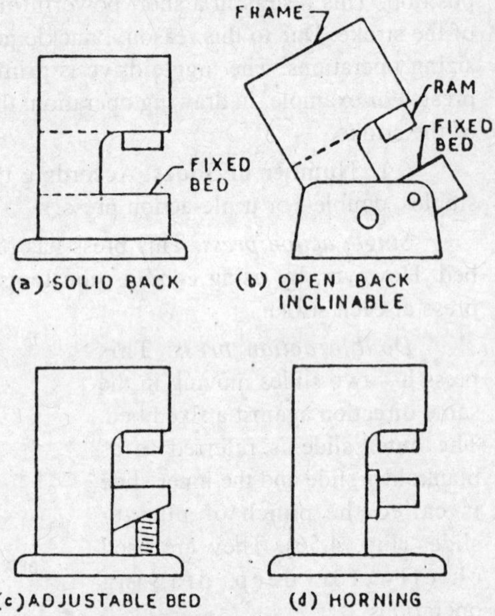

Fig. 4.58. Open Frame Press.

An inherent disadvant-age of *C*-frame type presses is that the top and bottom of the frame swing apart under load. This can prove detrimental to the work accuracy and tool life. To counteract this effect, tie rods can be fastened to the front. But this will nullify the main advantage of the press, that is, of open front.

Due to their construction, these presses are less rigid and strong and so are useful mainly for operations on smaller work. These presses are available upto 200-tonne capacity with strokes of 90 to 120 per minute.

The power flow diagram of a single column crank type press is shown in Fig. 4.59. The drive consists of an electric motor transmitting power through *V*-belts or gears to flywheel which rotates freely on crankshaft. The crankshaft, which is rotated by the flywheel when the clutch is engaged, reciprocates the slide through connecting rod (pitman). A brake is provided to stop the crank mechanism and the slide rapidly when the crankshaft and flywheel are disengaged by the clutch. Press operation is controlled by a pedal which engages or disengages the clutch.

Closed frame press. These presses have two upright columns on each side of the die. They are stronger, more rigid and balanced than *C*-type frame presses. The design is appropriate for wide beds and long strokes and so is suitable for heavier work. Work is fed either from the front or from the back of the press into the die area. Closed frame presses can be of two designs : arch frame press and straight side frame press, (Fig. 4.60). In arch press, the upper part is narrow as compared to the lower part. Straight-side presses are more rigid and strong than the arch press, since the heavy loads are taken up in a vertical direction by the massive side frame. Such presses are available upto at least 3000 tonnes capacity. Another design of a large mechanical press can be straight-side type with tie rods, Fig. 4.60 (*c*). The bed columns and crown are separate units but are keyed to hold alignment and are fastened together by tie rods. The tie rods are heated and shrunk in place to minimize straining under load. The pillar or open frame press, (Fig. 4.60 (*d*)), has four pillars. The pillars hold the crown to the bed and act as guides or ways for the ram.

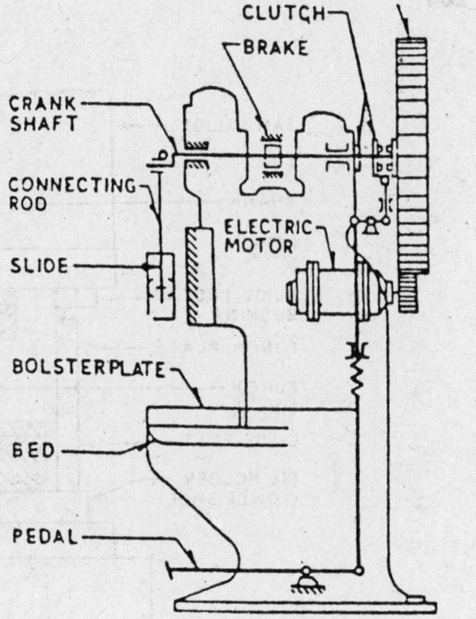

Fig. 4.59. Power Flow of Single Column Crank Press.

Hydraulic press frames can be of the *C*-type, straight side type with tie rods or of pillar type.

(*e*) **Type of Work.** The various types of work for which the press may be designed are, Punching, Blanking, Drawing, Bending, Forming, Coining, Embossing etc.

4.7.4. Press Selection. The factors which should be considered while selecting a press for a given job, are : the overall work size, the stock thickness and material, kind of operation to be performed, power required and speed of operation.

For punching, blanking and trimming operations, usually the crank or eccentric type mechanical press is used. This is due to their small working strokes and high production rates. In these operations, there is sudden release of load at the end of the cutting stroke. This sudden release of load is not advisable in hydraulic presses. So, hydraulic presses are not preferred for these operations. If however these are inevitable, then some damping devices are incorporated in the press design. For coining and other squeezing operations, which require very large forces, knuckle joint mechanical press is ideally suited. Hydraulic presses, which are slower and more powerful, can also be used for these operations. Hydraulic presses are also better adapted to pressing, forming and operations, which are basically slower processes.

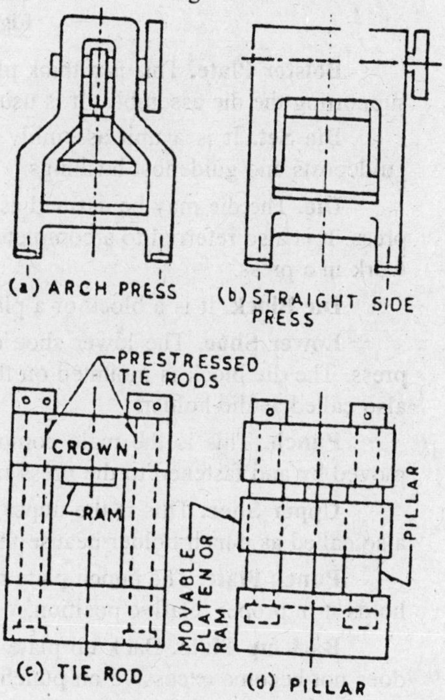

Fig. 4.60. Closed Frame Presses.

4.7.5. Press Working Terminology. A simple cutting die used for punching and blanking operations is shown in Fig. 4.61. Below, the definitions of the main components of the die and press are given :

Bed. The bed is the lower part of a press frame that serves as a table to which a bolster plate is mounted.

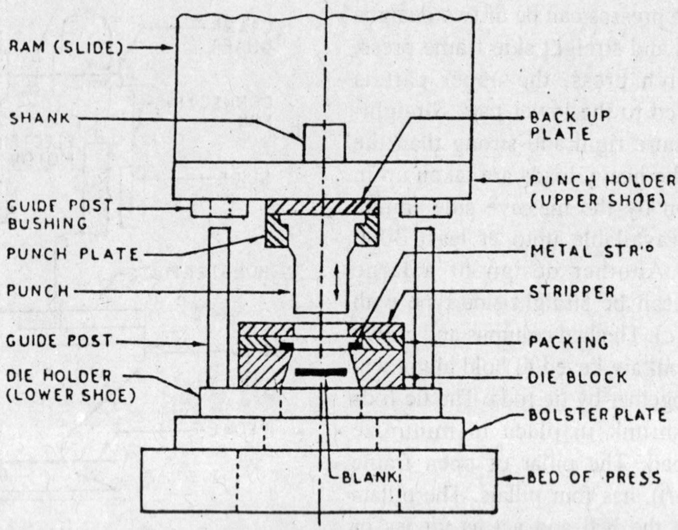

Fig. 4.61. A simple Cutting Die.

Bolster Plate. This is a thick plate secured to the press bed, which is used for locating and supporting the die assembly. It is usually 5 to 12.5 cm thick.

Die Set. It is a unit assembly which incorporates a lower and upper shoe, two or more guideposts and guidepost bushings.

Die. The die may be defined as the female part of a complete tool for producing work in a press. It is also referred to a complete tool consisting of a pair of mating members for producing work in a press.

Die Block. It is a block or a plate which contains a die cavity.

Lower Shoe. The lower shoe of a die set is generally mounted on the bolster plate of a press. The die block is mounted on the lower shoe. Also, the guide posts are mounted in it. It is also called as die-holder.

Punch. This is the male component of the die assembly, which is directly or indirectly moved by and fastened to the press ram or slide.

Upper Shoe. This is the upper part of the die set which contains guidepost bushings. It is also called as punch holder beause the punch is mounted on it.

Punch Plate. The punch plate or punch retainer fits closely over the body of the punch and holds it in proper relative position.

Back up Plate. Back up plate or pressure plate is placed so that the intensity of pressure does not become excessive on punch holder. The plate distributes the pressure over a wide area and the intensity of pressure on the punch holder is reduced to avoid crushing.

Stripper. It is a plate which is used to strip the metal strip from a cutting or non-cutting punch or die. It may also guide the sheet.

Knockout. It is a mechanism, usually connected to and operated by the press ram, for freeing a workpiece from a die.

Pitman. It is a connecting rod which is used to transmit motion from the main drive shaft to the press slide.

Shut height. It is the distance from top of the bed to the bottom of the slide, with its stroke down and adjustment up.

Stroke. The stroke of a press is the distance of ram movement from its up position to its down position. It is equal to twice the crankshaft throw or eccentricity of the eccentric drives but is variable on the hydraulic press.

Working of a Cutting Die, (Fig. 4.62). As is clear in the figure, the punch holder (upper shoe) is fastened directly to the ram of the punch press, and the die shoe (lower shoe) is fastened

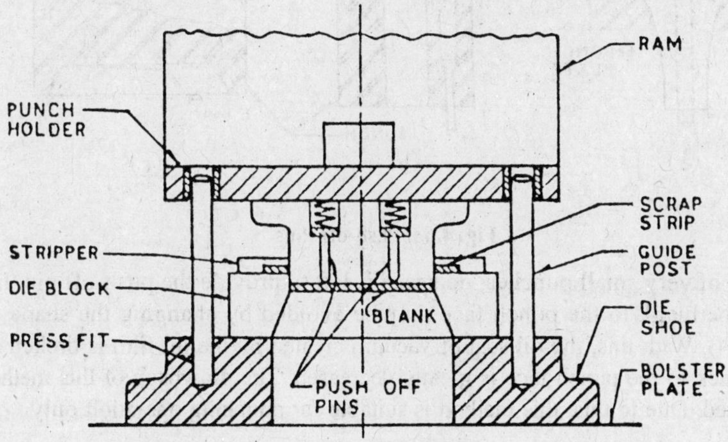

Fig. 4.62. A Cutting Die.

to the bolster plate of the press. Guide posts may be used to better align the punch holder with the die shoe. These three main components (punch holder, die shoe and guide posts) constitute what is known as the die set. A die set can be had with two guide posts located at the rear of the die set (known as back posts), diagonally; one at the back and the other in the front of the die set, or with four guide posts, one in each corner of the die set. The lower ends of the guide posts are press fitted into the die shoe. At the upper end, the guide posts have a slip fit with the guide bushings which are press fitted into the punch holder. With this, the guide posts have a free movement in the bushings. The punch is fastened to the punch holder and the die block is fastened to the die-shoe. The punch is aligned with the opening in the die-block. Since, both the punch and die block act as cutting tools, they are hardened.

The cutting action takes place during the downward movement of the punch into the die block. After the cutting action, the elastic recovery in the strip material takes place. Due to this, the size of the blank (cut portion from the strip) increases and that of the hole in the strip decreases. So, at the end of the cutting action, when the punch starts to move upwards, the scrap strip clings to the punch and the blank gets clogged in the die- opening. To remove the scrap strip from the punch surface, a stripper is used. In Fig. 4.62, a simplest type of stripper is used. It strips off the scrap strip from the punch surface when the scrap strip strikes the bottom surface of the stripper during the upward movement of the punch. To avoid clinging of the blank in the die opening, the walls of the die-opening are tapered.

In addition to the two above mentioned troubles, the blank may also adhere to the face of the punch. This usually happens with thin blanks or with blanks which have been treated with a lubricant. To help the blank free itself from the punch face, pushoff pins are provided which are fitted into the punch body, (Fig. 4.62).

The various basic designs of pushoff pins are shown in Fig. 4.63. The most popular design is a straight, round plunger with a sliding fit in the punch body, Fig. 4.63 (*a*). It is held in place by a light compression spring and a set screw. If at the end of the punch body, a pilot is to be provided, then the push- off pin is located off centre, (Fig. 4.63 (*b*)). When the space available for a compression spring is insufficient, a flat spring may be used, (Fig. 4.63 (*c*)).

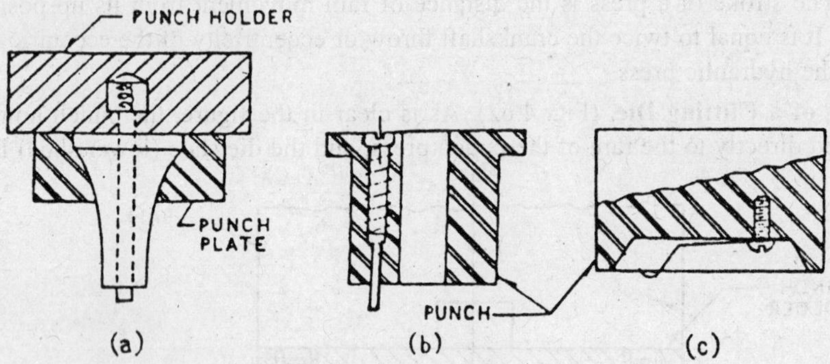

Fig. 4.63. Push-off Pins.

In the case of very small punches, no space is left to provide the push-off pin. In such cases, the sticking of the blank to the punch face may be avoided by changing the shape of the punch bottom, (Fig. 4.64). With this, the oil seal or vacuum created by the oil film is broken and tendency of the blank to stick to the punch face is greatly decreased. The drawback of this method is that the blank gets distorted. Due to this, this method is suitable for punching operation only.

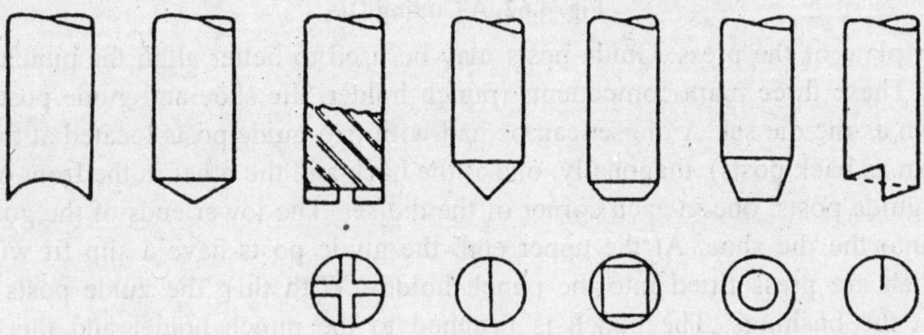

Fig. 4.64. Punch face contours to prevent blanks from sticking to very small punches.

4.7.6. Types of Dies. The dies may be classified according to the type of press operation and according to the method of operation.

4.7.6.1. Type of press operation. According to this criterion, the dies may be classified as : cutting dies and forming dies.

Cutting Dies. The dies are used to cut the metal. They utilize the cutting or shearing action. The common cutting dies are : blanking dies, piercing dies, perforating dies, notching, trimming, shaving and nibbling dies etc.

Forming Dies. These dies change the appearance of the blank without removing any stock. These dies include bending dies, drawing dies, squeezing dies etc.

4.7.6.2. Method of operation. According to this criterion, the dies may be classified as : Single-operation dies or simple dies, compound dies, combination dies, progressive dies, transfer dies, and multiple dies.

Simple Dies. Simple dies or single action dies perform single operation for each stroke of the press slide. The operation may be any of the operations listed under cutting or forming dies.

Compound Dies. In these dies, two or more operations may be performed at one station. Such dies are considered as cutting tools since, only cutting operations are carried out. Fig. 4.65 shows a simple compound die in which a washer is made by one stroke of the press. The washer is produced by simultaneous blanking and piercing operations. Compound dies are more accurate and economical in mass production as compared to single operation dies.

Combination Dies. In this die also, more than one operations may be performed at one station. It differs from compound die in that in this die, a cutting operation is combined with a bending or drawing operation. Fig. 4.66 explains the working of a combination blank and draw die. The die ring which is mounted on the die-shoe, is counterbored at the bottom to allow the flange of a pad to travel up and down. This pad is held flush with the face of the die by a spring. A drawing punch of required shape is fastened to the die shoe. The blanking punch is secured to the punch holder. A spring stripper strips the skeleton from the blanking punch. A knockout extending through the centre opening and through the punch stem ejects the part on the upstroke as it comes in contact with the knockout bar on the press. In operation, the blank holding ring descends as the part is blanked, then the drawing punch contacts and forces the blank into the drawing die which is made in the blanking punch.

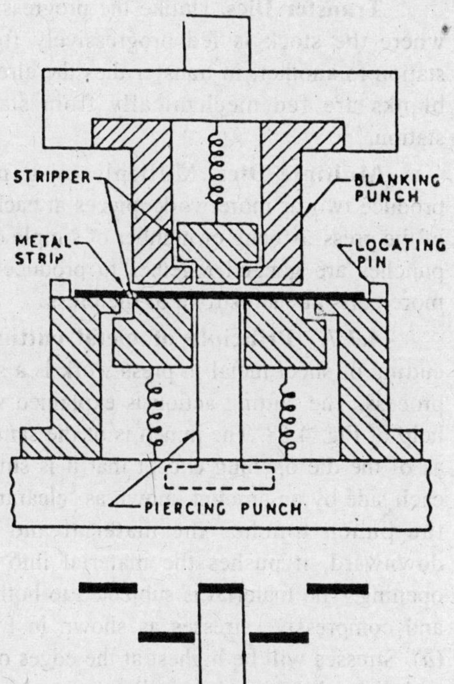

Fig. 4.65. Compound Die.

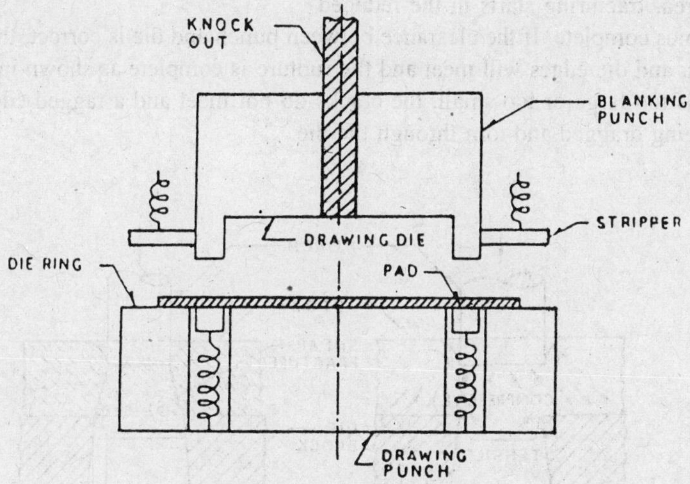

Fig. 4.66. Combination Die.

Progressive Dies. A progressive or follow on die has a series of stations. At each station, an operation is performed on a workpiece during a stroke of the press. Between stroke, the piece in the metal strip is transferred to the next station. A finished workpiece is made at each stroke of the press. A progressive die is shown in Fig. 4.67. While the piercing punch blanks out a portion of the metal in which a hole had been pierced at a previous station. Thus after the first stroke, when only a hole will be punched, each stroke of the press produces a finished washer.

Transfer Dies. Unlike the progressive dies where the stock is fed progressively from one station to another, in transfer dies the already cut blanks are fed mechanically from station to station.

Multiple Dies. Multiple or gang dies produce two or more work- pieces at each stroke of the press. A gang or number of simple dies and punches are ganged together to produce two or more parts at each stroke of the press.

4.7.7. Principle of metal cutting. The cutting of sheet metal in press work is a shearing process. The cutting action is explained with the help of Fig. 4.68. The punch is of the same shape as of the die opening except that it is smaller on each side by an amount known as 'clearance'. As the punch touches the material and travels downward, it pushes the material into the die opening. The material is subjected to both tensile and compressive stresses as shown in Fig. 4.68 (*a*). Stresses will be highest at the edges of punch and die and the material will start cracking there. The various steps in the rupture or fracture of the material can be written as : stressing the material beyond its elastic limit, plastic deformation, reduction in area, fracturing starts in the reduced area and becomes complete. If the clearance between punch and die is correct, the cracks starting from the punch and die edges will meet and the rupture is complete as shown in Fig. 4.68 (*b*). If the clearance is too large or too small, the cracks do not meet and a ragged edge results due to the material being dragged and torn through the die.

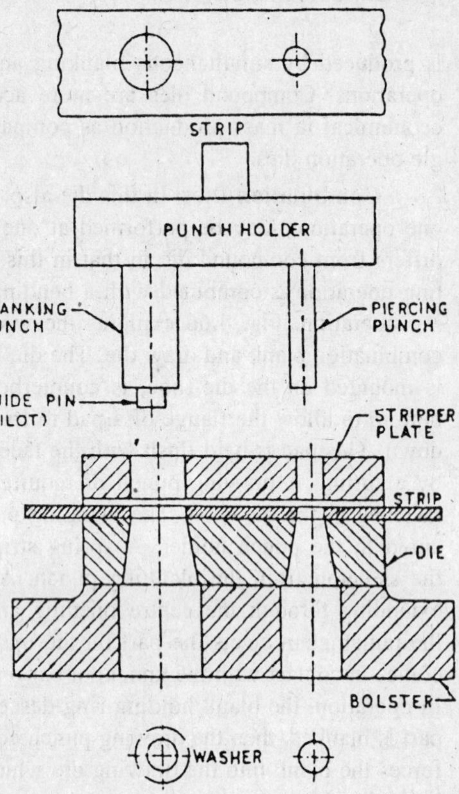

Fig. 4.67. Progressive Die.

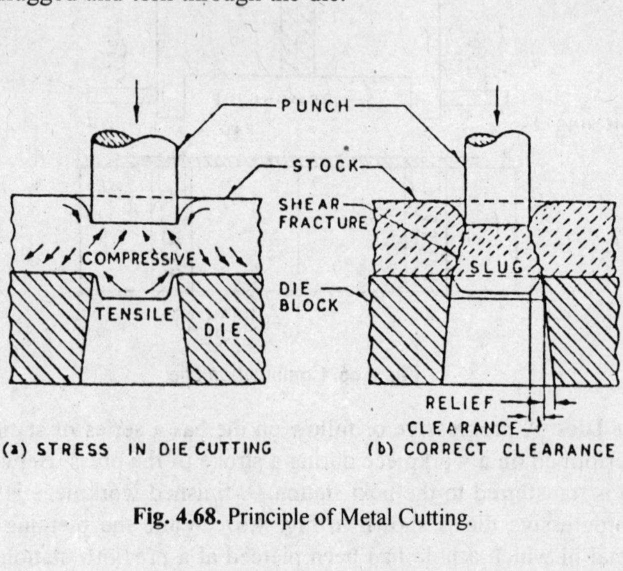

Fig. 4.68. Principle of Metal Cutting.

This is explained in Fig. 4.69.

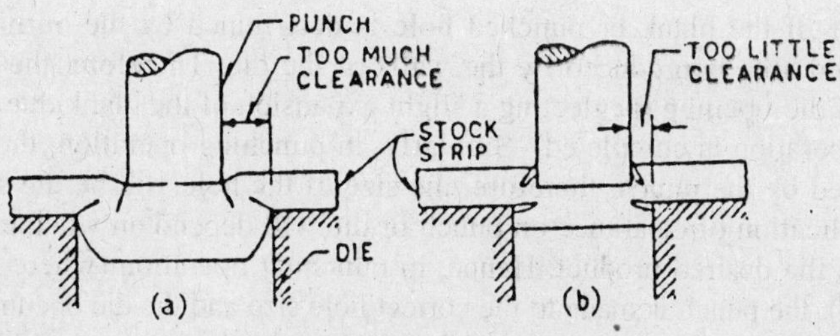

Fig. 4.69. Improper Clearance.

4.7.8. Clearance. As is clear in the previous article, the die opening must be sufficiently larger than the punch to permit a clean fracture of the metal. This difference in dimensions between the mating members of a die set is called 'clearance'. This clearance is applied in the following manner :

(*i*) When the hole has to be held to size, *i.e.*, the hole in the sheet metal is to be accurate (Punching operation), and slug is to be discarded, the punch is made to the size of hole and the die opening size is obtained by adding clearance to the punch size, (Fig. 4.70 (*a*)).

(*ii*) In blanking operation, where the slug or blank is the desired part and has to be held to size, the die opening size equals the blank size and the punch size is obtained by subtracting the clearance from the die-opening size, (Fig. 4.70 (*b*)).

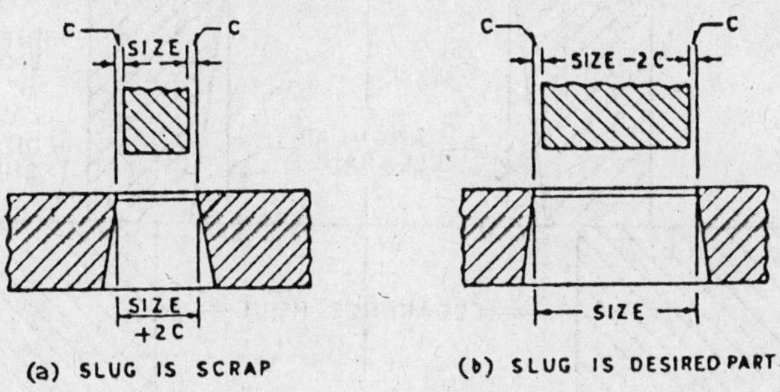

Fig. 4.70. Application of Clearance.

In Fig. 4.70 c, is the amount of clearance per side of the die opening. The clearance is a function of the kind, thickness and temper of the work material, harder materials requiring larger clearance than soft materials, the exception being Aluminium. The usual clearances per side of the die, for various metals, are given below in terms of the stock thickness, t :

For brass and soft steel, $c = 5\%$ of t

For medium steel, $c = 6\%$ of t

For hard steel, $c = 7\%$ of t

For aluminium, $c = 10\%$ of t

The total clearance between punch and die size will be twice these figures. These clearances are for blanking and piercing operations.

The clearance may also be determined with the help of the following equation :

$$c = 0.0032\, t \,.\, \sqrt{\tau_s}, \text{mm}$$

where τ_s is the shear strength of the material in N/mm^2.

The reason behind the application of clearance in the manner as given above, is explained below :

The diameter of the blank or punched hole is determined by the burnished area. On the blank, the burnished area is produced by the walls of the die. Therefore, the blank size will be equal to the size of die-opening (neglecting a slight expansion of the blank due to elastic recovery after the cutting operation is completed). Similarly, in punching operation, the burnished area in the hole is produced by the punch, therefore, the size of the hole will be the same as the punch. Therefore, the application of clearance on punch or die will depend on whether the punched hole or the cut blank is the desired product. Hence, in punching operation (where hole in the strip is the desired product), the punch is made to the correct hole size and the die opening is made oversize an amount equal to die clearance. Similarly, if the blank is the desired product, the die opening size is made to the correct blank size and the punch is made smaller an amount equal to die clearance. In other words, punch controls the hole size and die opening controls the blank size.

A section through blanking die is given in Fig. 4.71, showing clearance, land, straight and angular clearance.

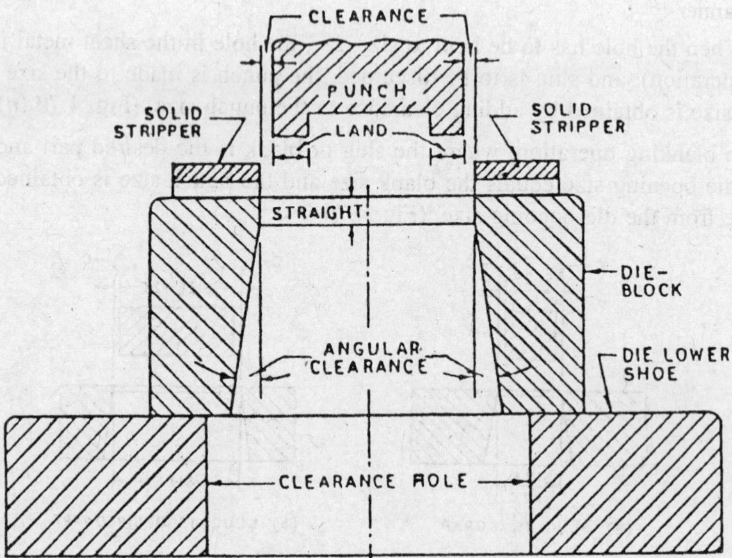

Fig. 4.71. Section through Blanking Die.

Land. It is the flat (usually horizontal) surface contiguous to the cutting of a die which is ground and reground to keep the cutting edges of the punch sharp.

Straight. It is the surface of a cutting die between its cutting edge and the beginning of the angular clearance. This straight portion gives strength to the cutting surface of the die and also provides for sharpening of the die. This straight portion is usually kept at about 3 mm for all materials less than 2 mm thick. For thicker materials, it is taken to be equal to the metal thickness.

Angular clearance. Angular clearance or relief is provided to enable the slug to clear the die. It is provided below the straight portion of the die surface. It is usually $\frac{1}{4}^{\circ}$ to $1\frac{1}{2}^{\circ}$ per side but occasionally as high as 2°, depending mainly on stock thickness and frequency of sharpening.

Punch and Die Clearance after considering the elastic recovery of the material. It has already been explained that after the cutting operation has been completed, elastic recovery of the strip material takes place. In blanking operation, after the release of blanking pressure, the blank expands slightly. The blanked part is thus actually larger than the die opening that has produced it. Similarly, in punching operation, after the strip is stripped off the punch, the material recovers and the hole contracts. Thus, the hole is actually smaller than the size of the punch which produced it. This difference in size due to elastic recovery will depend upon : blank size, stock thickness and stock material. It may be taken as between 0.0125 mm and 0.075 mm. If the stock thickness is upto 0.25 mm, this difference may be taken as zero. For stock thickness between 0.25 mm and 0.75 mm, it may be taken as equal to 0.025 mm and for stock thickness more than 0.75 mm, it may be taken as 0.050 mm. Thus to produce correct hole and blank sizes, the punch size should be increased and the die opening size should be decreased by an amount as explained above.

4.7.9. Cutting Forces. In cutting operation, as the punch in its downward movement enters the material, it need not penetrate the thickness of the stock in order to affect complete rupture of the part. The distance which the punch enters into the work material to cause rupture to take place is called 'penetration' and is usually given as the percentage of the stock thickness.

The percent penetration depends on the material being cut and also on the stock thickness. When a hard and strong material is being cut, very little penetration of the punch is necessary to cause fracture. With softer materials, the penetration will be greater. For example, for soft aluminium, it is 60% of 't' ; for 0.15% carbon steel annealed, it is 38% of 't' ; and only 24% of 't' for 0.5 percent carbon steel annealed. The percentage penetration also depends upon the stock thickness, being smaller for thicker sheets and greater for thinner sheets, as shown in Table 4.1.

Table. 4.1. Penetration

Stock Thickness t. mm	25	20	15	12.5	10	8	6	5	3	2.5	1.6	Below 1.6
Penetration % of t	25	31	34	37	44	47	50	56	62	67	70	80

The maximum force F_{max} in newtons needed to cut a material is equal to the area to be sheared times the shearing strength, τ_s in N/m m^2, for the material. For a circular blank of diameter D mm and of thickness t, mm., the cutting force will be given as,

$$F_{max} = \pi D t \, \tau_s = P.t.\tau_s \ldots \quad \ldots (4.1)$$

where 'P' is the perimeter of the section to be blanked

For rectangular blanks with length L and width b, it is

$$F_{max} = 2(L + b)\, t \,.\, \tau_s \ldots \quad \ldots (4.2)$$

The shear strengths of the various metals are given in Table 4.2.

Table 4.2. Shear strengths of various metals

Metal	τ_s *N/mm²*
Carbon steels :	
0.10% C	245 to 311
0.20% C	308 to 385
0.30% C	364 to 469
High strength low-alloy steels	315 to 446

Silicon steels	420 to 490
Stainless steels	399 to 903
Aluminium alloys	49 to 322
Copper and bronze	154 to 490
Lead alloys	12.8 to 41
Magnesium alloys	119 to 203
Nickel alloys	245 to 812
Tin alloys	20.5 to 77
Titanium alloys	420 to 490
Zinc alloys	98 to 266

Energy in Press work :

Energy in press work or the work done to make a cut is given ideally as,

$$E = F_{max} \times \text{punch travel}$$
$$= F_{max} \times K \times t$$

where K = percentage of penetration required to cause rupture.

To allow for energy lost in machine friction and in pushing slugs through the die etc., the above equation gets modified as,

$$E = F_{max} \times K \times t \times C$$

where the factor C accounts for the amount of extra energy required. This depends upon the circumstances in each case. For general purpose, a factor of 1.16 is recommended.

4.7.10. Methods of reducing cutting forces. For calculating the cutting forces in the last article, it has been assumed that the bottom of the punch and the top of the die block lie in parallel planes and that the blank is severed from the sheet metal by shearing it simultaneously along the whole perimeter. This process is characterized by very high punch forces exerted over a very short time, resulting in shock or impulse conditions. It is usual, however, to reduce cutting forces and to smooth out the shock impact of heavy loads. This is achieved by arranging for a gradual cut instead of sudden cut of the stock. For this, two methods are generally used :

(1) Shear. The working faces of the punch or die are ground off so that these don't remain parallel to the horizontal plane but are inclined to it. This angle of inclination is called 'shear'. This has the effect of reducing the area in shear at any one time and the maximum force is much less. It may be reduced by as much as 50%. This is made clear in Fig. 4.72, which shows the relation of cutting forces to amount of shear.

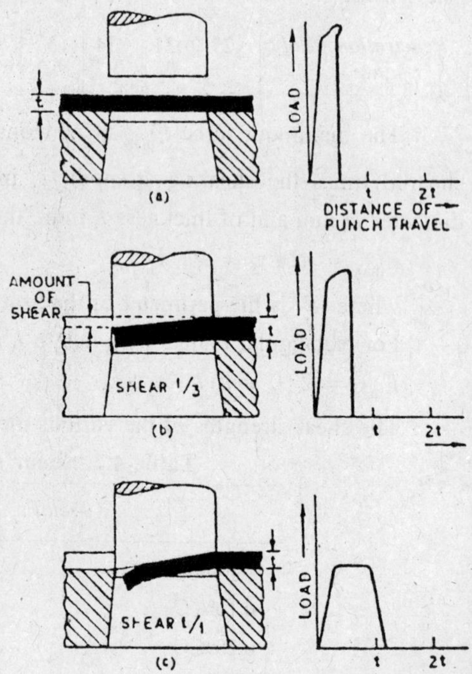

Fig. 4.72. Effect of shear on cutting force.

In fig. 4.72 (*a*), the shear is zero, *i.e.*, the cutting edges are parallel. The material is cut at once on the entire perimeter resulting in maximum load. The force diagram shows a steep rise at maximum load and then sudden load release sometimes severe on both press and dies, as the cut is completed. In Fig. 4.72 (*b*), the face of the punch is ground off so that shear = t / 3. The cutting action will start at the leading edge of the punch and then it will gradually spread to the rest of the punch. With this, only a part of the punch would be cutting at any one instant. While the maximum force will decrease, the energy needed to complete the cut is unchanged. So, the punch travel in this case will be more than in case (*a*). In case 4.72 (*c*), shear = t / 1. When the leading edge has travelled through the stock a distance 't', the trailing edge will state making contact with the material. Maximum force would be at this position of the punch and since the cut is complete, the maximum load would be about half of that when shear is zero. The punch travel will be still greater.

The provision of shear distorts the material being cut. When shear is on the face of the punch, the blank cannot be flat and when shear is on the die, the piercing cannot be flat. So, for blanking operation, shear is provided on the die face and for punching or piercing operation shear is provided on the punch face. Fig. 4.73 shows the various methods of applying shear on the punch and die face. Wherever possible, double shear should be used so that the two shear faces neutralise the side thrusts which each sets up.

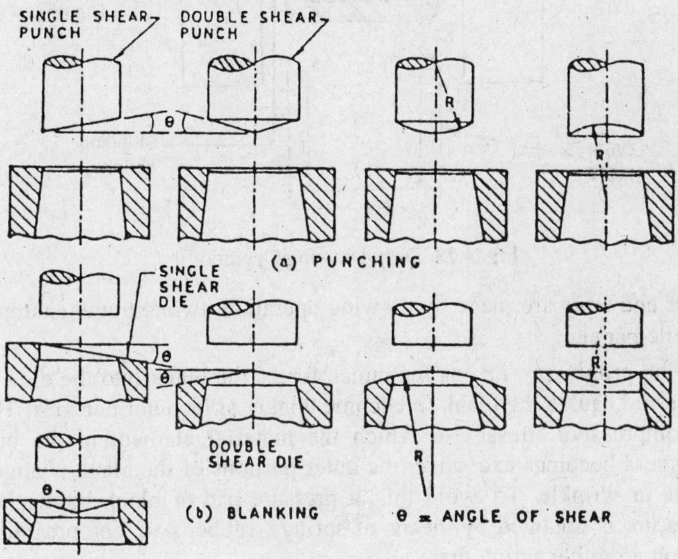

Fig. 4.73. Application of Shear.

It is clear from the above analysis that it is advisable to provide shear on punches and dies. The amount of shear to be applied is a matter of compromise. If the shear is quite big, say $2t$ or $3t$, they the cutting edges of the tools will become too acute and liable to break away easily. However, the shear must be at least equal to the percentage penetration.

(2) Staggering of Punches. An effect similar to shear can be obtained by staggering two or more punches, that is, making them of unequal lengths, that all operate in one stroke of the press. The punches are arranged so that one does not enter the material until the one before it has penetrated through. Like this, the cutting force can be reduced approximately 50%. (Fig. 4.74)

4.7.11. Deep Drawing. Drawing operation is the process of forming a flat piece of material (blank) into a hollow shape by means of a punch which causes the blank to flow into the die- cavity, Fig. 4.75.

The depth of draw may be shallow, moderate or deep. If the depth of the formed cup is upto half its diameter, the process is called "Shallow drawing". If the depth of the formed cup exceeds the diameter, it is termed as "Deep Drawing". Parts of various geometries and sizes are made by drawing operation, two extreme examples being bottle caps and automobile panels.

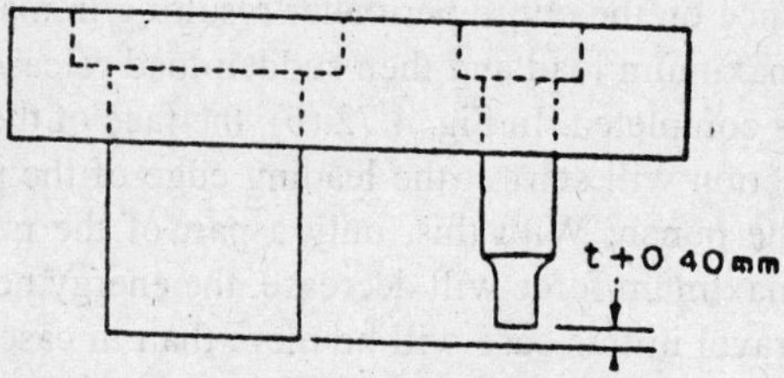

Fig. 4.74. Staggering of Punches.

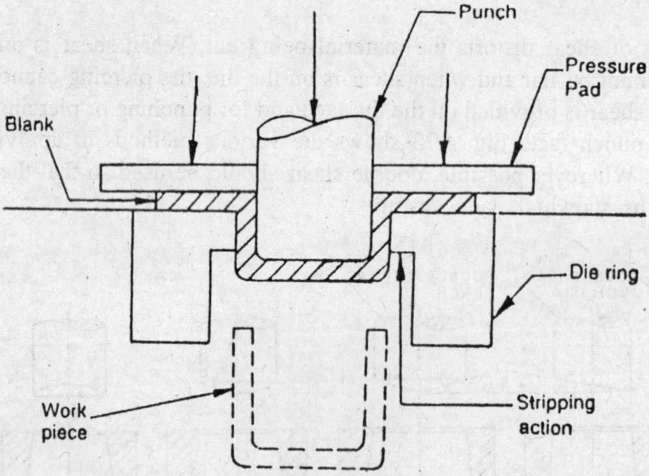

Fig. 4.75. Deep Drawing Operation.

As the drawing progresses, *i.e.*, as the punch forces the blank into the die cavity, the blank diameter decreases and causes the blank to become thicker at its outer portions. This is due to the circumferential compressive stresses to which the material element in the outer portions is subjected. If this stress becomes excessive, the outer portions of the blank (flange) will have the tendency to buckle or wrinkle. To avoid this, a pressure pad or blank holder is provided. The holding down pressure is obtained by means of springs, rubber pad, compressed air cylinder or the auxiliary ram on a double action press.

The portion of the blank between the die wall and the punch surface is subjected to nearly pure tension and tends to stretch and become thinner. The portion of the formed cup, which wraps around the punch radius is under tension in the presence of bending. This part becomes the thinnest portion of the cup. This action is termed as 'necking' and in the presence of unsatisfactory drawing operation, is usually the first place to fracture.

As noted above, the outer portions of the blank under the blankholder become thicker during the operation. When these portions are drawn into the die cavity, 'ironing' of this section will occur if the clearance between the punch die is not enough to accommodate this increased thickness of the workpiece. This ironing is useful if uniform thickness of the product is desired after the drawing operation.

Deep Drawability. Deep drawability or drawing ratio of a metal is defined as the ratio of the maximum blank diameter to the diameter of the cup drawn from the blank (usually taken to be equal to the punch diameter), *i.e.* D/d. For a given material there is a limiting drawing ratio (LDR), after which the punch will pierce a hole in the blank instead of drawing the blank. This ratio depends upon many factors, such as type of material, amount of friction present, etc. The usual range of the maximum drawing ratio is 1.6 to 2.3.

A simple push through drawing die is shown in Fig. 4.76. The drawing punch must be properly vented with drilled passages. Venting serves double purpose. It eliminates suction which would hold the cup on the punch and damage the cup when it is stripped from the punch. Secondly, venting provides passages for lubricants.

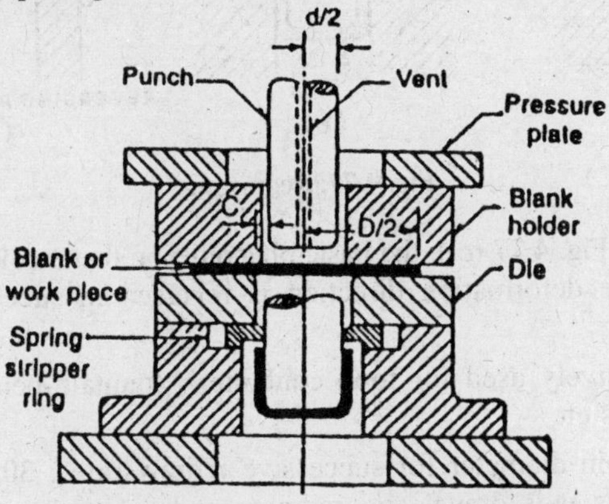

Fig. 4.76. A simple push through drawing die.

As already discussed, the presses used for deep drawing are : double action cam press or double action toggle press. Many presses used for deep drawing operations are hydraulically operated and these presses have an additional hydraulic cylinder and piston for the additional slide.

Product applications of deep drawing process are : Cups, shells, short tubes, automotive bodies, gas tanks, appliance covers and household hardware (spoons, utensils etc.), beverage cans, seamless can bodies.

Redrawing. In deep drawing, the percentage reduction in one draw is defined as :

$$\% \text{ reduction} = \frac{D - d}{D} \times 100$$

now
$$\frac{D}{d} = 1.6 \text{ to } 2.3$$

$\therefore$
$$\frac{d}{D} = 0.435 \text{ to } 0.625 = 0.50 \text{ (average)}$$

$\therefore$ The average reduction in deep drawing

$$= \left(1 - \frac{d}{D}\right) \times 100 = 50\%$$

To make tall cups of smaller diameter (such as cartridge cases and closed-end tubes), it is necessary to use successive drawing operations. Reducing a drawn cup to a smaller diameter and increased height is known as "Redrawing". The two methods used for redrawing are : direct redrawing and indirect or reverse redrawing.

Fig. 4.77 (*a* and *b*) are direct redrawing methods. In Fig. 4.77 (*a*), the metal bends twice inducing high strain hardening in it. In Fig. 4.77 (*b*), the metal again bends twice, but the angle through which it bends is less than 90°. This reduces strain hardening as compared to Fig. 4.77 (*a*) and punch load is reduced. But the drawback is that the initial drawn cup must be made with a taper corner.

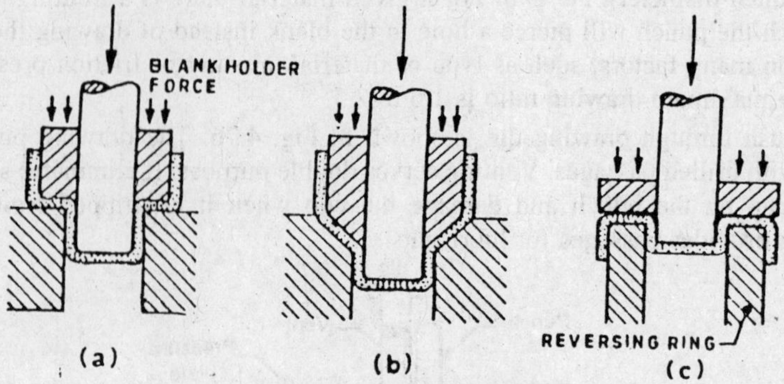

Fig. 4.77. Redrawing.

In reverse drawing, (Fig. 4.77 (*c*)), the basic phenomenon that a cold-worked material exhibits greater ductility when the deformation direction is reversed in successive operations (strain softening) is made use of.

Redrawing is extensively used for food containers, fountain-pen caps, oil filter housings and shock absorber pistons etc.

Practical reductions in diameter for successive redraws are : 30, 25, 16 and 13%. Metal becomes work-hardened as it is drawn and redrawn and must be annealed to prevent failure before it reaches its limit.

Ironing. It is the process of smoothening or thinning the wall of a shell or cup (cold or hot) by forcing the shell through a die with a punch. If the thickness of the sheet is more than the clearance between the punch and die, the thickness will get reduced. This is known as ironing and it produces cups with thinner walls than the base, Fig. 4.78, as in the case of beverage cans and household saucepans.

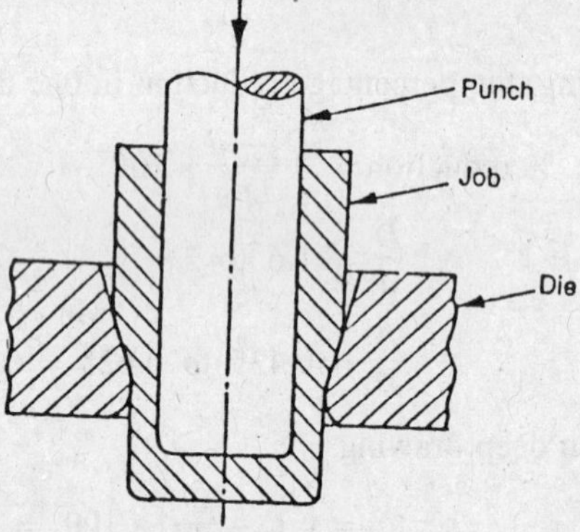

Fig. 4.78. Ironing.

4.7.12. Bending. Bending is the metal working process by which a straight length is transformed into a curved length. It is a very common forming process for changing sheet and plate into channels, drums, tanks etc. During the bending operation, the outer surface of the material is in tension and the inside surface is in compression. The strain in the bent material increases with decreasing radius of curvature. The stretching of the bend causes the neutral axis of the section to move towards the inner surface. In most cases, the distance of the neutral axis from the inside of the bend is 0.3 t to 0.5 t, where 't' is the thickness of the part. Bending terminology is illustrated in Fig. 4.79.

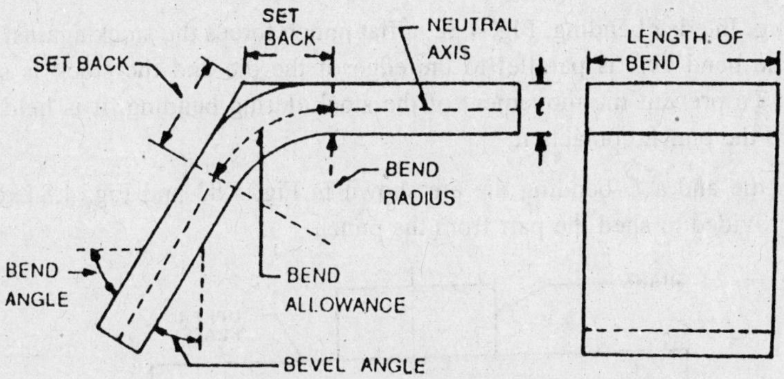

Fig. 4.79. Bending Terminology.

Bending Methods. The three bending methods commonly used are : V- bending, Edge bending, and U-bending (channel bending)

V– Bending. In *V*-bending, a wedge-shaped punch forces the metal sheet or strip into a wedge shaped die cavity, (Fig. 4.80). The bend angle may be acute, 90°, or obtuse. As the punch descends, the contact forces at the die corner produce a sufficiently large bending moment at the punch corner to cause the necessary deformation. To maintain the deformation to be plane-strain, the side creep of the part during its bending is prevented or reduced by incorporating a spring loaded knurled pin in the die. The friction between the pin and the part will help to achieve this, (Fig. 4.81). Plane strain conditions will also be established in the centre of the sheet if its width is more than 10 time its thickness.

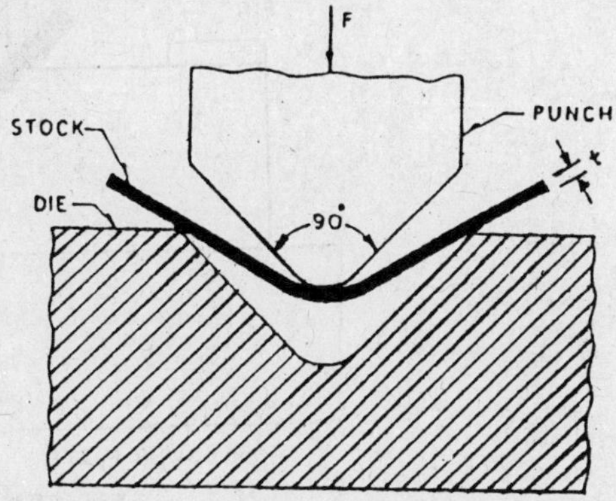

Fig. 4.80. V-Bending.

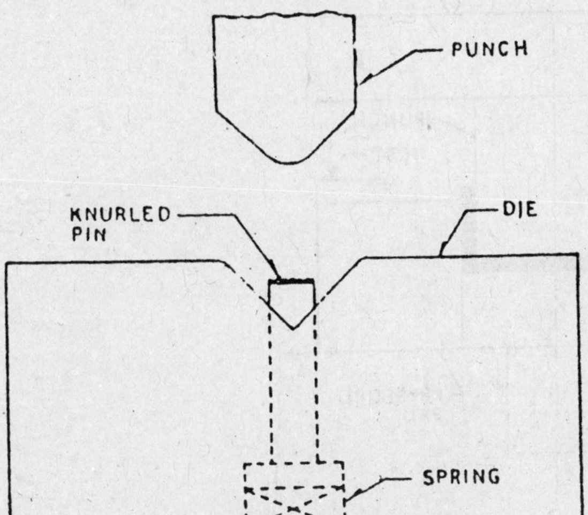

Fig. 4.81.

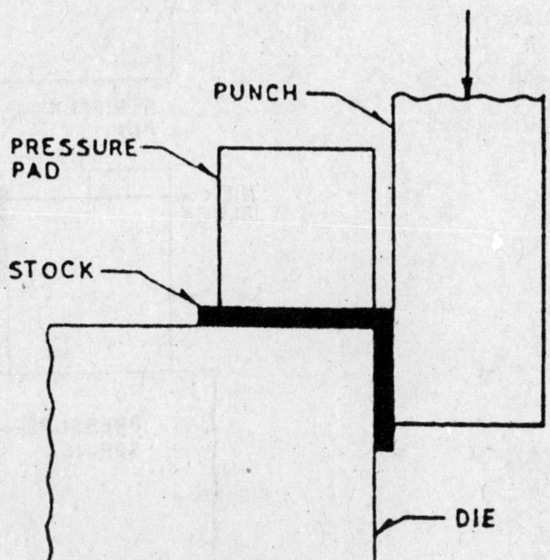

Fig. 4.82. Edge Bending.

Edge Bending. In edge bending, Fig. 4.82, a flat punch forces the stock against the vertical face of the die. The bend axis is parallel to the edge of the die and the stock is subjected to cantilever loading. To prevent the movement of the stock during bending, it is held down by a pressure pad before the punch contacts it.

A *V*-bending die and a *U*-bending die are shown in Fig. 4.83 and Fig. 4.84 respectively. Stripper pins are provided to shed the part from the punch.

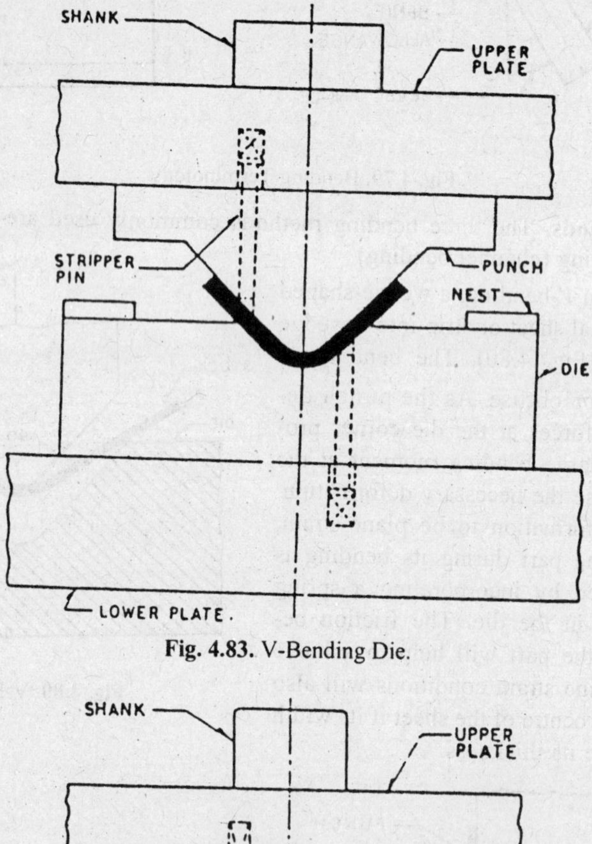

Fig. 4.83. V-Bending Die.

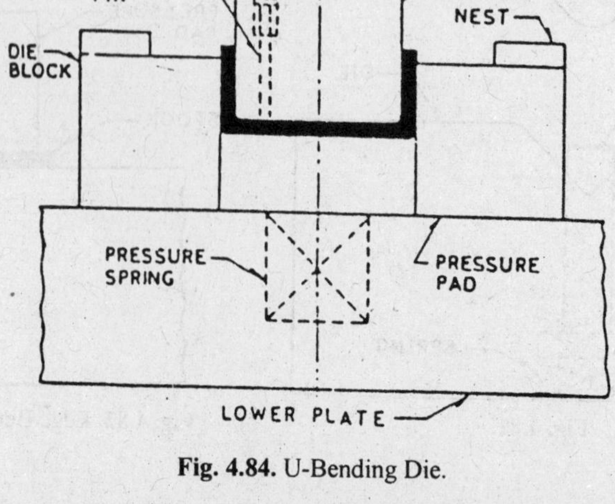

Fig. 4.84. U-Bending Die.

Spring back. At the end of the bending operation, when the pressure on the metal is released, there is an elastic recovery by the material. This causes a decrease in the bend angle and this phenomenon is termed as spring back. For low carbon steel, it can be 1° to 2° and for medium carbon steel, it can be 3° to 4°, for phosphor bronze and spring steel the spring back can be from 10 to 15°. To compensate for spring back, the wedge-shaped punches and the mating dies are made with included angles somewhat less than required in the formed component. Due to this, the component will be bent to a greater angle than desired, but it will spring back to the desired angle. For other types of bending, the part is overbent by an angle equal to spring-back angle by having the face of the punch undercut or relieved. The values of spring back given above are for 90° bends and are usually greater for greater angles.

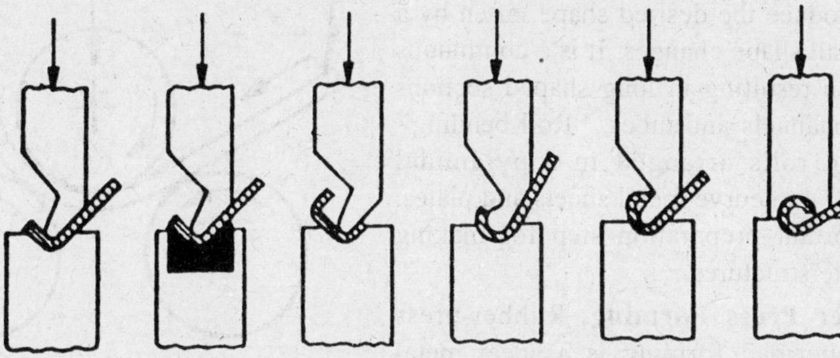

(a) (b) (c) (d) (e) (f)

Fig. 4.85. Press brake Forming.

4.7.13. Press Brake. The bending methods discussed under Art. 4.7.12, refer to bent parts with short lengths. These parts can be bent at high rates in mechanical presses in dies. For longer lengths, presses with very long and narrow beds (9 m long or even more) are used. Such a press is called as "Press brake". The press brake is chiefly used to form long, straight bends in sheets and plates to produce structural shapes such as angles, channels, automobile and truck chassis, and corrugated sheets. The metal is bent between interchangeable dies attached to the bed and the ram. By using different dies, many types of bends may be obtained, (Fig. 4.85).

Fig. 4.86 shows a typical mechanical press brake and a hydraulic press brake constructed by HMT.

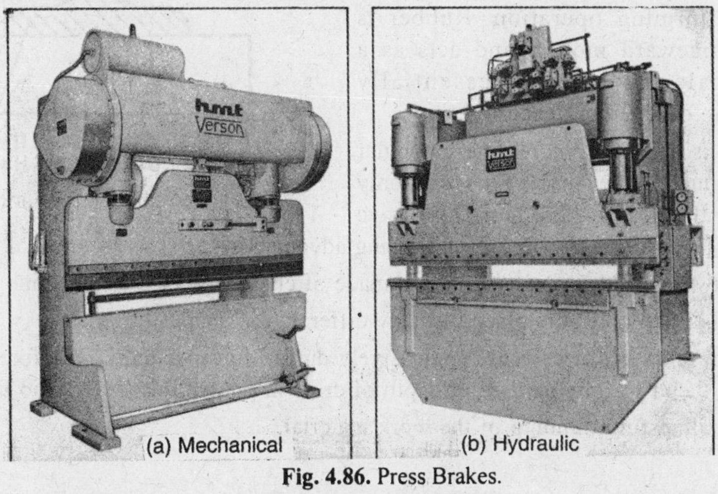

(a) Mechanical (b) Hydraulic

Fig. 4.86. Press Brakes.

4.7.14. Hot Press working of Metals. Hot press working of metals is resorted to for sheet steel over 6 or 8 mm thick, or even for thinner material of low ductility. Hot press working or hot stamping is employed in the manufacture of ship hull components, bottoms and other parts of tank cars, boilers, chemical machinery and apparatus etc. The plate is usually cut up into the required blanks mainly by oxygen cutting techniques. The blanks are preheated in batch-type furnaces. Hot press working is mainly done in hydraulic presses or friction screw presses. In some cases, crank type forming presses are employed for this purpose.

4.7.15. Miscellaneous Forming Operations

1. Roll Forming. On long parts, the punch and die can not be made full length, and forming can be accomplished by "Roll forming", where the material, usually strip stock, is fed through a series of rolls which produce the desired shape, often by a succession of small shape changes. It is a continuous bending operation resulting in long shaped sections such as angles, channels and tubes. "Roll bending" uses three large rolls arranged in a pyramidal fashion, (Fig. 4.87), to curve metal sheets and plates. This is an important preparation step for making large welded plate structures.

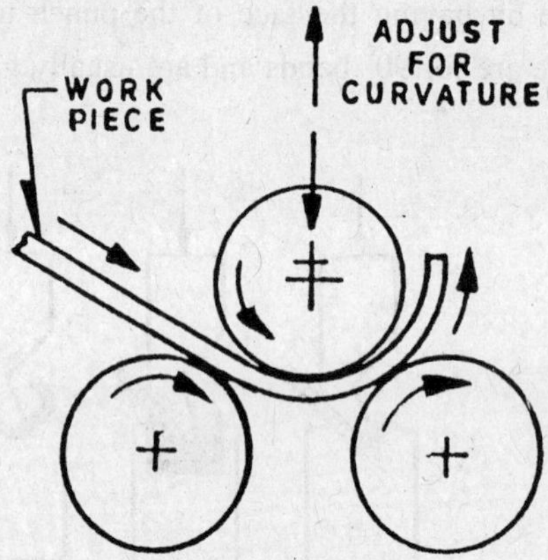

Fig. 4.87. Roll bending.

2. Rubber Press Forming. Rubber-press forming or Rubber-pad forming is a sheet metal forming (bending and stretching or drawing) operation in which either the punch or the die is a piece of constrained elastomer (rubber or polyurethane). The basic principle of the process is that if a block of rubber or polyurethane is placed in a cylinder and pressure is brought to bear upon it by applying a force by a ram of a press, any such force must develop a resultant reaction on every surface with which the elastomer comes in contact. This resultant reaction force is used for forming sheet metal to the required shape using form blocks or punches or dies. Fig. 4.88 shows a schematic view of Rubber-press forming operation. Rubber is restrained from sideward motion and acts as a deformable female die to exert essentially hydrostatic pressure on the workpiece.

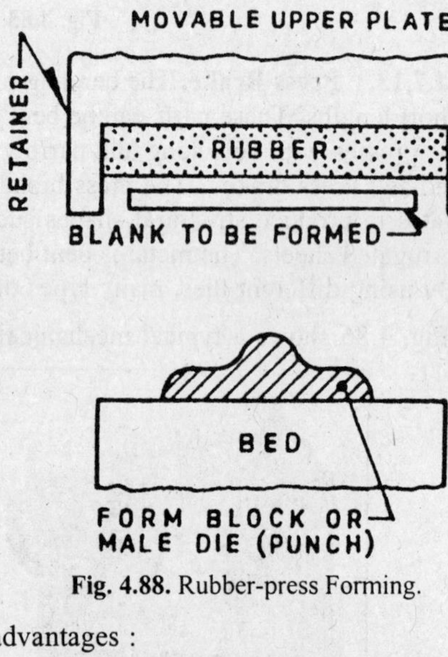

Fig. 4.88. Rubber-press Forming.

This process is becoming increasingly popular for short run production because of its economy since it eliminates the need for the more expensive mating steel dies. The process has got the following advantages :

1. Tools have fewer components and are made of cheap and easy to machine materials.
2. One rubber pad takes the place of many different die shapes.
3. The forming radius decreases progressively during forming unlike the fixed radius on conventional dies, thereby allowing greater depth of draw in one stroke (draw ratio upto 2.7).
4. There is almost no thinning of the work material.

5. Tool setting time is greatly reduced.

6. The variation in sheet thickness is not a limitation as in conventional drawing, since only one tool is used in rubber pad forming.

The limitations of the process are :

1. Rubber pads wear out rapidly.

2. Details of the components formed are less sharp as compared to conventionally formed components.

3. Rubber Hydroform Process. In this process, which is a variation of rubber-press forming process, a pressurized liquid behind the rubber pad is used to exert the force required for forming. Fig. 4.89 shows one such set up. The rubber pad or diaphragm acts as a seal at the end of the liquid container. The part to be formed is wrapped on to the punch due to high hydrostatic pressure developed which acts equally all over the surface of the sheet when the container descends over the punch. Since hydrostatic pressure acts equally all over the surface, deeper cups and complex shapes with sharper details can be achieved.

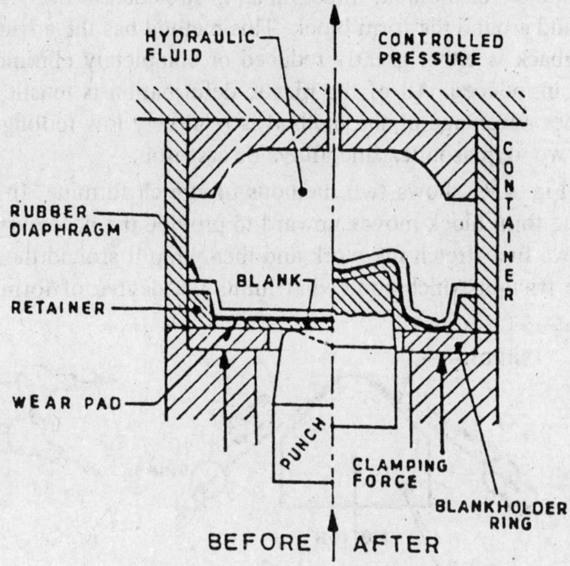

Fig. 4.89. Rubber Hydroform process.

4. Hydromechanical Forming. This process is a modification of rubber hydroform process. Here, the container containing the liquid (oil) is fixed on the bed of the press and the rubber pad or diaphragm acting as a seal is not used. The blank which is directly placed on the draw ring acts as the sealing element, Fig. 4.90. Blank holder descends on the blank and applies

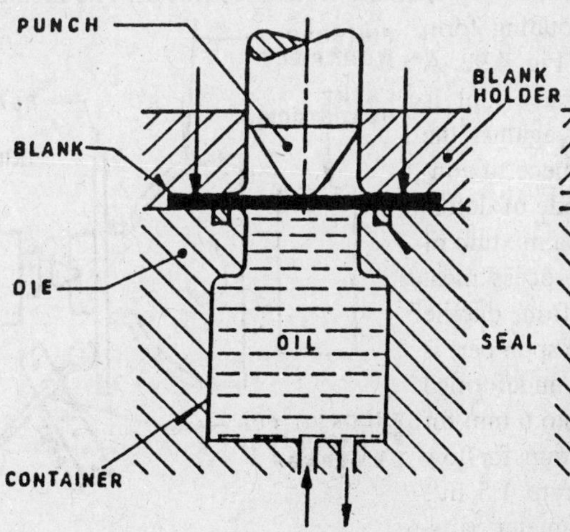

Fig. 4.90. Hydromechanical forming.

the blank holding pressure. As the punch travels into the container, the cup is formed against the hydrostatic pressure developed inside the container. This enables the formation of deeper cups and complex shapes. In this process, since the rubber is not used, the associated problems of wear and replacement etc. do not arise.

5. Stretch Forming. This method was developed during world war II as a means for producing large accurately contoured sheets for aircraft wing and fuselage parts. Since then, the method has been developing rapidly. Like rubber-press forming, stretch forming uses only the male die or form block. Sheet metal is stretched to the yield point in tension, and then wrapped over and around the form block. This method has the advantage over other forming methods that springback is either greatly reduced or completely eliminated, since direct bending stresses are never introduced. All of the plastic deformation is tensile extension in the direction of pulling. Another advantage of the method is relatively low tooling costs. Male dies or form blocks may be of wood, masonite, zinc alloys or cast iron.

Fig. 4.91 shows two methods of stretch forming. In Fig. 4.91 (*a*), the grips are stationary and the form block moves upward to provide the necessary tension and motion. In Fig. 4.91 (*b*), the jaws first stretch the work and then wrap it around the form block. This method avoids most of the friction which somewhat limits the degree of forming obtainable with the moving form block.

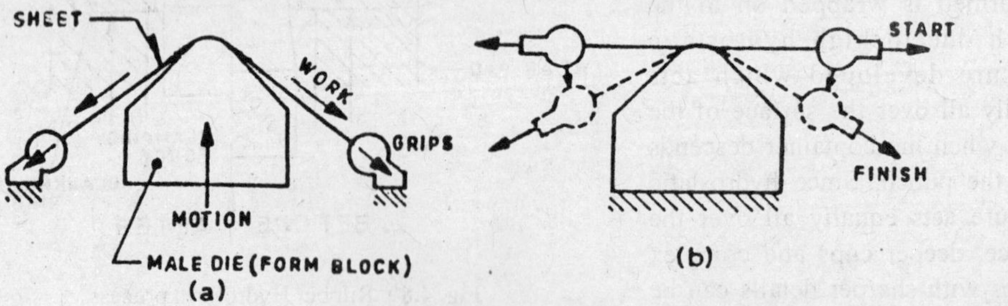

Fig. 4.91. Stretch Forming.

The process is most suitable for low volume production as is typical of air craft industry, for example, for the production of fuselage skins, wing skins and boat hulls etc.

6. Metal Spinning. Spinning is one of the oldest methods of sheet metal forming. Parts that have circular cross-sections can be made by spinning from sheet metal. The method involves the forming of a workpiece over a rotating form block or chuck held in a special lathe, Fig. 4.92. A smooth, hardened, rotating or stationary tool is held by the operator and is pressed against the blank to progressively bend the workpiece to conform to the chuck or mandrel. The mode of deformation of the metal during spinning is a mixture of bending and stretching, making the process most suitable for shaping of hollow parts from ductile metals and alloys. The thickness of the spun part is nearly the same as the thickness of the undeformed blank. The thickness of the blank is upto 6 mm for soft non - ferrous metals and upto 5 mm for low carbon steel. Spinning speeds vary from 1.5 m/s for small parts to 25 m/s for large- diameter parts. Spinning has been used to produce parts more than

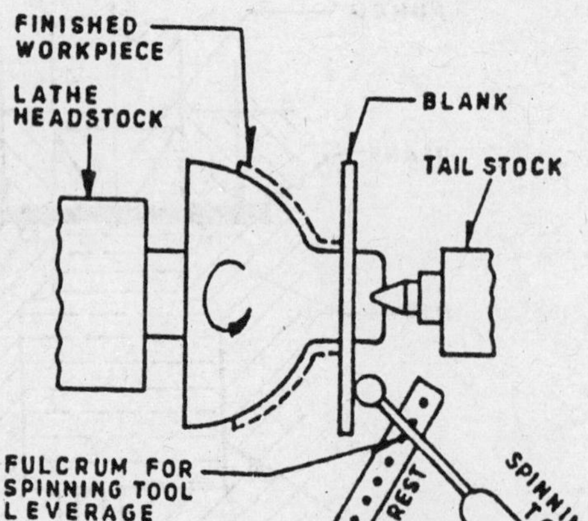

Fig. 4.92. Metal Spinning.

3.6 m in diameter. Before spinning, a suitable lubricant should be applied to the surface of the metal. Soap, beeswax, white lead and linseed oil are commonly used for this purpose. The degree of deformation obtainable by spinning depends upon the shape of the finished part, the material, the lubricant, and particularly the skill of the operator.

Product applications of spinning are : reflectors, kitchenware, bells on musical instruments, light fixtures, funnels and large processing kettles, radar dishes, roket motor cases.

The special features of metal spinning are :

1. Low equipment cost. The equipment consists of a speed lathe to the chuck of which the form block is attached. The blank is held between the formblock and a tailplate.

2. Low tool costs. The form block may be of plaster, wood or metal.

3. Some complex parts of re-entrant shape (kettles, pitchers) can be very economically produced by spinning.

The drawbacks of the method are :

1. The method depends to a large extent on the skill of the operator.

2. Finished parts are not always uniform and close tolerance cannot be obtained.

(*a*) **Power Spinning.** In manual spinning, the skill of the operator plays an important role in getting components of required quality. Also, manual spinning of metals having thickness of more than 1 mm and with complicated contour is difficult to shape. Hence, in order to obtain the economic advantage of spinning and at the same time to reduce the skill required of the operator and to spin thicker metals, spinning machines have been evolved. In these machines, the action of the spinning roller is controlled mechanically (tracer mechanism) or under NC control. This is called as Power spinning.

HMT has developed a Flow Turning Lathe FTL 40, which can be used for power spinning also. It is a semiautomatic spinning and flow turning machine with a hydraulic copying attachment. The forming spinning cycle of the machine is automatic, but the loading and unloading of work pieces is manual. The machine consists of a head stock, a hydraulically operated tailstock, and slides mounted on an auxiliary bed which is fixed to the mainbed. A circular blank is held between the tailstock quill and the spinning mandrel (form block) and the metal is shaped to the required form in a single pass or with a number of successive passes, using a single or multiple templates respectively. There is very little change in the metal thickness and no limitations in the shape of the parts. When the roller control mechanism is set for a particular part, each successive part will then be spun in the same manner.

(*b*) **Comparison of Spinning and Deep Drawing.** Metal spinning is both supplementary and competitive with drawing in presses. Due to its low tooling and equipment costs, spinning is normally used for low volume production. But drawing is preferred for mass production (due to high cost of dies and presses) to reduce the production cost of the component. Labour costs are higher for manual spinning than for press work. Also, production rates are much less. However, as the size and complexity of the part increase, the spinning process becomes more competitive.

(*c*) **Hot Spinning.** Hot spinning of metals is used commercially to dish or form thick circular plates to some shape over a revolving form block. A blank of sheet metal is clamped on centre against a form block or chuck, which is revolved on the spindle of a lathe. A rounded stick or roller is pressed against this revolving piece and moved in a series of sweeps. This displaces the metal in several steps to conform to the shape of the chuck. The pressure may be applied by hand or mechanically. A hot spinning machine is a combination of a vertical press and a large vertical spinning lathe which rotates the blank about a vertical axis. Power actuated rollers do the forming, as the part is rotated. Metal upto 150 mm thick is routinely hot spun into dished pressure vessels and tank shapes. Thinner plates of hard to form metals like Titanium are also shaped by hot spinning.

4.8. SHOT PEENING

Shot peening is mainly employed to increase the fatigue strength of work pieces subjected to impact and/or fatigue loads. The other functions of shot peening are to prevent the cracking of work pieces in corrosive media and to improve the oil retaining properties of the processed surfaces. The process is based on plastic deformation of the surface layer and consists in subjecting the surface to impacts of a jet of shots. Many overlapping indentations are made, causing localized compressive deformation of the surface. Since bulk of the material is not affected, compressive residual stresses are set up. Since fatigue failure occurs due to tensile stresses, the compressive residual stresses greatly offset any tendency to fatigue failure. The surface also gets slightly hardened and strengthened by shot peening (a cold working process). The shots are made of cast iron, steel, aluminium or glass. Cast iron or steel shot is used in peening steel work pieces, and aluminium or glass shot for non ferrous alloys. The depth of the workhardened layer obtained does not usually exceed 2 mm. This depth increases with the diameter of shot (0.4 to 2 mm) and its velocity (60 to 100 m/s), and decreases with an increase in the initial hardness of the workpiece. The efficiency of the process also depends upon the angle between the path of the shot and the surface being peened, and the duration of the peening.

Shot peening is performed in special equipment consisting of a workpiece chamber and a shot blasting device. The most widely used shot blasting devices are :

1. Air - nozzle
2. Centrifugal - wheel type.

In the air-nozzle device, the shot is thrown by compressed air at a pressure of 6 to 8 atm. from several nozzles simultaneously.

The centrifugal wheel device (airless blast equipment) consists of a work piece chamber with a mechanism for moving the work piece under the jet of shot and a centrifugal wheel. The latter is a rapidly rotating bladed wheel which throws the shot centrifugally in the required direction. This process of shot peening is done on the following types of products :- torsion bars, the tension side of automobile springs, other types of leaf springs, all sorts of shafts and axles (sometimes only in key ways), splines and fillets, gear teeth roots after carburizing and case hardening, oil well drill pipe (inside or outside, or both), and automobile connecting rods and crankshafts etc.

Characteristics of the shot peened surfaces are given below :

Increase in hardness = 20 to 50% over the original

Residual compressive stresses set up in the surface layer = 500 to 800 N/mm^2.

Increase in Fatigue life :

Gear wheels = 2.5 times

Coil springs = 1.5 to 2 times

Leaf springs = 10 to 12 times

Roughness of shot peened surface = 3.2 to 0.8 μm Ra.Finish of originally rough-machined surfaces becomes finer and that of finish machined surfaces coarser.

Another peening method is with pneumatic hammer. It is shaped like a spherical striker, its action results indents on the surface worked. The process is used for cold working of stressed areas in large parts prior to finish machining.

Peen Forming : Peen forming is a sheet metal working process utilizing the method of shot peening which has been discussed in detail under Art. 4.8 along with its advantages and applications in industry. Another very useful application is to produce curvatures on thin metal sheets by striking shots on one side of the sheet. The main product application is the forming of smooth and complex curvatures on aircraft wing skins. When the sheet metal is subjected to shot peening, the surface layer expands, while the bottom layer remains rigid. The shots may be of cast sheel about 2.5 mm in diameter striking the sheet surface at about 60 m/s. For thicker sections, the shots diameter can be as large as 6 mm

4.9 SUPER PLASTIC FORMING (SPF)

'Superplasticity' refers to large neck-fee extensions before fracture, (between 100 and 1000%) when deforming certain metals and alloys. Extreme examples of superplastic materials are : Bubble gum and molten glass which can be drawn from melt into glass fibres without the fibres necking down. The preconditions for super plastic behaviour of a metarial are listed under Art. 4.3.12 on "Syer-plastic forging", Materials a high strain-rate sensitivity ($0.3 < m < 1.0$) exhibit pronounced resistance to necking (Here m is stran-rate sensitivity index in the equation $\sigma = \varepsilon^m$; ε is the strain-rate, For glass, m = 1). Generatlly, this condition is satisfied in materials with a very fine grain size (of the order of 1 μm) and at deformation temperatures of > 0.4 Tm. Also, for super-plastic materials, there is a limiting strain rate (generally below $0.0.1 s^{-1}$), at which they should be formed.

Advantages of SPF : – 1. High ductility and Low flow stress (of the order of 5 to 30 MPa) exist at superplastic condition. These lead to lower tooling costs and forging of difficult-to work superalloys (see Art. 4.3.12) or in embossing of fine details in other applications.

2. Forming of complex shapes with fine details and close tolerances.

3. Savings in materials as secondary operations are eliminated.

4. Little or no residual stresses in components.

Draw backs : – (*i*) Very low strain-rates, resulting in large forming times. So, best suited for batch production.

(*ii*) The component material must not be super-plastic at service temperature.

Common super-plastic materials used are : – Zn – 22Al and Ti – 6Al – 4V. Their bulk deformation can be done by – forging, coining, compression moulding (see chapter 5), extrusion and hubbing. Sheet forming of these materials can be done by : thermo-forming, vacuum forming and Blow moulding (see Chapter 5)

4.10. HIGH ENERGY RATE FORMING (HERF, HVF)

These methods have been developed to form parts made of, new temperature-resistant, high strength metals, that require processing in short production runs and that, often, could not be formed by conventional methods. These methods can be grouped under two categories :

1. **High Energy Rate Forming Processes (HERF).** These methods are so called because the energy (mainly chemical energy in explosive materials and electric/magnetic energy) needed for processing the material is released in a very short time, usually in milli- seconds or micro-seconds. The important processes in this category are : Explosive forming (using chemical energy in explosive materials), Electric spark forming or Electro-hydraulic forming, and electro-magnetic forming (using electric energy)

2. **High Velocity Forming Processes (HVF).** In these processes, the metal is deformed by using very high velocities (movement of ram/die). Since kinetic energy is proportional to velocity square, high energy (mechanical) can be delivered to the metal in these methods with relatively small weights (of ram or die). This reduces the cost and size of the machine. Also, due to very high accelerations, high velocities can be obtained by using short strokes of the ram, which makes it possible to increase production rate. In these methods, the energy stored in air/gas (at high pressure) or the stored chemical energy of a hydrocarbon (petrol or diesel) is used to move the ram/die. So, the difference between HERF and HVF is that whereas in former, the energy stored in some medium is used directly to deform the metal, in HVF, it is converted into mechanical energy which imparts high velocities to ram/die. The common processes under HVF are :- petro-forge hammers, Dynapac, clearing hammers, trans- energy and linear induction motor. In HVF, the forming velocities usually range from 30 to 300 m/s, or sometimes higher.

This should be compared with velocity of ram in punch press where it is less than 3 mps and is slightly higher in drop hammers. The velocity of forming cannot be increased indefinitely, as all metals have a critical strain rate above which deformation is seriously reduced. For many high strength materials, the critical velocity is in the range of 300 m/s.

A common feature of the above methods is a high rate of strain (conventional forming and forging processes take place at relatively low velocities and strain rates). With increased rate of forming, the formability of most metals is improved. Also, spring-back is minimal in these methods. Although, these methods are mostly applied to forming of sheet and plate products, often of very large size, these can be applied to certain-die forgings.

4.8.1. Explosive Forming. In explosive forming, the chemical energy stored in explosives is used to process the material. A charge, detonated either above or within a pressure transmitting medium generates very high pressures, which can be used to form parts of large size. The pressure (shock waves) developed will depend upon the compressibility of the energy-transmitting medium. Water, being less compressible than air is a better medium. There are two basic systems of explosive forming :-

(*a*) Confined system, (*b*) Unconfined system

1. **Confined System, (Fig. 4.93).** This system is basically used for small and tubular parts for flaring and bulging operations. This system is not used much due to the following reasons :

(*i*) Being a closed-die operation, there is a greater hazard of die failure due to high pressure genera-ted.

(*ii*) Die erosion is a big problem.

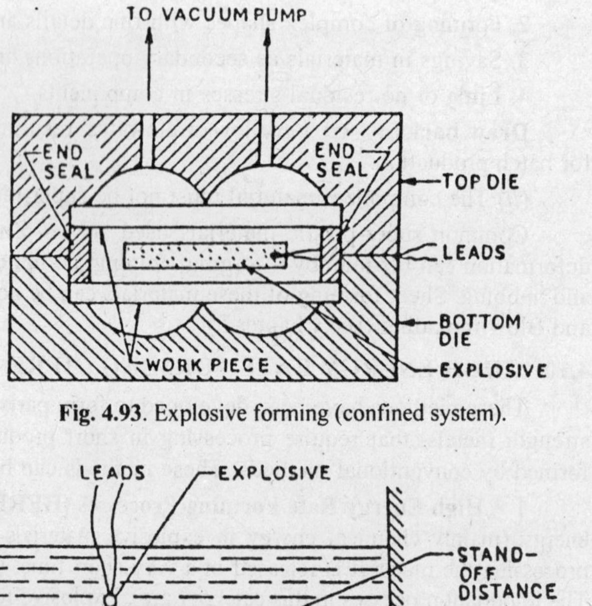

Fig. 4.93. Explosive forming (confined system).

In this system, a die in two or more pieces is used which completely encloses the workpiece. When the explosive is detonated, a shock wave is produced in the energy transmitting medium (in this case, air) which forces the workpiece into the die. Vacuum should be created in between the die and the workpiece, otherwise, the trapped air will be subjected to virtually adiabatic compression, resulting in very high temperatures which can burn or melt the workpiece or the die. The trapped air will also prevent the complete forming of the part.

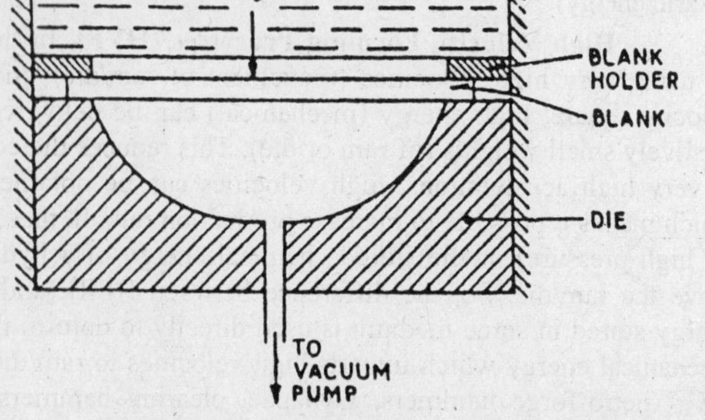

Fig. 4.94. Explosive Forming (Unconfined system)

2. **Unconfined System, (Fig. 4.94).** This system is used for

forming large parts from very thin sheet metal upto large ship plates about 25 mm thick. This method has been used for many years in ship building. The workpiece or blank is clamped to the die and the entire die-work assembly is lowered into a swimming pool. An explosive is denoted above or within the water. The shock waves are produced in the body of water (energy transmitting medium) which in turn forces the workpiece into the die. Again, vacuum should be drawn in between the workpiece and the die.

The important process varia-bles are : type and amount of explo-sive, the distance between the explo-sive and workpiece, the type of energy transmitting medium, and the workpiece.

Use of water as the trans-mitting medium has the following advantages :

1. Since air is very much more compressible than water, the shock pre-ssures produced are very much lower.
2. With water, the noise level of explosion is reduced.
3. Possibility of damage to the workpiece from particles of explosive and detonator is greatly reduced.

Basically two types of explosives are used :

(*a*) Low pressure explosives or firing explosives, such as smokeless powder and black powder. With low pressure explosives, known as cartridge system, the expanding gas is confined (confined system) and the pressure may built upto 700 MPa.

(*b*) High pressure explosives or detonating explosives. The most widely used detonating materials are :- dynamite, amatol, TNT (tri-nitrotoluene), RDX, Research and Developement Explosive (Cyclotrimethyl- enetrinitramine), Tetryl and PETN (Pentaerythriol tetranitrate) and their mixtures. With detonating materials which are preferred for explosive forming process (unconfined system), the gas pressure may be upto 20 times of that for low pressure explosives. The approximate deformation velocities range from 9 to 230 m/s.

The distance between workpiece and explosive charge (called as "Stand off" distance) or that between the explosive and the water level is also important, since, if the latter distance is small, the shock waves could vent to the surface and lose much of their energy. Greater force can be exerted on the workpiece if the "stand off" distance is smaller. Thus, small stand offs are used for deep drawing and large stand offs for shallow drawing. A too small a stand-off distance will result in excessive local deformation and thinning of the workpiece.

The evaluation of the effects of stand-off distance are usually (normally) based on the criterion that the optimum distance is that at which a central charge will produce the maximum deformation in free forming. Since the optimum distance depends on several factors such as the type of explosives used and the geometry of the tank-workpiece system, it is not possible to define a universal optimum distance. However, as a general guide, where a single point charge is used, the stand-off distance is given as,

S = Diameter for circular jobs of relatively small dimensions, say less than 60 cm.

= should be decreased to the order of 0.5 diameter for larger jobs.

The distance 'S' has a marked effect on the shape of freely formed parts. Generally, the greater S, the more nearly conical is the formed shape. As S is decreased to the order of 0.2 diameter of the workpiece, a more rounded shape approaching an ellipsoid or segment of a sphere is produced.

It has been suggested that for high efficiency, the depth of water over the explosive should be at least twice the stand-off distance.

In explosive forming, spring back is minimised. Less spring back occurs with :

(*i*) the use of sheet explosives close to the workpiece.

(*ii*) high clamping forces on the hold down areas.

(*iii*) the absence of lubricant.

Thick materials exhibit less spring back than do parts made from thin blanks.

Depending on the number of parts to be made, the die-material can be : concrete, wood, epoxy, plastic, plaster of paris, C.I. steel, aluminium alloys and various composite materials. Because of the nature of the process, safety is a very important consideration.

The various die materials with their areas of application are given below :

Concrete	:	Medium pressure and large parts.
Cast Iron (ductile)	:	High pressure and many parts.
Concrete and epoxy resin	:	Low pressure and large parts.
Concrete and fibre glass	:	Low pressure and large parts.
Kirksite	:	Low pressure and few parts.
Kirksite and fibreglass	:	Low pressure and few parts.

Explosive forming has got the following fields of application :

1. For the production of components having a wide-range of sizes and shapes, like radar dishes, and rocket motor cases etc.

2. The capital cost being low and the rate of output limited, the method is economically suitable only for low volume production and for the production of prototype parts.

3. For the production of very large parts with re-entrant sections, which are not possible by conventional methods.

4.10.2. Electro-Hydraulic Forming. The effect produced in explo sive forming (generation of high intensity shock waves), can also be obtained by discharging stored-electrical energy in a bank of condensers, across electrodes submerged in an electrolyte (usually water). This is the principle of operation of "electro- hydraulic forming" which is also known as "electric spark forming", "electric discharge forming" or "underwater spark forming".

The stored electric energy can be discharged either through a wire or across a gap. A potential difference of 50 kV can jump a gap of about 25 mm. When this discharge takes place under water, the arc (spark) produced converts water into steam. This generates high pressures which are utilized for forming the workpiece. Again, if a potential difference of 30 kV is discharged through a wire of 1 mm diameter, in water, the centre of the wire will instantaneously be raised to about 5100°C. The wire melts and vaporizes resulting in very high pressures and formation of water bubbles. The flow of current is temporarily disrupted due to formation of

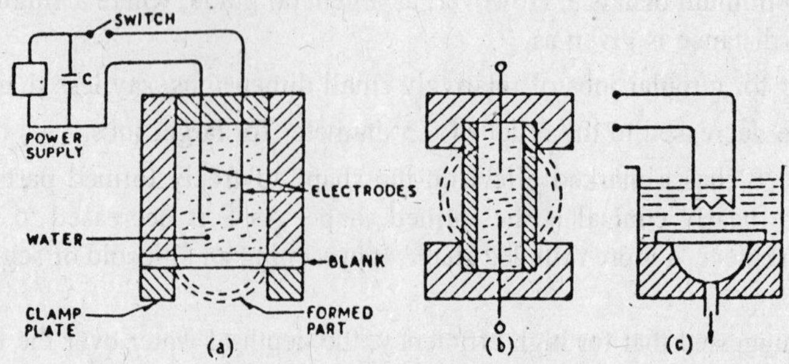

Fig. 4.95. Electro-Hydraulic Forming.

vapour. As the water bubbles expand, vapour pressure drops and an arc is struck between the electrodes. Due to this, high pressure shock waves are generated, which are used to form a workpiece. The second method is more efficient than the first, and also low voltage is needed. Typical set ups of "electro-hydraulic forming', are shown in Fig. 4.95. In Fig. 4.95(*a*) and (*b*) the parts are formed freely in air. However, a female die can be used, Fig. *c*, to produce a specific shape. In this case, arrangements must be made to remove air from in between the workpiece and die. This process can be used for : forming, bulging, beading and drawing. The main advantage of this process is that large amounts of energy can be directed to isolated areas of the workpiece. The main limitation of the process is the critical impact velocity of the material to be formed. Materials with critical impact velocities less than 30 m/s are not suitable for this method.

The die material can be as given below :

For small quantity production (upto about 12) : Kirksite, epoxy resin, plaster of Paris, provided the die is well supported to prevent cracking or deformation,

For large. quantity output : Steel.

Electro-hydraulic forming has got the following fields of application :

1. It is a versatile process and can be used for the forming of tubular and dished shapes. Shapes impractical by conventional methods can be made by this method.
2. Since only female die is needed, tooling costs are substantially reduced. The die materials being also of low cost, the method is economically more attractive for short run and prototype production.
3. Production rate is higher as compared to explosive forming.
4. Only relatively small components can be made.

4.10.3. Electro - Magnetic Forming. Electro-magnetic forming, also called magnetic pulse forming, is based on the fundamental principles of electrical technology *viz.*, when a current flows in a conductor, a magnetic field is set up around it. If the current and hence the magnetic field changes, a current is induced in any other conductor placed in the magnetic field. The direction of this induced current is such that the induced magnetic field opposes the magnetic field producing it. The two conductors get repelled due to the interaction of the two magnetic fields. In electro-magnetic forming, one conductor is a coil, through which current is suddenly discharged, giving rise to a rapidly changing magnetic field. The work-piece is the second conductor, in which the eddy current are set up. The repelling forces between the coil and the work-piece are high (of the order of 350 N/mm^2 for several micro/seconds), and since the coil is held rigidly in place, the work-piece is repelled and forced against a die. Fig. 4.96 shows the

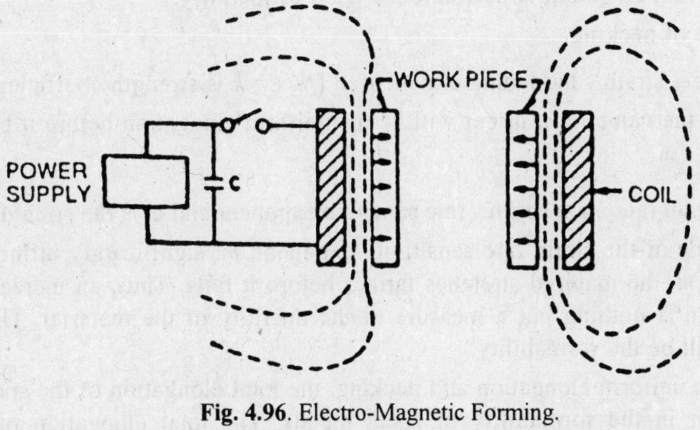

Fig. 4.96. Electro-Magnetic Forming.

basic circuit for electro-magnetic forming. The energy source is a bank of capacitors, charged to a predetermined voltage. The electric energy is discharged through a forming coil. By suitably designing the coil, the process can be used for compression, expansion or for forming contours from sheet metal and tubes. For example, for compressing a tube, the coil is wound around the tube which is contracted onto a mandrel and for expanding, it is inserted inside of a tube surrounded by die. The product applications of this process are : it works effectively on relatively thin materials and is particularly suitable for ; bulging of tubes, shrinking of tubes, flaring and swaging tubes over rods etc. The work-piece must be an electrical conductor, but it need not be magnetic. In this process, since the pulse duration is very short, sufficient pressures are not generated for forming thick material; so, its use is only limited to sheet metal working.

To minimise the magnetic cushion effect, the die materials should be of low electrical conductivity. Due to this, the dies are made usually of steel or epoxy resin. To ensure good reproduction of detail and prevent distortion, air is evacuated from the die. Steel dies have longer life than resin dies but with these, the cushion effect is not entirely prevented.

Electro-magnetic forming has got the following characteristics :

1. The process is simple, clean and safe to operate. The operating costs are low. A high cycle rate can be attained and the process is suitable for automation.

2. Deep drawing is not possible, because of the very short duration of the pressure pulse.

3. Asymmetrical parts are not suitable for this method.

4. Irregularities such as slots, holes and slits present problems, because the magnetic field is disrupted.

4.11. SHEET-METAL FORMABILITY :-

In sheet metal forming process, a blank (cut from a larger sheet) is formed into various shapes. All sheet metal forming operations involve a pair of tools (punch and die) to form the blank into the desired shape. During sheet metal forming operation, the material undergoes two basic modes of deformations : mainly stretching and also drawing or bending. Before selection a forming process, it is important to determine whether the blank can be formed into the desired shape withut failure or not. This is decided on the basis of what is known as ''Sheet-metal formability''. It may be defined as the ability of the sheet metal to undergo the desired shape change without failure, such as by necking or tearing.

The sheet metal formability depends upon certain sheet metal characteristics as discussed below : -

Elongation :- Since the material undergoes stretching and drawing/bending, during forming, large uniform elogation is desirable for good formability.

At the oneset of necking.

True strain, ε = strain - hardening exponent, n ($k. \varepsilon^n$; k is strength coefficient)

Thus, higher the value of n, larger will be the uniform elongation before it begins to neck.

Again, $\sigma = C\varepsilon^m$.

where ε is strain rate, m is strain - rate sensitive exponent and C is the strength co-efficient.

The magnitude of the strain rate sensitivity exponent, m, significantly influences necking. Wth an increasing m, the material stretches farther before it fails. Thus, an increasing m delays necking. Elongation is nothing but a measure of the ductility of the material. Thus higher the ductility, higher will be the formability.

In addition to uniform elongation and necking, the total elongation of the specimen is also a significant factor in the formability of sheet metals. The total elongation of the meterial increases with increasing values of n and m.

Sheet metal grain size is also important. Grain size affects mechanical properties and it influences the surface appearance of the formed part. The coarser the grains, the rougher is the surface appearance.

Formability is also dependent upon; thickness of sheet, friction and lubrication at the punch-die-sheet metal interface, anistropy and yield point elongation.

It has been proved that the thicker the sheet, the larger is its formability. However, in actual practice, a thick blank may not bend as easily around small radii without cracking.

With well lubricated interface between sheet and punch/die, the strains are more uniformly distributed over the punch. Surface scratches, deep cuts and blemishes can reduce formability and cause premature tearing and failure.

Again, the formability of a metal is a function of its tensile strength. The tensile strength and yield point must not be high or too much work will be required for forming the metal.

Sheet Formability Test:- During the various sheet metal forming processes, the stress and strain developed in the meterial are quite complex. So, any measure of the formability of the sheet metal on the basis of material properties derived from a simple tension test (for example, % elongation an index of ductility of the meterial) is not very accurate. Over the years, a number of tests have been developed to evaluate the formability of a sheet metal. These tests are discussed below :-

Cupping Tests :- These are the earliest developed tests. The various tests in this category are :-

1. **Erichsen Test :-** Here, the standard sheet metal specimen (90 mm wide) is clamped between two ring dies having 27 mm diameter openings. A stell ball or round punch (of 20 mm diameter) is then hydraulically forced against the sheet until a crack begins to appear on the bottom face of the specimen as it is stretched into the die opening (Fig. 4.97). The depth, d, is a measure of the formability of the sheet.

2. **Olsen Test :-** This test is just similar to the Erichsen test with one difference that the die opening size is 50 mm.

3. **Swift Test :-** Swift flat bottom cup test is a standard test for deep drawing process. The sheet formability is measured in terms of Limiting Draw Ratio, LDR (See Art. 4.7.11).

4. **Fukui Test :-** Erichsen and Olsen tests mainly evaluate the stretchability and the Swift test the deep drawability of the sheet metal. However, the actual conditions during any forming process are quite complex (mainly a combination of stretching and drawing/blending). The Fukui test uses a hemispherical punch and produces a conical cup. Thus, this test provides a combination of both stretching and deep drawing.

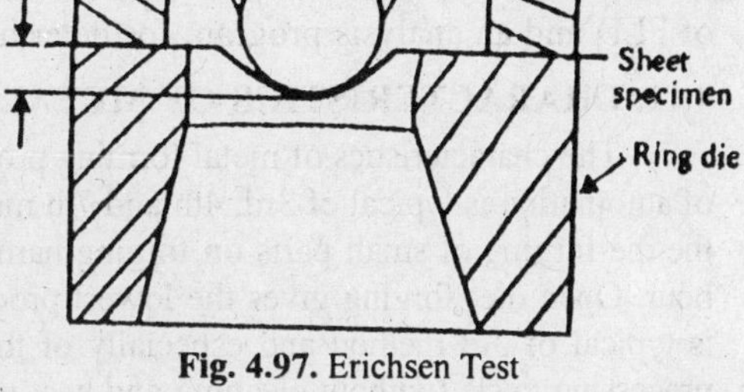

Fig. 4.97. Erichsen Test

The latest test methods which nearly relate to the actual conditions are discussed below :-

5. **Forming Limit Diagrams, FLD :-** This is a very useful technique for controlling failure in sheet metal forming and thus giving an accurate forming limit criteria. The surface of the sheet metal specimen is covered with a grid-pattern of circles (typically 2.5 mm to 5 mm in diameter), using electrochemical or photoprinting techniques. During testing, when the

sheet gets formed, the circles distort into ellipses. The major and minor axes of an ellipse give the two principal strain directions during forming. These strains are determined as the % change in the lengths of major and mior axes. These strains at any point on the surface of the sheet are then compared with the Keeler Good win diagram for the material (See Fig. 4.98). Strain position

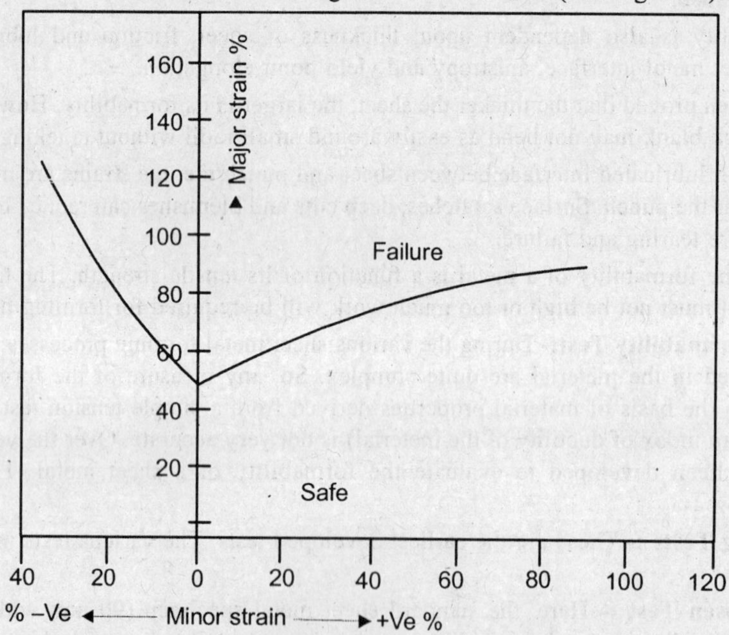

Fig. 4.98. Forming Limit Diagram

above the curve indicate failure. For enhanced accuracy, the diameter of circles in the grid can be reduced. (The failure curve for the tension-tension region was determined by Keeler and that for the tension-compression region was determined by Goodwin). The strain distribution can be altered (for example, by changing the die-radius) so that the strain position moves to the safe region.

6. **Stretch-Draw Shape analysis :-** Here, the forming limit of the sheet metal is determined by the Olsen and Swift cupping tests. Then the specimen is broken into simple shapes and the % of stretch and draw are determined from the geometry. Then, with the help of FLD for the material, the degree of severity of the part can be determined.

7. **Formability Charts :-** Very useful formability charts have been developed for many processes, such as, brake forming, linear stretch forming, rubber forming, dimpling, beading and metal spinning etc.

8. **Computer-analysis Program :-** Just like the bulk deformation processes, computer-aided design and analysis have been applied in determining the sheet metal forming limits. The total program includes : Constitutive relations for the material, a program for analysis of FLD and an analysis program for determining the critical strains.

4.12. CHARACTERISTICS OF METAL FORMING PROCESSES

The characteristics of metal forming processes are given in tables 4.3 to 4.5. The greatest extent of automation is typical of 3rd, 4th and 7th methods and to a lesser extent of 1st and 2nd methods. In the die-forging of small parts on forging hammers and presses, upto 1000 parts can be obtained per hour. Open die- forging gives the lowest production rates. The greatest metal utilization factor (0.9) is typical of 3rd method and especially of the 4th method, where this factor is unity. The shortest processing cycle (without cleaning and heat treatment) is provided by the 3rd and 4th methods.

PROBLEMS

1. Define the process of Mechanical working of metals.
2. What is the starting material for mechanical working of metals ?
3. Enumerate the advantages of mechanical working of metals over other manufacturing processes.
4. Differentiate between hot working and cold working of metals. Bring out the advantages and disadvantages of each of these techniques.
5. Discuss "worm working" of metals.
6. Define the Rolling process.
7. What is the difference between a bloom and a billet ?
8. What is the difference between plate, sheet, strip and foil ?
9. What is the difference between wire and rod ?
10. Define "Slab".
11. Why is metal cold worked ?
12. Explain what happens to a metal, when it is cold worked.
13. Why a cold worked metal is annealed ?
14. What is angle of bite in rolling ? On what factors does its value depend ?
15. Describe and specify the merits and limitations of the different kinds of rolling mills.
16. Sketch and explain the working of "universal rolling mill" and "Planetary rolling mill".
17. What is "Steckel rolling" ?
18. Why are a number of passes required to roll a steel bar ?
19. Describe what occurs in metal when it is rolled.
20. How is cold rolling done ? What benefits are obtained from cold rolling metal ?
21. Why for cold rolling, a four high rolling mill is usually used ?
22. Define forging process.
23. Give the advantages and drawbacks of forging process.
24. Describe the common types of forging hammers.
25. How the size of a forging hammer is specified ?
26. How are a forging hammer and drop hammer alike and how do they differ ?
27. What is the difference between hammer forging and drop forging ?
28. What is Press forging ? How does it differ from drop forging ?
29. What is upset forging and how it is done ?
30. Discuss the rules which govern the flow of metals in upset forging process.
31. How are the sizes of presses and upset forging machines specified ?
32. What is roll forging and how it is done ?
33. What is rotary forging or swaging and how it is done ?
34. With the help of a suitable sketch, explain the working of "Board drop hammer".
35. Describe the common defects in forgings and write about the causes of each.
36. Describe the various methods for cleaning the forgings.
37. Describe the various methods for heat treatment of steel forgings.
38. What is the purpose of heat treatment of forgings ?
39. Define "Forgeability". On what factors does it depend ?
40. List the advantages and limitations of liquid forging.

Table 4.3 Metal Forming Process—Design Considerations

S. No.	*Process*	*Choice of Materials*	*Complexity of part*	*Maximum size*	*Minimum size*	*Mechanical Properties*
1.	Drop Forging	Medium-many alloy are forgeable	Moderate	Large	Small even fraction of a kg.	High
2.	Press Forging	Medium-Best for non-ferrous alloys	Limited, but better than drop forging	Moderate : 0.012 to 0.014 t	Smaller than drop forging	High
3.	Upset Forging	Medium	Limited to cylindrical shapes	Medium 250 mm bar about largest	Moderate	High
4.	Cold headed parts.	Narrow-Steel and highly ductile alloys	Limited-less than forgings	Small-12.5 cm by 1.25 cm diameter usual maximum	Small : 3 mm dia parts	High
5.	Extrusion	Restricted-Light metals, some steels, Cu and Til.	Limited-can be complex in cross-section only	Medium 20 to 25 cm dia maximum	Small-1.25 mm sections in Al and Mg	Good
6.	Impact Cold Extrusion	Narrow-Al, Tin, Mg, Lead	Limited-must be concentric	Small-15 cm dia in soft alloys, 10 cm in hard.	Moderate : 18 mm dia smallest	High
7.	Roll Forming	Narrow-Cold rolled steel; Some Al and stainless steel	Restricted to thin sections and uniform cross-section	Large	Sections from 3mm up	Good
8.	Press Working	Wide-Includes all workable metals	Limited-many design restrictions	Large	Small sections as thin as 0.075 mm possible	Fair to High
9.	Metal Spinning	Wide-Many sheet metals can be spun	Limited-cylindrical or concentric shapes	Large : Upto several m in dia	Moderate-6 mm in dia in gauges less than 1 mm	Good

Table 4.4 Metal Forming Process-Design Considerations (Cont.)

S.No	Process	Precision and Tolerance	Special Structural properties	Surface smoothness	Surface Details	Getting into production	Rate of output	Remarks
1.	Drop Forging	± 0.25 to 0.75 mm	Grain flow provides toughness	Fair	Fair	Slow-dies require much work	Medium-120 per hour on small parts	Used for high strength
2.	Press Forging	Medium-Better than drop forging	Same	Same	Same	Same	Medium-Slower than drop forging	Greater complexity than drop forging
3.	Upset forging	Medium-comparable to press forging	Toughness	Medium	Medium	Slow	Medium	Best Suited to Small parts
4.	Cold Headed parts	± 0.25 mm common æ 0.05 mm possible	Toughness	High	Fair	Fair-dies relatively simple	Extremely high	One of fastest Process
5.	Extrusion	± 0.125 to 0.5 mm	Grain flow improves properties	Good	Only as part of Contour	Moderate-dies relatively simple	High	Sometimes used as blanks for other processes
6.	Cold Impact Extrusion	± 0.25 mm	Same	Good	Good	Same	High-Upto 2000/h	Can eliminate machining.
7.	Roll Forming	± 0.05 to 0.4 mm	Cold work improves properties	Good	None	Slow-roll takes long to be made	High	Requires large quantities
8.	Press Working	± 0.25 mm common	None	High	Fair	Slow-dies require several weeks	High-Upto Several thousand	Low cost for mass production
9.	Metal Spinning	± 0.40 to 1.5 mm	Grain flow and Cold work improves properties	Good	None	Fast-forms can be made quickly	Slow 12 to 30 per h.	Except for large piece best in short runs

Table 4.5. Metal Forming Process-Cost Consideration

S.No.	*Process*	*Raw Material Costs*	*Tool and die costs*	*Direct Labour Costs Medium*	*Finishing Costs*	*Scrap loss*	*Optimum lot size*
1.	Drop Forging	Low to Moderate	High	Medium	Medium	Moderate	Large, > 1000
2.	Press Forging	Low Moderate-equal to drop forging	High, usually less than drop forging	Medium-less than drop forging	Medium	Moderate, usually less than drop forging	Medium to High
3.	Upset Forging	Low to Moderate	High	Medium less than other forging process	Medium-less than other forging process	Medium-lowest of Forging processes	Same
4.	Cold Headed Parts	Low to Moderate-chiefly Steel wire	Medium	Low-almost completely automatic	Low	Low-practically nil	Large, not suited to small quantity
5.	Extrusion	Moderate, primarily non-ferrous metals, some alloy steels	Moderate	Moderate	Low	Low	Moderate
6.	Impact Extrusion	Moderate-Primarily Tin, Lead, Aluminium	Medium	Low-Little skilled labour needed	Low-often nil	Low-most scrap in blank scrap	Wide, from hundreds to thousands
7.	Roll Forming	Low to Moderate-mostly low carbon steel sheet	High	Moderate	Low-cutting done	Low	High
8.	Press Forming	Low to Moderate-ranging from carbon steel to stainless stell	High	Medium	Low-cleaning and trimming most frequent	Low to Moderate	Large, > 10,000
9.	Metal Spinning	Low to Moderate	High-skilled operators needed	Low	Low-only cleaning and trimming	Low-practically no scrap	Low

41. List the steps of operations in liquid forging process.
42. Write a short note on isothermal forging.
43. Write a short note on Powder metal forging.
44. What is "No-draft forging" ?
45. Write a short note on "Orbital forging".
46. Write a short note on "Hot isostatic pressing".
47. Write short notes on : Incremental forging, Gatorizing and Super-plastic forging.
48. Define extrusion process.
49. Compare extrusion and rolling processes,
50. Describe the common ways of extruding metals.
51. Compare direct extrusion and indirect extrusion.
52. What is Tubular extrusion ?
53. Define "extrusion ratio" and give its common values for various metals.
54. What is impact extrusion ? Give its product applications.
55. What is "The Hooker method" of cold extrusion ?
56. Describe cold hydrostatic extrusion and give its advantages.
57. List the advantages and disadvantages of extrusion process.
58. Compare open and closed die forging.
59. How is seamless tubing pierced ?
60. What are the final operations in producing finished steel tubes ?
61. Describe the three ways of making butt welded pipe.
62. Give the product applications of extruded pipe, pipe made by piercing technique and the welded pipes.
63. Describe the operation of hot drawing and cupping of metals.
64. How the oxygen gas cylinders are made by this method ?
65. Describe wire drawing and rod drawing.
66. Define "Degree of drawing". What are its normal values ?
67. Describe Tube drawing.
68. How the hypodermic tubes are produced ?
69. How the metal is prepared for drawing operation ?
70. What is the material of the dies used for drawing operation ?
71. Give the advantages of Sheet metal working processes.
72. What is elastic recovery or spring back ?
73. What are the basic units of a press and what does each do ?
74. How do hydraulic presses compare with mechanical presses ?
75. What are knuckle joint and toggle process and what is the purpose of each ?
76. What are single action, double action and triple action presses ?
77. Compare open and closed frame presses.
78. Sketch a Blanking die and label it properly. Write the function of each part.
79. With the help of a sketch, describe the working of a Cutting Die.
80. Describe a progressive, a combination and a compound die.
81. Describe what happens when sheet metal is sheared ?

82. Write the formula for calculating the cutting force in shearing operation.
83. How can the cutting force be reduced ?
84. Define "deep drawing process".
85. With the help of a neat diagram, describe a "deep drawing process". Give its product applications.
86. What is LDR ? What is its usual range ?
87. What is "redrawing" ?
88. Explain the "Bending terminology" with the help of a suitable sketch.
89. What are the various methods of bending ? Sketch and describe each.
90. What is "Press brake" ?
91. Describe "Roll forming".
92. Describe : "Rubber press forming", "Rubber hydro forming", and "Hydromechanical forming".
93. Describe "Stretch forming".
94. Describe "Metal spinning". Write its product applications. Differentiate between cold and hot metal spinning.
95. What is "Power spinning" ?
96. What is "Flow turning" ?
97. Differentiate between "coining" and "embossing".
98. Compare "metal spinning" with "deep drawing".
99. What is the difference between "riveting" and "staking" ?
100. What does "hubbing" do ?
101. Describe "shot peening".
102. What is "cold heading" ? What are its uses and advantages ?
103. Name two metals which are hot worked at room temperature.
104. Name six metals or alloys which may be cold worked at room temperature
105. What are the effects of forging on porosity and segregation ?
106. What is the effect of forging on grain structure and why is this benefitial to the properties of the metal ?
107. What is the relationship between carbon cantent of a steel and its maximum allowable forging temperature ?
108. What will be the consequences if a steel is forged :
    (*a*) at too high a temperature ?
    (*b*) at too low a temperature ?
109. Explain the difference between cold working and hot working of metals. (UPSC, DU)
110. How annealing is related to cold working ?
111. What is the effect of hot working on the structure and mechanical properties of metals?

Lubrication:

1. Rolling : – Hot rolling of ferrous metals is normally done without a lubricant. If need be, graphite can be used. Water-based solutions are used to cool the rolls and to break the scale on the rolled material. For hot rolling of non-ferrous metals, thrse is a wide choice; a wide variety of compounded oils, emulsions and fatty acids.

For cold rolling, the common lubricatnts are : – water-soluble oils, low-viscosity lubricants, such as mineral oils, emulsions, paraffins and fatty acids.

2. Forging : – Lubricants greatly influence : friction, wear, deforming forces and flow of material in die-cavities. They act as thermal barriers between hot workpiece and the relatively cool dies, thereby, slowing the rate of cooling of the workpiece and improving metal flow. Also, they serve as partting agents which inhibit the forging from sticking to the dies and so help in its release from the dies.

For hot forging, the common lubricants are : graphite, MOS_2 and sometimes molten glass. For cold forging, these are : mineral oil and soaps. In hot forging, the lubricant is applied to the dies, but in cold forging, it is applied to the workpiece.

3. Extursion :– For hot extrusion, glass is an excellent lubricant with steels, stainless steels and high temperature metals and alloys. For cold extrusion, lubrication is critical, especially with steels, ebcause of the possibility of sticking (siezure) between the workpiece and the tooling if the lubrication breaks down. Most effective lubricant is : application of a phosphate conversion coating on the workpiece followed by a coating of soap or wax.

5

Processing of Plastics

5.1. GENERAL

Plastics belong to the family of organic materials. Organic materials are those materials which are derived directly from carbon. They consist of carbon chemically combined with hydrogen, oxygen and other non- metallic substances, and their structures, in most cases, are fairly complex. The large and diverse organic group includes the natural materials : wood, coal, petroleum, natural rubber, animal fibers and food, which have biological origins. Synthetics include the large group of solvents, adhesives, synthetic fibers, rubbers, plastics, explosives, lubricants, dyes, soaps and cutting oils etc. which have no biological origins. Of them, plastics and synthetic rubbers are termed as " polymers ".

5.2. POLYMERS

The term " polymer " is derived from the two Greek words : *poly*, meaning "many", and *meros* meaning "parts" or "units". Thus polymers are composed of a large number of repeating units (small molecules) called monomers. The monomers are joined together end-to-end in a polymerization reaction. A polymer is, therefore, made up of thousands of monomers joined together to form a large molecule of colloidal dimension, called macromolecule. The unique characteristic of a polymer is that each molecule is either a long chain or a network of repeating units all covalently bonded together. Polymers are molecular materials and are generally noncrystalline solids at ordinary temperature, but pass through a viscous stage in course of their formation when, shaping is readily carried out.

The most common polymers are those made from compounds of carbon, but polymers can also be made from inorganic chemicals such as silicates and silicones. The naturally occuring polymers include : protein, cellulose, resins, starch, shellac and lignin. They are commonly found in leather, fur, wool, cotton, silk, rubber, rope, wood and many others. There are also synthetic polymers such as polyethylene, polystyrene, nylon, terylene, dacron etc., termed under plastics, fibers and elastomers. Their properties are superior to those of the naturally occuring counterparts. Our concern, here, is therefore, with synthetic polymers, also called plastics (again from Greek *plastikos*, derived from *plassein* : to form, to mould) or resins.

5.3. POLYMERIZATION

The process of linking together of monomers, that is, of obtaining macromolecules is called "polymerization". It can be achieved by one of the two processing techniques :

(a) *Addition Polymerization*. In addition or chain polymerization under suitable conditions of temperature and pressure and in the presence of a catalyst called an initiator, the polymer is produced by adding a second monomer to the first, then a third monomer to this dimer, and a fourth to the trimer, and so on until the long polymer chain is terminated. Polyethylene is produced by the addition polymerization of ethylene monomers. This linear polymer can also be converted to a branched polymer by removing a side group and replacing it with a chain. If many such branches are formed, a network structure results.

" Co - polymerization " is the addition polymerization of two or more different monomers. Many monomers will not polymerize with themselves, but will copolymerize with other compounds.

(*b*) **Condensation Polymerization.** In this process, two or more reacting compounds may be involved and there is a repetitive elimination of smaller molecules, to form a by-product. For example, in the case of phenol formaldehyde (bakelite), the compounds are : formaldehyde and phenol. Metacresol acts as a catalyst and the by-product is water. The structure of the 'mer' is more complex. Also, there is the growth perpendicular to the direction of chain. This is called 'cross-linking'.

Size of a Polymer. The polymer chemist can control the average length of the molecules by terminating the reaction. Thus, the molecular weight (the average weight, in grams, of 6.02×10^{23} molecules) or degree of polymerization, D.P., (the number of mers in the average molecule) can be controlled. For example, the length of molecules may range from some 700 repeat units in low-density polyethylene to 1,70,000 repeat units in ultrahigh-molecular-weight polyethylene.

5.4. ADDITIONS TO POLYMERS

The properties of polymers can be further modified by the addition of agents which are basically of two types. Those that enter the molecular structure are usually called " additives ", whereas those that form a clearly defined second phase are called " fillers ".

1. Plasticizers. Plasticizers are liquids of high boiling point and low molecular weight, which are added to improve the plastic behaviour of the polymer. The broad role of a plasticizer is to separate the macro-molecules, thus decreasing the inter-molecular forces and facilitating relative movement between molecules of the polymer, that is, making deformation easier. They are essentially oily in nature. Organic solvents, resins and even water are used as plasticizers.

2. Fillers. A filler is used to economize on the quantity of polymer required and / or to vary the properties to some extent, for example, mechanical strength, electrical resistance etc. A filler, whose function is to increase mechanical strength, is termed a " reinforcing filler ". A filler is commonly fibrous in nature and is chemically inert with respect to the polymer with which it is to be used. Common fillers are wood flour, cellulose, cotton flock, and paper (for improving mechanical strength) ; mica and asbestos (for heat resistance) ; talc (for acid resistance). Other filler materials are : fabric, chipped-wood moulding compound, wood veneer, textile or glass fibres.

Wood flour is a general purpose filler. It improves mouldability, lowers the cost with fairly improved strength of the plastics. Mica also imparts excellent electrical properties to plastics and results in low moisture absorption. The commonly used "reinforcing filler agents" with plastics are : fibres/filaments of glass, aramid, graphite or boron. Reinforcing by metal and glass fibres make plastics strong, flexible and light materials such as used in bullet proof vests. Cotton fibres improve toughness. Carbon fibres are used for high performance installations such as aircrafts etc. requiring high strength and stiffness.

3. Catalysts. These are usually added to promote faster and more complete polymerization and as such they are also called 'accelerators' and 'hardeners' *e.g.*, ester is used as a catalyst for Urea Formaldehyde.

4. Initiators. As the name indicates, the initiators are used to initiate the reaction, that is, to allow polymerization to begin. They stabilize the ends of the reaction sites of the molecular chains. H_2O_2 is a common initiator.

5. Dyes and Pigments. These are added, in many cases, to impart a desired colour to the material.

For example, titanium dioxide is an excellent white pigment; iron oxides give yellow, brown or red colour; carbon black is not only a pigment but also a UV light absorbent. Finely divided calcium carbonate dilutes (extends) the colour and is used in large quantities as low-cost filler.

6. **Lubricants.** Lubricants are added to the polymers for the following purposes: to reduce friction during processing, to prevent parts from sticking to mould walls, to prevent polymer films from sticking to each other and to impart an elegant finish to the final product. Commonly used lubricants include : oils, soaps and waxes.

7. **Flame retardants.** Most plastics will ignite at sufficiently high temperatures. The non-inflammability of the plastics can be enhanced either by producing them from less inflammable raw materials or by adding "flame retardants". The common flame retardants are : compounds of chlorine, bromine and phosphorous.

8. **Solvents.** Solvents are useful for dissolving certain fillers or plasticizers and help in manufacturing by allowing processing in the fluid state. For example, alcohol is added in cellulose nitrate plastics to dissolve Camphor. However, subsequently, the solvents must be removed by evaporation.

9. Stabilisers and anti-oxidants are added to retard the degradation of polymers due to heat, light and oxidation.

10. Elastomers are added to plastics to enhance their elastic properties.

Note. Above, excepting fillers, all other materials used fall under the category of "Additives".

5.5. PLASTICS

Polymers can be divided into three broad divisions : plastics, fibers and elastomers (polymers of high elasticity, for example, rubber). Synthetic resins are usually referred to as plastics. Plastics derive their name from the fact that in a certain phase of their manufacture, they are present in a plastic stage (that is, acquire plasticity), which makes it possible to impart any desired shape to the product. Plastics fall into a category known chemically as high polymers.

Thus, " Plastics " is a term applied to compositions consisting of a mixture of high-molecular compounds (synthetic polymers) and fillers, plasticizers, stains and pigments, lubricating and other substances. Some of the plastics can contain nothing but resin (for instance, polyethylene, polystyrene).

5.5.1. Types of Plastics. Plastics are classified on the broad basis of whether heat causes them to set (thermosetting) or causes them to soften and melt (thermoplastic).

1. **Thermosetting Plastics.** These plastics undergo a number of chemical changes on heating and cure to infusible and practically insoluble articles. The chemical change is not reversible Thermosetting plastics do not soften on reheating and can not be reworked. They rather become harder due to completion of any left-over polymerization reaction. Eventually, at high temperatures, the useful properties of the plastics get destroyed. This is called degradation. The commonest thermosetting plastics are : alkyds, epoxides, melamines, polyesters, phenolics and ureas.

2. **Thermoplastic Plastics.** These plastics soften under heat, harden on cooling, and can be resoftened under heat. Thus, they retain their fusibility, solubility and capability of being repeatedly shaped. The mechanical properties of these plastics are rather sensitive to temperature and to sunlight and exposure to temperature may cause thermal degradation. Common thermoplastic plastics are : acrylics, poly tetra fluoro ethylene (PTFE), polyvinyl chlorides (PVC), nylons, polyethylene, polypropylene, polystryrene, etc.

Thermosetting plastics are cross-linked polymers and the thermo- plastics are linear and branched-linear polymers. The method of process-ing a plastic is determined largely by whether a plastic is thermosetting or thermoplastic.

Thermosetting plastics are usually harder, stronger, and more brittle than thermoplastics. Since they do not soften when heated and roasted or charged at high temperatures, thermosetting plastics have applications only in moderate service temperatures, for example, saucepan handles. Thermoplastics are reclaimable while thermosetting plastics are not.

5.5.2. Properties of plastics. Their great variety of physico- chemical and mechanical properties, and the ease with which they can be made into various articles have found plastics their wide application in the engineering and other industries.

1. Their comparatively low density (1 to 2 g/cm^3), substantial mechanical strength, higher, shength-to-weight ratio and high anti-friction properties have enabled plastics to be efficiently used as substitute for metals, for example, non-ferrous metals and alloys-bronze, lead, tin, babbit etc., for making bearing.

2. With certain special properties (šilent operation, corrosion resistance etc.), plastics can sometimes replace ferrous metals.

3. From the production point of view, their main advantage is their relatively low melting points and their ability to flow into a mould.

4. Simple processing to obtain machine parts. Generally there is only one production operation required to convert the chemically manufactured plastic into a finished article.

5. In mass production, plastics substituted for ferrous metals allow the production costs to be reduced by a factor of 1.5 to 3.5 and for non-ferrous metals by a factor of 5 to 20.

6. Good damping capacity and good surface finish of the product.

7. The high heat and electric insulation of plastics permits them to be applied in the radio and electrical engineering industries as dielectrics and as substitutes for porcelain, ebonite, shellac, mica, natural rubber, etc.

8. Their good chemical stability, when subjected to the action of solvents and certain oxidizing agents, water resistance, gas-and steam - proof properties, enable plastics to be used as valuable engineering materials in the automobile and tractor, ship building and other industries.

With the development of high-performance engineering plastics, they are successfully competing with other engineering materials. They are replacing sheet metal parts, zinc and Al alloy castings and C.I. They are being increasingly used in automobile industry, where weight reduction is one of the means of fuel economy. The initial use of plastics in this direction was for fascia panels, interior fittings, and other non-load bearing components. Now structural parts such as bumpers, brake fluid tanks and some body parts are also being made of plastics

Disadvantages :

1. Comparatively higher costs of materials.
2. Inability of most plastics to withstand even moderately high temperatures.

The factors which have determined the rapid growth of polymer materials in the recent past, are :

1. Ready availability of the basic raw chemical materials in large quantities and, in general, at a low cost.
2. The large number of available starting materials for their production provide us with an almost continuous spectrum of composition and structure, and hence of mechanical, optical, electrical and thermal properties of the resulting polymers.
3. The engineer now has at his disposal many well developed processes and machines to convert these materials (as they come from the factory and that he can choose according to the specification of the ultimate product) into useful goods.

5.6. PROCESSING OF THERMOPLASTIC PLASTICS

The common forms of raw materials for processing plastics into products are : – pellets, powders, sheet, plate, rod and tubing. Liquid plastics are used especially in the fabrication of reinforced–plastic parts.

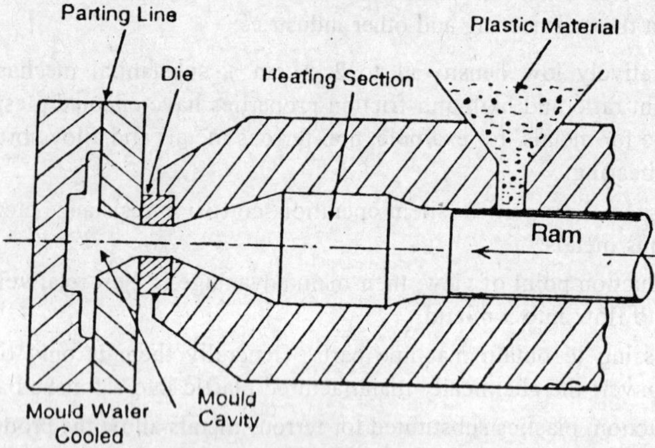

Fig. 5.1. Injection Moulding Process.

Thermoplastics can be processed to their final shape by moulding and extrusion processes. However, extruding is often used as an intermediate process to be followied by other processes, for example, vacuum forming or machining.

5.6.1. Injection Moulding. An important industrial method of producing articles of thermoplastics is injection moulding, (Fig. 5.1). The process is essentially as follows :

The moulding material is loaded into a hopper from which it is transferred to a heating section by a feeding device, where the temperature is raised to 150°C – 370°C. and pressure is built up. The material melts and is forced by an injection ram at high pressure through a nozzle and sprue into a closed mould which forms the part.

The mould is in at least two sections, so that it may be split in order to eject the finished component. For the process to be competitive, the mould must be fairly cool (between ambient temperature and the softening point of the plastic) and consequently the mould must be cooled by circulating water.

The improvement to the ram type injection moulding machine lies in the separation of the plasticizing and filling actions. The single-screw pre-plasticizer is probably the most successful design for injection moulding machines, (Fig. 5.2). The rotation of the screw provides the plasticizing action by shearing and frictional effects and the axial motion of the screw provides the filling action.

Injection moulding machines have a high production capacity : some can produce from 12 to 16 thousand parts per shift. This method is suitable for making parts with complex threads and intricate shapes, thin-walled parts etc. Typical parts include : Cups, containers, housings, tool handles, toys, knobs, plumbing fittings, electrical and communication components such as telephone receivers etc.

The limitations of the process are :- Equipment of cylinder and die should be non-corrosive. Also, reliable temperature controls are essential.

Injection moulding machines range in size from an injection capacity of 12,000 mm^3 to 2.2×10^6 mm^3. The locking forces are applied to the mould usually by hydraulic means, and may vary from 0.1 MN to 8.0 MN or even more. The injection pressure may range from 100 MPa to 150 MPa.

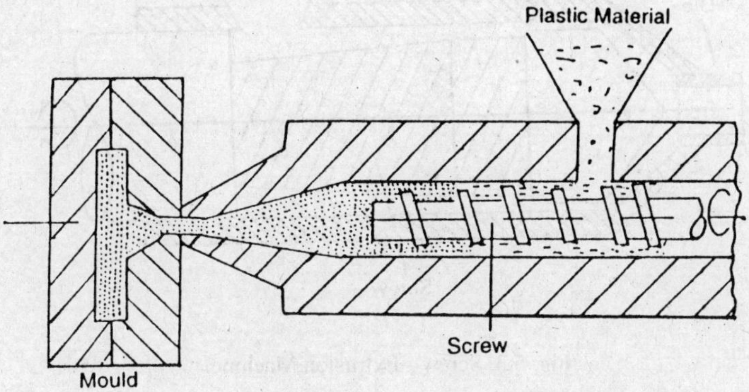

Fig. 5.2. Screw Injection Moulding Machine.

5.6.2. Extrusion process. The extrusion process, in many cases, produces material in an intermediate form for subsequent reprocessing to its final component form. The process is the same as for metals, that is, the expulsion of material through a die of the required cross-section. The earliest extrusion machines were of the ram type. The cylinder of the machine (container) is filled with prepared plastic and extruded through a die under the pressure of the ram. The advantages of this machine are : simplicity in operation and a controlled pressure which can be virtually as high as required. If the polymer can be plasticized by pressure, then the ram extruder is advantageous in view of its simplicity. But for plastics which require heat, the separate pre-processing may be regarded as a draw back. Another major drawback of this type of machine is the reciprocating action of the ram which is time wasting since the ram must be withdrawn after its power stroke and a new dolly of material inserted in the container. Also, with many materials, the die orifice must be cleaned between each working stroke.

Nowadays, the ram machine is mainly used for " wet extrusion " that is for extruding plastics which have been softened by the addition of solvents. Although useful in homogenizing materials which contain hard inclusions, wet extrusion has the disadvantage of producing a component from which the solvent has to be removed.

For the extrusion of plastics, single-screw machine has completely replaced the ram type machine. There are two basic types of screw extruders : the melt extruder and the plasticizing extruder. In the former, the material is delivered to the extruder already melted and thus the function of the extruder is merely to push the material to the die and through the orifice. In the plasticizing extruder the material is in the form of granules or particles and so the extrude has to compress and work it until it melts before delivering it, under pressure, to the die orifice.

Fig. 5.3. illustrates a screw type extrusion machine. It consists of : a water cooled screw having a special thread form to suit the material being extruded ; a barrel in which the screw rotates (including a form of heating in the case of a plasticizing extruder); and an extrusion die.

The material is fed from a hopper through a port in the cylinder where the rotation of the screw imparts both axial and rotary motion to the particles. The restricting effect of the die at the far end builds up a pressure in the particulate mass which is then worked by shearing and heated by frictional effects until it is in a plastic state and can be extruded.

Complex shapes with constant cross-sections can be extruded with relatively inexpensive tooling. The extruded product can be coiled or cut into desired lengths. Typical product applications are: – rods, pipes channels, window frames, sheets, architectural components, plastic–coated wire, cable or strips for electrical applications.

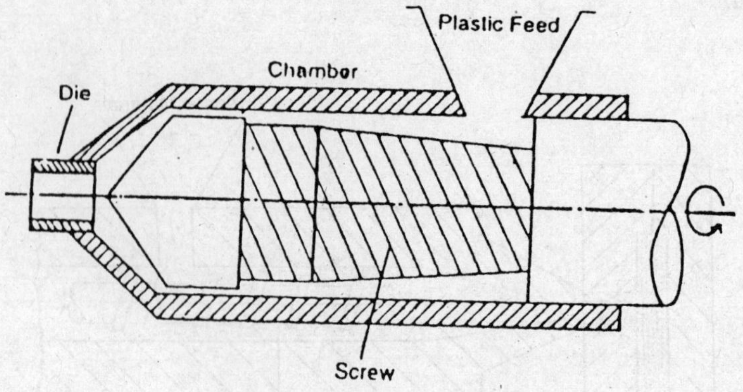

Fig. 5.3. Screw - Extrusion Machine.

5.6.3. Sheet-forming process. Many plastic articles are formed from sheet. The processes resemble those for metals, but require very low forces. Even atmospheric pressure may be sufficient. In "Drape forming", the sheet is heated to a moderate temperature. It is then clamped at the edges and stretch-formed over a die (See Fig. 4.99). One of the problems encountered is that the portions of the sheet first touching the die will be chilled and remain thicker than the rest. This is overcome or minimised by blowing hot air between the sheet and the die. Vacuum forming is a process, in which a heated plastic sheet is changed to a desired shape by causing it to flow against the mould surface by reducing the air pressure between one side of the sheet and the mould surface. The process consists of clamping the heated plastic sheet over a mould in such a way that the air between the sheet and mould can be evacuated (Fig. 5.4). This vacuum, of increasing intensity, draws the sheet against the surface of the mould, where it cools and solidifies. The solidification will take place earliest in those regions which touch the mould first (Fig. 5.5 *a*). This will cause differential

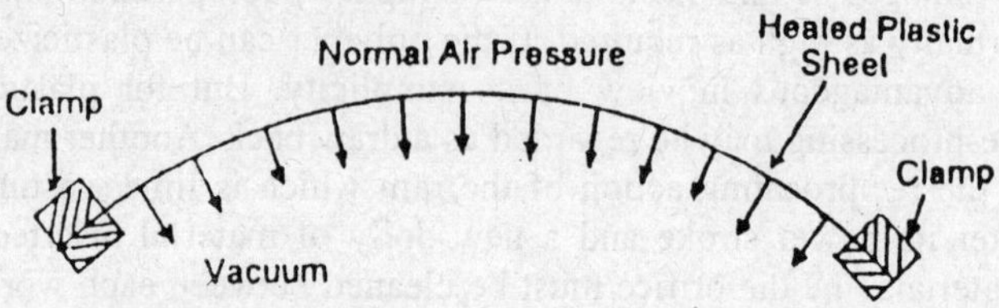

Fig. 5.4. Principle of Vacuum Forming.

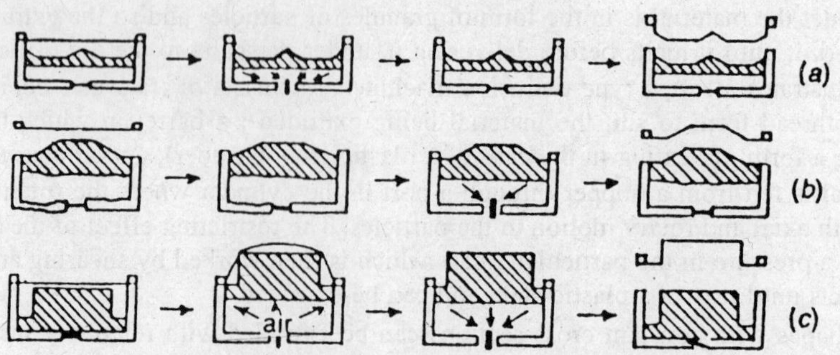

Fig. 5.5. Vacuum Forming Processes.

cooling and, as a result of non uniform temperature distribution, there will be a marked change in thickness along any given section of the component.

Thinning is clearly worst where the sheet contacts the mould first (high spots on the mould) and near the peripheral clamping ring. This effect is reduced in drapeforming by replacing the clamping ring by a movable ring or drape, so that part of the profile is achieved before evacuation occurs (Fig. 5.5 *b*). There is still a definite depth limitation, however, and additional steps must be taken to eliminate thinning in very deep components. Where one of the difficulties is the restriction of movement at a right-angle corner (Fig. 5.5 *c*), the same circuit which is used to extract air can be used earlier in the cycle, to blow hot air between the sheet and the mould. The air pressure acts as a cushion and the air temperature delays cooling of the sheet.

This method is also known as "Thermo-forming". Since high strength of the moulds is not a prerequisite, these are usually made of aluminium. So, tooling cost is quite less. Product applications are : Refrigerator liners, Packaging, Appliance housings, Panels for shower stalls and advertising signs.

5.7. PROCESSING OF THERMOSETTING PLASTICS

Compression moulding and Transfer moulding are the most common methods of processing thermosetting plastics. Although, suitable for thermoplastics also, the main application of these methods is to thermosets.

5.7.1. Compression Moulding.

Compression moulding, (Fig. 5.6), is the equivalent of closed-die forging. In this process, a premeasured quantity of plastic in the form of particles or briquettes, is placed in a heated mould and compressed at suitable pressure and temperature. Hydraulic presses are usually employed to provide the pressure (which may range from 20 to 30 MPa or even higher upto 80 MPa in some cases) for compressing the plastic compound. Other equipment, such as friction and screw presses, can also be used. The object of compression moulding is to bring the plastic to virtually molten state. Thus the process is, effectively, forming from the liquid state, the material being held in the mould until the curing stage is over when polymerization is complete. The process is rather slow with the phenolics and urea resins, but some of the newer resins have shorter curing times and this has improved the production rates appreciably.

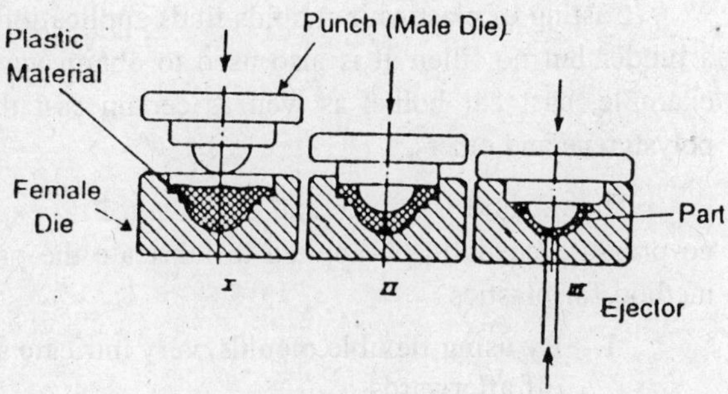

Fig. 5.6. Compression Moulding.

When the plastic is completely trapped between the male and female die, it is called as "positive mould". Closer tolerances can be held if a small flash is allowed to extrude, usually along the male die perimeter in "Semipositive moulds". More plastic is lost in " flash moulds ", similar to those used in impression-die forging.

Typical product applications are : Disches, handles, container caps, fittings, electrical and electronic components, washing machine agitators and housings etc.

5.7.2. Transfer Moulding. Transfer moulding (Fig. 5.7) is a modification of compression moulding in which the material is first placed in a separate chamber (transfer pot), from which it is pushed through an orifice (sprue) into the mould cavity (by the action of a punch) as the mould closes. The material to be moulded is often pre-heated by radio-frequency methods and, where it is desired to improve toughness and strength, reinforcing fillers may be used. The process has got the following advantages :

1. There is little pressure inside the mould cavity until it is completely filled, at which stage the full liquid pressure is transmitted.
2. The plastic acquires uniform temperature and properties in the transfer pot prior to transfer. The plastic is further heated by shearing through the orifice, viscosity is reduced, and the plastic fills the intricate mould cavities.
3. It scores over normal compression moulding in that cold presses can be used, since, heating of the plastic is affected, not by press itself, but by a simple heating jacket round the transfer chamber.

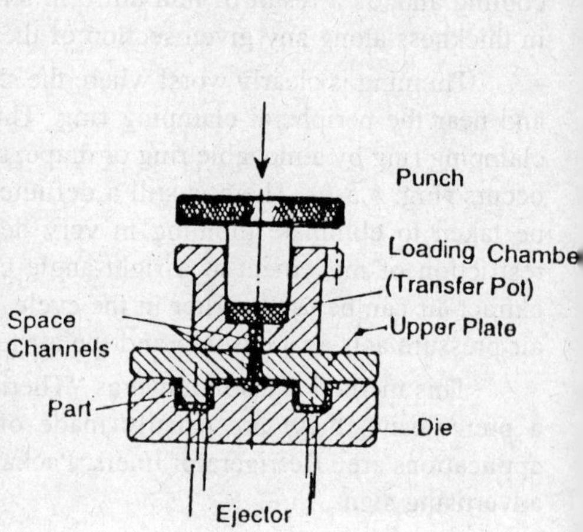

Fig. 5.7. Transfer Moulding.

Field of application :- It is used less for mass production. Its chief application is in short runs where the shape of the moulding would make readjustment of the injection moulding machine profitable.

5.8. CASTING OF PLASTICS

Casting of plastics in moulds finds application when making parts of a plastic material with a binder but no filler. It is also used to obtain various kinds of cast thermosetting plastics, for example, cast car bolite, as well as certain cast thermoplastic materials, such as organic glass, polystyrene and others.

The method is simple and cheap since no expensive tooling or equipment is required, and no pressure needs to be applied to fabricate the part. There are many variations of the casting method for plastics :

1. By using flexible moulds, very intricate shapes can be fabricated. The mould is peeled off afterwards.
2. Thick plastic sheets are produced by using plate glass moulds.
3. Thinner plastic sheets are produced by using moving stainless steel belts which contain and cool the resin.
4. Hollow shapes can be obtained by centrifugal casting of the molten plastic material.
5. **Potting.** In this method, the plastic material is cast around an electrical component which gets embedded in the plastic material. This is achieved by pouring the molten plastic material in a housing or case which is an integral part of the component and in which the component is prepositioned before pouring the plastic.
6. **Encapsulation.** Here, the component is covered with a layer of cooled and solidified plastic.

   Both potting and encapsulation are very important to the electrical and electronics industry. The plastic material serves as a dielectric.
7. **Foam moulding/casting.** In this method a foaming agent is mixed with the plastic resin. The mixture is placed in a mould and heated. The foaming agent makes the

material to expand (even upto 50 times the original size) to take up the shape of the mould. The amount of expansion can be controlled through temperature and time. Both rigid and flexible foamed plastics can be obtained from thermoplastics and thermosetting plastics. Rigid construction is used for structural purposes and flexible for cushioning. Product applications include : Shaped packaging materials for cameras, appliances and electronics etc., insulating blocks, food containers and styrofoam cups.

5.9. MACHINING OF PLASTICS

Plastics can be machined, but in most cases, machining of plastics is not required. Acceptable surface quality and dimensional accuracy can be obtained by moulding and forming methods. However, there are certain plastics like PTFE (Polytetra fluoroethylene) which are sintered products and are not mouldable by usual techniques, as they do not melt. For such "thermo stable plastics" machining is a viable alternative to moulding.

The machining of plastics (by operations such as turning, drilling and milling) has special features due primarily to the structure of the material. It also depends upon the binder and the filler and the method of moulding the component. For example, the machining of thermosetting plastics allows optimum cutting variables and tool geometry to be employed because these do not soften on heating, whereas thermoplastic resins soften under heat. The permissible maximum temperature in the cutting zone is 160°C for thermosetting resins and only 60°C to 100°C for thermoplastics.

Special features of the machining of plastics are :

1. The tendency of certain plastics to splitting.
2. High elasticity (40 times as much as that of steels). Therefore, they must be carefully supported, to avoid their deflection during machining.
3. Non-homogeneous structure of the material, with components of different hardness. This results in poor surface finish after machining.
4. Plastics have a strong abrading action on cutting tools.
5. Their low thermal conductivity results in poor heat dissipation from the cutting zone and in over-heating of the cutting edges.
6. The intense dust formation, especially for thermosetting plastics, makes it necessary to use special dust-removing devices.
7. The hygroscopicity of plastics excludes the use of liquid cutting fluids. Compressed air is commonly used for cooling.
8. Reinforced plastics are very difficult to machine.

Plastics can be machined with H.S.S. and cemented-carbide tools. In machining a plastic material with a filler of glass, quartz or mica type, a satisfactory tool life can be obtained only with carbide-tipped tools. Only diamond tools are suitable for turning high-strength plastics of this type. The strength of cast parts of laminate plastics is 40 to 50 per cent less than that of the parts made by compression moulding. Therefore, higher cutting speeds and feeds can be used in their machining than for strong thermosetting plastics. The main trouble in turning laminated plastics is the peeling of the surface layer.

The cutting variables are also influenced by the life of the cutting tool which is subject to abrasive wear in machining most engineering plastics. Dulling of the cutting tool leads to a poor surface finish and to breaking out of the material at the points the cutting tool enters and leaves the cut. This makes it necessary to use more keenly sharpened cutting tools for plastics. The need for sharp cutting edges also follows from the high elasticity of plastics.

The selection of cutting variables is also influenced by the low heat conductivity of the plastics, since, in machining the tool may be within a closed volume (as in drilling) with no cooling facilities. This may lead to charring of the machined surface.

The cutting tool angles for machining plastics are made somewhat different than those of tools for ferrous and non-ferrous metals. The rake angles are positive and relatively larger. Because of the viscoelastic behaviour of thermoplastics, some of the local elastic deformation is regained when the load is off. Therefore tools must be made with large relief angles (20° to 30°).

Abrasive machining of plastics has many advantages over machining with metal cutting tools. These include the absence of splitting and crack formation, and the better surface finish that can be obtained.

In grinding, the contact between the wheel and the surface being ground, should be as short as possible, to avoid burns. Organic glass is commonly ground with coated abrasive, applying an ample amount of water as a coolant. If possible, however, grinding should be replaced by polishing with a felt, broadcloth or flannel wheel charged with lapping paste, the process is known as "Buffing". The buffing wheels are of diameter 250 mm, 40 to 60 mm wide and of speed 2000 rev/min. Meduim and fine lapping pastes are used as the buffing compound for plastics. Laminate fabric base, asbestos-fibre and glass-fibre laminate can be cut with abrasive wheels (SiC) of grain size 24 to 46 and with a 5% emulsion as the cooling fluid.

5.10. OTHER PROCESSING METHODS FOR PLASTICS

1. Calendering. It is an intermediate process where the extruded plastic sections are reduced to sheet which may or may not, then be formed to final shape by vacuum forming. It is clear that the calendering process can be used only for thermoplastics and not for thermosetting plastics.

Calendering is in some ways similar to rolling process in that the material is compressed between rolls and emerges as sheet, (Fig. 5.8 *a*). However, there are differences. There is appreciable thickening after the material has reached minimum thickness at the roll gap and the pre-calendered material is not in sheet form, but of indefinite shape. The method of producing wide sheet and foil is illustrated in Fig. 5.8 *b*. The thermoplastic melt is fed to a multiroll

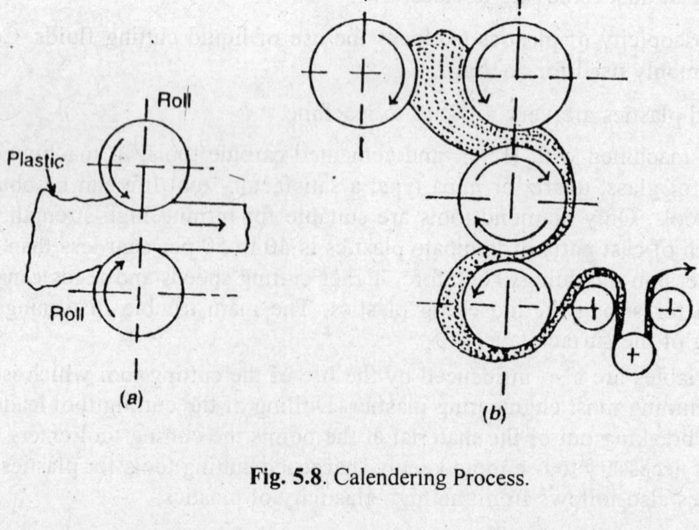

Fig. 5.8. Calendering Process.

calendar. The first roll gap serves as a feeder, the second as a metering device, and the third roll gap sets the gauge of the gradually cooling plastic which is then wound, with about 25% stretching onto a drum.

Calendering is a high-production rate (typically 100 m/min) process, mostly for flexible PVC, for example, upholstery, rainwear, shower curtains, tapes, etc. and rigid PVC, for example, trays, credit cards, laminations. PVC is also calendered into the well known transparent film widely used for packaging.

2. Rotational Moulding. In this process, also called "roto- moulding", large relatively thin-walled hollow (open or closed) parts are made. A measured quantity of polymer powder is placed in a thin-walled metal mould. The mould is closed and is rotated about two mutually perpendicular axes as it is heated. This causes the powder to sinter against the mould walls, building up the wall thickness of the component. At the end of the heating and sintering operations, the mould is cooled while it is still rotating. Cooling is done by applying cold water and air to the outside of the rotating mould. The rotation is then stopped and the component is removed. To increase production rates, three moulds at the end of three arms joined together to the central spindle (just like centrifuge casting) are used, with one mould for each stage of the process, that is, load-unload, heat and cool positions.

The process is simple as no pressure is employed and the part is free of moulded in stresses. The technique is extensively used for the production of toys in P.V.C. such as boats, horses etc. Large containers of polyethylene (or upto 20,000 litre capacity) and large components like laminated petrol tanks for motor cars are made from polythene (outer shell) and nylon (inner shell). Other products include : Trash cans, boat hulls, buckets, housings, and carrying cases etc.

3. Blow Moulding. In this process, a hot extruded tube of plastic, called a parison, is placed between the two part open mould, (Fig. 5.9 *a*). The two halves of the mould move towards each other so that the mould closes over the tube. The tube gets pinched off and welded at the

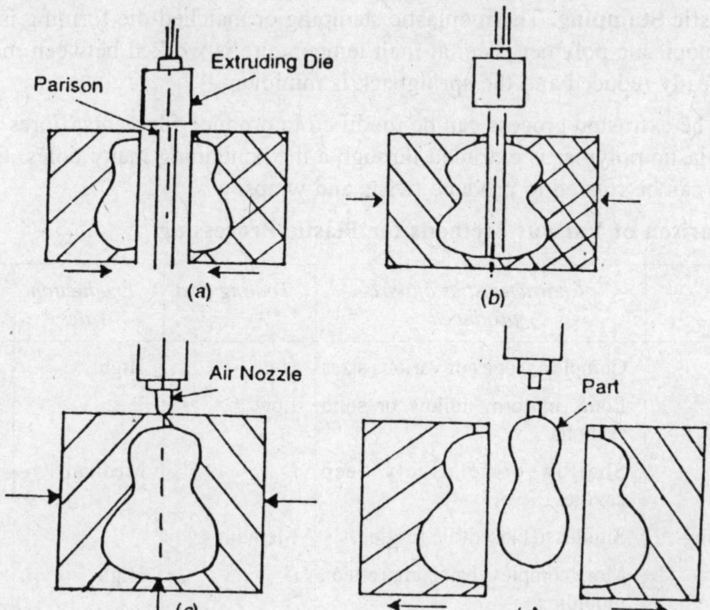

Fig. 5.9. Blow Moulding.

bottom by the closing moulds, (Fig. 5.9 *b*). The tube is then expended by internal pressure, usually by hot air, which forces the tube against the walls of the mould, (Fig. 5.9 *c*). The component is cooled and the mould opens to release the component, (Fig. 5.9 *d*). Typical product applications are : Plastic beverage bottles and hollow containers.

4. Reaction Injection Moulding (RIM). This method differs from the conventional injection moulding process in the sense that it is not the molten polymer which is injected into a mould, but a mixture of two or more monomers (reactants) are forced into a mould cavity. Chemical reaction takes place between the constituents of the mixture giving off heat to form a plastic polymer, which solidifies producing a thermoset component. The major product applications include : Automotive bumpers, and fenders, thermal insulation for refrigerators and freezers and stiffners for structural components.

5. **Solid State Forming.** The term is a misnomer because the temperature of the polymer is just (10°C to 20°C) below the melting point of the polymer. The main operations involved are : Sheet metal techniques such as stretching, bending and deep drawing. Many food-packaging tubs and containers are fabricated from Polypropylene. Forging is also used mainly for producing gears.

6. **Cold Forming.** All the cold working methods used for metals can be used for polymers. Filaments and fibres are produced by "Cold drawing", that is, continuous stretch drawing. Conventional rolling can also be used for producing fibres. In "Cold pressing" or "Cold moulding", the raw thermosetting material (or mixed plastic compounds) are put in the mould and pressed to shape at room temperature. The formed part is then removed from the mould and cured in an oven. Pressures applied by the press range from 14 MPa to 84 MPa. The moulds are made of abrasion resistant tool steel. The process is quite economical and the process cycle is relatively short. However, the surface quality and dimensional accuracy of the part is not very good.

7. **Thermoplastic Stamping.** Thermoplastic stamping or matched-die forming is a method in which filled thermoplastic polymer sheet at melt temperature is worked between mating dies. The cycle time is greatly reduced and the springback is minimum.

8. **Spinning.** The extrusion process can be modified to produce filaments, fibres and yarns. The molten thermoplastic polymer is extruded through a die containing many holes. For obtaining strands, the dies can be rotated to produce twists and wraps.

5.10.1. Comparison of Various Methods for Plastic Processing

Process	*Characteristics of parts produced*	*Tooling cost*	*Production rates*	*Surface Quality*
Injection Moulding	Complex shapes of various sizes	High	High	Good
Extrusion	Long, uniform hollow or solid sections	Low	"	tolerances
Thermoforming	Shallow or relatively deep cavities	"	Medium	Good
Compression Moulding	Similar to closed-die forging.	Medium	"	Good
Transfer Moulding	More complex than compression moulding	"	High	Good
Casting	Intricate with flexible moulds	Low	Low	Good
Calendering	Sheets	Medium	High	Good

Rotational Moulding	Large hollow parts of relatively simple shapes	Low	Low	Good
Blow Moulding	Hollow thin walled parts of various sizes.	Low	High	Good
Structural Foam Moulding	Large parts with high stiffness to weight ratio	Meduim	Low	Good

5.11. WELDING OF PLASTICS

Most plastics are available in the form of mould powders or granules, which are then fabricated into finished articles. The principal processes for the manufacture of articles from plastic materials have been discussed earlier. Plastics differ drastically from metals in their behaviour, when deformed at room and elevated temperatures. Even at their flow temperature, they behave like a highly viscous liquid. Therefore, some sort of pressure has to be applied to make a welded joint for plastic articles.

As we know that heating has no softening effect on thermosetting plastics, therefore, they can not be welded by conventional processes. Thermosetting plastics are available as semi-finished products, which can be either joined mechanically or cemented. On the other hand, when heated, thermoplastics can repeatedly pass into a highly elastic and then into a plastic state, without losing their original properties on cooling again. These plastic materials are worked by the application of heat and pressure and so can readily be welded.

For making permanent connections between plastic parts and components, welding is superior to cementing and riveting, in many respects, as it offers : high speed, low labour requirements, economy, improved working conditions, stronger and tighter joints.

The various welding processes for plastics can be grouped into two broad groups, depending upon the source of heat used :

(*a*) Welding processes, utilizing heat from an external source, such as, a stream of hot gas, a hot extruded filler metal, or a hot tool. In all these processes, heat is transferred to the surfaces being welded by convection, conduction and partly by radiation.

(*b*) Process in which heat is generated within the workpiece through conversion of some other form of energy, such as, r.f. current, ultrasound, friction, infrared light, chemical reactions, or nuclear welding.

The common welding processes for plastics are discussed below. The application of heat and pressure during welding, causes the plastic workpieces to undergo autohesion.

1. **Hot-Gas Welding.** In this process, the edges to be joined are heated by a stream of hot gas from a torch, (Fig. 5.10 *a*). The hot gas can be : air, nitrogen, products of combustion of some

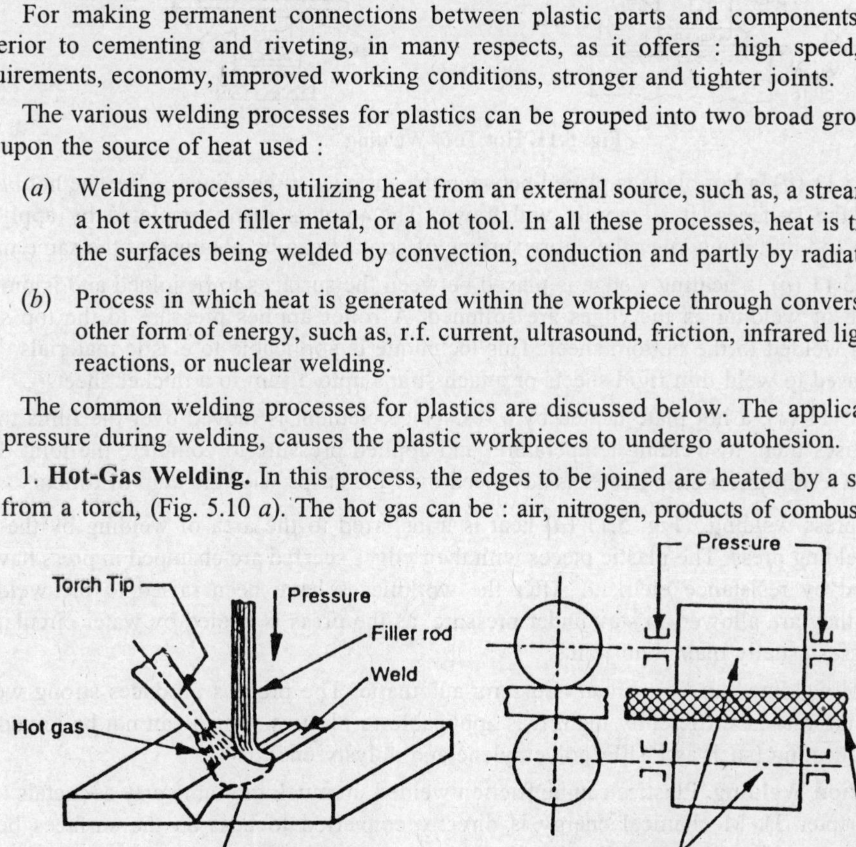

Fig. 5.10. Hot Gas Welding.

fuel gas, such as acetylene or hydrogen. The hot gas heats the edges and the filler rod to viscous fluid state. As the filler rod is forced down by hand, it welds to the softened edges and forms a weld. This technique is like manual gas welding of metals. The drawbacks of the process are : Low joint strength, reduced plasticity in the weld and near-weld area, low welding rate, especially in thick sheets, danger of overheating, and dependence on the operators skills.

Hot gas welding, without filler rods, (Fig. 5.10 *b*) speeds up the process and enhances the mechanical properties of the point. The edges are scarfed and sheets are fitted up for welding, and the edges are uniformly heated by hot gas. The hot gas jet is followed by cold rollers, which complete the weld by pressure. The process is most often applied to make lap joints in films.

2. **Hot Tool Welding.** In this process, a hot tool transfers heat to the plastic workpiece by direct contact. There are many variations of this process.

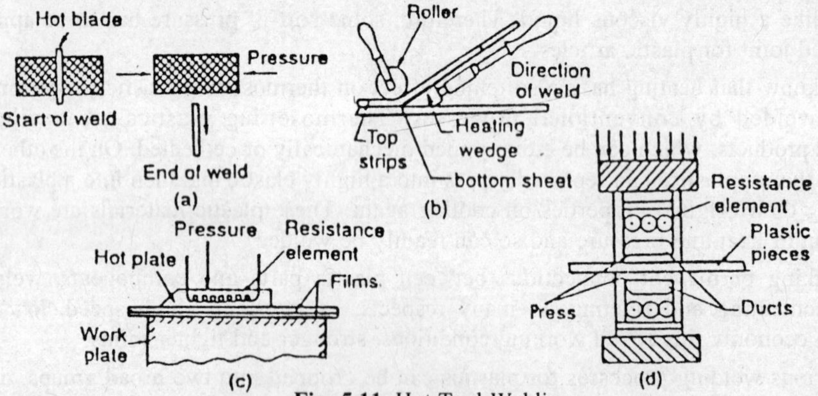

Fig. 5.11. Hot Tool Welding.

In Fig. 5.11 (*a*), a hot blade is placed between the surfaces to be joined. After the hot blade has softened the surfaces, it is rapidly withdrawn. The joint is then completed by applying pressure. Lap and butt joints over the entire surface of contact can be obtained at the same time.

In Fig. 5.11 (*b*), a heating wedge is placed between the surfaces to be joined and is moved along the line of welding as the edges are softened. A roller applies pressure to the top strip which is then welded to the bottom sheet. This technique is applicable to elastic materials. The technique is used to weld thin rigid sheets or attach straps upto 5 mm to a thicker sheet.

In Fig. 5.11 (*c*), a hot plate heated by a resistance element is moved over the films to be lap joined, raises them to welding temperature; and applied pressure to complete the joint. In a variant of this technique, a strip heater is advanced over the strips with the help of rollers.

In hot press welding, (Fig. 5.11 (*d*) heat is transferred to the area of welding by the hot platen of a welding press. The plastic pieces with their edges scarfed are champed in press having platens heated by resistance element. After the workpieces have been raised to the welding temperature, they are allowed to stay under pressure, as the press is cooled by water circulating in ducts. Presses usually make butt joints.

Hot tool welding can be manual and semi-automatic. The process produces strong welds and the welding rate is sufficiently high. It is applicable to plastics, which can not be joined by r.f. induction heating (such as PTFE, polyethylene and polystyrene).

3. **Friction Welding.** Plastics can be friction welded, in much the same way as metals (See Fig. 3.37, chapter 3). Mechanical energy is directly converted to heat on the surfaces being welded. Friction welding has the following limitations :

(*i*) One of the pieces must be a body of revolution, and its section at the joint must be a circle or an annuals :

(*ii*) When pressure is applied to the pieces to complete the joint, flash is formed at the connection.

4. **Ultrasonic Welding.** An ultrasonic plastic welder has the same elements as one for metals (See Fig. 3.40 chapter 3). Ultrasonic welding is applicable to acrylics, PVC, polystyrene and synthetic textile. Lap and tee spot joints are made best of all. Satisfactory joints are also made in the case of lap welds in static jigs. In all cases, neither edge preparation nor filler material is needed. Ultrasonic welds can also be made in dissimilar plastics.

5. **High frequency induction welding.** The high-frequency induction welding of plastics is a widely used process. The principle is the same as of the process of induction welding of metals (See Fig. 3.36 b, Chapter 3). The process consists in that the workpiece is placed in a high-frequency electric field, set up between two metal electrodes. After the plastics have been heated in the r.f. field, the electrodes apply pressure to complete the joint.

6. **Extruded-filler Welding.** In this method, the filler fed into the joint is in a viscous fluidic state. The hot filler material melts the edges of the plastic being joined, and a strong bond is formed between the filler and parent material. The process makes satisfactory welds in both films and heavy-gauge sheets.

7. **Nuclear Welding.** In this method, the workpieces to be joined are irradiated with a steam of neutrons. The surfaces to be welded are bombard these elements, nuclear reactions take place, resulting in the generation of heat. Due to this, the surfaces to be joined turn into a polyethylene, polystyrene, quartz, aluminium and some other metals. The method can not be applied to materials which become strongly radioactive, when irradiated with neutrons.

8. **Infra-red Welding.** In this process, welding heat is supplied by a source of infra-red light, such as, a sylite glower, a chrome-steel is carried out on a black backing plate of a formed plastic, sponge rubber or thick rubberized fabric. The resistence of the back up plate held firmly against the workpiece supplies the necessary welding pressure. The process is satisfactory for joining polyethylene films.

As noted above since the thermoplastics soften and melt as the temperature is increased, the methods discussed above, welding (fusion methods) and other methods are used to join thermoplastics.

Thermoplastics can also be joined by adhesive bounding.

Thermosets do not soften on heating. So, the above methods can not be used to join thermosets. Thermosets are usually joined by using :–

(*i*) threaded or other moulded in inserts.

(*ii*) Mechanical fasteners.

(*iii*) Solvent bonding. This method involves the following steps :–

(*a*) The surfaces to be joined are roughened with an abrasive.

(*b*) Wiping the surfaces with a solvent.

(*c*) Pressing the surfaces together and holding the pressure until sufficient bond strength is obtained.

PROBLEMS

1. Define the term " Polymer ".
2. What is " polymerization " ? Discuss the various methods of polymerization.
3. What is degree of polymerization (D.P.) ?
4. What are the various agents added to polymers to modify their properties ? Discuss their functions.

5. What are plastic materials ?
6. What are thermoplastics and thermosetting materials ?
7. Discuss the properties of plastic materials and give their limitations.
8. Sketch and explain injection moulding process.
9. Why is screw injection moulding machine better than a ram type injection moulding machine ?
10. Sketch and explain extrusion process for plastics.
11. Sketch and explain the principle of vacuum forming process for plastics.
12. Sketch and explain the "Compression moulding" and "Transfer moulding" processes.
13. What special cares are taken when machining plastics ?
14. Sketch and explain calendering process for plastics.
15. What is "rotational moulding" of plastics ?
16. Sketch and explain "Blow moulding" process for plastics.
17. Discuss the various design factors for plastic parts.
18. Write short notes on : (*i*) RIM (*ii*) Cold forming (*iii*) Solid state forming (*iv*) Spinning (v) Thermoplastic stamping.
19. What are "Potting" and "Encapsulation" ?
20. What is foam moulding ?
21. Discuss the various methods of welding of plastics.
22. Discuss the factors which have led to the rapid growth of polymers.
23. Write the product applications of engineering plastics.
24. List the typical product applications of compression moulding process.
25. What are the advantages of Transfer moulding process.
26. What is the difference between "Fillers" and "Additives" as used with polymers ?
27. List the advantages of cold forming of plastics.
28. What is the difference between potting and encapsulation ?
29. List several products that can be produced by Injection moulding process.
30. Give the advantages of thermoplastic forming process. Where do the thermoformed parts find their largest market ?
31. How are the injection moulding machines rated ?
32. What is "Parison"?
33. How is thin plastic film produced ?
34. What are the differences among three types of compression moulding methods ?
35. List the field of application of "Transfer Moulding" process.
36. What is "Wet extrusion" of plastics ?
37. What is the drawback of ram type extrusion machine ?
38. What are the forms of raw materials for processing plastics into products ?
39. List the product applications of "RIM".
40. List the product applications of Calendering process.
41. List several products that can be made by Rotational moulding process.
42. List the product applications of Blow moulding process.
43. Discuss : Design of Plastic parts.
44. Sketch and label a schematic of mould for compression moulding.

45. Sketch and label a schematic of mould for injection moulding.
46. Discuss the various cavity designs for plastic parts.
47. Write a note on "Materials for Moulds" for plastic parts.
48. How the moulds for plastic parts are manufactured ?
49. What additives are used in plastics and why ?
50. What are the differences tween thermoplastic and thermosetting plastics ?
51. List the effects a plasticizing agent have on a polymer.
52. What is an elastomer ?
53. How can you tell whether a component is made of a thermoplastic ?
54. Explain why thermoplastics are easy to recycle than thermosetting plastics.
55. The various plastic components in an auto-mobile are made of thermoplastics or thermosetting plastics.
56. List the limitations of injection moulding process.
57. List the products made by thermo-forming process.
58. List the similarities between compression moulding and closed-die forging.
59. Write the advantages of thermo-forming process.
60. List the product applications of foam moulding process.
61. List the advantages of casting process for plastics.
62. Compare the various methods for processing plastics.
63. Give the specifications of Plastic Processing Machines :

Machine		Main Specifications
(i) Plastic Injection Moulding Machine	:	Tonnage
(ii) Plastic Compression Moulding Machine	:	Tonnage
(iii) Plastic Blow Moulding Machines	:	Tonnage
(iv) Plastic Extruders	:	Screw Size

APPENDIX

Problems From Competitive Examinations
(GATE, IES, IAS)

CHAPTER 2

1. In a green-sand moulding process, uniform ramming leads to
   (A) less change of gas porosity
   (B) uniform flow of molten metal into the mould cavity
   (C) greater dimensional stability of the casting
   (D) less sand expansion type of casting defect [*GATE 1992*] (**Ans.** C)
2. Match the following moulding casting processes with the product:

Moulding/Casting processes	*Product*
(A) Slush casting	(P) Turbine blades
(B) Shell moulding	(Q) Machine tool bed
(C) Dry sand moulding	(R) Cylinder block
(D) Centrifugal casting	(S) Hollow castings like lamp shades
	(T) Rain water pipe
	(U) Cast iron shoe brake

[*GATE 1992*] (**Ans.** A – S, B – P, C – R, D – T)

3. Centrifugally cast products have
   (A) large grain structure with high porority
   (B) fine grain structure with high density
   (C) fine grain structure with low density
   (D) segregation of slag towards the outer skin of the casting [*GATE 1993*] (**Ans.** B)
4. List I gives a number of processes and List II gives a number of products. Match correct pairs:

List I	*List II*
(A) Investment casting	1. Turbine rotors
(B) Centrifugal casting	2. Turbine blades
(C) Die-casting	3. Connecting rods
(D) Drop forging	4. Galvanized iron pipe
(E) Extrusion	5. Cast iron pipes
(F) Shell moulding	6. Carburettor body

[*GATE 1994*]

(**Ans.** A – 2, B – 5, C – 6, D – 3, E – 4, F – 1).

5. Light impurities in the molten metal are prevented from reaching the mould cavity by providing a
   (A) Strainer (B) Bottom well (C) Skim bob (D) All of the above
   [*GATE 1996*] (**Ans.** C)
6. Chills are used in moulds to
   (A) achieve directional solidification
   (B) reduce the possibility of blow holes
   (C) reduce freezing time
   (D) smoothen metal flow for reducing splatter [*GATE 1998*] [*IAS 1994*] (**Ans.** A)

7. Which of the following materials requires the largest shrinkage allowance, while making a pattern for casting?

(A) Aluminium (B) Brass (C) Cast iron (D) Plair carbon steel

[*GATE 1999*] (**Ans.** D)

8. Disposable patterns are made of

(A) Wood (B) Rubber (C) Metal (D) Polystyrene (**Ans.** D)

9. Match List I with List II and select the correct answer using the codes given below the lists:

List I (Equipments)	*List II (Functions)*
A. Hot chamber machine	1. Cleaning
B. Muller	2. Core making
C. Dielectric baker	3. Die casting
D. Sand blasting	4. Annealing
	5. Mixing

Codes:

	A	B	C	D		A	B	C	D
(*a*)	3	5	2	1	(*b*)	4	2	5	3
(*c*)	4	2	3	1	(*d*)	3	5	1	2

[*IES 1993*] (**Ans.** *a*)

10. Which of the following materials can be used for making patterns?

1. Aluminium 2. Wax 3. Mercury 4. Lead

Select the correct answer using the codes given below:

Codes:

(A) 1, 3 and 4 (B) 2, 3 and 4 (C) 1, 2 and 4 (D) 1, 2 and 3

[*IES 1994*] (**Ans.** D)

11. Which of the following materials will require the largest size of riser for the same size of casting?

(A) Aluminium (B) Cast iron (C) Steel (D) Copper

[*IES 1995*] (**Ans.** A)

12. Directional solidification in castings can be improved by

(A) Chills and chaplets (B) Chills and padding

(C) Chaplets and padding (D) Chills, chaplets and padding

[*IES 1995*] (**Ans.** B)

13. Match List I with List II and select the correct answer taking the help of codes given below the lists:

List I (Products)	*List II (Process of Manufacture)*
A. Automobile piston in aluminium alloy	1. Pressure die-casting
B. Engine crankshaft in spherioidal graphite iron	2. Gravity die-casting
C. Carburettor Housing in aluminium alloy	3. Sand casting
D. Cast titanium blades	4. Precision investment casting
	5. Shell moulding

Codes:

	A	B	C	D		A	B	C	D
(*a*)	2	3	1	5	(*b*)	3	2	1	5
(c)	2	1	3	4	(*d*)	4	1	2	3

[*IES 1995*] (**Ans.** *A*)

14. Consider the following ingredients used in moulding:
    1. Dry silica sand 2. Clay 3. Phenol formaldehyde 4. Sodium silicate
    Those used for shell mould casting include:
    (A) 1, 2 and 4 (B) 2, 3 and 4 (C) 1 and 3 (D) 1, 2, 3, 4
    [*IES 1996*] (**Ans.** C)
15. Which of the following methods are used for obtaining directional solidification for riser design:
    1. Suitable placement of chills 2. Suitable placement of chaplets 3. Employing padding
    Select the correct answer:
    (*a*) 1 and 2 (*b*) 1 and 3 (*c*) 2 and 3 (*d*) 1, 2 and 3
    [*IES 1996*] (**Ans.** B)
16. Misrun is a casting defect which occurs due to
    (A) very high pouring temperature of the metal
    (B) insufficient fluidity of the molten metal
    (C) absorption of gases by the liquid metal
    (D) improper alignment of the mould flasks [*IES 1996*] (**Ans.** B)
17. Which of the following pairs are correctly matched?
    1. Pit moulding .... For large jobs 2. Investment moulding .... lost wax process
    3. Plaster moulding ...... Mould prepared in gypsum
    (A) 1, 2 and 3 (B) 1 and 2 (C) 1 and 3 (D) 2 and 3
    [*IES 1996*] (**Ans.** A)
18. Which one of the following pairs is not correctly matched?
    (A) Aluminium alloy Piston ...... Pressure die casting
    (B) Jewellery .... Lost wax process
    (C) Large pipes .... Centrifugal casting
    (D) Large bells ...... Loam moulding [*IES 1997*] (**Ans.** D)
19. If the melting ratio of a cupola is 10 : 1, then the coke requirement for one ton melt will be
    (A) 0.1 ton (B) 10 tons (C) 1 ton (D) 11 tons
    [*IES 1997*] (**Ans.** B)
20. Which one of the following are the requirements of an ideal gating system?
    1. The molten metal should enter the mould cavity with as high a velocity as possible.
    2. It should facilitate complete filling of the mould cavity
    3. It should be able to prevent the absorption of air or gases from the surroundings of the molten metal while flowing through it.
    Select the correct answer using the codes given below:
    (A) 1, 2 and 3 (B) 1 and 2 (C) 2 and 3 (D) 1 and 3
    [*IES 1998*] (**Ans.** C)
21. A spherical drop of molten metal of radius 2 mm was found to solidify in 10 seconds. A similar drop of radius 4 mm will solidify in
    (A) 14.14 s (B) 20 s (C) 28.30 s (D) 40 s
    [*IES 1998*] (**Ans.** D)

(**Hint:** Solidification for sphere is $\alpha\left(\frac{V}{A}\right)^2$)

22. In solidification of metal during casting, compensation for solid contraction is
    (A) provided by the oversize pattern
    (B) achieved by properly placed risers
    (C) obtained by promoting directional solidification
    (D) made by providing chills. [*IES 1999*] (**Ans.** A)
23. Disk-shaped components are cast by
    (A) True Centifigural casting (B) Semi-Centrifugal casting
    (C) Centifuge casting (**Ans.** B)
24. Consider the following ingredients used in moulding:
    1. Dry silica sand 2. Clay 3. Ethyl silicate 4. Phenol formaldehyde
    Those used for Lost Wax casting method include:
    (A) 1, 2 and 4 (B) 2, 3 and 4 (C) 1 and 3 (D) 1, 2, 3 and 4
    (**Ans.** *C*)
25. Metal patterns require ...... (more/less) draft allowance than wooden patterns. (**Ans.** more)
26. The molten metal is poured from the pouring basin to the gate with the help of a
    (A) Riser (B) Sprue (C) Runner (D) Core (**Ans.** B)
27. In cold chamber die-casting process, only non-ferrous alloys with ...... (high melting point/low melting point) are cast. (**Ans.** High melting point)
28. In the casting of large pipes by true centrifugal casting
    (A) Core is of sand (B) Core is of metal (C) no core is used (**Ans.** C)
29. Core prints are provided on patterns
    (A) to support the core
    (B) to locate the core in the mould
    (C) to support as well as locate the core in the mould (**Ans.** C)
30. Contraction allowance is provided on a pattern:
    (A) for machining of castings (B) for contraction in metal on cooling
    (C) for making a good casting (**Ans.** B)
31. To provide the machining allowance, a pattern should be made ...... (larger/smaller) than the size of the finished casting required. (Ans. Larger)
32. A gate
    (A) acts as a reservoir for the molten metal
    (B) delivers molten metal into the mould cavity
    (C) delivers molten metal from the pouring basin to runner
    (D) all of the above (**Ans.** B)
33. A riser
    (A) acts as a reservoir for the molten metal
    (B) delivers molten metal into the mould cavity
    (C) delivers molten metal from the pouring basin to runner
    (D) feeds the molten metal to the casting in order to compensate for solidification shrinkage
    (**Ans.** D)
34. The function of a core is:
    (A) to improve mould surface

(B) to form internal cavities in the casting

(C) to form a part of a green sand mould

(D) none of the above

(E) all of the above **(Ans.** E)

35. When the molten metal flows into the cavity of a metallic mould by gravity, the method of casting is known as:

(A) Die-casting (B) Centrifugal casting

(C) Permanent mould casting (D) Plaster mould casting **(Ans.** C)

36. The toys and ornaments of non-ferrous alloys are made by:

(A) Die-casting (B) Lost-wax method

(C) Permanent mould casting (D) Slush casting **(Ans.** D)

37. Two castings of the same metal have the same surface area. One casting is in the form of a sphere and the other is a cube. What is the ratio of the solidification time for the sphere to that of a cube. [*GATE 1998*]

Solution. According to Chvorinov's rule, solidification time is $\propto \left(\frac{V}{A}\right)^2$, where

V = Volume of casting

A = Surface area of casting

Now $A_s = A_c$

$$\therefore \quad \text{Ratio of solidification time} = \left(\frac{V_s}{V_c}\right)^2$$

$$= \left(\frac{\frac{4}{3}\pi R^3}{a^3}\right)^2 = \frac{16\pi^2 R^6}{9a^6}$$

Where R = Radius of sphere, and

a = side of the cube.

38. Risering is not needed when casting Gray cast iron, True/False. **(Ans.** True)

39. Calculate the ratio of the solidification times of two steel cylindrical risers of sizes 30 cm in diameter by 60 cm height and 60 cm in diameter by 30 cm in height subjected to identical conditions of cooling. [*GATE 1992*]

Solution. According to Chvorinov's rule solidification time is $\propto \left(\frac{V}{A}\right)^2$ (See problem 37)

Riser 1.

$$\frac{V}{A} = \frac{\frac{\pi}{4} d^2 \times h}{\pi d h + 2 \times \frac{\pi}{4} d^2}$$

$$= \frac{\frac{1}{4} \times 900 \times 60}{30 \times 60 + \frac{1}{2} \times 900} = \frac{13500}{2250} = 6$$

Riser 2: $$\frac{V}{A}=\frac{\frac{1}{4}\times 3600\times 30}{60\times 30+\frac{1}{2}\times 3600}=\frac{27000}{3600}=7.5$$

$\therefore$ Ratio of solidification times, $=\left(\frac{7.5}{6}\right)^2=1.5625$

40. An aluminium cube of 10 cm side has to be cast along a cylindrical riser of height equal to its diameter. The riser is not insulated on any surface. If the volume shrinkage of aluminium during solidification is 6%, calculate
(*i*) Shrinkage volume of cube on solidification, and
(*ii*) minimum size of the riser so that it can provide the shrinkage volume.

[*GATE 1993*]

Solution:

(*i*) Volume of casting = 10 × 10 × 10 = 1000 cm^3

Shrinkage volume = 6% = 60 cm^3 (Depending upon metal, this shrinkage varies from 2.5 to 7.5%)

A riser should be designed with the minimum possible volume while having a longer solidification time than for casting.

Now from practice,

Minimum volume of riser is approximately three times the shrinkage volume,

$\therefore$ Minimum volume of riser = 3 × 60 = 180 cm^3

$$\therefore \qquad \frac{\pi}{4}d_r^2\times h_r=180$$

now $h_r = d_r$

$\therefore$ From here, $d_r = 5.2$ cm

Now for sound casting, the metal in the riser should be the last to cool, that is, the riser should have a longer solidification time than the casting, so

$$\left(\frac{A}{V}\right)_r\le\left(\frac{A}{V}\right)_c \quad \text{or} \quad \left(\frac{V}{A}\right)_r\ge\left(\frac{V}{A}\right)_C$$

Now $$\left(\frac{A}{V}\right)_r=\frac{6}{d_r}=\frac{6}{5.2}$$ (it is simple to prove this, with $d_r = h_r$)

$$\left(\frac{A}{V}\right)_c=\frac{6\times 10\times 10}{10\times 10\times 10}=\frac{3}{5}$$

As is clear $\left(\frac{A}{V}\right)_r$ is > $\left(\frac{A}{V}\right)_c$, which is not desirable.

$$\therefore \qquad \left(\frac{A}{V}\right)_r\le\left(\frac{A}{V}\right)_c$$

$$\le 0.6$$

or $\frac{6}{d_r}\le 0.6$, $\quad \therefore \frac{10}{d_r}\le 1$, $\quad \therefore d_r\ge 10$.

$\therefore$ Minimum size of riser = 10 cm diameter × 10 cm height

$$\therefore \qquad V_r=\frac{\pi}{4}\times 100\times 10= 785.4 \text{ cm}^3.$$

41. With cylindrical riser, prove that for a longer solidification time, diameter of riser = height of riser.

Solution: We know that for longer solidification time, the ratio $\left(\frac{V}{A}\right)$ should be maximum or the ratio, $\left(\frac{A}{V}\right)$ should be minimum.

Now, $$V = \frac{\pi}{4} d^2 \times h$$

$\therefore$ $$h = \frac{4V}{\pi d^2}$$

$$A = \pi d.h + 2.\frac{\pi}{4} d^2$$

$\therefore$ $$A = \pi d.\frac{4V}{\pi d^2} + \frac{\pi}{2} d^2$$

$\therefore$ $$A = \frac{4V}{d} + \frac{\pi}{2} d^2$$

For A to be minimum for a given V, $\partial A / \partial d = 0$, and it will be seen that for this, $d = h$.

i.e. $$-\frac{4V}{d^2} + \pi d = 0$$

or $$d^3 = \frac{4V}{\pi}$$

Also, $$\frac{4V}{\pi} = hd^2$$

$\therefore$ $$d^3 = hd^2$$

or $$d = h$$

This optimum ratio is true only for a side riser, Fig. 3.22 (*a*). For a top riser,

$$A = \pi dh + \frac{\pi}{4} d^2$$

Doing the analysis as above, it will be seen that

$$d = 2h$$

or $$\frac{h}{d} = \frac{1}{2}$$

42. Shrinkage allowance on pattern is provided to compensate for shrinkage when
(*a*) the temperature of liquid metal drops from pouring to freezing temperature
(*b*) the metal changes from liquid to solid state at freezing temperature
(*c*) the temperature of solid phase drops from freezing to room temperature
(*d*) the temperature of metal drops from pouring to room temperature

[*GATE 2001*] (**Ans.** *c*)

43. Two solid workpieces (*i*) Sphere with radius R, (*ii*) a cylinder with diameter equal to its height, have to be sand cast. Both workpieces have the same volume. Show that the cylindrical workpiece will solidify faster than the spherical work piece. [*GATE 2001*]

Solution. Volume of sphere = $\frac{4}{3}\pi R^3$

Volume of cylinder = $\frac{\pi}{4}d^2.h$

$$= \frac{\pi}{4}d^3$$

$$\frac{\pi}{4}d^3 = \frac{4}{3}\pi R^3$$

$$\therefore \quad d^3 = \frac{16}{3}R^3$$

Now solidification time is $\alpha\left(\frac{V}{A}\right)^2$

Now, $$\left(\frac{V}{A}\right)_{cyl} = \frac{\frac{\pi}{4}d^2.h}{\pi dh + 2.\frac{\pi}{4}d^2} = \frac{d}{6} \text{ (with } d = h)$$

$$\therefore \quad \left(\frac{V}{A}\right)_{cyl} = \frac{1}{6}.\left(\frac{16}{3}\right)^{\frac{1}{3}}.R = 0.29 \text{ R}$$

Again, $$\left(\frac{V}{A}\right)_{sphere} = \frac{\frac{4}{3}\pi R^3}{4\pi R^2} = \frac{R}{3} = 0.33R$$

Now, it is known that risers with a higher value of $\left(\frac{V}{A}\right)$ loose heat at a slower rate. Thus, it is clear from above, that a solid casting of cylindrical section will solidify faster than the spherical work piece.

44. The primary purpose of sprue in a casting mould is to
    (a) feed the casting at a rate consistent with the rate of solidification
    (b) act as a reservoir for molten metal
    (c) feed molten metal from the pouring basin to the gate
    (d) help feed the casting until all solidification takes place (*GATE 2002*) (**Ans.** c)

45. In centrifugal casting, the impurities are
    (*a*) uniformly distributed
    (*b*) forced towards the outer surface
    (*c*) trapped near the mean radius of the casting
    (*d*) collected at the centre of casting [*GATE 2002*] (**Ans.** *a*)

CHAPTER 3

1. For resistance spot welding of 1.5 mm thick steel sheets, the current required is of the order of

   (A) 10 A (B) 100 A (C) 1000 A (D) 10,000A

   [*GATE 1992*] (**Ans.** D)

2. In an explosive welding process, the ..... (maximum/minimum) velocity of impact is fixed by the velocity of sound in the .... (flyer/target) plate material.

   [*GATE 1992*] (**Ans.** *maximum, flyer*)

3. In *d.c.* welding, the straight polarity (electrode negative) results in

   (A) lower penetration (B) lower depositing rate

   (C) less heating of work piece (D) smaller weld pod [*GATE 1993*] (**Ans.** B)

4. The electrodes used in arc welding are coated electrodes. The coating is not expected to

   (A) provide protective atmosphere to weld (B) stabilize the arc

   (C) add alloying elements (D) prevent electrode from contamination

   [*GATE 1994*] (**Ans.** D)

5. The ratio of acetylene to oxygen is approximately ...... for a neutral flame used in gas welding. [*GATE 1994*] (**Ans.** 1 : 1)

6. Preheating before welding is done to

   (A) make the steel softer

   (B) burn away oil, grease etc. from the plate surface

   (C) prevent cold cracks

   (D) prevent plate distortion [*GATE 1996*] (**Ans.** C)

7. Match 4 correct pairs between List I and List II

List I	*List II*
(A) Welding of aluminium alloy	1. Submerged arc welding
(B) Ship building	2. Electron beam welding
(C) Joining of HSS drill bit to carbon steel shank	3. TIG welding
(D) Deep penetration precision welds	4. Gas welding

   [*GATE 1996*] (**Ans.** A – 3, B – 1, C – 4, D – 2)

8. For butt-welding 40 mm thick steel plates when the expected quantity of such jobs is 5000 per month over a period of 10 years, choose the best suitable welding process out of the following available alternatives

   (A) Submerged arc welding (B) Oxy–acetylene gas welding

   (C) Electron beam welding (D) MIG welding [*GATE 1999*] (**Ans.** B)

9. Electron–beam welding can be carried out in

   (A) open air (B) a shielding gas environment

   (C) a pressurised inert gas chamber (D) vacuum

   [*IES 1993*] (**Ans.** D)

10. In gas welding of mild steel using an oxy-acetylene flame, the total amount of acetylene used was 10 litre. The oxygen consumption from the cylinder is

    (A) 5 litre (B) 10 litre (C) 15 litre (D) 20 litre

    [*IAS 1994*] (**Ans.** B)

    [**Hint:** For gas welding of M.S. neutral flame is used].

11. Which one of the following welding processes uses non-consumable electrodes:

(A) TIG welding (B) MIG welding

(C) Manual arc welding (D) Submerged arc welding

[*IES 1994*] (**Ans.** A)

12. Match List I with List II and select the correct answer using the codes given below the lists:

List I (Filler)	*List II (Joining process)*
A. Cu, Zn, Ag alloy	1. Braze welding
B. Cu, Zn, alloy	2. Brazing
C. Pb, Sn alloy	3. Soldering
D. Iron oxide and aluminium powder	4. TIG welding of Al

Codes:

	A	B	C	D		A	B	C	D
(*a*)	2	1	3	–	(*b*)	1	2	4	–
(*c*)	2	1	3	4	(*d*)	2	–	3	4

[*IES 1994*] (**Ans.** *a*)

13. Consider the following statements:

MIG welding process uses

1. consumable electrode 2. non-consumable electrode

3. D.C. power supply 4. A.C. power supply

Of these statements

(A) 2 and 4 are correct (B) 2 and 3 are correct

(C) 1 and 4 are correct (D) 1 and 3 are correct [*IES 1997*] (**Ans.** D)

14. In oxy-acetylene gas welding, for complete combustion, the volume of oxygen required per unit ton of acetylene is

(A) 1 (B) 1.5 (C) 2 (D) 2.5

[*IES 1998*] (**Ans.** D)

15. The voltage-current characteristics of a dc generator for arc welding is a straight line between an open circuit voltage of 80 V and short circuit current of 300 A. The generator setting for maximum arc power will be

(A) 80 V and 150 A (B) 40 V and 300 A (C) 40 V and 150 A (D) 80 V and 300 A

[*IES 1998*] (**Ans.** C)

16. Which of the following joining processes are best suited for manufacturing pipes to carry gas products?

1. Riveting 2. Welding 3. Nuts and bolts

Select the correct answer using the codes given below:

Codes :

(A) 1 and 2 (B) 1 and 3 (C) 2 alone (D) 1, 2 and 3

[*IES 1998*] (**Ans.** C)

17. The correct sequence of the given materials in descending order of their weldability is

(A) MS, Cu, C.I, Al (B) C.I., M.S., Al, Cu (C) Cu, C.I., MS, Al (D) Al, Cu, C.I., M.S.

[*IES 1999*] (**Ans.** A)

18. Failure of a bead weld between a heavy steel section and a thin section is mainly due to the formation of

(A) spheriodite

(B) bainite

(C) carbon free zone due to burning of carbon at high temperature

(D) martensite [*GATE 1998*] (**Ans.** C)

19. The open circuit voltage (OCV) in arc welding ranges from:

(A) 40 to 80 V (B) 100 to 150 V (C) 200 to 220 V (D) 400 to 440 V

(**Ans.** A)

20. In thermit welding, heat is generated

(A) from combustion of gas

(B) by an arc

(C) by chemical reaction between aluminium and iron oxide

(D) none of the above (**Ans.** C)

21. For resistance welding

(A) voltage is high, current is low (B) voltage is low, current is high

(C) both voltage and current are low (D) both voltage and current are high

(**Ans.** B)

22. In resistance welding, the greatest resistance, when the power is switched on, is:

(A) at the point of contact between the electrode and the job.

(B) at the point of contact of work pieces to be joined

(C) at the surface

(D) none of the above (**Ans.** B)

23. The maximum temperature produced by oxy-hydrogen flame is about:

(A) 3300°C (B) 2500°C (C) 2000°C (D) 1800°C (**Ans.** B)

24. Majority of the oxy-acetylene welding is done with:

(A) Neutral flame (B) Reducing flame (C) Oxidising flame (**Ans.** A)

25. Brasses and Bronzes are welded by:

(A) Neutral flame (B) Reducing flame (C) Oxidising flame (**Ans.** C)

26. Fluxes are used while welding :

(A) to increase the rate of welding

(B) to clean the joint

(C) to prevent oxidation of metal during welding

(D) all the above (**Ans.** D)

27. The selection of welding electrodes depends upon:

(A) the thickness of the metal to be joined (B) the type of metal to be joined

(C) position of welding (D) strength of joint

(E) all the above (**Ans.** E)

28. Low pressure OAW is the name given to the welding process in which

(A) oxygen is supplied at low pressure

(B) acetylene is supplied at low pressure

(C) Both oxygen and acetylene are supplied at low pressure (**Ans.** B)

29. In arc welding, the electric arc is produced between the job and the electrode due to

(A) voltage (B) flow of current (C) contact resistance (D) all of the above

(**Ans.** C)

30. In submerged arc welding, the arc is struck between
   (A) consumable coated electrode and job (B) bon-consumable electrode and job
   (C) consumable bare electrode and job (D) two tungsten electrodes and the job
   (**Ans.** C)

Note: For MIG, the answer will be C

TIG, the answer will be B

SMAW, the answer will be A

31. Resistance spot welding is performed on two plates of 1.5 mm thickness with 6 mm diameter electrode, using 15000 A current for a time duration of 0.25 s. Assuming the interface resistance to be 0.0001 ohm, the heat generated to form the weld is
   (A) 5625 W-sec (B) 8437 W-sec (C) 22500 W-sec (D) 33750 W-sec
   [*GATE 2001*] (**Ans.** A)
32. Two plates of the same metal having equal thickness are to be butt welded with electric arc. When the plate thickness changes, welding is achieved by
   (A) adjusting the current (B) adjusting the duration of current
   (C) changing the electrode size (D) changing the electrode coating
   [*GATE 2001*] (**Ans.** A, C).
33. Which of the following arc welding processes does not use consumable electrodes :
   (A) SAW (B) GMAW
   (C) GTAW (D) None of these (GATE 2002) (**Ans.** C)
34. The temperature of a carburising flame in gas welding is .... that of a neutral or oxidising flame.
   (A) Lower than (B) Higher than
   (C) Equal to (D) Unrelated to (*GATE 2002*) (**Ans.** A)
35. The arc length-voltage characteristic of a D.C. arc is given by the equation V = 24 + 4L, where V is voltage in volts and L is arc length in mm. The static volt-ampere characteristic of the power source is approximated by a straight line with a no load voltage of 80 V and a short circuit current of 600 A. Determine the optimum arc length for maximum power.
   (*GATE 2002*)

V = 24 + 4L

The static volt-ampere characteristic of power source is given as,

$$\frac{V}{80}+\frac{I}{600}=1$$

Now Power, $P = V \times I$

$$= (24 + 4L) \times \left(1-\frac{V}{80}\right) \times 600$$

$$= (24 + 4L) \times \left(\frac{80-24-4L}{80}\right) \times 600$$

$$= \frac{6+L}{20} \times (56 - 4L) \times 600$$

$$= 120\,(6 + L)\,(14 - L)$$

$$= 120\,(84 + 8L - L^2)$$

For Max. P, $\frac{\partial P}{\partial L} = 0$

$\therefore \; 0 + 120 \times 8 - 2L \times 120 = 0$

$\therefore \; L = 4$ mm

36. The purpose of a gas-welding flux is to:
(A) lower the melting point of the metal
(B) lower the melting point of the oxide
(C) remove oxides from the surface of the metal
(D) remove elements from parent metal *(L.U., BTE, M.U.)* **(Ans.** C)

37. The flux coating of an electric arc electrode has a melting point:
(A) Higher than the metallic core
(B) Lower than the metallic core
(C) The same as the metallic core
(D) The same as the metal being welded. **(Ans.** A)

38. Which gas from the following will form a hard constituent when it combines with molten steel and remains after solidification has taken place:
(A) O_2 (B) N_2
(C) H_2 (D) He *(BHU, BTE, AMIE, I.Mech.E)* **(Ans.** B)

39. The purpose of preheating low-alloy steel pipes, before they are electric arc welded is to:
(A) Refine grain structure
(B) Relieve internal streesses
(C) Retard rapid cooling
(D) Regulate excessive expansion *(L.U., AMIE, UPSC)* **(Ans.** C)

40. Hard-Zone cracking in low-alloy steel due to welding is the result of an absorption of:–
(A) N_2 (B) O_2
(C) H_2 (D) C *(BTE, UPSC, DU)* **(Ans.** C)

41. A low carbon steel plate is to be welded by the manual metal are welding process using a linear V – I characteristic D – C. Power source. The following data are available :
OCV of Power source = 62 V
Short circuit current = 130 A
Arc length, L = 4 mm
Traverse speed of welding = 15 cm/mm
Efficiency of heat input = 85%
Voltage is given as V = 20 + 1.5 L
Calculate the heat input into the workprice *(GATE 1992)*

Solution. Volt ampere characteristic of source is given as

$$\frac{V}{62} + \frac{I}{130} = 1$$

Now V = 20 + 1.5 × 4 = 26 Volts

$$\therefore \; I = 130 \left(1 - \frac{26}{62}\right) = 75.5 \text{ A}$$

$\therefore$ Power consumed, P = V × I = 75.5 × 26 = 1963 W

Efficiency of heat input = 85%

$\therefore$ Heat input into workpiece = 0.85 × 1963 = 1668.5 watts

CHAPTER 4

1. The thickness of the blank needed to produce, by power spinning a missile cone of thickness 1.5 mm and half cone angle 30°, is

   (*a*) 3.00 mm (*b*) 2.5 mm (*c*) 2.0 mm (*d*) 1.55 mm

   [*GATE 1992*] (**Ans.** *a*)

2. The true strain for low carbon steel bar which is doubled in length by forging is

   (*a*) 0.307 (*b*) 0.5 (*c*) 0.693 (*d*) 1.0

   [*GATE 1992*] (**Ans.** *c*)

3. The process of hot extrusion is used to produce

   (*a*) curtain rods made of aluminium (*b*) steel pipes for domestic water supply

   (*c*) stainless steel tubes used in furniture (*d*) Large size pipes used in city water mains

   [*GATE 1994*] (**Ans.** *a, c*)

4. Match 4 correct pairs between List I and List II

List I	*List II*
(*a*) Rivets for air craft body	1. Forging
(*b*) Carburettor body	2. Cold heading
(*c*) Crank shafts	3. Aluminium-based alloy
(*d*) Nails	4. Pressure die casting
	5. Investment casting

[*GATE 1996*] (**Ans.** *a*– 3, *b* – 4, *c* – 1, *d* – 2)

5. Match 4 correct pairs between List I and List II

List I	*List II*
(A) Sand casting	(1) Symmetrical and circular shapes only
(B) Plaster mould casting	(2) Parts have hardened skin and soft interior
(C) Shell mould casting	(3) Minimum post casting processing
(D) Investment casting	(4) Parts have a tendency to warp
	(5) Parts have soft skin and hard interior
	(6) Suitable only for non-ferrous metals

[*GATE 1998*] (**Ans.** A – 2, B – 6, C – 1, D – 3)

Match 4 Correct pairs between List I and List II

6.

List I	*List II*
(A) Aluminium brake shoe	(1) Deep drawing
(B) Plastic water bottle	(2) Blow moulding
(C) Stainless steel cups	(3) sand casting
(D) Soft drink can (aluminium)	(4) Centrifugal casting
	(5) Impact extrusion
	(6) Upset forging

[*GATE 1998*] (**Ans.** A – 3, B– 2, C – 1, D – 5).

7. In sheet metal work, the cutting force on the tool can be reduced by

   (A) grinding the cutting edges sharp (B) increasing the hardness of tool

   (C) providing shear angle on tool (D) increasing the hardness of die

   [*IES 1993*] (**Ans.** C)

8. Tandem drawing of wires and tubes is necessary because
   (A) it is not possible to reduce in one stage
   (B) annealing is needed between stages
   (C) accuracy in dimensions is not possible otherwise
   (D) surface finish improves after every drawing stage [*IES 1993*] (**Ans.** A)
9. In order to get uniform thickness of plate by rolling process, one provides
   (A) camber on the rolls (B) offset on the rolls
   (C) hardening of the rolls (D) antifriction bearings [*IES 1993*] (**Ans.** A)
10. A moving mandrel is used in
    (A) wire drawing (B) tube drawing (C) metal cutting (D) forging
    [*IES 1993*] (**Ans.** B)
11. Which of the following methods can be used for manufacturing 2 metre long seamless metallic tubes?
    1. Drawing 2. Extrusion 3. Rolling 4. Spinning [*IAS 1994*]
    Select the correct answer using the codes given below:
    Codes.
    (A) 1 and 3 (B) 2 and 3 (C) 1, 3 and 4 (D) 2, 3 and 4 (**Ans.** B)
12. In blanking operation, the clearance provided is
    (A) 50% on punch and 50% on die
    (B) on die
    (C) on punch
    (D) on die or punch depending upon designers choice [*IAS 1995*] (**Ans.** C)
13. Consider the following states of stress:
    1. Compressive stress in flange 2. Tensile stress in the wall
    3. Tensile stress on the bottom part
    During drawing operation, the states of stress in cup would include
    (*a*) 1 and 2 (*a*) 1 and 3 (*c*) 2 and 3 (*d*) 1, 2 and 3
    [*IAS 1995*] (**Ans.** *a*)
14. The following operations are performed while preparing the billets for extrusion process:
    1. Alkaline cleaning 2. Phosphate coating
    3. Pickling 4. Lubrication with reactive soap
    The correct sequence of these operations is
    (A) 3, 1, 4, 2 (B) 1, 3, 2, 4 (C) 1, 3, 4, 2 (D) 3, 1, 2, 4
    [IAS 1995] (**Ans.** D)
15. Match List I with List II and select the correct answer using the codes given below the lists:

List I (Metal forming process)	*List II (A similar process)*
A. Blanking	1. Wire drawing
B. Coining	2. Piercing
C. Extrusion	3. Embossing
D. Cup drawing	4. Rolling
	5. Bending

Codes:	A	B	C	D		A	B	C	D
(*a*)	2	3	4	1	(*b*)	2	3	1	4
(c)	3	2	1	5	(*d*)	2	3	1	5

[*IES 1994*] (**Ans.** *b*)

16. In sheet metal blanking, shear is provided on punches and dies so that

(A) press load is reduced (B) good cut edge is obtained

(C) warping of sheet is minimised (D) cut blanks are straight

[*IES 1994*] (**Ans.** A)

17. Which of the following pairs of process and draft are correctly matched?

1. Rolling ....2 2.Extrusion.........50 3.Forging......4

Select the correct answer using the codes given below:

Codes:

(A) 1, 2 and 3 (B) 1 and 2 (C) 1 and 3 (D) 2 and 3

[*IES 1994*] (**Ans.** A)

18. The mode of deformation of the metal during spinning is

(A) bending (B) stretching

(C) rolling and stretching (D) bending and stretching

[*IES 1994*] (**Ans.** D)

19. In drop forging, forging is done by dropping

(A) the work piece at high velocity

(B) the hammer at high velocity

(C) the die with hammer at high velocity

(D) a weight on hammer to produce the requisite impact [*IES 1994*] (**Ans.** C)

20. Metal extrusion process is generally used for producing

(A) uniform solid sections (B) uniform hollow sections

(C) uniform solid and hollow sections (D) varying solid and hollow sections

[*IES 1994*] (**Ans.** C)

21. Which one of the following is an advantage of forging?

(A) Good surface finish (B) Low tooling cost

(C) Close tolerance (D) Improved physical property

[*IES 1996*] (**Ans.** D)

22. In wire drawing process, the bright shining surface on the wire is obtained if one

(A) does not use a lubricant (B) uses solid powdery lubricant

(C) uses thick paste lubricant (D) uses thin fluid lubricant

[*IES 1996*] (**Ans.** D)

23. Match List I with List II and select the correct answer

List I (Metal forming process)	*List II (Associated force)*
A. Wire drawing	1. Shear force
B. Extrusion	2. Tensile stress (force)
C. Blanking	3. Compressive force
D. Bending	4. Spring back force

Codes:	A	B	C	D		A	B	C	D
(*a*)	4	2	1	3	(*b*)	2	1	3	4
(*c*)	2	3	1	4	(*d*)	4	3	2	1

[*IES 1996*] (**Ans.** *c*)

24. In metals subjected to cold working, strain hardening effect is due to
(A) slip mechanism (B) twining mechanism
(C) distocation mechanism (D) fracture mechanism [*IES 1997*] (**Ans.** A)

25. Which one of the following processes is most commonly used for the forging of bolt heads of hexagonal shape?
(A) Closed die drop forging (B) Open die upset forging
(C) Closed die press forging (D) Open die progressive forging
[*IES 1998*] (**Ans.** D)

26. The bending force required for V-bending, U-bending and Edge bending will be in the ratio of
(A) 1 : 2 : 0.5 (B) 2 : 1 : 0.5 (C) 1 : 2 : 1 (D) 1 : 1 : 1
[*IES 1998*] (**Ans.** A)

27. Match List I with List II and select the correct answer using the codes given below the lists:

List I	*List II*
A. Drawing	*1. Soap solution*
B. Rolling	*2. Camber*
C. Wire drawing	*3. Pilots*
D. Sheet metal operations using progressive dies	*4. Crater*
	5. Ironing

Codes:

	A	B	C	D		A	B	C	D
(*a*)	2	5	1	4	(*b*)	4	1	5	3
(*c*)	5	2	3	4	(*d*)	5	2	1	3

[*IES 1999*] (**Ans.** *D*)

28. Which one of the following is the correct temperature range for hot extrusion of aluminium?
(A) 300 – 340° C (B) 350 – 400°C (C) 430 – 480°C (D) 550 – 650°C
[*IES 1999*] (**Ans.** B)

29. Major operations in the manufacture of steel balls used for ball bearings are given below:
1. Oil lapping 2. Cold heading 3. Annealing 4. Hardening 5. Rough grinding
The correct sequence of these operations is
(A) 3, 2, 4, 1, 5 (B) 3, 2, 1, 4, 5 (C) 2, 3, 4, 5, 1 (D) 2, 3, 5, 4, 1
[*IAS 1994*] (**Ans.** C)

30. In metal working processes, the stresses induced in the metal are
(A) less than the yield strength of the metal
(B) greater than the yield strength of the metal
(C) less than the breaking strength of the metal
(D) greater than the breaking strength of the metal (**Ans.** B, C)

31. In cold working of metals, the working temperature is
(A) room temperature (B) below the recrystallisation temperature
(C) above the recrystallisation temperature (D) less than the room temperature (**Ans.** B)

32. In hot working of metals, the working temperature is
(A) below the recrystallisation temperature

(B) above the recrystallisation temperature

(C) equal to the melting point of the metal

(D) 150°C (**Ans.** B)

33. Which mechanical property a metal should possess to enable it to be mechanically formed?

(A) Ductility (B) Malleability (C) Elasticity (D) Machinability (**Ans.** A)

34. Mass production of cooking utensils is usually done by

(A) metal spinning (B) deep drawing (C) coining (D) embossing (**Ans.** B)

35. The equipment generally used for metal spinning is

(A) mechanical press (B) hydraulic press (C) drop hammer (D) speed lathe (**Ans.** D)

36. Collapsible tubes are made by

(A) direct extrusion (B) indirect extrusion

(C) cold impact extrusion (D) piercing (**Ans.** C)

37. Large size Rivet heads are made by

(A) swaging (B) hammering (C) upset forging (D) drop forging (**Ans.** C)

38. In blanking operation, the clearance is provided on .... (**Ans.** *Punch*)

39. In punching operation, the clearance is provided on .... (**Ans.** *Die*)

40. In punching operation, shear is provided on .... (**Ans.** *Punch*)

41. In blanking operation, shear is provided on .... (**Ans.** *Die*)

42. In punching operation, the size of hole is dependent on the size of punch or die? (**Ans.** *Punch*)

43. In blanking operation, the size of blank is dependent on the size of punch or die? (**Ans.** *Die*)

44. For the same thickness of sheet, the clearance is more for hard or ductile materials? (**Ans.** *Hard*)

45. Blanking and punching operations can be performed simultaneously on

(A) combination die (B) compound die (C) progressive die (D) simple die (**Ans.** B)

46. Cutting and forming operations can be performed simultaneously on

(A) combination die (B) compound die (C) progressive die (D) simple die (**Ans.** A)

47. Cold working of metal increases

(A) tensile strength (B) hardness (C) yield strength (D) all of these (**Ans.** D)

48. The cutting force in punching and blanking operations mainly depends on

(A) the modulus of elasticity of metal (B) the shear strength of metal

(C) the bulk modulus of the metal (D) the yield strength of metal

[*GATE 2001*] (**Ans.** B)

49. Hot rolling of mild steel is carried out

(A) at recrystallisation temperature. (B) below 100°C and 150°C

(C) above recrystallisation temperature (D) below recrystallisation temperature

(*GATE 2001*) (**Ans.** C)

50. When a steel undergoes a cold working process, it becomes progressively

(A) Softer (B) Harder

(C) Ductile (D) Malleable (*UPSC, AMIE*) (**Ans.** B)

51. When a cold worked metal is heated upto its recrystallization temperature, it becomes :-
(A) Harder (B) Softer
(C) Stays unchanged is hardness (D) Staysunchanged in softness
(*L.U., A.M.I.E.*) (**Ans.** B)

CHAPTER 5

I. Which of the following pairs are correctly matched?
1. Cellulose nitrate .... Table tennis ball
2. Phenol furfurol ...... Brake linings
3. Epoxies .... Jigs and Fixtures

Select the correct answer using the codes given below:

Codes:
(A) 1 and 2 (B) 2 and 3 (C) 1 and 3 (D) 1, 2 and 3
[*IES 1993*] (**Ans.** A)

2. To reduce consumption of synthetic resins, the ingredient added is
(A) accelerator (B) elastomer (C) modifier (D) filler
[*IAS 1994*] (**Ans.** D)

3. Which of the following pairs of plastics and their modes of formation are correctly matched?
1. Polythene .... Condensation polymerisation
2. Polycarbonate .... Addition polymerisation
3. Polystyrene .... Addition polymerisation
4. Polyamide .... Either by addition or by condensation polymerisation

Select the correct answer using the codes given below:
(A) 1 and 2 (B) 2 and 4 (C) 1 and 4 (D) 3 and 4
[*IES 1994*] (**Ans.** D)

4. Match List I (materials) with List II (applications) and select the correct answer using the codes given below the lists:

List I	*List II*
A. Engineering Ceramics	1. Bearings
B. Fibre reinforced plastics	2. Control rods in nuclear reactors
C. Synthetic carbon	3. Aerospace industry
D. Boron	4. Electrical insulator

Codes:

	A	B	C	D		A	B	C	D
(*a*)	1	2	3	4	(*b*)	1	4	3	2
(*c*)	2	3	1	4	(*d*)	4	3	1	2

[*IES 1994*] (**Ans.** *D*)

5. Consider the following statements:
Thermosetting plastics are
1. Formed by addition polymerisation
2. Formed by condensation polymerisation
3. Softened on heating and hardened on cooling for any number of times

4. Moulded by heating and cooling of the statements

(*A*) 1 and 3 are correct (*B*) 2 and 4 are correct

(*C*) 1 and 4 are correct (*D*) 2 and 3 are correct

[*IES 1996*] (**Ans.** *B*)

6. Match List I with Ljst II and select the correct answer

List I (Material)	*List II (Nature of product)*
A. Polyethylene	1. Adhesive
B. Polyurathane	2. Film
C. Cyano–acrylate	3. Wire
D. Nylon	4. Foam

Codes	A	B	C	D		A	B	C	D
(*a*)	2	4	3	1	(*b*)	4	2	3	1
(*c*)	2	4	1	3	(*d*)	4	2	1	3

[*IES 1996*] (**Ans.** *c*)

7. Consider the following statements:

Fibre Reinforced Plastics are:

1. made of thermosetting resins and glass fibre
2. made of thermoplastic resins and glass fibre
3. anisotropic
4. isotropic

Oof these statements

(A) 1 and 4 are correct (B) 1 and 3 are correct

(C) 2 and 3 are correct (D) 2 and 4 are correct [*IES 1996*] (**Ans.** B)

8. Match List I with List II and select the correct answer using the codes given below the Lists:

List I	*List II*
A. Neoprene	1. Electric switches
B. Bakelite	2. Adhesive
C. Foamed polyurethane	3. Thermal insulator
D. Araldite	4. Oil seal

Codes:

	A	B	C	D		A	B	C	D
(*a*)	4	1	2	3	(*b*)	1	4	2	3
(*c*)	4	1	3	2	(*d*)	1	4	3	2

[*IES 1997*] (**Ans.** *c*)

9. Which one of the following materials is used for car tyres as a standard material?

(A) Styrene-Butadiene Rubber (SBR) (B) Butyl rubber

(C) Nitrile rubber (D) Any of the above depending upon the need

[*IES 1997*] (**Ans.** A)

10. Which one of the following refractory materials is recommended for steel furnaces containing CaO slags?

(A) Alumina (B) Silica (C) Magnesia (D) Fireclay

[*IES 1997*] (**Ans.** B)

11. Consider the following statements:

The strength of the fibre reinforced plastic product

1. depends upon the strength of the fibre alone
2. depends upon the fibre and the plastic
3. is isotropic
4. is anisotropic

Which of these statements are correct?

(A) 1 and 3 (B) 1 and 4 (C) 2 and 3 (D) 2 and 4

[*IES 1999*] (**Ans.** B)

12. Consider the following pairs of plastics and their distinct characteristics:

1. Acrylics .... Very good transparency to light
2. Polycarbonate .... Poor impact resistance
3. PTFE .... Low co-efficient of friction
4. Polypropylene .... Excellent fatigue strength

Which of these pairs are correctly matched?

(A) 2 and 3 (B) 1 and 3 (C) 1 and 4 (D) 2 and 4

[*IES 1999*] (**Ans.** C)

13. The bottles from thermo-plastic materials are made by:

(A) Compression moulding (B) Extrusion
(C) Injection moulding (D) Blow moulding (**Ans.** D)

14. The long plastic rods and tubes are produced by:

(A) Compression moulding (B) Extrusion
(C) Injection moulding (D) Blow moulding (**Ans.** B)

15. The process of producing thin sheets by squeezing a thermo-plastic material between revolving cylinders, is known as:

(A) Transfer moulding (B) Injection moulding
(C) Blow moulding (D) Calendering (**Ans.** D)

16. The process of producing plastic components in moulds without the application of pressure, is known as:

(A) Moulding (B) Laminating (C) Calendering (D) Casting (**Ans.** D)

IES 2002

1. Consider the following statements related to piercing and blanking :

1. Shear on the punch reduces the maximum cutting force.
2. Shear increases the capacity of the press needed.
3. Shear increases the life of the punch.
4. The total energy needed to make the cut remains unaltered due to provision of shear.

Which of the above statements are correct ?

(*a*) 1 and 2 (*b*) 1 and 4
(*c*) 2 and 3 (*d*) 3 and 4 (**Ans.** : *b*)

2. Consider the following steps involved in hammer forging a connecting rod from bar stock:

1. Blocking 2. Trimming
3. Finishing 4. Fullering

5. Edging

Which of the following is the correct sequenc of operations ?

(*a*) 1, 4, 3, 2 and 5 (*b*) 4, 5, 1, 3 and 2

(*c*) 5, 4, 3, 2 and 2 (*d*) 5, 1, 4, 2 and 3 **(Ans.** ***b*****)**

3. Match List I (Parts) with List II (Manufactiring processes) and select the correct answer using the codes given below the lists :

List I (Parts)	List II (Manufacturing Process)
A. Seamless tubes	1. Roll forming
B. Accurate and Smooth tubes	2. Shot peening
C. Surface having higher hardness and fatigue strength.	3. Forging
	4. Cold forming

Codes :-

	A	B	C
(*a*)	1	4	2
(*b*)	2	3	1
(*c*)	1	3	2
(*d*)	2	4	1

(Ans. ***a*****)**

4. Which of the following are fabricated using engineering plastics ?

1. Surface plate 2. Gears

3. Guideways for machine tools 4. Foundry patterns

Select the correct answer using the codes given belos Codes

(*a*) 1, 2 and 3 (*b*) 1 (*c*) 2, 3 and 4 (*d*) 1, 2, 3 and 4 **(Ans.** ***c*****)**

5. Which one is not a method of reducing cutting forces to prevent the overloading of press ?

(*a*) Providing shear on die (*b*) Providing shear on punch

(*c*) Increasing die clearance (*d*) Stepping punches. **(Ans.** ***c*****)**

6. In which one of the following welding techniques is vacuum envrionment required ?

(*a*) Ultrasonic welding (*b*) Laser beam welding

(*c*) Plasma arc welding (*d*) Electron beam welding **(Ans.** ***d*****)**

7. Match list I (Ingredients) with List II (Welding functions) and select the correct answer using the codes given below the lists :

List I (*Ingredients*)	List II (*Welding functions*)
A. Silica	1. Arc stabilizer
B. Potassium silicate	2. De-oxidizer
C. Ferro-silicon	3. Fluxing agent
D. Cellulose	4. Gas forming material

Codes :

	A	B	C	D
(*a*)	3	4	2	1
(*b*)	2	1	3	4
(*c*)	3	1	2	4
(*d*)	2	4	3	1

(Ans. ***c*****)**

8. In rolling a strip between two rolls, the position of the neutral point in the arc of contact does not depend on
   (*a*) amount of reduction (*b*) diameter of the rolls
   (*c*) co-efficient of friction (*d*) material of the rolls **(Ans. *d*)**
9. Compare die-casting with investment casting w.r.t. production rate, melting point of work materials and ability of producing complex shaped products.
10. Name three processes of manufacturing long low carbon steel tubes.
11. What are the advantages of submerged arc welding over conventional open air manual arc welding and why ?

GATE 2003

1. Hardness of green sand mould increases with
   (A) increase in moisture content beyond 6 per cent
   (B) increase in permeability (C) decrease in permeability
   (D) increase in both moisture content and permeability **(Ans. C)**
2. In oxyacetylene gas welding, the temperature at the inner cone of the flame is around
   (A) 3500°C (B) 3200°C
   (C) 2900°C (D) 2550°C **(Ans. A)**
3. Cold working of steel is defined as working
   (A) at its recrystallisation temperature (B) above its recrystallisation temperature
   (C) below its recrystallisation temperature
   (D) at two thirds of the melting temperature of the metal **(Ans. C)**
4. With a salidification factor of 0.97×10^6 s/m^2, the solidification time (in seconds) for a spherical casting of 200 mm diameter is
   (A) 539 (B) 1078
   (C) 4311 (D) 3233 **(Ans. B)**

   Hint : In Chvorinov's equation

$$t = C \cdot \left(\frac{V}{SA}\right)^2, \quad C = 0.97 \times 10^6 \ s/m^2$$

5. Match the following

Work material	Type of joining
P Aluminium	1. Submerged Arc Welding
Q Die Steel	2. Soldering
R Copper wire	3, Thermit welding
S Titanium sheet	4. Atomic Hydrogen Welding
	5. Gas Tungsten Arc Welding
	6. Laser Beam Welding

(A)	(B)	(C)	(D)
P – 2	P – 6	P – 4	P – 5
Q – 5	Q – 3	Q – 1	Q – 4
R – 1	R – 4	R – 6	R – 2
S – 3	S – 4	S – 2	S – 6

(Ans. D)

IES – 2004

1. Write the Chemical equation for the reaction going on in the oxy-acetylene flame. Where do you get the most part of the oxygen; is it from the cylinder or the atmosphere ?
2. Clearances have to be provided on the press tools. What is the order of clearances for shearing operation and deep drawing operation ? On what tool will you provide the clearance is :
   (*i*) Punching operation (*ii*) Blanking operation

3. What is the three high rolling mill ? Indicate the movement of rolls and the workpiece.
4. What are the step involved in making of carbide tool bits ? What do you understand by the green density of a metal powder compacted part ? What are the chief merits of a bush made by metal powder compaction.
5. Match List – I (Type of Moulding) with List – II (Mechanism involved) and select the correct answer using codes given below the lists :

List – I **(Type of moulding)**	**List – II** **(Mechanism involved)**
A. Compression moulding	1. Mould cavity must be heated to cure the plastic forced into it
B. Injection moulding	2. Similar to hydraulic extrusion
C. Jet moulding	3. Analogous to hot pressing of powdered metals
D. Extrusion moulding	4. Analogous to die casting of metals.

Codes :

	A	B	C	D
(*a*)	2	4	1	3
(*b*)	3	1	4	2
(*c*)	2	1	4	3
(*d*)	3	4	1	2

(Ans. : d)

6. Consider the following statements :

   In comparison to hot working, in cold working,

   1. higher forces are required
   2. no heating is required
   3. less ductility is required
   4. better surface finish is obtained

   Which of the statements given above are correct ?

   (*a*) 1, 2 and 3 (*b*) 2 and 4
   (*c*) 1 and 3 (*d*) 2, 3 and 4 **(Ans.** : b)

7. Consider the following factors :

   1. Size and shape that can be produced economically
   2. Porosity of the parts produced
   3. Press Capacity
   4. High density

   Which of the above are the limitations of powder metallurgy ?

   (*a*) 1, 3 and 4 (*b*) 2 and 3
   (*b*) 1, 2 and 3 (*d*) 1 and 2 **(Ans.** : c)

8. Match List – I (Welding problems) with List – II (Causes) and select the correct answer using codes given below the Lists :

List – I (Welding problems)	**List – II (Causes)**
A. Cracking of weld metal	1. Excessive stresses
B. Cracking of base metal	2. High joint rigidity
C. Porosity	3. Failure to remove slag from previous deposit
D. Inclusions	4. Oxidation

5. Excessive H_2, O_2, N_2, in the welding atmosphere

Codes :

	A	B	C	D
(*a*)	2	1	5	3
(*b*)	3	4	2	1
(*c*)	2	4	5	3
(*d*)	3	1	4	2

(**Ans.** : a)

9. Consider the following statements :
The size of heat affected zone (HAZ) will increase with
1. increased starting temperature
2. increased welding speed
3. increased thermal conductivity of the base metal
4. increase in base metal thickness
Which of the statements given above are correct ?
(*a*) 1, 2 and 3 (*b*) 1 and 3
(*c*) 1 and 4 (*d*) 2 and 3 (**Ans.** b)

10. Consider the following statements :-
The magnitude of residual stresses in welding depends upon
1. Design of weldment
2. Support and clamping of components
3. Welding process used.
4. amount of metal melted/deposited.
Which of the statements given above are correct ?
(*a*) 1, 2 and 4 (*b*) 1, 2 and 3
(*c*) 1 and 3 (*d*) 2 and 3 (**Ans.** a)

GATE – 2005

1. Match the items of List I (Equipment) with the items of List II (Process) and select the correct answer using the given codes.

List – I (Equipment)	**List – II (Process)**
P – Hot Chamber Machine	1 – Cleaning
Q – Muller	2 – Core making
R – Dielectric Baker	3 – Die Casting
S – Sand Blaster	4 – Annealing
	5 – Sand mixing

(A) P–2, Q–1, R–4, S–5 (B) P–4, Q–2, R–3, S–5
(C) P–4, Q–5, R–1, S–2 (D) P–3, Q–5, R–2, S–1 (**Ans.** D)

2. When the temperature of a solid metal increases,
(A) strength of the metal decreases but ductility increases.
(B) both strength and ductility of the metal decrease
(C) both strength and ductility of the metal increase
(d) strength of the metal increases but ductility decreases (**Ans.** A)

3. The strength of a brazed joint
(A) decreases with increase in gap between the two joining surfaces.
(B) Increases with increase in gap between the two joining surfaces.
(C) decreases upto certain gap between the two joining surfaces beyond which it increases.
(D) increases upto certain gap between the two joining surfaces beyond which it decreases
(**Ans.** : D)

INDEX

A TEXTBOOK OF HYDRAULIC MACHINES

(Fluid Power Engineering)

R.K. Rajput

10 194 **2nd Edn. 2004** **ISBN : 81-219-1668-2**

A TEXTBOOK OF HYDRAULICS

R.K. Rajput

CONTENTS: Properties of Fluids ●·PRESSURE MEASUREMENT Pressure Measurement Pressure ● Hydrostatic Forces on Surfaces ●·Bouyancy and Floatation ● Fluid Kinematics ● Fluid Dynamics ●·Dimensional and Model Analysis ● Flow Through Orifices and Mouthpieces ●·Flow over Notiches and Weirs ● Laminar Flow ● Flow Through Pipes ●·Flow Around Submerged Bodiesadrag and Lift ● Flow in Open Channels ●·Impact of Free JETS ● Hydrauli Turbines ●·Centrifuga Pumps ●·Reciprocating Pumps ● Miscellaneous Hydraulic Machines ● Experiments

Code : 10 198 **1st Edn. 1998** **ISBN : 81-219-1731-X**

ISO 9000 CONCEPTS, METHODS AND IMPLEMENTATION

Tapan P. Bagchi

CONTENTS: ● Quality Management Practices Worldwide ● Quality, Customers and ISO 9000 ● ISO 9000: A Management Overview ● Writing the Company's Quality Policy ● Interpretation of Key ISO 9000 Clauses ● Aspects of Manufacturing and ISO 9000 ● ISO 9000 Standards and Production ● Process Control by Control Charts ● Problem Solving with 7 Tools ● Data, Records and Traceability ● Procurement System Standards in ISO 9000 ● Inspection Test Standards and Calibration ● An Overview of Acceptance Sampling and Documenting the Quality System ● What is a Quality Audit? ● Design and Change Control Standards ● What are Taguchi Methods? ● Implementing ISO 9000 ● ISO 9000 Registration: Two Case Studies ● Total Quality Management: Concepts and Accessories ● A Prototype Quality Manual ● References

AHW 0112 **ISBN : 81-7544-011-2**

ENGINEERING MATERIALS

R.K. Rajput

CONTENTS: Introduction ● Building Stones ● Bricks and other Clay Products ● Lime ● Cement ● Mortar ● Concrete ● Timber and Wood-Based Products ● Metals and Alloys ● Paints, Varnishes, Distempers and Anti-Termite Treatment ● Asphalt, Bitumen and Tar ● Asbestos, Adhesives and Abrasives ● Plastics and Fibres ● Glass ● Insulating Materials ● Fly-Ash, Gypsum and Gypsum Plaster ● Rubber and Composite Materials ● Lubricating, Belting and Packing Materials ● Cutting Tool Materials ● Electrical Engineering Materials ● Material Science of Metals

Code : 10 210 **2nd Edn. 2004** **ISBN : 81-219-1960-6**

PRESS TOOLS: DESIGN AND CONSTRUCTION

P.H. Joshi

CONTENTS: ● Tool and Workpiece Materials ● Cutting Action ● Chip Formation ● Tool Wear ● Lubrication and Surface Finish ● Cutting Power ● Machining Economics ● Single Point Turning and Threading Tools ● Tools for Holes ● Milling ● Broaching ● Gear Cutting Tools ● Index

AHW 4465 **ISBN : 81-85814-46-5**

STRENGTH OF MATERIALS (MUTLI COLOUR)

(Machanics of Solids S.I. Units)

R.K. Rajput

CONTENTS: Simple Stresses and Strains ● Principal Stresses and Strains ● Centroid and Moment of Inertia ● Bending Stresses ● Combined Direct and Bending Stresses ● Shear Stresses in Beam ● Thin Shells ● Thick Shells ● Riveted and Welded Joints ● Torsion of Circular and Non Circular Shafts ● Springs ● Stain Energy and Deflection due to Shear & Bending ● Columns and Struts ● Analysis of Framed Structures ● Theories of Failure ● Rotating Discs and Cylinders ● Bending of Curved Bars ●·Unsymmetrical Bending ●·Material Testing Experiments ●·Index

Code : 10 174 **5th Edn. 2006** **ISBN : 81-219-1381-0**

CUTTING TOOLS

P.H. Joshi

CONTENTS: ● Tool and Workpiece Materials ● Tool Geometry ● Chip Formation ● Tool Wear ● Lubrication and Surface Finish ● Cutting Power ● Machining Economics ● Single Point Turning and Threading Tools ● Tools for Holes ● Milling ● Broaching ● Gear Cutting Tools ● Standards for Cutting Tools ● Index

AHW 4538 **ISBN : 81-85814-53-8**

A TEXTBOOK OF ENGINEERING MECHANICS

(Applied Mechanics)

R.S. Khurmi

CONTENTS: Introduction ● Composition and Resolution of Forces ● Moments and their Applications ● Parallel Forces and Couples ●· Equilibrium of Forces ●·Centre of Gravity·●·Moment of Inertia ●·Friction ● Principles of Lifting Machines ● Simple Lifting Machines ● Support Reactions ●·Analysis of Perfect Frames (Analytical Method & Graphical Method) ● Equilibrium of Strings ● Virtual Work ● Plane Motion ●·Motion Under Variable Acceleration ●·Relative Velocity ● Projectiles ●·Motion of Rotation ●·Combined Motion of Rotation and Translation ●·Simple Harmonic Motion ●·Laws of Motion ●·Motion of Connected Bodies ●·Helical Springs and Pendulums ●·Collision of Elastic Bodies ●·Motion Along a Circular Path ●·Balancing of Rotating Masses ●·Work, Power and Energy ●·Kinetics of Motion of Rotation ●·Motion of Vehicles ●·Transmission of Power by Belts and Ropes ●·Gear Trains ●·Hydrostatics ●·Equilibrium of Floating Bodies ●·Index.

Code : 10 023 **20th Edn. 2005** **ISBN : 81-219-0651-2**

APPLIED MECHANICS AND STRENGTH OF MATERIALS

R.S. Khurmi

CONTENTS: Introduction ● Composition and Resolution of Forces ● Moments and their Application ● Parallel Forces and Couples ● Equilibrium of Forces● Centre of Gravity ● Moment of Inertia ● Principles of Friction ● Applications of Friction ● Principles of Lifting Machines ●Simple Lifting Machines ● Linear Motion ●·Circular Motion ● Projectiels ●Laws of Motion ● Work, Power and Energy ● Simple Stresses and Strains ●Thermal Stresses and Strains ●·Elastic Constants ●·Strain Energy and Impact Loading ●·Bending Moment and Shear Force ●·Bending Stresses in Beams ●·Shearing Stresses in Beams ●·Deflection of Beams ●·Deflections of Cantilevers ●Torsion of Circular Shafts ●·Riveted Joits ●·Thin Cylindrical and Spherical Shells ● Analysis of Perfect Frames (Analytical Method & Graphical Method)

Code : 10 025 **13th Edn. 2005** **ISBN : 81-219-1077-3**

A TEXTBOOK OF APPLIED MECHANICS

R.S. Khurmi

CONTENTS: Introduction ● Composition and Resolution of Forces ● Moments and their Applications ● Parallel Forces and Couples ● Equilibrium of Forces ● Centre of Gravity ● Moment of Inertia ● Principles of Friction ● Applications of Friction ● Principls of Lifting Machines ● Simple Lifting Machines ● Linear Motion ● Circular Motion ● Projectiles ● Laws of Motion ● Work, Power and Energy ● Analysis of Perfect Frames (Analytical Method) ● Analysis of Perfect Frames Graphical Method ● Index

Code : 10 191 **13th Edn. 2004** **ISBN : 81-219-1643-7**

STRENGTH OF MATERIALS

R.S. Khurmi

CONTENTS: Introduction ●·Simple Stresses and Strains ● Stresses and Strains in Bars of Varying Sections ● Stresses and Strains in Statically Indeterminate Structures ● Thermal Stresses and Strains ● Elastic Constants ● Principal Stresses and Strains ● Strain Energy and Impact Loading ● Centre of Gravity ● Moment of Intertia ● Analysis of Perfect Frames (Analytical Method & Graphical Method) ● Analysis of Perfect Frames (Graphical Method) ● Bending Moment and Shear Force ● Bending Stresses in Simple Beams ● Bending Stresses in Composite Beams ●·Shearing Stresses in Beams ● Direct and Bending Stresses ● Dams and Retaining Walls ● Deflection of Beams ● Deflection of Cantilevers ● Deflection By Moment Area Method ●Deflection by Conjugate Beam Method ●Propped Cantilevers and Beams ●·Fixed Beams ●Theorem of Three Moments ●·Moment Distribution Method ●Torsion of Circular Shafts ●·Springs ●·Riveted Joints ●·Welded Joints ●·Thin Cylindrical and Spherical Shells ●·Columns and Struts ●·Introduction to Reinforced Concrete ●·Index

Code : 10 024 **23rd Edn. 2005** **ISBN : 81-219-0533-8**

A TEXTBOOK OF HYDRAULICS, FLUID MECHANICS AND HYDRAULIC MACHINES

R.S. Khurmi

CONTENTS: Introduction ● Fluid Pressure and its Measurement ● Hydrostatics ● Applications of Hydrostatics ● Equilibrium of Floating Bodies ● Hydrokinematics ● Bernoulli's Equation and its Applications ● Flow through Orifices (Measurement of Discharge & Time) ● Flow through Mouthpieces ● Flow over Notches ● Flow over

Weirs● Flow through Simple Pipes ● Flow through Compound Pipes ●Flow through Nozzles ● Uniform Flow through Open Channels ● Non-uniform flow through Open Channels ● Viscous Flow ● Viscous Resistance Fluid Masses Subjected to Acceleration ● Vortex Flow ● Mechanics of Compressible Flow ● Compressible Flow of Fluids ● Flow Around Immersed Bodes ●·Dimensional Analysis ●·Model Analysis (Undistorted Models and Distorted Models ●·Non-Dimensional Constants ●·Impact of Jets ●·Jet Propulsion ●·Water Wheels ●·Impulse Turbines ●·Reaction Turbines ● Performance of Turbines ●·Centrifugal Pumps ●·Reciprocating Pumps ●·Performance of Pumps ●·Pumping Devices ●·Hydraulic Systems ●·Index

Code : 10 026 **19th Edn. 2004** **ISBN : 81-219-0162-6**

STEAM TABLES

(With Mollier Diagram in S.I. Units)

R.S. Khurmi

CONTENTS: Rules for S.I. Units ● Introduction to Steam Tables (Temperature) and Mollier Diagrams ● Saturated Water & Steam Tables ● Saturated Water & Steam Tables (Pressure) ● Specific Volume of Super-heated Steam ● Specific Enthalpy of Super-heated Steam ● Entropy of Super-heated Steam ● Specific Volume of Super-critical Steam ● Specific Enthalpy of Superecritical Steam ● Specific Entropy of Super Critical Steam

Code : 10 044 **8th Edn. 2005** **ISBN : 81-219-0654-7**

A TEXTBOOK OF PRODUCTION ENGINEERING

P.C. Sharma

CONTENTS: Jigs and Fixtures ●·Press Tool Design ●·Forging Die Design ●·Cost Estimation ●·Economics of Tooling ●·Process Planning ●·Tool Layout for Capstans and Turrets ●·Tool Layout for Automatics ●·Limits, Tolerances and Fits ●·Gauges and Gauge Design ●·Surface Finish ●·Measurement ●·Analysis of Metal Forming Process ●·Theory of Metal Cutting ● Design and Manufacture of Cutting Tools ● Gear manufacturing ● Thread manufacturing ● Design of Machine Tool elements and machine tool testing ● Machine Tool installation and maintenance ● Design of product for economical production ● Statistical quality control ● Kinematics of machine tools ● Production planning and control ● Manufacturing systems and automation ● Computer integrated manufacturing ● Plant Layout ● Production and Productivity ● Appendix ● Index

Code : 10 038 **10th Edn. 2005** **ISBN : 81-219-0421-8**

A TEXTBOOK OF THERMAL ENGINEERING

R.S. Khurmi & J.K. Gupta

CONTENTS: Introduction ● Properties of Perfect Gases ● Thermodynamic Processes of Perfect Gases ● Entropy of Perfect Gases ● Thermodynamic Air Cycles ● Formation and Properties of Steam ● Entropy of Steam ● Thermodynamic Processes of Vapour ● Thermodynamic Vapour Cycles ● Fuels ● Combustion of Fuels ● Steam Boilers ● Boiler Mountings and Accessories ● Performance of Steam Boilers ● Boiler Draught ● Simple Steam Engines ● Compound Steam Engines ●Performance of Steam Engines ●Steam Condensers ● Steam Nozzles ● Impulse Turbines ●·Reaction Turbines ●·Performance of Steam Turbines ●·Modern Steam Turbines ●·Internal Combustion Engines ●·Testing of Internal Combustion Engines ●·Reciprocating Air Compressors ●·Air Motors ●·Gas Turbines ●·Performance of Gas Turbines ●·Introduction to Heat Transfer ●·Air Refrigeration Systems ●·Vapour Compression Refrigeration ●·Psychrometry Air Conditioning Systems ●·General Thermodynanuc Relation ●·Variable Specific Heat ●·Index

Code : 10 172 **15th Edn. 2004** **ISBN : 81-219-1337-3**

A TEXTBOOK OF HYDRAULICS

R.S. Khurmi

CONTENTS: Introduction ● Fluid Pressure and its Measurement ● Hydrostatics ● Applications of Hydrostatical Equilibrium of Floating Bodiesl Hydrokinematics ● Bernoulli's Equation and its Applications ● Flow Through Orifices (Measurement of Discharge) ● Flow Through Orifices (Measurement of Time) ● Flow Through Mouthpieces ● Flow Over Notches ● Flow Over Weirs ● Flow Through Simple Pipes ● Flow Through Compound Pipes ● Flow Through Nozzles ● Uniform Flow Through Open Channels ● Non Uniform Flow Through Open Channels ●·Viscous Flow ●·Viscous Resistance ●·Impact of Sets ●·Hydraulic Turbines ●·Hydraulic Pumps ●·Pumping Devices ●·Hydraulic System ●·Index

Code : 10 027 **19th Edn. 2004** **ISBN : 81-219-0135-9**

THEORY OF MACHINES

R.S. Khurmi & J.K. Gupta

CONTENTS: Introduction • Kinematics of Motion • Kinetics of Motion • Simple Harmonic Motion • Simple Mechanisms • Velocity in Mechanisms (Instantaneous Centre Method)• Velocity in Mechanisms (Relative Velocity Method) •·Acceleration in Mechanisms •·Mechanisms with Lower Pairs • Friction •·Belt, Rope and Chain Drives •·Toothed Gearing •·Gear Trains •·Gyroscopic Couple and Precessional Motion •·Inertia Forces in Reciprocating Parts •·Turning Moment Diagrams and Flywheel •·Steam Engine Valves and Reversing Gears •·Governors •·Brakes and Dynamometers •·Cams •·Balancing of Rotating Masses •·Balancing of Reciprocating Masses •·Longitudinal and Transverse Vibrations •Torsional Vibrations •·Computer Aided Analysis and Synthesis of Mechanisms •·Index

Code : 10 013 **13th Edn. 2005** **ISBN : 81-219-0132-4**

MATERIAL SCIENCE

R. S. Khurmi & R.S. Sedha

CONTENTS: Part I: Science of Metals : •·Introduction •·Structure of Atom •·Crystal Structure •·Bonds in Solids •·Electron Theory of Metals •·**Part II: Mechanical Behaviour of Metals:** •·Mechanical Properties of Metals •·Mechanical Tests of Metals •·Deformation of Metals Fracture of Metals •·**Part III: Enginering Metallurgy:** Iron Carbon Alloy System •·Heat Treatment •·Corrosion of Metals •·**Part IV: Engineering Materials** •·Ferrous and Non-Ferrous Alloys •·Organic Materials and Ceramics • Composite Materials • Semiconductors • Insulating Materials •·Magnetic Materials •·Appendices •·Index

Code : 10 109 **4th Edn. 2004** **ISBN : 81-219-0146-4**

A TEXTBOOK OF MATERIAL SCIENCE

K.G. Aswani

CONTENTS: Selection of Materials •·Structure of Atoms, Crystal Structure and Bonds in Solids •·Mechanical Properties of Materials •·Deformation of Metals •·Alloy Systems and Phase Diagrams •·Cast Iron •·Carbon and Alloy Steels •·Heat Treatment of Carbon and Alloy Steels •Non-Ferrous Metals and Alloys •·Polymers and Polymeric Products •·Corrosion and its Prevention •·Composites •·Electrical and Magnetic Properties of Materials •·Manufacturing Processes •·Refractories •·Cementitious Materials •·Tables •·Index

Code : 10 157 **2nd Edn. 2001** **ISBN : 81-219-1199-0**

A TEXTBOOK OF PRODUCTION TECHNOLOGY

(MANUFACTURING PROCESSES)

P.C. Sharma

CONTENTS: •·Introduction •·Engineering Materials and Heat Treatment • The Casting Process •·Mechanical Working of Metals •·The Welding Process •·The Machining Process •·Cutting Tools materials and Cutting Fluids •·Machine Tools •·Unconventional Manufacturing Methods •·Power Metallurgy •·Procesing of Plastic •·Special Processing Methods •·Ceramic Materials and Their Processing•·Composite Materials and Their Processing •·Tracer Controlled Machine Tools •·Numerically Controlled Machine Tools •·Appendix •·Index

Code : 10 160 **6th Edn. 2005** **ISBN : 81-219-1114-1**

ENGINEERING FLUID MECHANICS

K.L. Kumar

CONTENTS: •·Introduction •·Fluid Statics • Fluid Kinematics •·Fluid Dynamics •·Flow Measurement •·Ideal Fluid Flow •·Laminar Flow •·Boundary Layer Flow •·Flow Around Immersed Bodies •·Flow Through Pipes •·Flow Through Open Channels •·Compressible Flow •·Dimensional Analysis and Similitude •·Fluid Machines •·Appendices

Code : 10 020 **7th Edn. 2004** **ISBN : 81-219-0100-6**

INDUSTRIAL MAINTENANCE

H.P. Garg

CONTENTS: Part I: Restoration and Manufacture of Machine Parts •·Fits, Tolerances and Surface Finish •·Materials for Machine Parts and Heat Treatment of Steels •·Guide Surface •·Gear Transmission •·Bush Bearings, their Shank and Housing • Ball and Roller Bearings their Shank and Housing •·Key Fitting •·Spline Fitting •·Coupling •·Clutches •·Lead Screw and Nut •·Machine Spindle •·Vee Belt Drive •·Chains and Sproket Wheels •·Machine Hydraulics •·Tailstock Repairs •·Repair of Three Jaw Chucks •·Repair of Cracks in C.I. Body •·Restoration of Parts by Welding, Metalization, Chromium Plating •·Threads and Threaded Joints •·Seals and Packings •·Special Features of the Repair of Cranes, Hammers, Power Presses / Shears and Furnaces •·Fitters Common Tools, Appliances and Devices, Handling Facilities and Measuring Instruments

●·Typical Manufacture and Machining Norms ●·**Part II: Planned Maintenance** : Preventive Maintenance Planning ●·Repair Cycle ●·Repair Complexity Maintenance Stages ●·Machine Kinematics and Schedule of Complete Overhaul ●·Pert in Maintenance ●·Lubrication and Lubricants ●·Materials and Standard Spares Planning ●·Spare Parts Stock Pate Accuracy and Technological Test Charts ●·Organisation of Maintenance Department ●·Depreciation and Machine Life ●·Appendices ●·Index

Code : 10 021 **3rd Edn. 2002** **ISBN : 81-219-0168-5**

A TEXTBOOK OF REFRIGERATION AND AIR-CONDITIONING

R.S. Khurmi & J.K. Gupta

CONTENTS: Introduction ●·Air Refrigeration Cycles ●·Air Refrigeration Systems ●·Simple Vapour Compression Systems ●·Compound Vapour Compression Systems ●·Multi-evaporator and Compressor Systems ●·Vapour Absorption Refrigeration Systems ●·Refrigerants ●·Refrigerant Compressors ●·Condensers ●·Evaporators ●·Expansion Devices ●·Food Preservation ●·Low Temperature Refrigeration (Cryogenics) ●·Steam-Jet Refrigeration system ●·Psychrometry ●·Comfort Conditions ●·Air Conditioning Systems ●·Cooling Load Estimation ●·Ducts ●·Fans ●·Refrigeration Tables and Charts ●·Applications of Refrigeration and Air-Conditioning ●·Index

Code : 10 097 **3rd Edn. 2005** **ISBN : 81-219-0268-1**

A TEXTBOOK OF MACHINE DESIGN

R.S. Khurmi & J.K. Gupta

CONTENTS: Introduction ●·Engineering Materials and their Properties ●·Manufacturing Considerations in Machine Design ●·Simple Stresses in Machine Parts ●·Torsional and Bending Stresses in Machine Parts ●·Variable Stresses in Machine Parts ●·Pressue Vessels ●·Pipes and Pipe Joints ●·Riveted Joints ●·Welded Joints ●·Screwed Joints ●·Cotter and Knuckle Joints ●·Keys and Coupling ●·Screwed Joints ●·Levers ●·Columns and Struts ●·Power Screws ●·Flat belt Drives ●·Flat Belt Pulleys ●·V-belt and Rope Drives ●·Chain Drives ●·Flywheel ●·Springs ●·Clutches ●·Brakes ●·Sliding Contact Bearings ●·Rolling Contact Bearings ●·Spur Gears ●·Helical Gears ●·Bevel Gears ●·Worm Gears ●·Internal Combustion Engine Parts ●·Index

Code : 10 012 **13th Edn. 2005** **ISBN : 81-219-0501-X**

MECHANICAL ENGINEERING

(Objective Type)

R.S. Khurmi & J.K. Gupta

CONTENTS: ● Engineering Mechanics ● Strength of Materials ● Hydraulics and Fluid Mechanics ● Hydraulic Machines ●Thermodynamics ● Steam Boilers and Engines ● Steam Nozzles and Turbines ● I.C. Engines and Nuclear Power Plants ● Compressors, Gas Dynamics and Gas Turbines ● Heat Transfer, Refrigeration and Air Conditioning ● Theory of Machines ● Machine Design ● Engineering Materials ● Workshop Technology ● Production Engineering ● Industrial Engineering and Production Management ● Automobile Engineering ● Index

Code : 10 042 **5th Rev. Edn. 2004** **ISBN : 81-219-0628-8**

A TEXTBOOK OF ENGINEERING DRAWING

(Geometrical Drawing)

R.K. Dhawan

CONTENTS: Section I: Introduction & Drawing Instruments ●·Layout of Drawing Sheet ●·Conventions ●·Lettering ●·Dimensioning ●·Scales ●·Geometrical Constructions ●·**Section II:** Loci of Points ●·Conic Sections ●·Plane and Space Curves ●·**Section III:** Theory of Projection & Orthographic Projection ●·Orthographic Reading or Interpretation of Views ●·Identification of Surfaces ●·Missing Lines & Views ●·Sectional Views ●·Isometric Projections ●·Auxiliary Views ●·Freehand Sketching ●·**Section IV: Projections of Points** ●·Projections & Traces of Straight Lines ●·Projections of Planes ●·Projections of Solids ●·Sections of Solids ●·Intersection of Surfaces ●·Development of Surfaces ●·Perspective Projections.

Code : 10 147 **2nd Edn. 2005** **ISBN : 81-219-1431-0**

INDUSTRIAL MAINTENANCE MANAGEMENT

S.K. Srivastava

CONTENTS: Introduction and Objectives ●·Quality, Reliability and Maintainability (QRM) ●·Maintenance Jobs and Technologies ●·Defect List Generation & Defect/Failure Analysis ●·Maintenance Types/ Systems ●·Condition Monitoring (Equipment Health Monitoring) ●·Maintenance Planning and Scheduling ●·Systematic Maintenance ●·Computer Managed Maintenance System (CMMS) ●·Total Productive Maintenance (TPM) ●·Other Concepts of Maintenance Types/Systems ●·Maintenance Organisation ●·Maintenance Effectiveness & Performance Evaluation/Audit ●·Maintenance Budgeting, Costing and Cost-Control ●·Training (HRD) of Maintenance Personnel

•·Bibliography •·Index

Code : 07 300 3rd Edn. 2005 ISBN : 81-219-1663-1

A TEXTBOOK OF WORKSHOP TECHNOLOGY

(Manufacturing Processes)

R.S. Khurmi and J.K. Gupta

CONTENTS: Introduction •·Industrial Safety •·Fundamentals of Metals and Aloys •·Properties, Testing and Inspection of Metals •·Ferrous Metals and Alloys •·Non-ferrous Metals and Alloys •·Heat Treatment of Metals and Alloys •·Mechanical Working of Metals •·Carpentry and Joinery •·Pattern Making •·Foundry Tools and Equipments •·Moulding and Core Making •·Special Casting Processes • Smithy and Forging •·Welding •·Bench Work and Fitting •·Sheet Metal Work • Rivets and Screws •·Limit System and Surface Finish •·Measuring Instruments and Gauges •·Quality Control •·Powder Metallurgy •·Plastics •·Metallic and Non-metallic Coatings •·Pipes and Pipe Fittings •·Machine Tools (Introduction) •·Index

Code : 10 155 6th Edn. 2004 ISBN : 81-219-0868-X

THE AUTOMOBILE

Harbans Singh Reyat

CONTENTS: The Automobile •·Construction and Working of Automobile •·Suspension System •·Engine I • Engine II •·Engine Lubrication System •·Engine Cooling System • Fuels and Fuel Systems •·Fuel Injection System • Electrical Systems I • Electrical Systems II •·Transmission System •·Clutch •·Transmission •·Drive Shaft and Drive Axle •·Steering System •·Brakes •·Air Conditioning in Automobiles •·Maintenance of Automobiles •·Trouble Shooting •·Appendices •·Glossary of Technical Terms

Code : 10 019 6th Edn. 2004 ISBN : 81-219-0214-2

INDUSTRIAL ENGINEERING AND PRODUCTION MANAGEMENT

Martand Telsang

CONTENTS: Part I : Industrial Engineering •·Introduction to Industrial Engineering •·Productivity •·Work-Study •·Method Study •·Work Measurement •·Value Engineering •·Plant Location •·Plant Layout •·Material Handing •·Job Evaluation and Merit Rating •·Wages and Incentives •·Ergonomics •·**Part II: Production Management** •·Introduction to Production/Operations Management •·New Product Design •·Demand Forecasting •·Production Planning and Control •·Capacity Planning •·Material Requirement Planning (MRP) • Process Planning •·Project Scheduling with CPM and Pert •·Production Control •·Inventory Control •·Production Cost Concepts and Break-even Analysis •·Maintenance Management •·Make or Buy Decisions •·Planning and Control of Batch Production • **Part III: Advanced Topics in Production Management** •·Application of Linear Programming Technique in Production Management • TQM - Concept and Philosophy •·Business Process Reengineering •·Group Technology •·Just in Time (JIT) Manufacturing •·Operations Strategy •·Materials Management •·Project Management •·Service Management •·Product and Service Reliability •·Theory of Constraints (Toc) •·Advanced Manufacturing Technologies and Systems •·Supplements •·Additional Solved Problems •·Appendices

Code : 10 197 2nd Rev. Edn. 2004 ISBN : 81-219-1773-5

A TEXTBOOK OF MACHINE DRAWING

(In First Angle Projections)

R.K. Dhawan

CONTENTS: Section I: Introduction & Drawing Instruments • Layout of Drawing Sheet •Conventions •Lettering •Dimensioning •Scales •·**Section II:** Theory of Projection & Orthographic Projection •·Orthographic Reading or Interpretation of Views •·Indentification of Surfaces •·Missing Lines & Views •·Sectional Views •·Isometric Projections •·Auxiliary Views •·Freehand Sketching •·**Section III:** Detail & Assembly Drawings •·Limits, Fits & Machining Symbols •·Rivets and Riveted Joints •·Welding •·Screw Threads •·Fastenings •·Keys Cotters and Joints •·Shaft Couplings •·Bearings •·Brackets •·Pulleys •·Pipe Joints •·Steam Engine Parts •·I.C. Engine Parts •·Valves •·Gears •·Cams •·Jigs & Fixtures •·Miscellaneous Drawings

Code : 10 148 2nd Edn. 2005 ISBN : 81-219-0824-8

MECHANICS

(Dynamics & Statics)

P. Duraipandian, Laxmi Duraipandian & Muthamizh Jayapragasam

CONTENTS : • Introduction•Kinematics •Force• Equilibrium of a Particle•Forces on a Rigid Body • A Specific Reduction of Forces • Centre of Mass • Stability of Equilibrium • Virtual Work • Hanging Strings • Rectilinear

Motion Under Constant Forces • Work, Energy and Power • Rectilinear Motion Under Varying Force • Projectiles • Impact • Circular Motion • Central Orbits • Moment of Inertia • Two Dimensional Motion of a Rigid Body • Theory of Dimensions

Code : 14 199 **6th Edn. 2005** **ISBN : 81-219-0272-X**

INDUSTRIAL AND BUSINESS MANAGEMENT

Martand T. Telsang

CONTENTS : Part One : Business Environment • Introduction To Business • Forms of Business Organisation • Business Environment •Growth, Development and Internationalisation Of Business • Business Ethics and Social Responsibility **Part Two** : PRINCIPLES OF MANAGEMENT • Functions of Management • Evolution and Development of Management Thought • Planning • Decision Making • Organising • Staffing [Human Resource Management] • Communication • Motivation • Leadership • Controlling • **Part Three:** Human Behaviour At Work • Organisational Behaviour • Individual Behaviour • Group Behaviour • Job Satisfaction•Organisational Change And Development **Part Four** : Functional Areas of Management • Production Management • Materials Management • Marketing Management • Financial Management • Financial and Cost Accounting • Quality Management • Project Management **Part Five** : Entrepreneurship Development and Small Scale Industries • Entrepreneurship Development • Management of Small Scale Industries **Part Six** : Industrial Relations and Safety Management • Industrial Relations • Labour Legislations • Industrial Safety Management **Part Seven** : Industrial Economics • Managerial Economics • Engineering Economics **Part Eight** : Special Topics in Management • Management By Objectives (MBO) • Management Information System •·Office Management • Management of Technology •·Business Process Re-engineering • Introduction to Enterprise Resource Planning (ERP) • Kaizen—Continuous Improvement **Part Nine** : Quantitative Techniques for Managerial Decisions • Introduction to Operations Research • Pay-Off Matrix and Decision Trees •·Games Theory •·Linear programming Problems • Assignment Model • Transportation Model •·Sequencing Models • Queueing Models •·Simulation •·Replacement Models •·Break Even Analysis •·Forecasting Techniques **Part Ten** : Case Studies •·Case Method •·Case Problems •·APPENDICES •·Selected References

Code : 10 228 **1st Edition: 2001** **ISBN : 81-219-2056-6**

TRIBOLOGY IN INDUSTRIES

Sushil Kumar Srivastava

CONTENTS: • Tribology vis-A-vis Terotechnology • Friction • Wear • Lubrication •·Behaviour of Tribological Components • Other Tribological Measures • Tribo-Technical Systems •Tribological Monitoring of Equipment's Condition • Tribology in Metal Working Processes • Tribology in Steel Industries • Tribology in Mining Industries • Tribology in Paper and Pulp Industries • Tribology in Glass Fiber Industry • Tribology in Transport Sector • Bibliography • Index

Code : 10 209 **1st Edition : 2001** **ISBN : 81-219-2045-0**

PRODUCTION MANAGEMENT

Martand T. Telsang

CONTENTS: • Introduction to Production and Operations Management • Product Design and Process Selection • Managing the Supply Chain • Quality Management • Design of Facilities and Jobs • Supplement

Code : 07 389 **1st Edition : 2005** **ISBN : 81-219-2458-8**

A TEXT BOOK OF MACHINE TOOLS AND TOOL DESIGN

P. C. Sharma

CONTENTS: • Theory of Metal Cutting • Cutting Tool Materials and Cutting Fluids • Machine Tools • Machining Variables and Related Relations • Tracer Controlled Machine Tools • Gear Manufacturing • Tool Layout for Capstans and Turrets • Tool Layout for Automatics • Jigs and Fixtures • Press Tool Design • Appendix-I • Appendix -II • Index

Code : 10 285 **1st Edition : 2004** **ISBN : 81-219-2362-X**

MODERN POWER ELECTRONICS

P.C. Sen

Code : 10 287 **2nd Rev. Edn. 2005** **ISBN : 81-219-2425-1**

CONTENTS: • Characteristics of Semiconductor Power Devices • Linear Voltage Regulator • Switching Regulators • CUK Converter and Integrated Magnetics • Resonant Converter • Uninterruptible Power Supply • Stepper and Switched Reluctance Motor Drive • Solved Questions • Bibliography • Index

FUNDAMENTALS OF ELECTRICAL NETWORKS

B.R. Gupta & Vandana Singhal

CONTENTS: • Units and Definitions • Parameters of Electric Networks • DC Network Analysis and Network Theorems • Magnetic Circuits • Self and Mutual Inductance • Network Transients • Alternating Current and Voltage • Series AC Circuits • Parallel and Series Parallel Networks • Resonance • AC Network Theorems • State Variable Analysis • Non-Sinusoidal Waves • Polyphase Networks • Two-Port Networks • Filter Network • Network Functions: Poles and Zeros • Network Synthesis • Mechanical Analogies • Symmetrical Components • Appendices • Bibliography • Index

Code : 10 274 **Rep. Edn. 2005** **ISBN : 81-219-2318-2**

DIGITAL COMMUNICATIONS

S.K. Venkata Ram

CONTENTS: • Basics of digital communication • Digital Transmission of Analog signals • Digital Coding of Analog Waveforms • Intersymbol Interference And Its Cures • Digital Modulation Techniques • Optimum Receivers • Introduction to spread spectrum techniques • Optical Fiber communications • Satellite communication

Code : 10 247 **1st Edition : 2003** **ISBN : 81-219-2241-0**

A TEXTBOOK OF ELECTRICAL TECHNOLOGY

B.L. Theraja & A.K. Theraja, S.G. Tarnekar,
M.K. Khedkar, R.S. Sedha

CONTENTS: Vol. I: Electric Current and Ohm's Law •·D.C. Network Theorems •·Work, Power and Energy •·Electrostatics •Capacitance • Magnetism and Electromagnetism •·Electromagnetic Induction •·Magnetic Hysteresis •·Electronical Power Sources •·Electrical Instruments and Measurements •·A.C. Fundamentals •·Complex Numbers •·Series A.C. Circuits •·Parallel A.C. Circuits •·A.C. Network Analysis •·A.C. Bridges •·A.C. Filter Networks •·Circle Diagrams •·Polyphase Circuits •·Harmonics •·Fourier Series •·Transients •·Symmetrical Components •·Introduction to Electrical Energy Generation **•·Vol II:** • Elements of Electro-mechanical Energy Conversion • D.C. Generators • Armature Reaction and Commutation • Generator Characteristics • D.C. Motor • Speed Control of D.C. Motors •Testing of D.C. Machines •Transformer •Transformer-Three Phase • Induction Motors• Computation And Circle Diagrams•Single-Phase Motors • Alternators • Synchronous Motors • Special Machines **•·Vol III:** D.C. Transmission and Distribution •·A.C. Transmission •·Distriemtion Electric Traction • Industrial Applications of Electric Motors •·Distribution Automation •·Electronic Control of AC Motors •·Electric Heating •·Rating and Service Capacity •·Electric Welding •·Illumination •·Tariffs and Economic Considerations •·Semiconductor Physics •·P.N. Junction Diode •·Opto Electronic Devices •·Special Diodes **•·Vol IV:** • Semiconductor Physics • P-N Junction Diode • Optoelectronic Devices • Special Diodes • D.C. Power Supplies • Regulated Power Supply• Bipolar Junction Transistor • Load Lines and DC Biased Circuits •Transistor Equivalent Circuit and Models • Single-Stage Transistor Amplifiers• Multistage and Feedback Amplifiers • Feedback Amplifier• Field Effect Transistors• Breakdown Devices • Sinusoidal and Non-sinusoidal Oscillators • Modulation and Demodulation • Integrated Circuits • OP-AMP and its Applications • Number Systems and Codes • Logic Gates • Boolean Algebra & Logic Families • Electronic Instruments

(Also available in 4 individual parts)

Code : 10 291 **24th Rev. Edn. 2005** **ISBN : 81-219-0289-4**

VOL. I: (Basic Electrical Engineering)

B.L. Theraja, A.K. Theraja & S.G. Tarnekar

Code : 10 292 **23rd Rev. Edn. 2005** **ISBN : 81-219-0290-8**

VOL. II: (A.C. & D.C. Machines)

B.L. Theraja, A.K. Theraja & S.G. Tarnekar

Code : 10 293 | 23rd Rev. Edn. 2005 | ISBN : 81-219-2537-5

VOL. III: (Transmission, Distribution and Utilization)

B.L. Theraja, A.K. Theraja & M.K. Khedkar

Code : 10 004 | 23rd Rev. Edn. 2005 | ISBN : 81-219-1140-0

VOL. IV: (Electronic Devices & Circuits)

B.L. Theraja, A.K. Theraja & R.S. Sedha

Code : 10 005 | 23rd Rev. Edn. 2004 | ISBN : 81-219-1141-9

DIGITAL SIGNAL PROCESSING, TELECOMMUNICATIONS AND MULTIMEDIA TECHNOLOGY

V.K. Khanna

CONTENTS: • Communication and Information Systems Overview • Signal Processing Fundamentals • Digital Coding of Analog Waveforms • Speech and Audio Signal Processing • Image and Video Signal Processing • Digital Modulation Techniques • Networking Principles • Digital Satellite Communications • High Definition Television (HDTV) • Multimedia • Signal Processing and Communications • Bibliography • Review Exercises • Appendices • Index

Code : 10 276 | Rep. Edn. 2003 | ISBN : 81-219-2325-5

SENSORS AND TRANSDUCERS

D. Patranabis

CONTENTS: • Introduction: Definitions • Principles • Classification • Parameters • Environmental Parameters • Characterization; Mechanical and Electromechanical Sensors : Introduction • Resistive Potentiometers • Strain Gauge • Inductive Sensors • Capacitive Sensors; Thermal Sensors: Introduction • Gas Thermometric Sensors • Thermal Expansion Type Thermometric Sensors • Dielectric Constant and Refractive Index Thermosensors • Helium Low Temperature Thermometer • Nuclear Type • Magnetic Thermometer • Resistance Change Type Thermometric Sensors • Thermo emf Sensors • Junction Semiconductor Types • Thermal Radiation Sensors • Quartz Crystal Thermometric Sensors; Magnetic Sensors: Introduction • Types of Sensors Using the Principles • Magnetoresistive Sensors • Hall Effect Sensors • Inductance and Eddy Current Types • Angular/Rotary Movement Transducers • Eddy Current Sensors • Electromagnetic Flowmeter • Switching Magnetic Sensors • Squid Sensors; Radiation Sensors: Introduction • Basic Characteristics • Types of Photosensistors/Photodetectors • X-ray and Nuclear Radiation Sensors • Fibre Optic Sensors; Electroanalytical Sensors: Introduction • The Electrochemical Cell • The Cell Potential • Standard Hydrogen Electrode (SHE) • Liquid Junction and Other Potentials • Polarization • Reference Electrodes • Sensor Electrodes • Electroceramics in Gas Media • Chem FET; Smart Sensors: Introduction • Primary Sensors • Excitation • Amplification • Filters • Converters • Compensation • Information Coding/Processing • Data Communication • Review Questions • Bibliography • Index

AHW 004X | Rs. 170.00 | ISBN : 81-7544-004-X

CONCISE HANDBOOK OF ELECTRONICS AND ELECTRICAL ENGINEERING

V.K. Khanna

CONTENTS: • Mathematical and Physio-Chemical Foundations • Electromagnetics • Material Science • Network Analysis • Electrical Devices • Circuits and Systems • Telecommunication • Computers • Instrumentation • Appendices • Index

Code : 10 275 | Rep. Edn. 2003 | ISBN : 81-219-2319-0

FIBRE OPTIC COMMUNICATION

D.C. Agarwal

CONTENTS: • Principles of Fibre Optic Communication • Fibre Optical Source Devices • Mode Theory for Optical Fibres • Ray Theory for Optical Fibres • Fibre Optical Communication Components • Fibre Optical Communication Systems • Optical Fibre Measurements • Economics and Potential Applications of Optical Fibre Communication Systems • Appendix • Index

Code : 16 286 | Rep. Edn. 2004 | ISBN : 81-219-2317-4

FUNDAMENTALS OF ELECTRICAL ENGINEERING AND ELECTRONICS

B.L. Theraja

CONTENTS: Electric Current and Ohm's Law •·Division of Current • Network Analysis •·Work, Power and Energy •·Electrostatics •·Capacitance • Magnetism and Electromagnetism • Electromagnetic Induction •·Magnetic Hysteresis •·D.C. Generators •·Generator Characteristics •·D.C. Motor •·Speed Control of D.C. Motors •·Chemical Effects of Current •·Electrical Instruments and Measurements •·A.C. Fundamentals •·Series A.C. Circuits •·Parallel A.C. Circuits •·Complex Algebra and A.C. Circuits •·Three Phase Circuits •·Transformer •·Three Phase Induction Motor •·Single Phase Motors •·Alternators •·Synchronous Motor •·Q and A on Electric Machinery •·Semiconductor Physics •·Semiconductor Diodes •·Optoelectronic Devices •·Bipolar Junction Transistors •·Load Line and Biasing Circuits •·Transistor Equivalent Circuits and Models •·Transistor Amplifiers •·Field Effect Transistors •·Thyristors •·Digital Electronics •·Sinewave Oscillators •·Analog and Digital Communication •·Vacuum Tubes and Gas Valves •·Electron Ballistics •·Illumination •·Index

Code : 10 008 **28th Edn. 2004** **ISBN : 81-219-0099-9**

CONTROL SYSTEM DESIGN
The Optimal Approach

B. Sarkar

CONTENTS: • Introduction • Classical Design Approach • Modern Design Approach • Choice of Performance Criterion • Scaler Valued Performance Criterion • Vector Valued Performance Criterion • Equivalent Formulation of External Problems • Design Approach Using a Performance Criterion • Method of Optimization • Appendices: A Quadratic Performance; B. Digital Simulation for Optimization; C. Evaluation of Performance Criteria

AHW 0252 **Rs. 125.00** **ISBN : 81-7544-025-2**

A TEXTBOOK OF DIGITAL ELECTRONICS

R.S. Sedha

CONTENTS: •·Introduction •·Number Systems and Codes •·Logic Gates •·Boolean Algebra and Logic Simplification •·Flip-flops and Related Devices •·Arithmetic Oprations and Circuits •·Counters and Shift Registers •·Integrated Circuit Logic Families •·Medium-Scale Integrated Logic Circuits •·Appendix A, B, C, D, E •·Index

Code: 10 260 **1st Edition : 2004** **ISBN : 81-219-2378-6**

ELECTRICAL ELECTRONICS AND TELECOMMUNICATION ENGINEERING (OBJECTIVE TYPE)

B.L. Theraja

CONTENTS: Electrical Engineering: 1. Basic Electrical Engineering, 2. D.C,. Generators 3. D.C. Motors 4. Transformers 5. Three-phase Induction Motors 6. Single-phase Induction Motors, 7. Alternators 8. Synchronous Motors • Solid State: Electronics 1. Semiconductor Physics and Diodes 2. Tranistor Fundamentals 3. Transistor Parameters and Amplifiers 4. Fets, Diacs, Triacs and Scrs 5. Oscillators 6. Television 7. Charge Coupled Devices 8. Semiconductor Memories 9. Digital Electronics 10. Computer Engineering 11. Robotics 12. Electronic Instruments •·Telecommunications: 1. Engineering Materials 2. Applied Electronics and Circuits 3. Principles of Communication Engineering 4. Antenna and Wave Propagation 5. Electronic Measurements 6. Radio Communication Systems 7. Broadcasting and T.V. Systems 8. Microwave Communication Systems 9. Radar Engineering 10. Industral Electronics 11. Transmission Lines 12. Waveguides 13. Communication Satellites •·Standard Symbols and Vocabulary: 1. IS: 1885 Electrotechnical Vocabulary 2. IS: 2032 Graphical Symbols for Electrical Installatins in Buildings •·Index

Code : 10 037 **4th Edn. 2004** **ISBN : 81-219-0125-1**

BASIC ELECTRICAL ENGINEERING

V.K. Mehta & Rohit Mehta

CONTENTS: Basic Concepts •·D.C. Circuits •·Network Theorems (D.C.) •·Units-Work, Power and Energy •·Electrostatics and Capacitance •·Magnetism and Electomagnetism • Electomagnetic Induction •·Batteries • A.C. Fundamentals • Series A.C. Circuits •·Phasor Algebra •·Parallel A.C. Circuits •·Polyphase Circuits •·Index

Code : 10 113 **3rd Edn. 2004** **ISBN : 81-219-0871-X**